MATHEMATICS
with applications in
management and economics

**Irwin Series in
Quantitative Analysis for Business**

Consulting Editor
ROBERT B. FETTER   *Yale University*

MATHEMATICS
with applications in management and economics

EARL K. BOWEN
Professor of Mathematics
Babson College

 1976   Fourth Edition

RICHARD D. IRWIN, INC.   Homewood, Illinois   60430

Irwin-Dorsey Limited   Georgetown, Ontario   L7G 4B3

© RICHARD D. IRWIN, INC., 1963, 1967, 1972, and 1976

Fourth Edition

First Printing, May 1976
Second Printing, September 1976
Third Printing, May 1977
Fourth Printing, August 1977

ISBN 0-256-01839-1
Library of Congress Catalog Card No. 75–35096
Printed in the United States of America

LEARNING SYSTEMS COMPANY—
a division of Richard D. Irwin, Inc.—has developed a
PROGRAMMED LEARNING AID
to accompany texts in this subject area.
Copies can be purchased through your bookstore
or by writing PLAIDS.
1818 Ridge Road, Homewood, Illinois 60430.

To Steven and Pamela

Preface

The first edition of this text was published in 1963 with the objective of presenting mathematics at a level consistent with student preparation, and directed specifically toward applications in management and economics. The same objective motivates this fourth edition in which I have again enlarged the number of applications.

The basic explanation, example, exercise approach which has met with approval in earlier editions has been maintained. This approach makes the text semiprogrammed and adaptable to self-study. Students whose preparation is weak are given the opportunity to help themselves by working through the appendixes which review the elementary mathematics required in the text. If class time is spent on the appendixes, the prerequisite for the course should be no more than one year of secondary school algebra. I might add that the unit on linear systems makes relatively little demand upon previous preparation, so that work with this unit and the appendixes could proceed simultaneously.

Chapters 1–5 contain new sections on applications of linear systems, supply and demand analysis (shifts and their interpretations), tableaus of detached coefficients, nonnegativity constraints, optimization, interpretation of the outcome of successive steps in linear programming problem solutions, and sensitivity analysis. The beginning material of Chapter 6 has been rewritten, developing matrix operations in an applied setting. This chapter also contains new sections on Markov Chains, applications of the solution of $A x = b,$ the standard deviation, and the practical significance of the standard deviation (use of the Tchebysheff Theorem).

Additional examples of raising numbers to fractional powers have been added to Chapter 7. Chapter 8 has a new section on mortgage payments. A new high-accuracy table has been included for mortgage computations and other financial tables have been enlarged.

In response to suggestions from adopters and reviewers of the third edition, I have rewritten the calculus material (Chapters 9, 10, 11), reducing the theory and increasing the number of applications substantially. Space does not permit a detailed account of the new applications here, but I will mention some which can be identified without lengthy description. These are: inflation, the value of the dollar, and the Consumers' Price Index; perpetuities and capitalized cost; average cost and break-even analysis; cost curve inflection interpretations; operating areas for service organizations; oil well production; fuel consumption; expected value; continuous money flows compounded continuously; productivity; demand functions for competing products; the multiplier; and others. Additionally, I would like to call attention to the innovative discussion of parameterizing a model which appears in Section 10.6.

I should note here also that a brief table of derivatives and integrals has been placed at the end of the book for quick reference.

The thrust of Chapter 12, Probability, is the same as in the third edition. The chapter has been rewritten in major part to modernize the presentation and relate it more directly to everyday situations, rather than relying upon card and dice games to develop the concepts. A new illustrative table of Cumulative Binomial Probabilities ($n = 25$) has been included.

A brief review of sets is contained in Appendix 1. Appendixes 2 and 3, with minor additions and alterations, contain the same review of algebra that appeared in the third edition.

The book contains approximately 1,500 numbered problems, with answers, and an additional 660 review problems which can serve for examination and lecture purposes. Answers to the review problems, worked out in detail, are available in the *Solutions Manual* available to instructors. Additionally, several hundred examples and exercises are woven into the text. To the best of my knowledge, no other text in this field provides a comparable number and variety of problems and applications.

If the entire text, including work in the appendixes, is to be completed, there is sufficient material for a three-semester sequence of courses. However, the book is structured to be adaptable to a variety of courses, from one quarter to three semesters in duration, by making parts of chapters, whole chapters and groups of chapters independent. This structure permits omissions to be made without loss of continuity or prerequisite topics. In broad terms, one quarter or one semester courses in finite mathematics can be organized from the material in Chapters 1–8 and 12; courses of corresponding duration in calculus would cover Chapters 1, 7, 9–11, and 13. More specific detail on course design can be found in the introduction to the *Solutions Manual.*

I wish to acknowledge my debt to those who have taken the time to send me comments on earlier editions and the constructive suggestions which led to most of the major changes in this fourth edition. These are: R. Andres, P. Applebaum, W. Beatty, F. Benn, G. Bloom, R. Borman, A. Brunson, R. Carlson, W. Cassidy, D. Chesnut, T. Church, D. Cleaver, C. Crell,

R. Davis, W. Davis, E. Dawson, B. Dilworth, R. Dingle, D. Dixon, W. Etterbeek, J. Freigo, R. Fetter, J. Foster, R. Fox, Jr., R. Friesen, H. Frisinger, H. Fullerton, W. Furman, E. Goldstein, L. Goldstein, V. Heeren, J. Hindle, A. Ho, A. Hoffman, G. Horcutt, J. Hudson, and D. Isaacson.

Also: R. Jaffa, F. Jewett, C. I. Jones, R. J. Jones, H. King, R. Kizior, P. Latimer, R. Leezer, R. Leidig, J. Liff, S. Logan, G. Long, T. Lougheed, T. Lupton, M. Malchow, E. Marrinan, Jr., A. McLaury, E. Merrick, P. Merry, R. Moreland, J. Moreno, C. Murphy, D. Nichols, J. Papenfuss, R. Ralls, P. Randolph, G. Reeves, J. ReVelle, R. Salmon, F. Schwab, H. Sendek, P. Sgalla, R. Sheffield, L. Shumway, P. Siegel, B. Smith, J. Smith, W. Soule, Jr., M. Spinelli, H. Stein, D. Stoller, M. Tarrab, T. Taylor, O. Thomas, R. Tibrewalla, T. Tsukahara, E. Underwood, T. Vasper, G. Waldron, B. Walker, and M. Williamson.

Additionally, I wish to acknowledge the encouragement and contributions of my colleagues at Babson: W. Carpenter, D. Kopcso, H. Kriebel, W. Montgomery, M. Riskalla, J. Saber, A. Shah, and M. Weinblatt. Finally, I thank my wife, Dorothy Holmes Bowen, for her editorial assistance and for protecting the quiet of my workplace.

April 1976 EARL K. BOWEN

Contents

TO THE READER

Short exercises have been woven into textual discussions to help you pick up pertinent points as they occur. For example, on page 2 you will see the following:

Exercise. A worker's weekly earnings, y dollars, consist of a fixed salary of $50 plus $2 for each hour worked during a week. Letting x be the number of hours worked, write the equation for y in terms of x. Answer: $y = 2x + 50$.

To make the best use of these exercises, you should determine the answer to the question without reference to the given answer (it would help to cover up the given answer) and then compare your answer with the given answer.

1

Linear Equations

1.1 INTRODUCTION

MATHEMATICIANS, economists, statisticians, and others have applied their skills to management problems in some degree for many years, but the first concerted effort in this area occurred during World War II when these specialists were formed into Operations Analysis Groups to assist in the planning of military operations. The analysts used mathematics and statistics extensively in their studies, and the resulting recommendations were an important contribution to the war effort. Following the war, some analysts, soon joined by others, turned their attention to problems of management operations and accomplished major improvements in inventory control, quality control, warehouse location, oil industry operations, agriculture, purchasing decisions, scheduling of complex tasks such as building a shopping center, and a variety of other areas. Mathematics, old and *newly created*, coupled with innovative applications of the rapidly evolving electronic computer and directed toward management problems, resulted in a new field of study called Quantitative Methods (or Management Science or Operations Research) which in time became part of the curriculum of colleges of business. The importance of quantitative approaches to management problems is now widely accepted, and a course in mathematics, with management applications, is included in the core of subjects studied by almost all management students. This text, which has been used in many hundreds of classrooms, develops mathematics in the applied context required for an understanding of the quantitative approach to management problems.

Linear relationships are the subject matter of the first five chapters of the book. From the several pleasant comments which could be made about this material, we select three. First, the mathematics has direct management applications; indeed, at this moment many electronic computers are manipulating linear relationships and printing out information to aid in making lowest-cost or highest-profit decisions. Second, Chapters 4 and 5 introduce a relatively *new* and widely used mathematical technique known as linear programming. Third, most readers will be happy to learn that the mathematics of linear relationships, as we shall develop it, is quite easy. Only a minimal background, such as that provided in the appendixes,[1] is required to get started and to make progress. Every reader should scan these appendixes. Those who feel the need should work through them systematically and refer to them as often as necessary while studying the book.

In this chapter, we first consider in some detail the algebra and geometry of linear equations in two variables, that is, equations whose graphs are straight lines. Later in the chapter, linear equations in three or more variables will be introduced to lay the foundations for analysis of systems of linear equations.

Example.   A salesman receives a fixed salary of $100 per week, and, additionally, a commission of 10 cents for each dollar's worth of goods he sells. If we wish to express the relationship between his weekly earnings and the amount he sells, we would write

$$\text{Earnings} = 0.10(\text{Sales}) + 100$$

or, letting E be earnings and S be sales,

$$E = 0.10S + 100.$$

The last is a *linear* equation in the two variables, E and S, because its terms involve only constants or constants and the *first* power of *one* variable. The graph of this and any other linear equation in two variables is a straight line. Two points completely determine a straight line, so to graph a line we plot any two points satisfying its equation and draw a straight line through them. In the present case, let us arbitrarily let S be 0, then 100. The corresponding values of E are found by substitution in the equation to be, respectively, 100 and 110, so we have (0, 100) and (100, 110) as the coordinates for two points. These points and the line through them are shown in Figure 1–1.

Exercise.   A worker's weekly earnings, y dollars, consist of a fixed salary of $50 plus $2 for each hour worked during a week. Letting x be the number of hours worked, write the equation for y in terms of x. Answer: $y = 2x + 50$.

---

[1]For a review of the minimal background, read Appendix 2. In Appendix 3 read sections A3.1–3.5 and the parts of A3.6–3.11 that deal with linear equations.

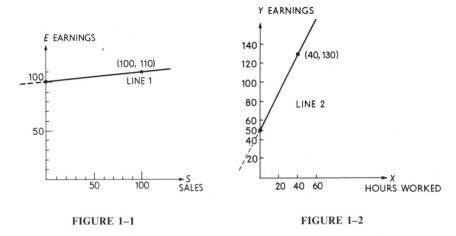

FIGURE 1–1   FIGURE 1–2

The answer to the exercise is a linear equation in the two variables x and y. Its graph is shown in Figure 1–2 which has been drawn to the same scale as Figure 1–1. In both cases, the vertical variable increases as the horizontal variable increases, so the lines slant upward to the right. Observe that line 2 is steeper than line 1, which means that for a *given* horizontal change, the vertical change is greater for line 2 than for line 1. We shall see that the ratio of vertical to horizontal change has numerous applied interpretations. Inasmuch as these changes are horizontal and vertical distances, we turn first to a consideration of such distances.

1.2 VERTICAL AND HORIZONTAL DISTANCES

The distance between two points is the length of the straight-line segment which joins the points. In beginning (plane) geometry and in applied mathematical work, we use a ruler or some other measuring device to determine lengths of segments. In *analytic* geometry, also called coordinate geometry, end points of line segments are specified by their x- and y-coordinates, and algebraic procedures are applied to the coordinates to find the distance between the points.[2]

Distances on horizontal and vertical line segments play an important role in the mathematics of straight lines. Distance on a vertical segment is computed by subtracting the y-coordinates of the end points of the segment. Distance on a horizontal segment is computed by subtracting the x-coordinates of the end points of the segment. Consider, for example, a city which has avenues running east and west and streets running north and south, dividing the city into square blocks. To get from (4th St., 3d Ave.) to (10th St., 3d Ave.), we would walk on Third Avenue a distance of $10 - 4 = 6$ blocks.

---

[2]Broadly speaking, analytic (coordinate) geometry has two distinguishing characteristics: first, points are specified by their coordinates; second, methodology is essentially algebraic.

III

    Exercise.   How many blocks would we walk going from (10th St., 6th Ave.) to (10th St., 13th Ave.)? Answer: Seven blocks.

IIIIIIIIIIIIIIIIIIIIIIIIIIIIIIIIIIIIIIIIIIIIIIIIIIIIIIIIIIIIIIIIIIIIIIIIIIIIIIIIIIIIIIIIIIIIIIIIIIIIIIIIIIIIIIIIIIIIIIIIIIIIIIIIIIIIIIIIIIIIIIIIIIIIIIIII

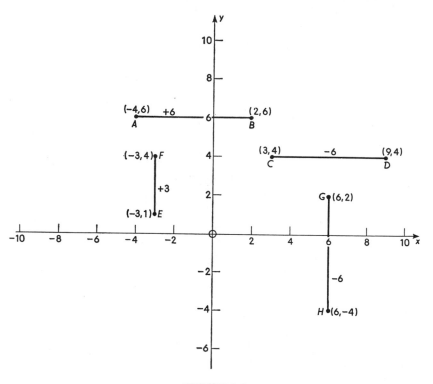

FIGURE 1–3

We shall require that *distance* be a positive number. Hence, if the subtraction of coordinates yields a negative number, the minus sign will be discarded. Consider Figure 1–3. The segment *CD* is horizontal. The distance *CD* is found by subtraction, thus:

$$CD = \text{horizontal distance} = \text{difference of } x\text{-coordinates} = 9 - 3 = 6.$$

If the calculation had been made as $3 - 9 = -6$, the minus sign would have been discarded.[3]

The segment *HG* is vertical. The distance *HG* is found by subtraction:

$$HG = \text{vertical distance} = \text{difference of } y\text{-coordinates} = 2 - (-4) = 6.$$

---

[3]The sign could be used to designate the direction of the segment. Thus, from *C* to *D* is $9 - 3 = +6$, plus meaning the segment goes from left to right, whereas *DC* is $3 - 9 = -6$, minus meaning the segment is taken from right to left. For example, if sales go from $100 to $150, then $150 - $100 = 50, or an increase of $50. On the other hand, if sales go from $200 to $100, then $100 - $200 = -$100$, or a decrease of $100.

Exercise. Find the distances AB and EF on Figure 1–3. Answer: $AB = 6$ and $EF = 3$.

Clearly, the ordinates (that is, the y-coordinates) of points on a horizontal segment are equal, and the abscissas (that is, the x-coordinates) of points on a vertical segment are equal. Thus, letting $P(a, b)$ mean the point P, whose coordinates are (a, b), and letting $Q(a, c)$ mean the point Q, whose coordinates are (a, c), it follows that the segment

$$P(a, b)\, Q(a, c)$$

is vertical. Similarly, the segment

$$R(a, b)\, S(c, b)$$

is horizontal because the points R and S have the same ordinates.

Exercise. Given $A(-2, -5)$; $B(2, 6)$; $C(3, 10)$; $D(2, -5)$. If each pair of points was connected by a straight-line segment, which segment would be parallel to the x-axis and which segment would be parallel to the y-axis? Answer: AD and BD, respectively.

Employing subscript notation, we may refer to any two points as (x_1, y_1) and (x_2, y_2). The symbols are read as "x-sub-one, y-sub-one" or simply "x-one, y-one," and similarly for (x_2, y_2). This notation provides a handy method for preserving the letters x and y to mean abscissa and ordinate, in general, with variations in the subscript serving to designate coordinates of different points. In this notation, a vertical segment would have (x_1, y_1) and (x_1, y_2) as end points. The distance between the points would be:

$$\text{Vertical distance} = |y_2 - y_1|.$$

The vertical lines, $|\ |$, are the *absolute value* symbol which means that the difference is to be taken as positive (or zero). Thus, $|5 - 2| = |+3| = 3$, but $|2 - 6| = |-4| = 4$.

The coordinates of the end points of a horizontal segment would be written in subscript notation as (x_1, y_1) and (x_2, y_1). The distance between the points would be:

$$\text{Horizontal distance} = |x_2 - x_1|.$$

1.3 PROBLEM SET 1

1. Given the points $A(3, 6)$, $B(-3, 6)$, $C(3, 9)$, $D(-3, 2)$, plot the points, then figure the following distances from the graph: AB, AC, DB.
2. Given the points $A(-5, -9)$, $B(-3, -9)$, $C(-5, 15)$, $D(12, -9)$, find the distances AB, AC, AD, BD.

3. Consider the segments AB and CD, where the coordinates are $A(p, q)$, $B(p, r)$, $C(n, q)$, and $D(m, q)$. If the two segments are extended, they will intersect in a right angle. Why?

4. Given $P_1(x_1, y_1)$, $P_2(x_2, y_2)$, $P_3(x_3, y_3)$, what relationships must exist among the coordinates if P_1P_2 is to be horizontal and P_1P_3 is to be perpendicular to P_1P_2?

5. Given P_1, P_2, P_3, P_4 with their respective coordinates in subscript form, what relationships must exist among the coordinates if P_1P_2 is to be horizontal and P_3P_4 is to be parallel to P_1P_2?

6. If cities A, B, and C are collinear (lie on the same straight line), how far would it be between B and C if:

   a) B is 60 miles east of A and C is 100 miles east of A?

   b) B is 30 miles west of A and C is 40 miles east of A?

7. If x represents sales and y represents selling expense, then (30, 22) would mean \$30 in sales were accompanied by \$22 of selling expense. Suppose last month's figures were (14, 10) and this month's are (30, 22).

   a) How much did sales increase?

   b) How much did selling expense increase?

   c) Make a graph showing the points and labeling the increases.

8. What is the advantage of subscript notation for coordinates of points?

9. A section of a city is divided into square blocks by streets and avenues. How many blocks would we walk on sidewalks going from (6th St., 12th Ave.) to (1st St., 5th Ave.)?

1.4 SLANT DISTANCE FORMULA

Figure 1–4 shows a building whose dimensions are 80 by 90 feet. A sidewalk is to be constructed from a building exit to a point which is 200 feet on the horizontal from the lower left corner of the building. We wish to determine the length of the dotted line which represents the sidewalk. We see that the sidewalk is the hypotenuse of a right triangle whose altitude is 90 feet and whose base is a horizontal line segment of length $(200 - 80) = 120$ feet.

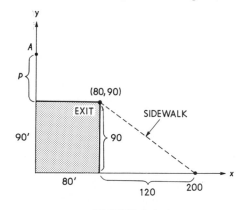

FIGURE 1–4

According to the Pythagorean theorem, the length of the hypotenuse is the square root of the sum of the squares of the sides. Hence, the desired length, L, is

$$L = \sqrt{(120)^2 + (90)^2} = \sqrt{22{,}500} = 150 \text{ feet.}$$

Exercise. If the sidewalk is continued to meet the vertical at point A, *a*) find the distance p from the properties of similar triangles. *b*) Find the length of the extension. Answer: *a*) p is to 80 as 90 is to 120, so p is 60. *b*) $\sqrt{(60)^2 + (80)^2} = \sqrt{10{,}000} = 100$ feet.

Now let us develop a formula for finding the distance between two points which are on a segment not parallel to an axis by reference to the segment AB on Figure 1–5. Dashed lines have been drawn to make a right angle at C, so that the distance AB is the length of the hypotenuse of a right triangle.

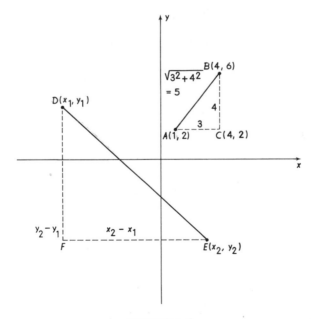

FIGURE 1–5

Inasmuch as AC is horizontal, C must have the same y-coordinate as A (namely, 2). Moreover, CB is vertical, so that C must have the same x-coordinate as B (namely, 4). The horizontal distance AC is $|4 - 1| = 3$ and the vertical distance CB is $|6 - 2| = 4$. Employing the Pythagorean theorem, the hypotenuse, AB, is

$$|\sqrt{(AC)^2 + (CB)^2}| = |\sqrt{3^2 + 4^2}| = |\sqrt{25}| = 5.$$

Applying the same methodology to the segment *DE* on Figure 1–5, we derive the general formula:

$$\text{Distance between two points} = |\sqrt{(x_2 - x_1)^2 + (y_2 - y_1)^2}|.$$

For example, the distance between (2, 3) and (5, 2) is

$$|\sqrt{(5-2)^2 + (2-3)^2}| = |\sqrt{9+1}| = |\sqrt{10}| = 3.162.$$

1.5 PROBLEM SET 2

1. In each case, the given point pairs are the end points of the diagonal of a rectangle whose base is parallel to the *x*-axis. What are the coordinates of the other corners of the rectangle, and what is the length of the diagonal?
   a) (3, 5), (7, 8).   *b*) (−1, −2), (5, 6).
   c) (−1, −2), (−10, −14).

2. Find the distance between the points.
   a) (5, 10), (11, 18).   *b*) (0, 0), (9, 12).   *c*) (−2, −5), (3, −4).
   d) (−2, 3), (6, 9).   *e*) (3, −5), (6, −5).   *f*) (4, 7), (4, 9).

3. A machine shop occupies a 100 foot by 50 foot rectangular area. The electrical outlets which supply the machines are located by reference to a grid with origin at one corner of the area and the axis of abscissas in the long direction of the area.
   a) What would be the coordinates of the outlet in the center of the shop?
   b) How long a line would it take to reach from the origin to the center of the shop?
   c) How long a line would it take to connect a machine at (10 feet, 5 feet) to a plug at (40 feet, 45 feet)?

4. With reference to an origin, City *A* is located at (3, 5), City *B* at (9, 13), and City *C* at (21, 4), all numbers being miles.
   a) Make a graph showing the positions of the cities.
   b) Compute the total distance covered traveling from *A* to *B*, then from *B* to *C*.

5. The coordinates of three cities are shown, in miles, on Figure A. What will be the total distance traveled if one goes from *A* to *B*, then *B* to *C*, then back to *A*? Note that *ABC* is not a right triangle.

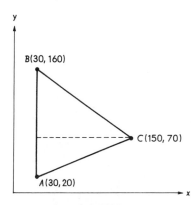

FIGURE A

1.6 SLOPE

The steepness of a ski slope, the pitch of a roof, and the steepness of the glide path of a descending airplane all are associated with the mathematical concept of the *slope* of a straight line or line segment. Numerically, the slope of a straight line is the ratio of the *rise (or fall)* to the *run* between two points on the line, where the rise or fall is the vertical separation and the run is the horizontal separation of the two points. In Figure 1–6, the slope of the segment *AB* is the ratio

$$\text{Slope} = \frac{\text{Rise}}{\text{Run}} = \frac{6}{2} = 3.$$

Clearly, the rise is a vertical segment and the run is a horizontal segment. Hence the slope (generally called *m*) of a straight line or line segment joining the points (x_1, y_1) and (x_2, y_2) is:

$$\text{Slope} = m = \frac{\text{Difference of } y\text{'s}}{\text{Difference of } x\text{'s}} = \frac{y_2 - y_1}{x_2 - x_1}.$$

Consider the segment *CD* in Figure 1–6:

$$m = \frac{\text{Rise}}{\text{Run}} = \frac{-3 - (-7)}{7.5 - 2.5} = \frac{4}{5} = 0.8.$$

It is important to distinguish between segments which rise to the right, such as *AB* in Figure 1–6, and those which fall to the right, such as *KL*. This is done by a convention which requires positive slopes for lines which rise to the right and negative slopes for those which fall to the right.

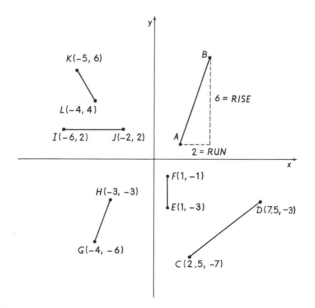

FIGURE 1–6

To adhere to the convention, it is necessary only to *use the same point as starting point* when computing the rise or fall and the run. For example, the slope of $C(2.5, -7)$, $D(7.5, -3)$ computed to be 0.8 in the foregoing starting with point D for both rise and run, is also 0.8 if point C is the starting point for both rise and run. Thus,

$$m = \frac{-7-(-3)}{2.5-7.5} = \frac{-4}{-5} = 0.8.$$

Exercise.   See Figure 1 — 6. Verify that the slopes of KL and GH are, respectively, -2 and 3.

Returning to Figure 1–6, observe the horizontal segment IJ. If we substitute the coordinates into the slope formula we find:

$$m = \frac{2-2}{-2-(-6)} = \frac{0}{4} = 0.$$

The slope is zero. Moreover, it is clear that the slope of any horizontal segment will be zero because the equality of y-coordinates on such a segment will make the numerator of the slope ratio zero, and zero divided by any (nonzero) number is zero.

Next consider the vertical segment EF (Figure 1–6). Substitution in the slope formula yields:

$$m = \frac{-3-(-1)}{1-1} = \frac{-2}{0}.$$

The expression $-2/0$ is not a number, nor is any expression with zero denominator a number. Inasmuch as every vertical segment will lead to a zero denominator in the slope expression, we conclude that there is no number representing the slope of such segments. If the slope formula, mechanically applied, leads to a zero denominator, we say, properly, "the segment is vertical; it has no slope number."

Exercise.   Which of the following segments is horizontal and which is vertical? *a*) The segment joining $(4, -6)$ and $(10, -6)$. *b*) The segment joining $(4, -6)$ and $(4, 10)$. Answer: *a*) is horizontal and *b*) is vertical.

Exercise.   Draw three segments through the point $(4, 7)$, one with no slope number, one with $m = 0$, and one with $m = \frac{2}{3}$. (A slope of $\frac{2}{3}$ is a risc of 2 for a run of 3.)

In applications, the slope of a line segment often is interpreted as the amount of change in the vertical for a unit change (that is, a change of one) in the horizontal, and a number of such slopes have been given names.

Example. This example uses the terms *disposable income, personal consumption expenditures,* and *savings.* It will be sufficient for our purposes here to think of disposable income as the amount of income left after taxes have been paid. The part of disposable income which is placed in a bank or otherwise invested is savings, and the remainder, personal consumption expenditure, is spent on food, clothing, housing, and so on.

A line segment fitted to points whose coordinates are in the order (disposable income,[4] personal consumption expenditures) for the United States in recent years passes through the points (312, 295) and (575, 537), the numbers being in billions of dollars. Thus, when income available for spending (disposable) rose from \$312 to \$575 billion, consumption expenditures rose from \$295 to \$537 billion. The slope of the segment is

$$\frac{\text{Change in consumption expenditures}}{\text{Change in disposable income}} = \frac{537 - 295}{575 - 312} = \frac{242}{263} = 0.92.$$

We see that consumption expenditures increased by \$0.92 when disposable income rose by \$1. The figure, 0.92, is called the *marginal propensity to consume,* or MPC for short. The word *marginal,* just used, means *extra,* and MPC is the extra consumption that accompanies a \$1 increase in income. Income not used for consumption expenditures is *saved.* We see from the foregoing that \$0.08 of each extra dollar of disposable income is saved, and 0.08 is called the *marginal propensity to save,* MPS. Clearly,

$$\text{MPS} = 1 - \text{MPC}.$$

The important bearing these two marginals have on the economic well-being of the nation can be understood by noting that savings are the source of investment in the factories and other economic activities that produce the income which becomes available to consumers. It is interesting to note that while many believe saving is a virtue, economics teaches us that when the nation has idle productive capacity, a high propensity to save (with the consequent low propensity to consume) is not a virtue for the national economy. What is needed in such times is a high propensity to consume, for this will lead to the increased demand which will bring idle capacity back into use and increase national income.

1.7 PROBLEM SET 3

1. If (x_1, y_1) and (x_2, y_2) are the coordinates of two points on a line,
   a) How is the rise computed?   *b)* How is the run computed?
2. *a)* What is the nature of the steepest line that can be drawn through a point?
   b) Why does such a line not have a slope number?

---

[4]Adjusted by removing interest consumers pay on loans.

3. If a set of stairs rises 8 inches for every horizontal run of 12 inches, what slope number (assumed positive) describes the steepness of the stairs?

4. If x is the ground path of an airplane and y is its altitude, and at one point in time the plane is at (500, 1000) and soon after it is at (500, 0), what was the path of the airplane during the time interval?

5. A ski slope whose fall line makes a 45° angle with the horizontal is said to have a 100 percent grade. What slope number represents a 100 percent grade?

6. If the total manufacturing cost, y dollars, of producing x units of a product is $500 at 50 units output and $900 at 100 units output, and the cost-output relation is linear,
   a) What is the slope of the cost-output line?
   b) How much does the production of one unit add to total cost?

7. A line segment fitted to points whose coordinates are in the order (disposable income, personal consumption expenditures) for a nation passes through the points (32, 30), (57, 54), the numbers being in billions of dollars. What are the values, names, and interpretations of the slope and (1 − slope) of this line?

Compute the slope of the segment joining each point pair:

8. (0, 0), (2, 2).   9. (−3, 5), (4, −2).
10. (6, −1), (−2, 0).   11. (−3, −2), (−3, −4).
12. (2 3), (6, 3).   13. (1/2, −1/4), (2, −1/4).
14. (−4 7), (−4, 2).   15. (1, 1), (3, 3).
16. (2/7,   1/3), (−2/9, −2/3).   17. (0, 3), (3, 0).
18. (12, −5), (3, 6).   19. (3, −7), (3, −15).
20. (−1, −2), (−3, −4).   21. (1.6, 3.8), (−3.6, 4.2).

22. Write the formula for the slope of a line segment; then discuss the formula by means of a numerical illustration.

23. Write the expression for the slope of the segment from $P(a, -2)$ to $Q(3, -b)$ in two equivalent forms.

24. Make and label sketches showing segments having positive slope, negative slope, zero slope, and no slope number.

25. Why is division by zero excluded from arithmetic calculations? (See Appendix 2.)

Mark (T) for true or (F) for false.

26. ( ) A line which rises to the right and is almost vertical does not have a slope number.

27. ( ) The slope of the x-axis is zero.

28. ( ) The segment $P(a, b) Q(a, c)$ is vertical.

29. ( ) A line segment of negative slope rises to the left.

30. ( ) The slope of the y-axis is not a number.

31. ( ) A line segment which is very, very close to the vertical has a slope number which has a large absolute value.

32. ( ) No matter how large a number we may write down, there is a line segment whose slope exceeds this number.

33. ( ) A line segment contained entirely in the second quadrant necessarily has a negative slope.
34. ( ) The quadrant in which a line segment lies has no necessary relation to the sign of the slope number of the segment.

1.8 DETERMINING EQUATIONS OF STRAIGHT LINES

We have been considering line segments and their slopes. We now turn to the infinite extension of a segment, the straight line.

A unique feature of a straight line is its constant slope. No matter what two points we consider on a given straight line, the slope number will be the same. A curve, on the other hand, changes "steepness" from point to point. In Figure 1–7, for example, the curve becomes steeper as we move along it to the right. In later work we shall speak of the slope of a curve *at a point* as the slope of the straight-line tangent to the curve at the point. In Figure 1–7 we see that the slope of the curve at Q is greater than the slope at P. Keep in mind that we may speak of *the* slope of a straight line, because the slope of a line is unique, but that we must refer to the slope of a curve at a point.

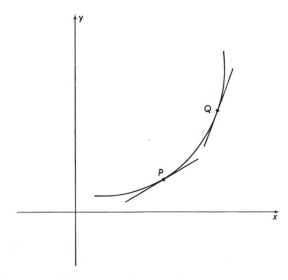

FIGURE 1–7

Slopes of lines tangent to curves are a fundamental consideration in the study of differential calculus later in this book. There we shall see that a point where the tangent to a profit curve is horizontal may identify for us a level of output which will lead to *maximum* profit. This and many other applications of linear relations make it worth our while to devote time to a study of linear equations.

1.9 POINT-SLOPE FORMULA FOR A STRAIGHT LINE

Suppose that a salesman earns a fixed weekly salary plus a 20 percent commission on each dollar's worth of goods he sells. Suppose further we are told that the salesman earned $180 for a week in which his sales were $500. Letting y be earnings and x be sales, we see that the given information specifies the point (500, 180) on a line, and the slope of the line, which is 0.20. Geometrically, the line passes through (500, 180) and rises by 0.2 when x (sales) increase by $1. See Figure 1–8 which shows a line of slope 0.2 passing through the given point.

Exercise. How can the slope number be used to draw the line? Answer: Starting from the given point, a run of, say, 100, will lead to a rise of 0.2(100) = 20, which will give a second point (600, 200). The line is then drawn through the two points.

We now wish to write the equation of the line in Figure 1–8. That is, we want an equation relating y and x which holds true for *any* point on the line.

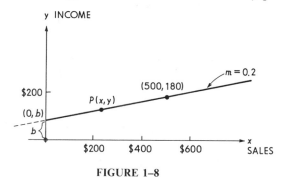

FIGURE 1–8

The word *any* in this context means *all,* because the equation must hold true for the x-, y-coordinates of all points on the line. Let us take $P(x, y)$ to represent *any* point on the line. To form the equation relating the coordinates x and y, we apply the fundamental property of a straight line; namely, that its slope is *constant.* We know the slope is $m = 0.2$. It follows that the slope computed by using any point, $P(x, y)$, and the given point (500, 180), must equal 0.2. Hence,

$$\frac{y - 180}{x - 500} = 0.2$$

from which

$$y - 180 = 0.2x - 100$$

or

$$y = 0.2x + 80.$$

This last expression is the equation of the line in Figure 1–8. It holds true for any point on the line. In particular, substitution of the given point (500, 180) leads to the true statement

$$180 = 0.2(500) + 80.$$

Next, let us derive a general formula for finding the equation of a line given its slope, m, and a point (x_1, y_1) on the line. See Figure 1–9. Note that

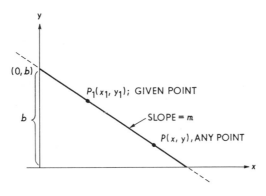

FIGURE 1–9

x_1 and y_1 in $P_1(x_1, y_1)$ and m are given numbers, but the x and y in $P(x, y)$ represent the variables in an equation which must hold true for all points on the line. Applying the criterion of constant slope, we see that

$$\frac{y - y_1}{x - x_1}$$

must equal the given slope, m. Hence, we have the *point-slope* formula for a straight line in either of the following forms:

$$\frac{y - y_1}{x - x_1} = m$$

or

$$y - y_1 = m(x - x_1).$$

As an example of application of the formula, suppose that we want the equation of a line of slope $\frac{2}{3}$ which passes through the point $(3, -2)$. We have that $m = \frac{2}{3}$, $x_1 = 3$, and $y_1 = -2$. Hence,

$$\frac{y - (-2)}{x - 3} = \frac{2}{3}$$

or

$$3y + 6 = 2x - 6$$

so that

$$3y - 2x + 12 = 0.$$

The last equation may be written in various forms, of course. Two equivalent equations are

$$3y = 2x - 12$$

and

$$y = \frac{2}{3}x - 4.$$

Exercise. Find the equation of the line of slope $-1/2$ which passes through the point $(1, 2)$. Answer: $x + 2y = 5$, or equivalent forms.

1.10 SLOPE-INTERCEPT FORMULA

Return now to the income-sales example shown in Figure 1–8. By the point-slope procedure we found the equation of the line to be

$$y = 0.2x + 80.$$

Note that if the salesman sells nothing during a week ($x = 0$), he still earns $y = \$80$. This represents his *fixed* salary.

Exercise. In Figure 1–8, what represents fixed salary? Answer: The vertical distance, b, shown where $x = 0$.

The y-intercept of a straight line is the value of y obtained from the equation when x is set equal to zero. It is designated commonly by the letter b. In the case of the line

$$y = 0.2x + 80,$$

b is 80, and the slope m is 0.2. Following the form of this equation we have the *slope-intercept* form of the equation of a straight line,

$$y = mx + b.$$

This form is a special case of the point-slope form in which the given point is the y-intercept, $(0, b)$. Applied problems often fall naturally into this form. For example, if it costs $\$1000$ to set up a plant to produce a product (before any units are made), and thereafter production costs amount to $\$5$ for each unit made, the total cost, C dollars, of making x units of the product is

$$C = 5x + 1000$$

because the given information establishes $\$1000$ as the value of C when x is zero (the vertical intercept), and the slope as 5.

Exercise. What is the equation of the line whose slope is 1 and whose *y*-intercept is 4? Answer: $y = x + 4$.

As another example, suppose C represents the cost of parking for H hours, and the cost is computed as $1 plus $0.50 per hour. We have

$$C = 1 + 0.5H$$

as the relationship. Generally, the number of hours is rounded up to the next higher whole number, so we have a series of points here rather than a complete line.

Exercise. The cost of renting a truck for a day is $40 plus 40 cents per mile driven. Write the equation of the relationship between cost, C, and number of miles driven, M. Answer: $C = 40 + 0.4M$.

1.11 FINDING THE SLOPE OF A LINE FROM ITS EQUATION

If the equation of a line is given as

$$3x + 6y = 5,$$

we may change it to slope-intercept form by solving for y, as follows:

$$y = -\frac{3}{6}x + \frac{5}{6} = -\frac{1}{2}x + \frac{5}{6}.$$

Associating this with the slope-intercept form

$$y = mx + b$$

we see that the slope of the line is $-1/2$ and the *y*-intercept is $5/6$.

Exercise. Find the slope and *y*-intercept of $2y - 3x - 10 = 0$. Answer: Solving for y, we find $y = (3/2)x + 5$, so the slope is $m = 3/2$ and the *y*-intercept is $b = 5$.

Linear equations in a form such as $ax + cy = d$, with a, c, and d constant, arise naturally in many applications. For example, suppose we have $12.60 to spend on pork and chicken. If we buy p pounds of pork at $0.90 per pound and c pounds of chicken at $0.72 per pound, our expenditures would be $0.9p + 0.72c$ dollars, and this must equal $12.60. Thus,

$$0.9p + 0.72c = 12.60.$$

In slope-intercept form, solving for p, this becomes

$$p = -\frac{0.72}{0.9}c + \frac{12.60}{0.9}, \quad \text{or} \quad p = -0.8c + 14.$$

The intercept tells us that we can buy 14 pounds of pork if we buy no chicken. The slope, -0.8, means that if we increase our purchase of chicken by one pound, we must decrease purchases of pork by 0.8 pounds. Thus, the *substitution rate* is 0.8 pounds of pork per pound of chicken.

Exercise. Solve the foregoing equation for c in terms of p in slope-intercept form, then interpret the intercept and the slope. Answer: $c = -1.25p + 17.5$. We can buy 17.5 pounds of chicken if we buy no pork. The substitution rate is 1.25 pounds of chicken per pound of pork.

1.12 EQUATION OF A LINE THROUGH TWO GIVEN POINTS

Two points completely determine a unique straight line. They also determine the slope of the line. Hence, we can find the desired equation by first finding the line's slope, then applying the point-slope procedure. For example, suppose that x is the number of units produced at a total cost of y dollars and that cost is linearly related to output. If we are told that cost goes from \$200 to \$350 when output increases from 100 to 200 units and we let y be cost and x be output, we have the two points (100, 200) and (200, 350) as shown in Figure 1–10. The slope of the line is

$$m = \frac{350 - 200}{200 - 100} = 1.5.$$

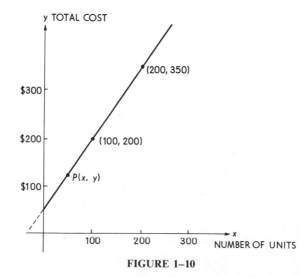

FIGURE 1–10

The point-slope formula now may be applied with either of the given points. Thus, with $P(x, y)$ representing any point on the line, and (100, 200) as the given point, we find

$$\frac{y - 200}{x - 100} = 1.5$$

so that

$$y - 200 = 1.5x - 150$$

or

$$y = 1.5x + 50.$$

Exercise. Find the equation of the line in Figure 1–10 using (200, 350) as the given point. Answer: $y = 1.5x + 50$, as before.

In general, then, the equation of a line through two given points can be found by first computing the slope, then applying the point-slope procedure using either of the given points.

Exercise. Find the equation of the line through $(-1, 7)$ and $(2, -2)$. Answer: $y = -3x + 4$.

1.13 LINES PARALLEL TO AN AXIS

Given the problem of finding the equation of the line passing through (3, 6) and (8, 6), we may write

$$\frac{y - 6}{x - 3} = \frac{6 - 6}{8 - 3}.$$

We find that

$$\frac{y - 6}{x - 3} = 0$$

$$y - 6 = (0)(x - 3)$$

$$y - 6 = 0$$

$$y = 6.$$

The two given points and the line joining them are shown in Figure 1–11. Inasmuch as the points have the same ordinate, the line is parallel to the *x*-axis. The line is completely and uniquely described by stating that it is the line containing all points whose *y*-coordinate is 6, and this is precisely what is stated by the linear equation $y = 6$. Because we have come to expect that

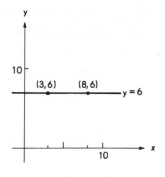

FIGURE 1–11

linear equations will have both the variables, *x* and *y*, among their terms, it may be helpful to point out that the equation at hand could be written in the form

$$0x + y = 6.$$

The point is that the coefficients of *x* and *y* in a straight-line equation may be any real numbers, provided both are not zero. The special case here is an equation where the coefficient of *x* is the real number zero. In general, viewed as an equation in *x* and *y:*

$$y = \text{a constant}$$

is a straight line, parallel to the *x*-axis.

As we might now expect, a line parallel to the *y*-axis will have an equation of the form

$$x = \text{a constant}.$$

Consider the problem of finding the equation of the line through (5, 2) and (5, 12). Applying the point-slope method:

$$\frac{y - 2}{x - 5} = \frac{12 - 2}{5 - 5}.$$

This leads to

$$\frac{y - 2}{x - 5} = \frac{10}{0}.$$

The zero in the denominator of the slope expression tells us that this is a vertical line. A glance at the coordinates (5, 2) and (5, 12) shows that the particular vertical line at hand is the one on which the *x*-coordinate of all points is 5. Its equation is

$$x = 5,$$

and its graph is shown on Figure 1–12.

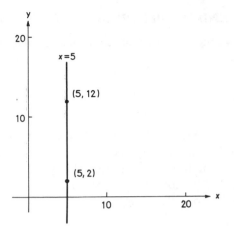

FIGURE 1–12

Slope ratios with zero numerator lead to an equation of the form

$$y = \text{a constant}$$

and ratios with zero denominator lead to

$$x = \text{a constant}.$$

Actually, of course, a glance at the two given points is sufficient to determine the equation. Thus, a line through $(4, -6)$ and $(5, -6)$ has the equation $y = -6$, and the line through $(4, -6)$ and $(4, 7)$ has the equation $x = 4$.

Exercise. What is the equation of the line through *a*) $(-4, 7)$ and $(-4, -3)$? *b*) $(3, 6)$ and $(2, 6)$? Answer: *a*) $x = -4$. *b*) $y = 6$.

In economics, a *demand curve* relates number of units demanded, q (for quantity), to p, the price per unit. Demand curves have negative slopes, indicating that demand goes down as price per unit goes up. In special cases, demand curves are horizontal or vertical lines which have equations $p = $ a constant or $q = $ a constant. Remembering that demand curves express a relationship between p and q, we would interpret $q = 1000$ to mean that 1000 units are demanded whatever the price may be.

Exercise. A college charges \$75 per credit hour. If q is the number of credit hours demanded by students, write and interpret the equation of the demand curve. Answer: $p = 75$, which means that the price per credit hour is \$75 no matter how many credit hours are demanded by students.

1.14 PARALLEL AND PERPENDICULAR LINES

Lines which have the same slope are parallel to each other. Figure 1–13 shows the lines

$$2y - x = 4$$

and

$$2y - x = 10$$

both of which have a slope of $1/2$. We shall have occasion to be concerned about parallel lines in coming chapters where we seek to find points of intersection of lines by simultaneous solution of systems of linear equations. There we shall see that the solution procedure leads to an inconsistent statement when applied to parallel lines.

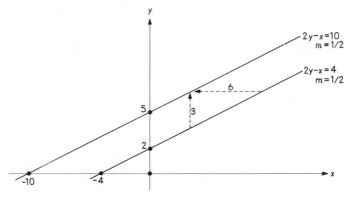

FIGURE 1–13

Distinct parallel lines which are not horizontal or vertical have a constant vertical separation and a (usually different) constant horizontal separation. These separations can be found easily from the intercepts. Thus, the parallel lines shown in Figure 1–13 have y-intercepts of 2 and 5, so the second line is 3 units above the first. The x-intercepts are -4 and -10, so that the second line is 6 units to the left of the first. Now suppose that the demand function for a product at one point in time is

$$p = 20 - 0.2q$$

where q is quantity demanded and p is price per unit. Suppose further that a competing product is taken off the market and that at a later point in time the demand function for our product is found to be

$$p = 22 - 0.2q.$$

Assuming p is on the vertical, we note that the p-intercepts increase by \$2 from \$20 to \$22, so that the price per unit is \$2 higher at any level of demand. We say that the demand curve has *shifted* upward by \$2 per unit.

Exercise. By how much and in what direction did the demand curve shift horizontally? Interpret the shift. Answer: The horizontal (q) intercept moved from 100 to 110, a shift of 10 units to the right. This means that demand is higher by 10 units at every level of price.

If two slant lines are *perpendicular,* the slope of one is the negative reciprocal of the slope of the other. Thus, if one has a slope of $2/3$, the other will have a slope of $-3/2$. The proof of this relationship and an application are introduced in Problem Set 4.

Exercise. Given line 1 is $2y = 4x - 7$ and line 2 is $2y + x = 4$, to which of these is $y - 2x = 4$ *a*) parallel *b*) perpendicular? Answer: *a*) 1. *b*) 2.

1.15 LINES THROUGH THE ORIGIN

Any equation in the variables x and y which does not have a constant term will have a graph which passes through the origin. For example, in the case of the straight line

$$3y - 2x = 0,$$

it is obvious that (0, 0), the coordinates of the origin, satisfy the equation. From an applied point of view, such lines are of interest because they are the mathematical expression of a *proportion.* That is, if we write the equation in the form

$$\frac{y}{x} = \frac{2}{3},$$

we may read the statement as "y is to x as 2 is to 3." An assumption that, say, output per man-hour is 3 units would translate to $y/x = 3$, where y is number of units of output and x is number of man-hours worked. Graphical expression of the assumption would be the straight line $y = 3x$, passing through the origin.

When considering revenue obtained from sale of a product, revenue, R, is 0 when the number of units sold, x, is 0. The revenue obtained by selling x units at a constant price of $5 per unit would be

$$R = 5x,$$

a line through the origin.

Exercise. If it costs C dollars to maintain a factory and produce x units of product, would it be reasonable to assume the cost curve goes through the origin? Answer: It would not be reasonable because at $x = 0$ (no output) costs of insurance, security, interest payments and other elements of fixed cost, which do not depend upon output, would still be incurred.

1.16 PLOTTING LINES BY INTERCEPTS

If we are asked to graph the straight line

$$3x + 4y = 12$$

we know that two points on the graph will be adequate. The simplest points to obtain are the *intercepts*. If we set $y = 0$ in the equation, we obtain the x-intercept; if we set $x = 0$, we obtain the y-intercept. In the case at hand the x-intercept is 4, and the y-intercept is 3. The two points are, of course, (4, 0) and (0, 3). The points and the line are shown in Figure 1–14.

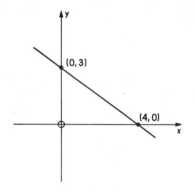

FIGURE 1–14

The term intercept refers to the point at which an axis is cut. We shall often refer to intercepts when graphing straight lines. Thus, a glance at

$$2x - 3y = 16$$

shows that this line cuts the x-axis 8 units to the right of the origin, and cuts the y-axis 16/3 units below the origin.

Exercise.   Verify that the x- and y-intercepts of $4x + 5y = 20$ are, respectively, 5 and 4. Then plot the line, labeling the co-ordinates of the intercepts.

1.17 PROBLEM SET 4

Find the equation of the line passing through the given point and having the given slope:

1. $(3, 4)$, slope 3.
2. $(-5, 6)$, slope -1.
3. $(-2, 6)$, slope $-2/3$.
4. $(1, -4)$, slope $1/2$.
5. $(-3, -8)$, slope 0.
6. $(2, 7)$, slope $-1/6$.
7. $(-5, -8)$, slope 13.
8. $(0, 0)$, slope 0.
9. $(5, 2)$, vertical.
10. $(0, 0)$, vertical.
11. $(3, 4)$, slope 0.
12. $(4, -3)$, slope 5.

Find the equation of the line passing through each of the given pairs of points:

13. $(4, 6)$, $(-3, 7)$.
14. $(-5, 3)$, $(2, 9)$.
15. $(1, 1)$, $(3, 3)$.
16. $(-2, -4)$, $(-1, 5)$.
17. $(0, 0)$, $(2, 3)$.
18. $(2, 4)$, $(-3, 4)$.
19. $(1/3, -1/2)$, $(2, -3)$.
20. $(-7, 2)$, $(-7, -8)$.
21. $(3, 5)$, $(1, 4)$.
22. $(3, 5)$, $(-4, 5)$.
23. $(6, 0)$, $(10, 0)$.
24. $(-3, -2)$, $(4, -7)$.

25. Write the equation of the x-axis.
26. Write the equation of the y-axis.
27. On the line passing through $(2, 3)$ and $(-5, 6)$, what is the ordinate of the point whose abscissa is 17?
28. On the line of slope -2 passing through $(3, 7)$, what is the abscissa of the point whose ordinate is 17?
29. What is the equation of the vertical line which passes through $(-6, 3)$?
30. *a*) What is the equation of the horizontal line which passes through $(-6, 3)$?
    b) A curve showing profit (vertical) and number of units produced and sold (horizontal) rises smoothly to a peak and then declines as we move to the right. The peak is at $(100, 500)$. What is the equation of the tangent line at the peak? What is the significance of this equation? Hint: Make a sketch labelling the axes, showing a curve which has a mound-shaped peak.
31. *a*) What is the equation of the line parallel to and 5 units above the x-axis?
    b) If q is the quantity demanded at price p per unit and the p axis is vertical, what is the geometric nature and interpretation of the demand curves $p = 2$ and $q = 500$?
32. As sales (x) change from \$100 to \$400, selling expense (y) changes from \$75 to \$150. Assume that the given data establish the relationship between sales and selling expense as the two change, and assume that the relationship is linear. Find the equation of the relationship.

33. As the number of units manufactured increases from 100 to 200, manufacturing cost (total) increases from $350 to $650. Assume that the given data establish the relationship between cost (y) and number of units made (x), and assume that the relationship is linear. Find the equation of the relationship.

34. If the relationship between total cost and number of units made is linear, and if cost increases by $3 for each additional unit made, and if the total cost of 10 units is $40, find the equation of the relationship between total cost (y) and number of units made (x).

35. *a*) If taxi fare (y) is 50 cents plus 20 cents per quarter mile, write the equation relating fare to number of miles traveled, m.

   b) The weekly earnings of a salesman are $50 plus 10% of the retail value of the goods he sells. Write the equation for earnings, E, in terms of sales volume, V. What is the slope of this line called?

Mark (T) for true or (F) for false.

36. ( ) The horizontal line through (5, 6) has the equation $x = 5$.

37. ( ) The y-coordinate is called the ordinate.

38. ( ) The slope of the line through (4, 6) and (5, 9) is greater than the slope of the line through (0, 0) and (1, 2).

39. ( ) The equation of the x-axis is $x = 0$.

40. ( ) The lines $x = 5$ and $x = 10$ are parallel to each other.

41. ( ) The lines $x = 5$ and $y = 10$ are perpendicular to each other.

Graph the following lines, using intercepts:

42. $x + y = 6$.   43. $3x - 2y = 12$.

44. $2x + 5y = 6$.   45. $x - y = 4$.

46. $7x + 6y - 3 = 0$.

What is the slope of each of the following lines?

47. $3x - 2y = 7$.   48. $x + y = 2$.

49. $2x - 6y = 5$.   50. $x - y = 0$.

51. *a*) A pound of food A contains 8 ounces of a nutrient, and a pound of B contains 12 ounces of the nutrient. Write the expression which must be satisfied if x pounds of A and y pounds of B are to provide 96 ounces of nutrient.

   b) What is the substitution rate of A per pound of B?

   c) What is the substitution rate of B per pound of A?

52. What is the equation of the line which has a slope of 2 and a y-intercept of -6?

53. If a straight-line equation is in the form

$$ax + by + c = 0$$

then the slope is $-a/b$; that is, the negative of the ratio of the coefficient of x to the coefficient of y. Why is this statement true?

54. If total cost is y and number of units is x, what expression represents cost per unit? What equation would replace the statement that cost per unit is $3? What is the slope and what are the intercepts of the line whose equation was just written?

55. *a*) Find the equation of the line through (2, 7) which is parallel to the line $3x - 2y = 7$.
   b) Find the equation of the line through $(-2, -6)$ which is perpendicular to the line $x = 3y - 4$.

56. If a demand function shifts from $p = 40 - 0.1q$ to $p = 35 - 0.1q$, what are the amounts and directions of the vertical and horizontal shifts? Interpret these shifts.

57. What is the equation of the line on which the ordinate of any point is twice the abscissa?

58. Find the equation of the line through the origin which is parallel to $4x - 5y = 10$.

59. *a*) Figure A shows two perpendicular lines, l_1 and l_2. Clearly, the slope of one is positive and the other negative. Prove that angle 1 = angle 2. Then, using similar triangle relationships, prove that the slopes of the lines are reciprocals.
   b) In what situation would slopes of perpendicular lines not be reciprocals?

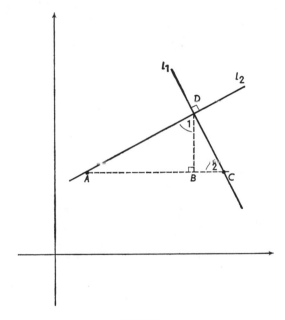

FIGURE A

60. See Figure B, which shows an existing pipeline passing through two points. Plant A is in existence at the point (5, 6) and a new plant, B, is to be located on the x-axis at a point such that the dotted pipeline which will be constructed to connect A and B to the existing line will be perpendicular to the existing line.
   a) Where should B be located?
   b) What is the advantage of having the new line meet the existing line at a right angle?

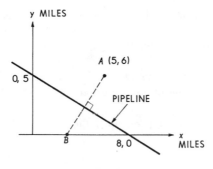

FIGURE B

61. Prove that every line whose equation is of the form $ax + by = 0$, where a and b are any numbers (not both zero), passes through the origin.

Mark (T) for true or (F) for false.

62. ( ) The x-intercept of $3x - 2y - 12$ is -6.
63. ( ) The slope of $2x + 3y = 6$ is $-2/3$.
64. ( ) If the graph of the equation $ax + by + c = 0$ is to pass through the origin, c must be 0.
65. ( ) If the ratio of y to x is constant (not zero) for an equation, the graph of the equation must pass through the origin.
66. ( ) $x + y = 6$ and $2x + 2y = 8$ are parallel but not identical lines.
67. ( ) $x + y = 6$ and $2x + 2y = 12$ are identical lines.
68. ( ) If the relationship between x and y can be expressed as $ax + by = 0$, then x and y are in proportion.
69. ( ) $x = 10$ has no y-intercept.
70. ( ) Any line parallel to $x - 2y = 7$ must have a slope of $\frac{1}{2}$.
71. ( ) If y is units of output and x is man-hours worked, and if $y = 3x$, then output per man-hour is constant.

1.18 INTERPRETIVE EXERCISE: COST-OUTPUT

The purpose of this exercise is to relate the mathematical terminology of linear equations to a real-world situation. To this end, we shall assume that C is the total factory (manufacturing) cost of production of a product when Q (for quantity) units of the product are made. We assume that the relationship between C and Q is linear,[5] as shown by the line segment LM in Figure 1–15.

When making interpretations from Figure 1–15, keep in mind that a vertical distance represents a cost and a horizontal distance represents a quantity (number of units). Thus, the segment PT represents the total cost of producing OP units of product, and UM is the total cost of producing OU units of product.

---

[5]When output interval (domain) is wide, the cost-output relationship may well be curved rather than straight. We shall consider the case of a curve in later chapters.

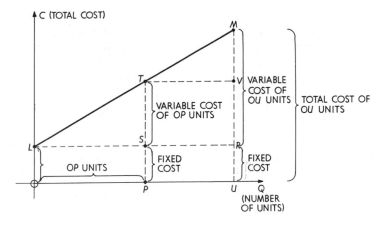

FIGURE 1–15

The distance *OL*, the vertical intercept, corresponds to the cost of operation when zero units are produced, the reasoning here being that some costs (such as insurance on the plant) exist even when no product is being made. Borrowing an accounting term, we interpret *OL* as *fixed cost,* that is, the component of cost which does not vary with the number of units made. In Figure 1–15, fixed cost is

$$OL = PS = UR.$$

If we make *OP* units of product, the total cost is *PT. PT,* in turn, is the sum of *PS* and *ST. PS* is fixed cost; *ST* we shall call the *variable cost* when *OP* units are made. The term variable cost refers to the component of cost which changes as the number of units produced changes. Thus, *ST* is variable cost when *OP* units are made; *RM* is the (larger) variable cost when *OU* units are made. At each level of output, total cost is the sum of fixed and variable cost.

Consider the ratio *RM/LR. RM* is the variable cost when *LR* (or *OU*) units are made. The ratio

$$\frac{\text{Variable cost}}{\text{Number of units made}}$$

is the variable cost *per unit* of product made. By definition, *RM/LR* is the slope of *LM,* and the slope of a straight-line segment is constant. Consequently, variable cost per unit is constant when cost and output are linearly related.[6]

Economic terminology leads us to another description of the slope of *LM* in Figure 1–15. Economists speak of the *extra* cost (the change in total cost)

---

[6]If, on a cost *curve,* variable cost is divided by number of units produced, the ratio is called *average* variable cost per unit and is *not* constant.

when one more unit is made as the *marginal cost* of that unit. This is the vertical change for a horizontal change of one. It is the slope of the line. We may say that when total cost is linearly related to output, marginal cost is constant.

For further practice, observe that *UM* is *VM* greater than *PT*, which means that it costs *VM* more dollars to produce *OU* units than to produce *OP* units. *TV*, on the other hand, represents how many more units can be produced for *UM* dollars than for *PT* dollars.

We have said that when total cost is linearly related to output, then variable cost per unit is constant. Average cost per unit, defined as total cost over number of units produced, is not constant. Rather than argue this statement from Figure 1–15, suppose that we interpret the equation

$$y = 3x + 2,$$

letting *y* be total cost of producing *x* units. The fixed cost is $2, and the variable cost per unit (the slope) is $3. However, by substitution, we find that total cost rises from $17 to $32 if *x* changes from 5 to 10 units. The average cost per unit declines from 17/5 to 32/10, that is, from $3.40 to $3.20. This reduction sometimes is referred to as being a consequence of spreading fixed cost over a larger number of units.

In order that we may differentiate clearly between the examples in the present section and those of the next section, it will be helpful to remind ourselves that the present section deals with total *factory* (or manufacturing) cost of making *x* units of a product. In the next section, *x* will be the dollar volume of sales of a retail firm and the cost expression will include fixed cost and a variable element which depends upon the level of sales.

1.19 PROBLEM SET 5

Mark (T) for true or (F) for false. (Assume that cost means factory cost.) Given that total cost, *y*, of making *x* units is $y = 5x + 10$:

1. (   ) The total cost of making 20 units is $110.
2. (   ) Average cost per unit is $6 if 10 units are made.
3. (   ) The marginal cost of the 11th unit is greater than $5.
4. (   ) The marginal cost of the 20th unit is $5.
5. (   ) The variable cost per unit decreases as the number of units made increases.
6. (   ) The variable cost incurred when making 10 items is $50.
7. (   ) Average cost per unit decreases as the number of units made increases.
8. (   ) Variable cost increases as the number of units made increases.
9. (   ) The marginal cost of every unit is the same.
10. (   ) The slope of the line is the variable cost per unit.
11. If the total factory cost, *y*, of making *x* units of a product is given by $y = 3x + 20$, and if 50 units are made:
    a) What is the variable cost?
    b) What is the total cost?

c) What is the variable cost per unit?

d) What is the average cost per unit?

e) What is the marginal cost of the 50th unit?

12. If total factory cost, *y*, of making *x* units of a product is given by $y = 2x + 25$:

   a) Graph the cost-units equation.

   b) Draw a line representing fixed cost on the graph for (*a*).

   c) Erect a vertical line at $x = 10$, intersecting the *x* axis at *R*, the fixed cost line at *S*, and the given equation at *T*.

   d) What are the numerical values and the interpretations of *RS*, *ST*, and *RT*?

13. In Figure A, the slant line represents the relation between total factory cost, *C*, of producing a number of units, and the number of units, *Q*, produced. What line segment(s), or ratios thereof, represent:

   a) Fixed cost?

   b) Total cost if *OA* units are made?

   c) Variable cost if *OA* units are made?

   d) Variable cost per unit made?

   e) Average cost per unit if *OA* units are made?

   f) Fixed cost per unit if *OA* units are made?

   g) How many more units can be made for *AD* dollars than for *AF* dollars?

   h) Marginal cost per unit?

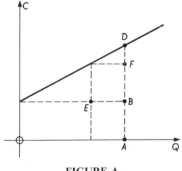

FIGURE A

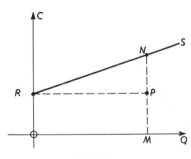

FIGURE B

14. Prior to making a number of units of a certain part, a machine must be made ready, the cost incurred being called the setup cost. The total machine shop cost, *C*, of making *Q* units of the part is shown in Figure B as the segment *RS*. What interpretation would be given to:

   a) *OR*?

   b) *PN*?

   c) *PN/OM*?

   d) *MN/OM*?

1.20 COMMENT ON MODELS

We call attention here to the distinction between the mathematics of the last discussion and the real-world cost-output situation to which the mathematics was related. Nothing was said about whether costs *are* or whether costs *should be* linearly related to output. The theme of the discussion was that *if* the relationship was linear, then certain interpretations suggested themselves. In particular, mathematical analysis of the linear equation showed that the ratio of y to x decreased as x increased, and we interpreted this as being a consequence of spreading fixed cost over an increasing output.

The linear equation under discussion is an example of a mathematical *model*. A model is a mathematical expression which seeks to capture the essential features of a real-world situation. If this objective can be achieved, then the full force of mathematical analysis can be applied to the model, and the outcome of the analysis may have important real-world interpretations. The key idea here is that mathematical analysis is a highly developed and powerful tool, but before it can be brought to bear on a real-world problem, it is necessary to construct an adequate mathematical representation (model) of the real-world situation. We do not expect a model to be an exact replica of the real world. We hope that the model will be adequate. In any event, we must be careful not to exceed the bounds of good sense when interpreting the outcome of mathematical analysis of a model. Thus, it probably would not be realistic to assume a linear relationship between cost and output no matter how large output became, and it certainly would make no sense to extend the linear model to negative outputs.

Physical scientists have had a high degree of success in formulating mathematical models. Witness, for example, the contributions of mathematical analysis of gravitational forces to man's achievements in space. Much effort now is being directed toward model building by workers in the social sciences. The linear programming model we shall encounter a few chapters hence is a good example of a relatively recent advance which has stirred interest in many parts of the world of business and economics.

A model should *fit* actual data reasonably well. To understand what is involved here, recall an earlier example in which a straight line was fitted to points whose coordinates were in the order (disposable income, personal consumption expenditures), and the line passed through (312, 295) and (575, 537). The actual data used in determining these points are shown in Figure 1–16, which also shows the plotted points (solid) and a line drawn freehand (with the aid of a stretched thread). Clearly, no line passes through all of the points, but the freehand line comes very close to the points, and the fit is very good—much better than is the case in other situations. Essentially the same line would be obtained if we had used points for all years from 1950 to 1970 and had applied the objective techniques of fitting developed in a study of statistics.

If we let C be personal consumption expenditures and D be disposable income, then use the two-point formula with (312, 295) and (575, 537), we obtain the linear model

$$C = 0.92D + 7.96.$$

The slope, 0.92, is called the marginal propensity to consume, as we learned earlier. What about the "intercept" or constant 7.96? Mechanically, this is consumption when disposable income is zero, but we should not infer any-

Linear Model Fitted to Consumption Expenditures and Disposable Income in the United States. Selected Years, 1950–70.

PERSONAL CONSUMPTION EXPENDITURES
(BILLIONS OF DOLLARS)

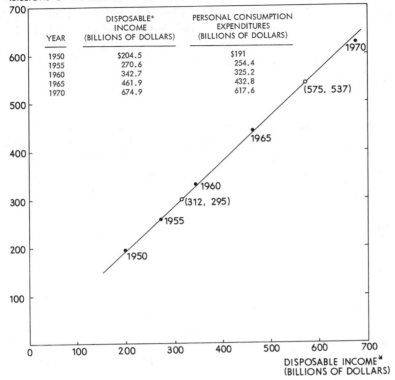

YEAR	DISPOSABLE* INCOME (BILLIONS OF DOLLARS)	PERSONAL CONSUMPTION EXPENDITURES (BILLIONS OF DOLLARS)
1950	$204.5	$191
1955	270.6	254.4
1960	342.7	325.2
1965	461.9	432.8
1970	674.9	617.6

DISPOSABLE INCOME*
(BILLIONS OF DOLLARS)

*Less interest paid by consumers.
Source: Economic Report of the President, 1973.

FIGURE 1–16

thing about consumption when D is zero because we have no data near zero and, in fact, D will never be zero. Moreover, in this changing economy we should be careful about extrapolating (going beyond the region of experience) very far. An appropriate interpretation of the constant 7.96 is that consumption expenditures are in major part determined by income, but that at any level of income, a part ($7.96 billion) is *independent* of income. This fixed part is called *autonomous* consumption. Thus, a common model of consumption in economics takes the form

$$C = A + MD$$

where the variables C and D are as defined earlier, the constant A is autonomous consumption, and the constant M is the marginal propensity to consume (about 0.92 in recent years).

Having warned about extrapolating beyond the region of experience, we must now admit that a major reason for developing a model is to make projections. Projections answer such questions as, What would consumption be if disposable income were $780 billion assuming the model $C = 0.92D + 7.96$ applies? Mechanical substitution leads to the projection $C = \$725.6$ billion. As a matter of record, disposable income did reach $780 billion in 1972, and in that year consumption expenditures were $726.5 billion, so the projection was very close to the actual figure. We see that extrapolation can lead to quite accurate results, but we have no assurance of accuracy, and our confidence lessens the further we depart from the region of experience.

As a final comment, we should point out that many of the equations used in textbooks are simplistic ones, with nice whole numbers as coefficients. Students ask, quite rightly, where they come from. The answer, of course, is that they were designed by the writer of the text to facilitate learning of mathematical techniques, for such is the objective of the text. One of the purposes of the foregoing discussion of models was to indicate where real-life equations come from, and to assure the student that simplified equations used for text illustration do have real-life counterparts.

1.21 BREAK-EVEN INTERPRETATIONS

In this section, we consider a retail firm which purchases and sells products. The firm plans to sell at prices which provide a markup of 35 percent on the retail value of the products, so that if x is the dollar value of sales, $0.35x$ is the markup and $0.65x$ is the cost of goods sold.[7] The cost, $0.65x$, is part of the variable cost of goods sold. Additionally, the firm expects other variable costs (for example, commissions) to be $0.10 per dollar of sales or $0.1x$, so that total variable cost is $0.65x + 0.10x$ or $0.75x$. Finally, the firm expects fixed cost of $12,000 in the period for which it is planning. Expressing total cost, y, as the sum of fixed and variable cost, we have

$$y = 12,000 + 0.75x.$$

Thus, if the firm's sales volume is $60,000, its cost will be

$$y = 12,000 + 0.75(60,000) = \$57,000,$$

and a profit, before taxes, of $60,000 - \$57,000 = \$3,000$ will be achieved.

---

[7] We wish to emphasize that in this section, x is the retail sales volume, in dollars. In order to achieve a markup rate of 0.35 or 35 percent on retail sales, the firm must mark up the cost of goods by $0.35/(1 - 0.35) = 0.35/0.65$. Thus, goods costing $650 would be priced at retail at $650 + [(0.35)/(0.65)](650) = \$1000$. The markup of $350 then is 35 percent of retail and the cost is 65 percent of retail. In general, if M_c and M_r are markup rates on cost and retail, then $M_c = M_r/(1 - M_r)$ and $M_r = M_c/(1 + M_c)$.

Exercise. Compute profit if sales are $40,000. Answer: Cost would be $42,000, so a loss of $2,000 would arise.

From the foregoing example of sales levels leading to profit and loss we are led to inquire what the *break-even* level of sales would be; that is, the level of sales which will equal cost. At this point, $x = y$. Call the point x_e the break-even level of sales, so that here $x = y = x_e$. Substituting x_e for both x and y in $y = 12{,}000 + 0.75x$ we have

$$x_e = 12{,}000 + 0.75x_e \quad \text{or} \quad x_e - 0.75x_e = 12{,}000 \quad \text{or} \quad 0.25x_e = 12{,}000,$$

so that

$$x_e = \frac{12{,}000}{0.25} = \$48{,}000.$$

Hence, sales above $48,000 lead to profit, sales below $48,000 lead to loss.

Exercise. If cost, y, is related to sales, x, by $y = 100 + 0.6x$, show that the break-even point is $250.

If the relationship between cost and sales is written in the general slope-intercept form

$$y = mx + b$$

then the interpretations are that b represents fixed cost, mx represents variable cost, and m itself represents variable cost per dollar of sales. At break-even:

$$y = x = x_e$$

so that, by substitution:

$$x_e = mx_e + b$$

$$x_e(1 - m) = b$$

$$x_e = \frac{b}{1 - m}.$$

Hence, the break-even level occurs when sales are equal to

$$\frac{\text{Fixed cost}}{1 - \text{Variable cost per dollar of sales}}.$$

Suppose that in making a budget for next year's operations, top management has set a sales goal of $200,000. Markup is to be 45 percent on retail (so cost is 55 percent of retail), and other variable cost is estimated at $0.05 per dollar of sales, so that variable cost is $0.55 + 0.05 = 0.60$ per dollar of sales. Fixed cost is projected at $56,000. Then the linear cost-sales model will be

$$y = 0.6x + 56,000.$$

The break-even level of sales will be

$$x_e - \frac{56,000}{1 - 0.6} = 140,000.$$

The calculations show the company will make a profit if its sales exceed $140,000. For example, if sales should turn out to be $190,000, net profit would be

$$190,000 - \text{Cost} = 190,000 - [0.6(190,000) + 56,000] = 20,000.$$

It is useful to keep in mind that a budget establishes *estimates* for coming operations, and that *actual* operations are not expected to be exactly equal to the estimates. However, assuming that the linear model applies, the estimates provide the information needed to predict results (say profit) if sales vary from the estimated value.

||||||||||||||||||||||||||||||||||||||||||||||||||||||||||||||||||||||||||||||||||||||||||||||||||||||||||||||||||||||||||||||||||||

Exercise.   Markup is to be 33 percent on retail and other variable cost is estimated at $0.13 per dollar of sales. Fixed cost is estimated at $4,000. a) What is the break-even point? b) Estimate profit if sales are $50,000. Answer: a) The equation is $y = 4,000 + 0.8x$ and the break-even point is $4000/(1 - 0.8) = \$20,000$. b) $6,000.

||||||||||||||||||||||||||||||||||||||||||||||||||||||||||||||||||||||||||||||||||||||||||||||||||||||||||||||||||||||||||||||||||||

A *break-even chart* can be constructed by plotting

$$y = mx + b$$

and the line where cost equals sales, $y = x$. The point of intersection of the last two lines establishes the break-even level of sales. Figure 1–17 shows the break-even chart for the foregoing example. The break-even point is seen to be $140,000. The separation of the lines to the right of break-even indicates profit; to the left, loss.

In passing, we should note that the break-even interpretations just discussed rest upon the assumption that total cost can be separated into two components, one fixed and the other varying directly in proportion to sales. These assumptions often are reasonably valid for a restricted range of sales. It is not realistic, however, to assume fixed cost as constant over all ranges of sales. If sales are proving to be considerably below expected levels, management may reduce salaries or take other actions to reduce "fixed" cost. It is not our purpose here to explore managerial action, but rather again to call for exercise of judgment when interpreting mathematical models.

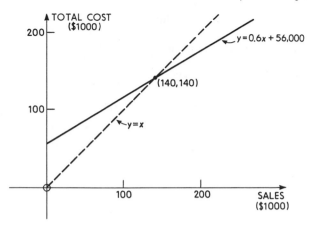

FIGURE 1–17

1.22 PROBLEM SET 6

Mark (T) for true or (F) for false.

Given that total cost, *y*, is related to sales volume, *x*, by the equation $y = 1000 + 0.2x$:

1. ( ) Variable cost will be $200 on sales of $1000.
2. ( ) Variable cost per dollar of sales is $0.20.
3. ( ) Fixed cost is $1000.
4. ( ) A loss would occur if sales were $1000.
5. ( ) Sales of $2000 would lead to a profit of $600.
6. ( ) The cost-sales line rises $2 for each increase of $10 in sales volume.
7. ( ) Average cost per dollar of sales is the same at various sales levels.
8. ( ) The slope of the line is interpreted as variable cost.
9. ( ) Variable cost is a constant.
10. ( ) Variable cost per dollar of sales is constant.
11. A company expects fixed cost of $22,800. Markup is to be 55 percent on retail. Variable cost in addition to costs of goods is estimated at $0.17 per dollar of sales.
    a) Find the break-even point.
    b) Write the equation relating cost and sales.
    c) What will net profit before taxes be on sales of $75,000?
    d) Make the break-even chart.
12. A company expects fixed cost of $36,000. Markup is to be 52 percent on retail, and variable cost in addition to cost of goods is estimated at $0.07 per dollar of sales.
    a) Find the break-even point.
    b) Write the equation relating sales and cost.
    c) What will net profit before taxes be on sales of $75,000?
    d) Make the break-even chart.

13. If total cost, y, is related to sales volume, x, by the equation $y = 0.47x + 29,786$, find:

   a) Variable cost per dollar of sales.
   b) Fixed cost.
   c) Total cost on sales of $72,000.
   d) The break-even point.
   e) Net profit before taxes on sales of $80,000.

14. If total cost, y, is related to sales volume, x, by the equation $y = 0.32x + 23,800$, find:

   a) Variable cost on sales of $40,000.
   b) Fixed cost.
   c) The break-even point.
   d) Net profit before taxes on sales of $30,000.

15. If variable cost per dollar of sales remains at last year's level, $0.40, but fixed cost this year is $3600 compared to $3000 last year, how much greater will this year's break-even point be than last year's?

16. Make the break-even chart for the equation in
   a) Problem 13.   *b*) Problem 14.

1.23 THREE-SPACE

The piece of paper or the blackboard upon which we plot points are segments of planes. On a given plane any point can be specified by an origin and two coordinates, and we refer to the points on a plane as points in space of two dimensions or, more briefly, points in *two-space*. If, now, we think of

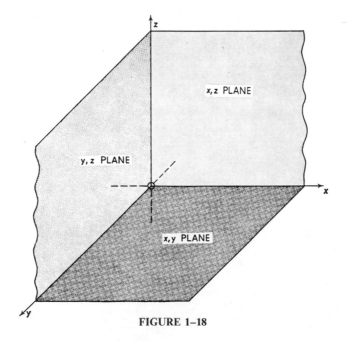

FIGURE 1–18

points in the interior of a room, referred to an origin which is the point at the lower left corner of the room, three coordinates are required to locate the point. The points in the room are points in *three-space.*

Figure 1–18 illustrates coordinates for three-space. The x-, y-, and z-axes are mutually perpendicular. The x- and z-axes are in the plane of the paper, and the y-axis is to be visualized as a perpendicular coming out of the page.[8] The dotted extensions represent the negative portions of the axes. Viewing the figure as a section of a room, the inner left corner of the room is the origin; the floor of the room is the plane of the x- and y-axes; one wall of the room is the plane of the x- and z-axes, the other is the plane of the y- and z-axes. We refer to these coordinate planes as the x, z plane; the x, y plane; and the y, z plane.

The z-axis is a straight line formed by the intersection of the two walls. The x- and y-axes similarly are formed by the intersection of two coordinate planes. In passing, it is worth noting for future reference that *the intersections here discussed are particular cases of the general observation that if two planes intersect, the intersection is a complete straight line.*

Figure 1–19 illustrates the plotting of points in three-space. The point (5, 3, 6), that is:

$$x = 5, \qquad y = 3, \qquad z = 6$$

is found by counting five units to the right along the x-axis, then three units out into the x, y plane in a direction parallel to the y-axis, then six units upward, parallel to the z-axis.

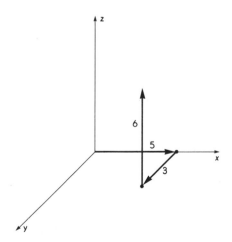

FIGURE 1–19

---

[8]The assignment we have chosen for the x, y, and z axes is fairly standard. However, some texts have the positive direction of the y-axis into, rather than out of, the page or interchange the x and y axes, or both.

⁣||||||||||||||||||||||||||||||||||||||||||||||||||||||||||||||||||||||||||||||||||||||||||||||||||||||||||||||||||||||||||||||||||||||||||||||||||

Exercise. Take three rectangular pieces of cardboard (3 by 5 inch cards will do) and by cutting slits and assembling, construct a model of three mutually perpendicular planes. The model should look like the interior walls, and separating floor, of a two-story house with four rooms up and four rooms down.

⁣||||||||||||||||||||||||||||||||||||||||||||||||||||||||||||||||||||||||||||||||||||||||||||||||||||||||||||||||||||||||||||||||||||||||||||||||||

⁣||||||||||||||||||||||||||||||||||||||||||||||||||||||||||||||||||||||||||||||||||||||||||||||||||||||||||||||||||||||||||||||||||||||||||||||||||

Exercise. The planes of the model referred to in the last exercise, when extended indefinitely, divide space into eight octants. Call the octant at the upper-front-right octant I, then going counterclockwise call the remaining octants on the upper floor II, III, and IV. Octant V is below I, and again going counterclockwise, the remaining octants on the lower side are VI, VII, and VIII. Verify that $(1, -2, 3)$ is in octant II, $(1, 2, -3)$ is in V, and $(-1, -2, -3)$ is in VII.

⁣||||||||||||||||||||||||||||||||||||||||||||||||||||||||||||||||||||||||||||||||||||||||||||||||||||||||||||||||||||||||||||||||||||||||||||||||||

1.24 LINEAR EQUATIONS IN THREE-SPACE

Earlier in the chapter, we saw that linear equations in two variables were represented geometrically by straight lines in two-space. The general form of such equations was $ax + by = c$. Special cases were those in which either a or b was 0, such as $y = 3$ or $x = -5$. We saw that these special cases were lines parallel to one of the coordinate axes (or perpendicular to the other). In what follows, we find that equations in three variables of the form

$$ax + by + cz = d$$

are represented geometrically in three-space as *planes,* and that special cases, where one or two of a, b, and c are zero, are represented by planes which bear parallel or perpendicular relationships to the coordinate planes and axes.

Consider the very special equation in three variables:

$$0x + 0y + z = 0$$

which we shall refer to as the equation $z = 0$ in three-space. Graphically, any point in the x, y plane has a z-coordinate of zero, and any point for which z is zero lies in the x, y plane. Hence, $z = 0$ is the equation of the x, y plane itself. Similarly, in three-space, $x = 0$ is the equation of the y, z plane, and $y = 0$ is the equation of the x, z plane. See Figure 1–20.

Consider next the plane which is parallel to the y, z plane, and four units to the right thereof. This plane will, of course, be perpendicular to the x, y

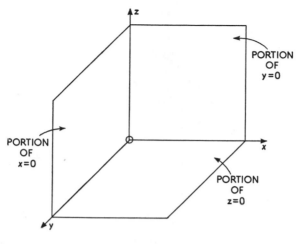

FIGURE 1–20

plane and to the x, z plane. See Figure 1–21. Clearly, x is 4 for any point on this plane, and any point for which x is 4 is on the plane. Hence, $x = 4$ is the equation of the plane. To emphasize that we are thinking in three-space, we could write the equation as $x + 0y + 0z = 4$. Similarly, in three-space, the equation $z = -5$ represents a plane which is parallel to, and five units below, the x, y plane. See Figure 1–22. Planes, of course, extend indefinitely, so that illustrations such as those in Figures 1–21 and 1–22 represent only finite portions of planes.

Exercise. Sketch a portion of the plane $y = 2$. Answer: This plane is parallel to, and two units "in front of," the x, z plane.

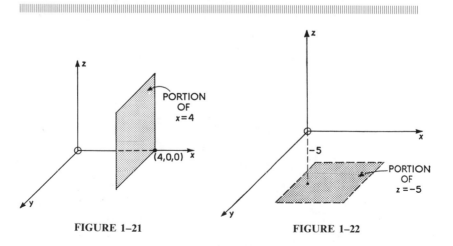

FIGURE 1–21   **FIGURE 1–22**

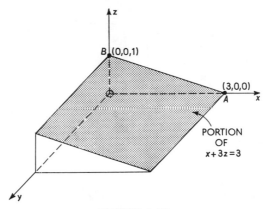

FIGURE 1–23

Figure 1–23 shows a segment of a plane which is perpendicular to the *x, z* plane (and parallel to the *y*-axis) and cuts the *x*- and *z*-axes at $A(3, 0, 0)$ and $B(0, 0, 1)$, respectively. The relationship among the coordinates at *A* and *B* is

$$x + 3z = 3.$$

Moreover, inasmuch as the plane is perpendicular to the *x, z* plane, the same relationship exists for every point on the plane, and the coordinates of every point on the plane will satisfy the equation. Hence:

$$x + 3z = 3$$

in three-space is the equation of a plane which is perpendicular to the *x, z* plane. Similarly:

$$2z + 5y = 10$$

in three-space is a plane which is perpendicular to the *y, z* plane.

Finally, think of a plane which slices up through the floor and cuts floor and walls in the manner shown in Figure 1–24. This plane is not parallel to

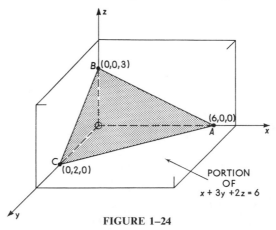

FIGURE 1–24

any of the coordinate planes. The section shown is of triangular appearance because the parts of the plane below the floor and behind the walls have not been shown. Suppose that the plane cuts the axes at the points *A, B,* and *C,* as shown. These three points completely determine a unique plane. By trial and error, we can determine that the points (6, 0, 0), (0, 0, 3), and (0, 2, 0) satisfy the linear equation

$$x + 3y + 2z = 6.$$

We state without further demonstration that any point on the plane containing *A, B,* and *C* satisfies the last equation and that any point whose coordinates satisfy the equation will lie on the plane.

Our purpose in the last few paragraphs has been to build an intuitive basis for the statement that *every* equation of the form

$$ax + by + cz = d$$

in three-space is represented by a plane. The statement can be justified by rigorous mathematical development, but we choose here to rely upon intuition.

For sketching purposes, points such as *A, B,* and *C* on Figure 1–24 can be found by setting pairs of coordinates equal to zero to obtain intercepts. Line segments drawn connecting intercepts (such as *AB* in Figure 1–24) represent the intersection of the plane with the coordinate plane and are called *traces;* thus, *AB* is the *x, z* trace of the plane $x + 3y + 2z = 6$.

Exercise.   Sketch the plane $6x + y + 2z = 12$ in the manner of Figure 1–24. Answer: The intercepts are (0, 0, 6), (0, 12, 0), and (2, 0, 0).

1.25 *n*-SPACE AND HYPERPLANES

The linear equation in four variables:

$$w + x + 2y + 3z = 12$$

is satisfied by the point (1, 5, 0, 2), by (0, 0, 0, 4), and by an unlimited number of other points. Each point has four coordinates (written in alphabetical order) and is a point in four-space. Inasmuch as we are limited to three-dimensional perception, the notion of four-space is purely a matter of terminology; we agree to call (1, 5, 0, 2) the coordinates of a point in four-space. Again, inasmuch as the last equation is linear, like that of a plane but with more than three variables, we agree to call the expression the equation of a *hyperplane.*

Although we are limited to three-space in geometrical constructions, there is no such limit on mathematical analysis. We may speak of 6-space,

150-space, or *n*-space, where *n* is any positive integer, and we may treat equations in *n* variables. It should be understood that our inability to graph equations in four or more variables in no way reflects upon the realism of such equations. Consider Table 1–1. Suppose that we wish to mix ingredients

TABLE 1–1

Ingredient	Number of Units of Ingredient	Weight per Unit
A . . . . *w*		2
B . . . . *x*		3
C . . . . *y*		1
D . . . . *z*		6

A, B, C, and D in amounts such that the total weight of the mixture will be 1000 pounds. The weight of *w* units of A, at two pounds per unit, is $2w$ pounds. Similar expressions hold for the weights of the other ingredients. The total weight is

$$2w + 3x + y + 6z$$

for any combination of ingredients, and the condition for the total weight is

$$2w + 3x + y + 6z = 1000.$$

If we decided to use only ingredients A and B in the mixture, the weight condition would be

$$2w + 3x = 1000.$$

Both of the last two equations describe a simple real-life situation. The fact that one is called a straight line and can be graphed, whereas the other is called a hyperplane and cannot be graphed, has no bearing on the question of realism.

Exercise.   If A, B, C, and D (Table 1–1) have unit costs of 2, 5, 7, and 11 cents, respectively, write the equation which specifies that the total cost of ingredients is to be $5. Answer: $2w + 5x + 7y + 11z = 500$.

To summarize the last two sections, we state that a linear equation in two-space is a straight line. In three-space a linear equation is a plane. A linear equation in *n*-space, where *n* exceeds 3, is called a hyperplane.

1.26 ARBITRARY VARIABLE TERMINOLOGY

The number of points on a straight line is without limit. For example, the equation

$$2x + y = 10$$

is satisfied by (0, 10), (1, 8), (−3, 16), and so on. For any arbitrarily stated value of x, the corresponding value of y is $10 - 2x$. We shall have occasion to state this last remark as follows: The *general* solution of the equation $2x + y = 10$ is

$$x \text{ arbitrary}$$
$$y = 10 - 2x.$$

Of course, we could have stated the general solution with y arbitrary; thus:

$$y \text{ arbitrary}$$
$$x = \frac{10 - y}{2}.$$

The term general solution is contrasted with a *specific* solution. Thus, (2, 6) is the specific solution when x is 2 (or when y is 6), whereas the general solution tells us how one variable is computed for any (every) stated value of the other variable.

If a single linear equation has n variables, the general solution will have $n - 1$ variables (that is, all but one) arbitrary. For example:

$$2x + 3y + 4z = 12$$

has three variables. We know that if specific values are assigned to two of the variables, we can compute the proper value of the third variable so that the equation will be satisfied. If values are assigned arbitrarily to x and y, then z must be assigned the value

$$\frac{12 - 2x - 3y}{4}$$

if the equation is to be satisfied. Hence, we may write the general solution as

$$x, y \text{ arbitrary}$$
$$z = \frac{12 - 2x - 3y}{4}.$$

Obviously, the solution could be written with x and z arbitrary, or with y and z arbitrary.

From a practical point of view, the statement that a variable may be assigned values arbitrarily means that we have freedom of choice. For example, suppose that

$$x + y = 1000$$

expresses the condition that a tank which holds 1000 gallons of fluid is to be filled using x gallons of fluid A and y gallons of fluid B. The general statement of the solution as

$$x \text{ arbitrary}$$
$$y = 1000 - x$$

means that we are free to choose the value for x as long as we then take y as $1000 - x$. Inasmuch as this is an applied problem, the quantities must not be negative; hence, x is restricted to positive values from 0 to 1000, inclusive.

Exercise.   In the last exercise, total cost was $2w + 5x + 7y + 11z = 500$. *a*) Write the general solution with x, y, and z arbitrary. *b*) If we buy 20 pounds of B, 10 pounds of C, and 4 pounds of D, so that (x, y, z) are (20, 10, 4), how many pounds of A can we buy? Answer: *a*) x, y, z arbitrary, $w = 250 - 2.5x - 3.5y - 5.5z$. *b*) 143 pounds.

In later chapters we shall describe problems which involve several linear conditions on the variables and discuss methods for choosing combinations of values of the variables which will satisfy the conditions and, at the same time, be a *best* combination in the sense of maximizing a profit or minimizing a cost. In these later chapters, we shall make extensive use of arbitrary variable terminology.

1.27 SET TERMINOLOGY

For reasons which will become clear in later applications, we have chosen to express general solutions in arbitrary variable terminology. It may be helpful for those who are familiar with sets (see Appendix 1) to show the relationship between set and arbitrary-variable terminology.

The coordinates of a point are an example of an *ordered pair* of numbers, that is, a pair in which order has significance. In the case of coordinates, the first number of the pair is the x value and the second is the y value. An equation in two variables has ordered pairs as members of its solution set. The expression

$$\{(x, y) : 2x + y = 12\}$$

is read as "the set of ordered pairs, x, y, such that $2x + y = 12$." If we interpret each ordered pair as a *point*, then the foregoing expression can be read as "the set of points such that $2x + y = 12$."

Exercise.   Write the symbolism which represents the solution set for the equation $y - 3x = 7$. Answer: $\{(x, y) : y - 3x = 7\}$.

Clearly, the solution set for an equation such as $y - 3x = 7$ is an infinite set. Or, to put the matter in another way, the set of points on the line $y - 3x$

$= 7$ is an infinite set of points. Similarly, the solution set of $x + 2y - 3z = 19$, which is symbolized

$$\{(x,\ y,\ z) : x + 2y - 3z = 19\}$$

is an infinite set. We cannot list all the elements of the solution set. We can, however, generate as many elements of the solution set as we wish. One convenient way to do this is to solve the equation for one of the variables in terms of the other variables, in which case the equation is said to be in *explicit* form. Thus,

$$x + 2y - 3z = 19$$

is in *implicit* form because it is not solved for one of the variables, whereas

$$x = 19 - 2y + 3z$$

is in explicit form. It explains how x is computed for stated values of y and z. We may construct as many members of the solution set as we wish by giving y and z any values, selected arbitrarily, then computing x as $19 - 2y + 3z$ to form the ordered triple which is a member of the solution set. Inasmuch as we are thinking about constructing solutions from arbitrarily selected values of y and z, we refer to these as arbitrary variables.

The terminology and symbolism in this section do not add to our ability to solve equations, but they do provide a fundamental understanding of what we mean when we refer to solutions. We do not have to come to grips with fundamentals when we deal with an equation such as $2x = 3x - 1$. We see immediately what the solution is and can prove it satisfies the equation. However, if we are asked to explain what is meant by "solution" with reference to the equation $y = x + 6$, we must get down to fundamentals and state that the solution is the (infinite) set of ordered pairs of numbers, (x, y) such that the second number is 6 more than the first. We express the same concept symbolically by

$$\{(x, y) : y = x + 6\}.$$

In this explicit form, we think of generating members of the solution set by assigning arbitrary values to x, then computing y as $x + 6$. It is clear that the statement of the general solution in terms of an arbitrary variable, shown earlier in the chapter, namely,

$$x \text{ arbitrary}$$
$$y = x + 6$$

is a satisfactory description of the solution set.

1.28 PROBLEM SET 7

Mark (T) for true or (F) for false:

1. ( ) Two planes may intersect at a single point.
2. ( ) $x + y = 7$ is a plane in three-space.

3. ( ) $x + y + z = 0$ is a plane in three-space.

4. ( ) In four-space, $x = 10$ would be called a hyperplane.

5. ( ) In three-space, $x = 0$ is the equation of the y-axis.

6. ( ) In three-space, $y = 0$ is the equation of the x, z plane.

7. ( ) The plane $3x + 2y = 6$ will never intersect the z-axis.

8. ( ) The plane $x - 3z = 7$ is perpendicular to the x, z plane.

9. ( ) In two-space, $y = 0$ is the equation of the y-axis.

10. ( ) In three-space, every linear equation in three variables is a plane.

11. ( ) It is not possible to graph a hyperplane.

12. ( ) Equations of hyperplanes have no applicability to real world situations.

13. ( ) The point $(-2, -3, 5)$ is in octant III.

14. ( ) The point $(1, 3, -4)$ is in octant V.

15. ( ) The origin of coordinates lies on the plane $x + 2y + 3z = 0$.

16. ( ) The general solution for a single equation which has five variables will have four arbitrary variables.

17. ( ) There are five ways of setting up the solution for the equation described in Problem 16.

18. ( ) The plane $2x + 3y - 5z = 14$ intersects all three coordinate planes.

19. ( ) The x-intercept of the plane in Problem 18 is 7.

20. ( ) The x, y trace of a plane is a line passing through the x- and y-intercepts of the plane.

21. Make a three-dimensional sketch showing the points:
a) $(5, 0, 0)$.
b) $(2, 7, 4)$.
c) $(-3, 0, 2)$.

22. Given that $x + 3z = 6$ is the equation of a plane in three-space, rewrite the equation, including the missing variable y.

23. Make graphs showing illustrative segments of each of the following planes:
a) $x = 10$.
b) $z = -3$.
c) $y = 2$.
d) $x + z = 5$.
e) $x + 2y = 4$.
f) $x + y + z = 3$.

24.

Ingredient	Number of Units	Weight per Unit (Pounds)	Cost per Unit (Dollars)
A . . . . w		1	$0.50
B x		3	0.30
C . . . . y		2	1.20
D . . . . z		4	0.80

a) A mixture is to be made with total weight 500 pounds. Write the equation whose solutions are the permissible numbers of units of each ingredient in the mixture.

b) A mixture is to be made costing $310. Write the equation whose solutions are the permissible numbers of units of each ingredient in the mixture.

25. Write in two different forms the general solution of

$$3x + 2y = 6.$$

26. Write in three different forms the general solution of

$$2x + 3y - 2z = 18.$$

27. What is meant by the statement that

$$3x - 4y = 6$$

has an unlimited number of specific solutions?

28. If the x, y trace of the plane $3x + 4y - 6z = 12$ were considered as a straight line in two-space, what would be the equation of the line?

29. What are the coordinates of the points which establish the x, y trace of the plane in Problem 28?

30. If $x + y = 1000$ expresses the condition that a tank which holds 1000 gallons of fluid is to be filled using x gallons of fluid A and y gallons of fluid B, show that the general solution, graphically, consists of the set of points on a line segment which is the hypotenuse of an isosceles right triangle.

31. (See Problem 30.) If A costs 20 cents a gallon and B costs 23 cents a gallon, and if the greatest amount of A that can be obtained is 600 gallons, what combination of x and y satisfies the requirement and also leads to minimum total cost?

32. (See Problems 30 and 31.) Under what conditions would the minimum cost combination be at one of the intercepts?

33. Shingle requirements (per house) for superdeluxe, top-grade, and regular houses are, respectively, 10, 8, and 7 bundles. Write the expression for the total number of bundles of shingles needed for x superdeluxe, y top-grade, and z regular houses.

34. (See Problem 33.) A freight car can hold 560 bundles of shingles.
    a) Write the equation whose solutions contain the numbers of houses which could be shingled with one freight car load of shingles.
    b) In addition to being nonnegative, what other requirements must be satisfied by the solutions in (a)?

Note: Problems 35–40 relate to Section 1.27.

Translate the set expressions of Problem 35–37 into words.

35. $\{(x, y) : y - 3x = 6\}$.   36. $\{(x, y, z) : 2x + y - 3z = 15\}$.

37. $\{(w, x, y, z) : 2w + x - y + 3z = 30\}$.

38. If the expression in Problem 35 were put into explicit form with x as the arbitrary variable, it would be $\{(x, y) : y = 3x + 6\}$. Write Problem 37 with x and z arbitrary.

39. See Problem 38. Write Problem 37 with all variables but y arbitrary.

40. The general solution for Problem 36 with x and z arbitrary may be written in the form

$$x, z \text{ arbitrary}$$
$$y = 3z - 2x + 15.$$

Write the general solution for Problem 37 in this form with all variables but x arbitrary.

1.29 REVIEW PROBLEMS

1. Given the points $A(3, -4)$, $B(5, -2)$, $C(3, -2)$, find the distances
   a) AC.   b) BC.   c) AB.

2. In Problem 1, which segment is vertical and which is horizontal?

3. In each case, the given points are the end points of the diagonal of a rectangle whose sides are parallel to the axes. What are the coordinates of the other corners of the rectangle, and what is the length of the diagonal?
   a) $(0, 0)$, $(3, 4)$.   b) $(-1, 2)$, $(8, 14)$.   c) $(1, 2)$, $(3, 4)$.

4. Find the perimeter of the triangle whose vertices are $(1, 2)$, $(7, 2)$, and $(7, 10)$.

5. a) If city B is 5 miles east and 12 miles north of city A, how far is it from A to B?
   b) A section of a city has square blocks with streets perpendicular to avenues. How far would we walk (in blocks) going by sidewalk from (2nd St., 7th Ave.) to (7th St., 19th Ave.)?
   c) See b). A subway is to be built under a straight line connecting the corners. If a block is 400 feet long, how long will the subway line be in feet?

6. If y and x represent, respectively, expenses and sales in thousands of dollars and last week's operations are characterized as $(10, 12)$, compared to this week's $(13, 16)$, by how much did expenses and sales change from last week to this?

7. If sales increased from $12 thousand to $15 thousand, the *percent* increase was (100 percent) $(15 - 12)/12 = 25$ percent.
   a) Write the expression for percent change if sales go from S_1 to S_2.
   b) What would it mean if the answer to a calculation such as that in (a) was negative?

8. Find the slope of the segment joining the following point pairs:
   a) $(-4, 7)$, $(-1, 3)$.   b) $(1, 2)$, $(5, 6)$.
   c) $(0, 0)$, $(0, 5)$.   d) $(5, -1)$, $(10, -1)$.
   e) $(5, 0)$, $(5, 3)$.   f) $(-2, -1)$, $(2, -4)$.
   g) $(1, -3)$, $(4, -1)$.   h) $(-2, 5)$, $(3, 5)$.
   i) If personal consumption expenditures increased from $254 billion to $618 billion when disposable income (less consumer interest on loans) increased from $270 billion to $675 billion, compute the marginal propensities to consume and to save.

9. Find the equation of the line passing through each of the point pairs in Problems 8a through 8h.

10. Find the equation of the line passing through the given point and having the stated slope:
    a) $(1, 3)$, slope $1/5$.   b) $(0, 0)$, slope 0.
    c) $(1, 1)$, slope 1.   d) $(1, 3)$, slope -2.
    e) $(-2, -4)$, vertical.   f) $(-3, -4)$, slope 0.

11. a) What is the equation of the line parallel to and five units below the x-axis?
    b) If p is vertical and q horizontal, what is the geometric nature and interpretation of the demand curves $p = 2$ and $q = 800$?

12. As sales, x, change from $300 to $600, selling expense changes from $250 to $400. Assume that the given data establish a linear relationship between sales and selling expense. Find the equation of the relationship.

13. If selling expense is $100 when sales are $150, and if expense increases $1 for

each increase of $3 in sales, write the straight-line relationship between expense and sales.

14. If a man's weekly pay is computed at $50 (whether or not he works) plus $5 per hour worked, what equation relates weekly pay, y, to hours worked, x?

15. Explain why the lines $y = -6$ and $x = 15$ are perpendicular.

16. What is the slope and y-intercept of the line in Problem 14?

17. Graph the following using the intercepts:
    a) $3x - 4y = 24$. b) $2x + y = 10$. c) $y = 15$.

18. Using the coefficients of x and y, determine the slope of:
    a) $x - y = 4$. b) $3x - 6y = 4$. c) $3x + 4y = 7$.
    d) An ounce of bourbon contains $\frac{1}{2}$ of an ounce of alcohol and an ounce of vermouth contains $\frac{1}{8}$ of an ounce of alcohol. Write the relation which exists if b ounces of bourbon and v ounces of vermouth are to be mixed to make a drink containing two ounces of alcohol. What is the substitution rate of bourbon per ounce of vermouth?

19. What is the equation of the line which has a slope of $-2/3$ and a y-intercept of -4?

20. a) Find the equation of the line through (3, 4) which is parallel to the line $x - 2y = 4$.
    b) A demand curve shifts from $p = 50 - 0.2q$ to $p = 60 - 0.2q$. If p is vertical, what are the amount and direction of the horizontal and vertical shifts? Interpret the shifts.

21. a) Find the equation of the line through $(-1, 15)$ which is perpendicular to the line $x - 2y = 4$.
    b) If an existing pipeline is described by $y = x$ and a plant at (14, 26) wishes to connect to the pipeline on a perpendicular, what is the equation of the perpendicular and how far will it be from the plant to the connecting point? Note: In the equation of the perpendicular, let $y = x$ to find the connecting point.

22. If y and x are in proportion so that y is to x as 2 is to 5, then:
    a) Write the equation relating y and x.
    b) Graph the equation in (a).

23. a) What is the equation of the line on which the abscissa of every point is $\frac{1}{4}$ of the ordinate?
    b) If a company sells x units of a product at $4 per unit, write the expression for R, the total revenue received for the x units. Interpret the intercepts of this line.

24. If the first of two lines has the form $ax + by + c = 0$, and the second the form $dx + ey + f = 0$, and if the first line is to pass through the origin and the second is to be parallel to the first, what relationships must exist among $a, b, c, d, e,$ and f?

25. The *productivity* (as contrasted to *production*) of a factory is often measured by the ratio *output per man-hour*. If a factory has a constant productivity of five units per man-hour, what is the equation relating units of output, y, to number of man-hours worked, x?

26. a) If y is a man's weekly pay in dollars and x is the number of hours the man worked, and if $y = 3.5x + 25$, interpret the numbers 3.5 and 25.

b) A machine purchased now ($t = 0$) for $10,000 depreciates in value by a constant amount per year for 20 years to a scrap value of $1,000. Write the equation for D, the depreciated value of the machine at time t years. Interpret the slope and intercept of this line.

27. If total factory cost, y, of making x units of a product is given by $y = 2x + 40$, and if 100 units are made:
    a) What is the variable cost?
    b) What is total cost?
    c) What is variable cost per unit?
    d) What is average cost per unit?
    e) What is the marginal cost of the 100th unit?
    f) What is the marginal cost of the 1st unit?

28. In Figure A, the slant line represents the total factory cost, C, of producing a number of units, Q. Write an interpretation of each of the following:
    a) *OT.*   *b*) *PL/WP.*   *c*) *RW/OR.*   *d*) *PL.*   *e*) *SW.*

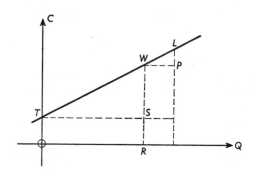

FIGURE A

29. A company expects fixed cost of $25,000. It plans to mark up goods 46 percent on retail, and to incur other variable costs of $0.06 per dollar of sales.
    a) Find the equation relating total cost to sales.
    b) Find the break-even point.
    c) Make the break-even chart.
    d) What will net profit before taxes be if sales are $100,000?

30. If the total cost of operations, y, is related to sales, x, by the equation $y = 10,500 + 0.58x$, find:
    a) Fixed cost.
    b) Total cost on sales of $60,000.
    c) The break-even point.
    d) Net profit before taxes on sales of $65,000.

31. If fixed cost is $10,000 and cost increases by $2 for each $3 increase in sales, find:
    a) The break-even point.
    b) The equation relating cost and sales.

32. In three-space, in what octant would each of the following points be?
    a) $(1, 7, 5)$.   b) $(-3, -5, -6)$.
    c) $(1, -3, -6)$.   d) $(2, 4, -3)$.

33. Make three-dimensional sketches showing the points:
    a) $(1, 3, 5)$.   b) $(2, 4, -6)$.   c) $(1, -2, 3)$.

34. Sketch the octant I portion of the plane $5x + 3y + 2z = 30$.

35. What are the coordinates of the points which establish the x, y trace of the plane $5x + 3y + 2z = 30$?

36. If a college has w freshmen, x sophomores, y juniors, and z seniors, what is the expression for the total number of students in the college?

37. (See Problem 36). If the college wishes to enroll exactly 1500 students, write the equation whose solutions are the permissible numbers of students in each class.

38. Write the general solution of $x + 2y + 3z = 6$ in three forms.

39. Translate the following set expressions:
    a) $\{(x, y) : 5x + 6y = 21\}$.   b) $\{(x, y, z) : x + 2y + 3z = 0\}$.

40. Express the solution space for the following in set notation:
    a) $3x - 5y = 15$.   b) $3x - 2y + 5z - 6 = 0$.

2

Systems of Linear Equations

2.1 INTRODUCTION

THE VARIABLES encountered in a problem may have to fulfill more than one condition. In a production problem, for example, the numbers of units of various products made will be restricted by conditions such as time available for production and money available for the purchase of raw materials. When each of the conditions can be expressed in the form of a linear equation, the mathematical description of the problem is a *system* of linear equations. The procedures for finding the values of the variables which satisfy all equations of the system *simultaneously* are the subject matter of this chapter.

Example. The break-even discussion of Chapter 1 introduced a system of two linear equations in two variables. There we saw that if sales, x, and cost, y, were related by the linear condition

$$y = 1200 + 0.4x$$

we could solve this equation simultaneously with the condition $y = x$ to find where sales revenue just equals cost. If we express the 2 by 2 system as

$$y = 1200 + 0.4x$$
$$y = x,$$

y may be *eliminated* by subtracting the second equation from the first, yielding

$$0 = 1200 - 0.6x$$

from which the break-even point is at $x = 2000$. Graphically, the simultaneous solution of the system, which is $x = y = 2000$ in this case, is the point of intersection of the graphs of the two equations.

54

The number of relevant variables in a system, and the number of equations representing conditions to be fulfilled, vary widely. The requirement that *all* conditions expressed by the system be satisfied sometimes cannot be met. In other cases only one set, or many sets, of values for the variables will satisfy all conditions. In this chapter we shall learn systematic procedures for solving systems with varying numbers of equations and variables, how to determine if no solution is possible, and how to express the solution if the system is satisfied by more than one set of values. We shall consider the effect of the requirement that values in the solution set not be negative, and also show how the best set can be chosen when the system has more than one solution set. Two of the five sections on applications are concerned with supply and demand analysis.

We turn first to a consideration of the relation between solutions and points of intersection.

2.2 NUMBER OF SOLUTIONS POSSIBLE

A specific solution of a system of linear equations is a set of values, one for each variable, which simultaneously satisfy all equations of the system. Geometrically, a set of values for the variables is represented by a point, and a set satisfying all the equations is represented by a point which lies on the graphs of all the equations; that is, a solution is a point of intersection of all the graphs. We shall appeal to the geometry of intersections of lines and planes to illustrate that *the number of solutions of a linear system is either zero, one, or unlimited.*

2.3 INTERSECTIONS OF STRAIGHT LINES

The graphs of linear equations in two-space are straight lines. The parts of Figure 2–1 show the intersection possibilities for such lines. If a system has two equations in two variables, either the corresponding lines intersect in a single point, or they are parallel and have no intersection point, as shown in Figures 2–1A and 2–1B. The three lines representing a system of three

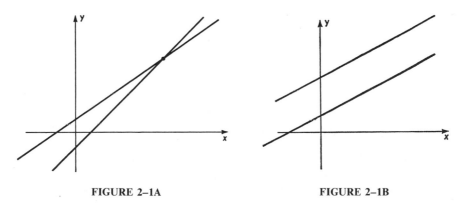

FIGURE 2–1A   FIGURE 2–1B

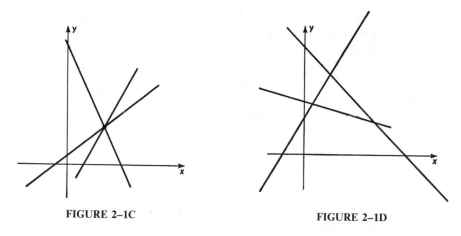

FIGURE 2–1C   FIGURE 2–1D

equations in two variables may intersect in a single point, as in Figure 2–1C, or there may be no point which lies on all three lines, as in Figures 2–1D and 2–1E. The same intersection possibilities exist if more than three lines are plotted, and we conclude that two or more different lines have one point in common or no points in common. As a special case, however, we note that if the system has two equations, and the terms of one equation are a constant multiple of the corresponding terms of the other, then both equations have the same graph. All of the (unlimited) points on one line also lie on the "other" line, so that in this special sense the system has an unlimited number of solutions.

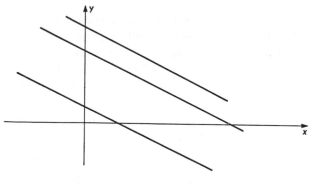

FIGURE 2–1E

2.4 INTERSECTIONS OF PLANES

Linear equations in three-space are represented by planes. A system of two equations in three variables is represented geometrically by two planes. Any points which lie on both planes will satisfy both equations and therefore be

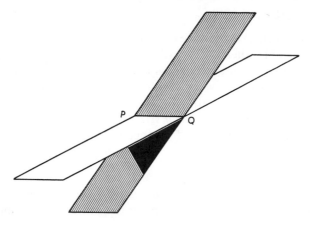

FIGURE 2–2A

solutions of the system. Figures 2–2A and 2–2B show that two different planes either intersect in a complete straight line *PQ*, and so have an unlimited number of points in common (Figure 2–2A), or they do not intersect at all (Figure 2–2B).

The three planes representing a system of three equations in three variables may have either zero, one, or an unlimited number of points in common. Observe Figure 2–2A, and visualize the intersection possibilities if a third plane cut into the figure. The third plane might contain the line *PQ*, so that the three planes would have an unlimited number of points in common, or it might slice across *PQ* at an angle, hitting this line at a single point. Finally, the third plane might not cut *PQ* at all, so that the three planes would have no point in common.

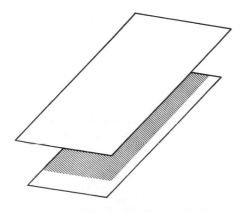

FIGURE 2–2B

2.5 A GENERALIZATION

Geometrical sketches cannot be made to aid us in determining the number of solutions possible when the equations of a system have more than three variables. However, intuition suggests that what has been said for systems with two variables and for systems with three variables carries over for systems with more than three variables, and intuition turns out to be correct. *The number of solutions for any linear system is either zero, one, or unlimited.*[1] We shall see many illustrations of this generalization as we move through the chapter.

Exercise. Imagine that two successive pages of this book are planes which meet in a straight line at the binding. Use a piece of paper as a third plane.

a) Arrange the third plane so that the three planes have a single point in common.

b) Arrange the third plane so that the three planes have an unlimited number of points in common.

c) Find two substantially different orientations of the third plane such that the three planes have no points in common.

2.6 LINEAR COMBINATIONS

If we have two equations, such as

$$x + y - 4 = 0$$
$$2x + 3y - 9 = 0$$

and we combine them into the equation

$$p(x + y - 4) + q(2x + 3y - 9) = 0$$

where p and q are any numbers, the last equation is called a linear combination of the two given equations. As examples of linear combinations of the given equations, we may write

$$3(x + y - 4) + 2(2x + 3y - 9) = 0$$
$$-2(x + y - 4) + 1(2x + 3y - 9) = 0.$$

Now, a solution of the first given equation is a number pair which makes $(x + y - 4)$ equal zero, and a solution of the second is a number pair

---

[1] In the language of sets, each specific solution is an element of the solution set. In this language, the statement referred to means that the solution set may be the empty set, it may have a single element, or it may be an infinite set. Each element of the solution set is a group of numbers. If the equations have two variables, the solution set elements are ordered pairs of numbers; if three variables, the solution set elements are ordered number triples; if n variables, the solution set elements are ordered *n-tuples* of numbers.

which makes $(2x + 3y - 9)$ equal zero. A solution of the system of two equations is a number pair which makes both parenthetical expressions equal zero, in which case

$$p(x + y - 4) + q(2x + 3y - 9)$$

will be zero, no matter what the numbers p and q are. In other words, the number pair which satisfies both given equations will satisfy any linear combination of the two equations.

Suppose that we construct the linear combinations

$$2(x + y - 4) - 1(2x + 3y - 9) = 0$$
$$3(x + y - 4) - 1(2x + 3y - 9) = 0.$$

Removing the parentheses and collecting terms, we find these linear combinations are simply

$$y = 1$$
$$x = 3.$$

Inasmuch as the common solution of the two given equations must satisfy every linear combination, it must satisfy the last two statements. We conclude that the solution of the system of two equations is the point $(3, 1)$.[2]

2.7 ELIMINATION PROCEDURE

The discussion in the last section suggests a method of attack when seeking solutions of a linear system; namely, form linear combinations of pairs of equations in a manner such that a variable is *eliminated*. For example, given the system

$$2x + 3y - 2 = 0$$
$$5x + 4y - 12 = 0$$

we see that multiplication of the first equation by 5 and the second by 2, followed by subtraction, will result in the elimination of x. Thus:

$$5(2x + 3y - 2) - 2(5x + 4y - 12) = 0$$

becomes, upon simplification:

$$7y + 14 = 0$$

from which we see that y must equal -2. Rather than forming a linear

---

[2] A set made up of the elements common to two sets is called the intersection of the two sets and is symbolized by $\cap$. The expression A $\cap$ B could be read as "A intersection B", or "the intersection of sets A and B." If, for example, $\{1, 3, 7\}$ and $\{1, 5, 7\}$ are two sets, then $\{1, 3, 7\} \cap \{1, 5, 7\} = \{1, 7\}$. In the equations under discussion, the solution set for the first is the set of ordered pairs (x, y) such that $x + y - 4 = 0$, and the solution set for the second is the set of ordered pairs (x, y) for which $2x + 3y - 9 = 0$. The solution set for the system is the set of ordered pairs for which both equations hold. We may write this symbolically as

$$\{(x, y) : x + y - 4 = 0\} \cap \{(x, y) : 2x + 3y - 9 = 0\} = \{(3, 1)\}.$$

See Appendix 1.

combination eliminating y in order to obtain the required value for x, we may substitute -2 for y in either of the original equations and find that x is 4. The solution of the system is $(4, -2)$.

We shall use the elimination-substitution method to solve linear systems in most of this chapter. The symbolism illustrated in the next example will be employed to keep track of the operations performed.

$$e_1: 3x - 2y = 4$$
$$e_2: 2x - 4y = 1.$$

The letter e with a subscript denotes an equation. Here we have given two equations, e sub 1 and e sub 2. Elimination of y will occur if we multiply e_1 by minus 2 and add to e_2. We obtain

$$e_3: -4x = -7 \qquad -2e_1 + e_2.$$

The procedure for obtaining e_3 is shown at the right of the equation. It follows from e_3 that $x = 7/4$ and substitution of this value into e_1 or e_2 yields $y = 5/8$, so the solution of the system is $(7/4, 5/8)$. The geometric nature of the system is shown in Figure 2–3, wherein the solution is seen to be the point of intersection of the two lines.

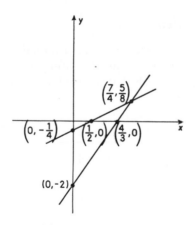

FIGURE 2–3

Exercise.   Find the solution of the following system of two equations in two unknowns, then plot the graph of the lines and label the coordinates of the point of intersection. Answer: $(3, 5)$.

$$4x - 3y = -3$$
$$5x - y = 10.$$

2.8 APPLICATIONS–1

Consider a buyer who wants to combine x gallons of X quality gasoline which costs \$0.50 per gallon with y gallons of Y quality at \$0.60 per gallon to obtain 2000 gallons of mixture worth \$0.53 per gallon. We find two equations by noting that the total amount of gas is $x + y$ which must equal 2000 and by noting that the value of x gallons at \$0.50 plus y gallons at \$0.60, which is $0.5x + 0.6y$, must equal the total value of 2000 gallons at \$0.53, which is \$1,060. Hence,

$$\begin{aligned} e_1: \quad & x + \quad y = 2000 \\ e_2: \quad & 0.5x + 0.6y = 1060 \end{aligned}$$

and

$$e_3: 0 + 0.1y = 60 \qquad -0.5e_1 + e_2$$

so that

$$y = 600 \text{ gallons}$$

and from e_1

$$x = 2000 - y = 1400 \text{ gallons.}$$

The buyer should combine 1400 gallons of X with 600 gallons of Y.

Exercise. We wish to invest x dollars in security X which pays 7 percent interest and y dollars in Y which pays 10 percent interest. We will invest \$10,000 and require that we receive \$820 interest. How much should be invested in each security? Answer: The equations are $0.07x + 0.1y = 820$ and $x + y = 10,000$. The solution is $x = \$6000$ and $y = \$4000$.

2.9 PROBLEM SET 1

Solve the following systems:

1. $x + y = 5$
   $2x + y = 7.$

2. $2x + 3y = 10$
   $3x + \ y = 1.$

3. $5x - 2y = 3$
   $2x + \ y = 3.$

4. $2x + 3y = 9$
   $4x - 2y = 2.$

5. $4x + 3y = 4$
   $2x + 6y = 5.$

6. $3x + 5y = 9$
   $4x + 2y = 5.$

7. If x gallons of quality X gasoline costing \$0.50 per gallon are to be mixed with y gallons of Y quality at \$0.66 per gallon to obtain 1000 gallons worth \$0.60 per gallon, how much of X and Y should be used?

8. We wish to invest x dollars in security X which pays 6.5% interest and y dollars in Y which pays 9% interest. We will invest \$50,000 and require that we receive \$4000 interest. How much should be invested in each security?

9. It takes 20 minutes and costs $2 to make a unit of X and the unit figures for Y are 30 minutes and $1. If 600 minutes are available and total cost is to be $40, how many units of X and Y can be made?

10. It takes 10 minutes to make and 20 minutes to paint one unit of X. Unit figures for Y are 5 minutes to make and 8 minutes to paint. If 300 minutes are available for making and 500 for painting, how many units of X and Y can be made?

2.10 APPLICATIONS–2: SUPPLY AND DEMAND ANALYSIS

Demand curves relate the demand (q units) for a product to the price per unit of product, p. Typical demand curves have negative slopes because demand usually *rises* when price per unit *falls*. Demand curves reflect consumer attitude toward price changes. On the producer's side, a *supply curve* reflects producer attitude toward price changes. Typically, supply *rises* when price per unit *rises*, so supply curves usually have positive slopes. Consider Figure 2–4, which shows the supply and demand curves (lines) labeled *SS*

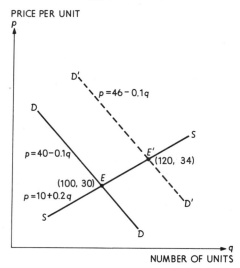

FIGURE 2–4
(not to scale)

and *DD* respectively. The equations are

$$e_1: (DD)\ p = 40 - 0.1q \quad \text{(negative slope, demand)}$$
$$e_2: (SS)\ \ p = 10 + 0.2q \quad \text{(positive slope, supply)}.$$

Point E, the intersection of *DD* and *SS*, is found by

$$e_3:\ 0 = 30 - 0.3q \quad e_1 - e_2$$

so that q is 100 units and, from e_1, p is $30 per unit. E is the *equilibrium point*. It shows the market price per unit which will equate the quantity consumers are willing to buy to the quantity producers are willing to supply.

Now suppose that due to population growth over a period of time, consumer demand is 60 units higher at every price level. The demand curve now shifts 60 units to the right, yielding $D'D'$ which is parallel to DD. We can obtain the equation of $D'D'$ easily by noting that at q units demand on $D'D'$ the price is the same as it was at $q - 60$ on DD. Thus, replacing the q of DD by $q - 60$ gives us

$$D'D' : p = 40 - 0.1(q - 60) \quad \text{or} \quad p = 46 - 0.1q.$$

Or, reasoning in a different manner, we observe that a right shift in DD is an upward $(+)$ shift in p. The amount of the shift in p is the absolute value of the slope of DD times the shift in DD; that is, $|-0.1(60)| = 6$. Hence, with DD as $p = 40 - 0.1q$, an upward shift of 6 yields $D'D'$ as $p = 46 - 0.1q$.

Solving the new system

$$e_1: (D'D') \; p = 46 - 0.1q$$
$$e_2: (SS) \quad p = 10 + 0.2q$$

we obtain

$$e_3: \; 0 = 36 - 0.3q \qquad e_1 - e_2$$

so that $q = 120$ and, from e_1, $p = \$34$. The new equilibrium point is $E'(120, 34)$.

Our example shows that an increase in demand (with an unchanging supply curve) brought more product to market at a higher unit price. Such equilibrium analysis to determine the outcome of shifting supply and demand curves is a fundamental tool in economics. The analysis is simple in the cases of parallel shifts of linear supply and demand relations. In such cases we can find the new supply or demand equation by computing the absolute value of the slope times the shift and adding this to (or subtracting it from) the original equation, keeping in mind that a shift to the right is an upward shift for a demand curve but a downward shift for a supply curve. We reverse the words upward and downward if the shift is to the left. For example, if we start with SS as $p = 5 + 0.05q$, and market changes lead producers to supply 100 units *less* at any price per unit, the supply curve shifts to the left and therefore upward. The amount of the shift is $|0.05(100)| = 5$, so we have $S'S'$ as $p = 10 + 0.05q$. The directions of the shifts can be kept in mind easily by making a sketch like Figure 2–4 and comparing the resultant $S'S'$ or $D'D'$ with the original SS or DD if shifts are made to the left and right.

Exercise. Given DD as $p = 100 - 0.5q$. Due to changing market conditions, consumers demand 80 units less at any price per unit. Find $D'D'$. Answer: The shift is downward by $|-0.5(80)| = 40$, so $D'D'$ is $p = 60 - 0.5q$.

2.11 PROBLEM SET 2.

In Problems 1–3, write the equation for $D'D'$ or $S'S'$ after the stated shift:

1. Given DD as $p = 75 - 0.25q$ and demand increases by 60 at each price.
2. Given SS as $p = 12 + 0.15q$ and 40 more units are supplied at every price.
3. Given DD as $p = 100 - 0.75q$ and demand decreases by 40 at every price.
4. Given DD as $p = 112 - 0.35q$ and SS as $p = 12 + 0.15q$:
   a) Find the equilibrium point.
   b) If, due to changing market conditions, producers supply 80 units less at every price, and the demand curve is unchanged, find the new point of equilibrium.
5. Given DD as $p = 200 - 0.08q$ and SS as $p = 10 + 0.02q$:
   a) Find the equilibrium point.
   b) If, due to changing market conditions, consumers demand 100 units less at any price, find the new equilibrium point.

2.12 DEFINITION, *m* BY *n* SYSTEM

The designation *m* by *n* means that the number of equations in a system is *m* and the number of variables is *n*. The number of variables in a system is the total number appearing in the equations of the system. It is not required that all variables appear in each of the equations. As an example:

$$2x + y - z = 4$$
$$x \quad\quad + 2z = 7$$

has two equations and three variables. It is a 2 by 3 system.

2.13 SOLVING SYSTEMS OF LINEAR EQUATIONS

We know that linear systems will have either zero, one, or an unlimited number of solutions, and we have discussed the elimination procedure as a method of finding solutions. Attention now turns toward relating the outcome of steps in elimination to the number of solutions, and to the method of describing the general solution when the number of solutions is without limit. It will be convenient to separate the examples into the two categories, *m* equals *n* and *m* does not equal *n*.

2.14 NUMBER OF EQUATIONS EQUALS NUMBER OF VARIABLES

Systems in this category can be described as *n* by *n* systems. The general approach to the solution is to form $n - 1$ linear combinations of pairs of equations, each linear combination eliminating the *same* variable, making sure that every given equation is used in at least one of the linear combinations. The outcome of these eliminations is an $n - 1$ by $n - 1$ system. The reduction procedure is repeated as many times as is necessary to learn the nature of the solution of the system. Consider the following example:

$$e_1: 2x + y - z = 2$$
$$e_2: x + 2y + z = 1$$
$$e_3: 3x - y + 2z = 9.$$

Two linear combinations, each eliminating x, carry us from the given 3 by 3 to the following 2 by 2:

$$e_4: -3y - 3z = 0 \qquad -2e_2 + e_1$$
$$e_5: -7y - z = 6 \qquad -3e_2 + e_3.$$

A linear combination of the last two equations leads to

$$e_6: 18y = -18. \qquad -3e_5 + e_4$$

The last statement says that y must be -1. We now use backward substitution to find x and z: Setting y equal to -1 in e_5 (or e_4) yields $z = 1$. Finally, putting $y = -1$ and $z = 1$ into e_2 (or e_1 or e_3) leads to $x = 2$.

The solution of the system is the point $(2, -1, 1)$. The reader should check this solution by showing that the number triplet satisfies every one of the original three equations. The system has exactly one solution. Geometrically, we have the case of three planes which have a single point in common.

||||||||||||||||||||||||||||||||||||||||||||||||||||||||||||||||||||||||||||||||||||||||||||||||||||||||||||||||||||||||||||||||

Exercise. Find the solution of the following system by use of the elimination procedure. Answer: $(1, 1, 2)$.

$$x + y + z = 4$$
$$2x - y + z = 3$$
$$x - 2y + 3z = 5.$$

||||||||||||||||||||||||||||||||||||||||||||||||||||||||||||||||||||||||||||||||||||||||||||||||||||||||||||||||||||||||||||||||

Consider next the system

$$e_1: x + y + z = 4$$
$$e_2: 5x - y + 7z = 25$$
$$e_3: 2x - y + 3z = 8.$$

Two linear combinations, eliminating y, lead to

$$e_4: 6x + 8z = 29 \qquad e_1 + e_2$$
$$e_5: 3x + 4z = 12 \qquad e_1 + e_3.$$

A linear combination of the last two equations leads to

$$e_6: 0 + 0 = 5 \qquad -2e_5 + e_4.$$

The last line is a contradiction. Zero cannot equal 5. We conclude that the system has no solutions. The argument here is that a solution of the system must satisfy the linear combinations. If a linear combination leads to a false statement, the system can have no solutions.

|||||||||||||||||||||||||||||||||||||||||||||||||||||||||||||||||||||||||||||||||||||||||||||||||||||||||||||||||||||||||||||||||||||||||||||||||

Exercise. Prove that the following system has no solutions.

$$2x - y + 3z = 5$$
$$x + 2y - z = 6$$
$$3x + y + 2z = 8.$$

|||||||||||||||||||||||||||||||||||||||||||||||||||||||||||||||||||||||||||||||||||||||||||||||||||||||||||||||||||||||||||||||||||||||||||||||||

The first example of the section had a single solution, the second no solution. Let us turn next to a system which has an unlimited number of solutions and learn how to write the general solution in arbitrary variable terminology.

$$e_1: \quad x + y + z = 4$$
$$e_2: \quad 5x - y + 7z = 20$$
$$e_3: \quad 2x - y + 3z = 8.$$

Proceeding in the usual manner, we find:

$$e_4: 6x + 8z = 24 \qquad e_1 + e_2$$
$$e_5: 3x + 4z = 12 \qquad e_1 + e_3.$$

A linear combination of the last equations yields

$$e_6: 0 + 0 = 0 \qquad -2e_5 + e_4.$$

The appearance of the true statement, zero equals zero, means simply that this linear combination arose from two equivalent equations. Looking back at

$$e_4: 6x + 8z = 24$$
$$e_5: 3x + 4z = 12$$

we see that e_4 could be divided by 2 and then would be identical to e_5. Any number pair satisfying one of these equations will satisfy the other. Moreover, any solution of the original system must satisfy these last equations. Solving (either) one for z yields

$$e_7: z = \frac{12 - 3x}{4} \quad \text{from } e_5.$$

To repeat, any solution of the original system requires that z be related to x according to the last equation.

Returning to e_1, or any of the three original equations, and replacing z by

$$\frac{12 - 3x}{4}$$

we obtain

$$e_8: \quad x + y + \frac{12 - 3x}{4} = 4 \qquad e_7 \text{ into } e_1.$$

Solving the last equation for y leads to

$$e_9: \quad y = \frac{4 - x}{4} \quad \text{from } e_8.$$

Both z and y have now been expressed in terms of x, and we state the *general* solution of the system as follows:

$$x \text{ arbitrary}$$

$$y = \frac{4 - x}{4}$$

$$z = \frac{12 - 3x}{4}.$$

The general solution just written shows not only that the number of specific solutions for the system is without limit, but also provides convenient formulas for finding specific solutions. To illustrate the matter of convenience, suppose that we are asked to find the specific solutions for values of x equal to 0, 1, 2, 3, 4, 5, 6, 7, 8, and 9. We could substitute $x = 0$ in the three *original* equations, then solve to get the corresponding y, then substitute to get the corresponding z. Next, we could substitute $x = 1$ in the original equations and repeat the procedure to obtain the corresponding values for y and z. The 10 specific solutions found in this manner would require 10 separate (duplicate) manipulations of the original equations. If, on the other hand, we substitute in the formulas of the general solution, we see immediately that if $x = 0$, then $y = 1$, and $z = 3$. If $x = 1$, then $y = 3/4$, and $z = 9/4$, and so on.

The general solution also can be of use in real-world problems to determine what conditions must be satisfied if the variables in solutions are not permitted to take on negative values. In the foregoing general solution, for example, it is clear that if y and z are to be nonnegative, then x must not exceed 4; that is, x must be in the interval whose limits are 0 to 4, inclusive.

The tactics of the last example are worth reviewing. First, observation of the identity, zero equals zero, directed attention to the equations from which the statement arose. One of these equations was solved for z in terms of x. This automatically relegated x to the role of arbitrary variable, and we sought to express y also in terms of the same arbitrary variable. To do so, we returned to one of the original equations and replaced z by its x equivalent, leaving an expression which was solved for y in terms of x.

||||||||||||||||||||||||||||||||||||||||||||||||||||||||||||||||||||||||||||||||||||||||||||||||||||||||||||||||||||||||||||||||||||||||||||||

Exercise. Start by eliminating z, then make y arbitrary and prove the general solution of the system can be expressed as y arbitrary, $x = 2y + 3$, $z = y + 1$.

$$e_1: \quad x - 3y + z = 4$$
$$e_2: \quad x - y - z = 2$$
$$e_3: \quad 2x - 5y + z = 7.$$

||||||||||||||||||||||||||||||||||||||||||||||||||||||||||||||||||||||||||||||||||||||||||||||||||||||||||||||||||||||||||||||||||||||||||||||

||||||||||||||||||||||||||||||||||||||||||||||||||||||||||||||||||||||||||||||||||||||||||||||||||||||||||||||||||||||||||||||||||

Exercise. The general solution just obtained can be checked by substitution into the original equations. For example, substitution into e_1 yields

$$(2y + 3) - 3y + (y + 1) = 4.$$

Removing parentheses, we find the statement is the identity

$$4 = 4.$$

Show that substitution into e_2 and e_3 also leads to identities.

2.15 MORE THAN ONE VARIABLE ARBITRARY

The following two examples provide practice with arbitrary variables and, at the same time, make clear that more than one variable may prove to be arbitrary in a general solution.

$$
\begin{array}{ll}
e_1: & w + x \qquad\quad + 3z = \quad 4 \\
e_2: & w - x + 2y + z = -2 \\
e_3: & -2w + x - 3y - 3z = \quad 1 \\
e_4: & w + 2x - y + 4z = \quad 7.
\end{array}
$$

The attack on the 4 by 4 system starts by reducing it to a 3 by 3; thus:

$$
\begin{array}{lll}
e_5: & 2w + 2y + 4z = 2 & \qquad e_1 + e_2 \\
e_6: & 3w + 3y + 6z = 3 & \qquad e_1 - e_3 \\
e_7: & w + y + 2z = 1 & \qquad 2e_1 - e_4.
\end{array}
$$

The next step would be to eliminate another variable, reducing the 3 by 3 to a 2 by 2. However, every linear combination which eliminates a variable will lead to zero equals zero. We solve (any) one of the three equations for one variable, say w:

$$e_8: \quad w = 1 - y - 2z \quad \text{from } e_7.$$

The last statement automatically makes y and z arbitrary variables, so x must be expressed in terms of y and z. Returning to the original equations and selecting, say, the first, we have

$$e_9: \quad (1 - y - 2z) + x + 3z = 4 \quad e_8 \text{ into } e_1.$$

Solving the last equation for x yields

$$x = y - z + 3.$$

The general solution, therefore, can be written as

$$
\begin{array}{l}
y, z \text{ arbitrary} \\
x = \quad y - z + 3 \\
w = -y - 2z + 1.
\end{array}
$$

||||||||||||||||||||||||||||||||||||||||||||||||||||||||||||||||||||||||||||||||||||||||||||||||||||||||||||||||||||||||||||||||||||

Exercise.   Return to equation e_7 of the foregoing example. Let w and z be arbitrary. Show that the general solution then is

$$w, z \text{ arbitrary}$$
$$y = 1 - w - 2z$$
$$x = 4 - w - 3z.$$

||||||||||||||||||||||||||||||||||||||||||||||||||||||||||||||||||||||||||||||||||||||||||||||||||||||||||||||||||||||||||||||||||||

The next example shows the introduction of an arbitrary variable during the backward substitution process.

$$
\begin{aligned}
e_1&: 2w + x + 2y + 3z = 4 \\
e_2&: w + 3x + y + 5z = -2 \\
e_3&: w - 2x + y - 2z = 6 \\
e_4&: 3w - x + 3y + z = 10.
\end{aligned}
$$

In this case the linear combinations chosen eliminate two variables:

$$
\begin{aligned}
e_5&: -5x - 7z = 8 && -2e_2 + e_1 \\
e_6&: -5x - 7z = 8 && -e_2 + e_3 \\
e_7&: -10x - 14z = 16 && -3e_2 + e_4.
\end{aligned}
$$

Any linear combination of a pair of the last three equations which eliminates a variable leads to zero equals zero. Solving any of the three for x, we find

$$e_8 : x = \frac{-7z - 8}{5} \quad \text{from } e_7.$$

Here, z has been delegated the role of an arbitrary variable.

$$e_9 : 2w + \frac{-7z - 8}{5} + 2y + 3z = 4 \quad e_8 \text{ into } e_1$$

$$w = \frac{14 - 4z - 5y}{5}.$$

The last equation makes y arbitrary along with z. The solution of the system is

$$y, z \text{ arbitrary}$$

$$x = \frac{-7z - 8}{5}$$

$$w = \frac{14 - 4z - 5y}{5}.$$

It may be helpful to think of a solution stated in general form as a set of formulas for finding specific solutions to the system of equations. In the present case, for example, if we are asked what the solution of the system is when $y = 2$ and $z = 6$, we can compute quickly from the formulas of the general solution that x must be -10 and w must be -4. The reader may verify that $(-4, -10, 2, 6)$ satisfies every one of the original equations.

In summary, the general approach to the solution of an n by n system is to form $n - 1$ linear combinations of pairs of equations, each linear combination eliminating the same variable, making sure that every equation at hand is used in at least one of the linear combinations. The outcome of these eliminations is an $n - 1$ by $n - 1$ system. The reduction process is repeated until a contradiction is encountered (no solutions), or until an identity such as zero equals zero is encountered (general solution required). If neither of the last two circumstances arises, the reduction will lead ultimately to a unique specific solution. The reader may wish to follow through the next examples before proceeding.

Example 1.   Solve the 4 by 4 system:

e_1: $2w + 3x + y + z = 8$
e_2: $w - 2x + 3y - 2z = -7$
e_3: $3w + 7x - y + 3z = 15$
e_4: $w - 3x + 2y + z = 9.$

We proceed as follows:

e_5: $7x - 5y + 5z = 22 \quad e_1 - 2e_2$
e_6: $9x - 3y - z = -10 \quad e_1 - 2e_4$
e_7: $16x - 7y = -12 \quad e_3 - 3e_4$

e_8: $16x - 7y = -12 \quad e_7$
e_9: $52x - 20y = -28 \quad e_5 + 5e_6$

e_{10}: $\dfrac{11}{4}y = 11 \quad e_9 - \dfrac{13}{4}e_8$

$\phantom{e_{10}:52x -}y = 4 \quad$ from e_{10}
$\phantom{e_{10}:52x -}x = 1 \quad$ substituting $y = 4$ in e_7
$\phantom{e_{10}:52x -}z = 7 \quad$ substituting $y = 4$, $x = 1$ in e_6
$\phantom{e_{10}:52x -}w = -3 \quad$ substituting $y = 4$, $x = 1$, $z = 7$ in e_4.

The solution of the system is $(-3, 1, 4, 7)$.

Example 2.   Solve the 2 by 2 system:

$$e_1: 2x + 3y = 5$$
$$e_2: 6x + 9y = 15.$$

We proceed as follows:

$$e_3: 0 + 0 = 0 \qquad e_2 - 3e_1$$

$$e_4: y = \frac{5 - 2x}{3} \quad \text{from } e_1.$$

The equations in this case represent the same straight line. The general solution is

$$x \text{ arbitrary}$$

$$y = \frac{5 - 2x}{3}.$$

Example 3. Solve the 3 by 3 system:

$$e_1: \qquad 5y - 2z = \quad 5$$
$$e_2: 2x + 3y + 4z = \quad 20$$
$$e_3: \quad x - \quad y + 3z = \quad 15.$$

We proceed as follows:

$$e_4: \qquad 5y - 2z = -10 \quad e_2 - 2e_3$$
$$e_5: \qquad 5y - 2z = \quad 5 \quad e_1$$
$$e_6: \qquad 0 - 0 = -15 \quad e_4 - e_5.$$

The last statement is a contradiction, so the system has no solutions.

2.16 TABLEAU OF DETACHED CONSTANTS

The reader will find the following to be an efficient way to carry out the elimination procedure. Consider the system

$$e_1: 2x - \quad y + 3z = \quad 9$$
$$e_2: \quad x + 2y + 2z = 11$$
$$e_3: \qquad 5y + \quad z = 13.$$

We detach the constants and arrange them in a table as follows:

	x	y	z	
e_1:	2	-1	3	9
e_2:	1	2	2	11
e_3:	0	5	1	13

Tableau (1)

Now we may carry out the elimination without having to write the variable names and the equality signs each time. (Note that the vertical line before the last column of numbers represents the equality signs.)

Let us use the -1 in the y column of the first row to eliminate y (obtain zeros) in the second and third rows.

Tableau (2)

e_4:	5	0	8	29	$2e_1 + e_2$
e_5:	10	0	16	58	$5e_1 + e_3.$

Elimination is completed by the third tableau.

Tableau (3) e_6:

0	0	0	0	$-2e_4 + e_5.$

From e_6 we have $0 = 0$, so the system has an unlimited number of solutions. Going back to Tableau (2), we have

$$e_4: 5x + 8z = 29$$

from which

$$x = \frac{29 - 8z}{5}, \; z \text{ arbitrary.}$$

Then from

$$e_1: 2x - y + 3z = 9$$

$$y = 2x + 3z - 9$$

$$= 2\left(\frac{29 - 8z}{5}\right) + 3z - 9 = \frac{13 - z}{5},$$

and the solution can be written as

$$z \text{ arbitrary}, \; x = \frac{29 - 8z}{5}, \; y = \frac{13 - z}{5}.$$

Exercise. Write the tableau of constants for the system

$$
\begin{aligned}
e_1: &\; w + x + y + z = 4\\
e_2: &\; w + 2x - 3y + z = 0\\
e_3: &\; 2w + 3x + y - z = 1\\
e_4: &\; 3w \quad\quad + y + 2z = 8.
\end{aligned}
$$

Answer: See the following.

We shall now run through the solution of the foregoing system.

		w	x	y	z		
	e_1:	1	1	1	1	4	
Tableau (1)	e_2:	1	2	-3	1	0	
	e_3:	2	3	1	-1	1	
	e_4:	3	0	1	2	8	

		w	x	y	z		
	e_5:	0	1	-4	0	-4	$-e_1 + e_2$
Tableau (2)	e_6:	0	1	-1	-3	-7	$-2e_1 + e_3$
	e_7:	0	-3	-2	-1	-4	$-3e_1 + e_4$.

Tableau (3)	e_8:	0	0	3	-3	-3	$-e_5 + e_6$
	e_9:	0	0	-14	-1	-16	$3e_5 + e_7$.

Tableau (4)	e_{10}:	0	0	45	0	45	$-3e_9 + e_8$.

From Tableau (4) we have $y = 1$. Using $y = 1$ in Tableau (3) yields $z = 2$. Proceeding to Tableau (2) we find $x = 0$, then from Tableau (1), $w = 1$, so the solution is $(w, x, y, z) = (1, 0, 1, 2)$.

Readers planning to omit the next optional section should now turn to Problem Set 3.

2.17 n BY n SOLUTION BY DETERMINANTS (Optional)

The elimination method we have been applying thus far to find solutions of systems of linear equations requires the exercise of judgment in the selection of appropriate multipliers which will result in elimination of a variable when equations are added or subtracted. A purely mechanical method which does not require judgment can be derived if we solve a system for the variables in terms of the constants which appear in the equations. Thus, in the 2 by 2 system

$$e_1: a_1x + b_1y = k_1$$
$$e_2: a_2x + b_2y = k_2$$

the constants are represented by the a's, b's, and the k's. If we multiply e_2 by a_1 and e_1 by a_2 and subtract, we eliminate x and find

$$e_3: a_1b_2y - a_2b_1y = a_1k_2 - a_2k_1$$

from which we have

$$y = \frac{a_1k_2 - a_2k_1}{a_1b_2 - a_2b_1}.$$

Exercise. Write the expression for $(b_2e_1 - b_1e_2)$ and solve for x. Answer: $x = (k_1b_2 - k_2b_1)/(a_1b_2 - a_2b_1)$.

In passing we should note that the denominator in the solution for both x and y is the same.

The calculations involved in the evaluation of x and y can be conveniently summarized in the conventions of evaluating *determinants*. Consider the determinant

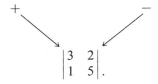

To evaluate the determinant, we follow the plus arrow and find the product $(3)(5) = 15$. The minus arrow yields $-(2)(1) = -2$. The value of the determinant is $15 - 2 = 13$.

Exercise.   Evaluate the determinant

$$\begin{vmatrix} 4 & -2 \\ 3 & 8 \end{vmatrix}.$$

Answer: $(4)(8) - (-2)(3) = 32 + 6 = 38$.

Now notice that the solutions for x and y which we obtained earlier can be expressed in determinant form as

$$x = \frac{\begin{vmatrix} k_1 & b_1 \\ k_2 & b_2 \end{vmatrix}}{\begin{vmatrix} a_1 & b_1 \\ a_2 & b_2 \end{vmatrix}} = \frac{k_1 b_2 - b_1 k_2}{a_1 b_2 - b_1 a_2}; \quad y = \frac{\begin{vmatrix} a_1 & k_1 \\ a_2 & k_2 \end{vmatrix}}{\begin{vmatrix} a_1 & b_1 \\ a_2 & b_2 \end{vmatrix}} = \frac{a_1 k_2 - k_1 a_2}{a_1 b_2 - b_1 a_2}.$$

(This method of expressing the solution of a system of equations in terms of determinants is known as Cramer's Rule.) Next observe how the determinants are constructed from the original equations

$$a_1 x + b_1 y = k_1$$
$$a_2 x + b_2 y = k_2.$$

The denominators of both solutions are simply the a and b numbers on the left. The numerator for x is obtained by replacing the a's by the k's, and the numerator for y has the b's replaced by the k's. Let us now apply the determinant method to solve

$$3x + y = 7$$
$$4x + 2y = 11.$$

For y we have

$$y = \frac{\begin{vmatrix} 3 & 7 \\ 4 & 11 \end{vmatrix}}{\begin{vmatrix} 3 & 1 \\ 4 & 2 \end{vmatrix}} = \frac{33 - 28}{6 - 4} = \frac{5}{2}.$$

Exercise.   Compute x for the last pair of equations. Answer: $x = (14 - 11)/(6 - 4) = 3/2$.

If we learn how to evaluate a 3 by 3 determinant, the method just developed can be extended to a 3 by 3 system. Consider the system

$$a_1 x + b_1 y + c_1 z = k_1$$
$$a_2 x + b_2 y + c_2 z = k_2$$
$$a_3 x + b_3 y + c_3 z = k_3.$$

Following the previous method, the denominator determinant is the set of coefficients on the left and the numerator of the solution for x replaces the a's by the k's. Thus,

$$x = \frac{\begin{vmatrix} k_1 & b_1 & c_1 \\ k_2 & b_2 & c_2 \\ k_3 & b_3 & c_3 \end{vmatrix}}{\begin{vmatrix} a_1 & b_1 & c_1 \\ a_2 & b_2 & c_2 \\ a_3 & b_3 & c_3 \end{vmatrix}}.$$

Exercise. How would the numerator for the solution for y be obtained? Answer: Write the denominator determinant with the b's replaced by the k's.

We must now learn how to compute the value of a 3 by 3 determinant. A convenient way to do this is to repeat the first two columns of the determinant at the right, then proceed according to the arrows, as shown next.

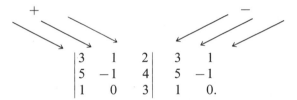

We have

$$+(-9 + 4 + 0) - (-2 + 0 + 15) = -18.$$

Exercise. Evaluate the following determinant:

$$\begin{vmatrix} 1 & 0 & 1 \\ 2 & 1 & 3 \\ 1 & 2 & 4 \end{vmatrix}.$$

Answer: $+(4 + 0 + 4) - (1 + 6 + 0) = 1.$

Now let us apply the procedure to solve the system

$$\begin{aligned} x + y + z &= 6 \\ 2x - y + z &= 3 \\ 2x + 3y - 2z &= 2. \end{aligned}$$

We have

$$x = \frac{\begin{vmatrix} 6 & 1 & 1 \\ 3 & -1 & 1 \\ 2 & 3 & -2 \end{vmatrix} \begin{matrix} 6 & 1 \\ 3 & -1 \\ 2 & 3 \end{matrix}}{\begin{vmatrix} 1 & 1 & 1 \\ 2 & -1 & 1 \\ 2 & 3 & -2 \end{vmatrix} \begin{matrix} 1 & 1 \\ 2 & -1 \\ 2 & 3 \end{matrix}} = \frac{+(12 + 2 + 9) - (-2 + 18 - 6)}{+(2 + 2 + 6) - (-2 + 3 - 4)} = \frac{13}{13} = 1.$$

Exercise. Find y and z for the foregoing system. Answer: $y = 26/13 = 2, z = 39/13 = 3$.

The determinant method of organizing the solution for a system of n equations in n variables can be extended to 4 by 4 and higher order systems but the evaluation procedures become somewhat involved and we shall not pursue this topic. We should mention in passing that if the denominator determinant is zero and any numerator determinant is not zero, the system has no solutions. However, if the denominator determinant and all numerator determinants are zero, it is not possible to state the nature of the system. It may be *consistent* (one solution), *inconsistent* (no solutions), or *dependent* (an unlimited number of solutions).

2.18 PROBLEM SET 3

1. Define:
   a) An m by n system of equations.
   b) An n by n system of equations.
   c) A 3 by 2 system of equations.
   d) The number of variables in a system of equations.
   e) A linear combination of two equations.
2. Write a general solution of
   a) $3x + 4y = 3$. b) $x + 2y - 4z = 15$.
3. What developments in the solution of a system indicate
   a) No solutions?
   b) An unlimited number of solutions?
4. State the situations which may arise in the solution of a 2 by 2 system, and discuss the geometrical interpretation of each.
5. State the situations which may arise in the solution of a 3 by 3 system, and discuss the geometrical interpretation of each.
6. What rules should be observed when reducing an n by n system to an $n - 1$ by $n - 1$ system?

7. Find the solution of each of the following. Plot the graph of each equation, and label the point of intersection of the lines.

a) $x - y = 5$
$x + y = 1.$

b) $x - 2y = 7$
$2x + y = 4.$

c) $2x + 7y - 5 = 0$
$3x + 2y = 6.$

8. Give a geometrical interpretation to the outcome of your attempt to solve

a) $3x + 2y = 4$
$6x + 4y = 6.$

b) $x - y - 4 = 0$
$2y - 2x = -8.$

Solve the following systems:[3]

9. $2x + y + 2z = 5$
$x + y - z = 0$
$3x - 2y + z = 1.$

10. $x + z = 5$
$y + z = 3$
$x - y = 2.$

11. $5x + y + z = 8$
$x + 2y - z = 1$
$2x + y = 3.$

12. $x - 2y + z = 7$
$x - y + z = 4$
$2x + y - 3z = -4.$

13. $x + y + z = 10$
$3x - y + 2z = 14$
$2x - 2y + z = 8.$

14. $x = 4$
$x - y - z = 7$
$x + y + z = 2.$

15. $2x - y + z = 5$
$x + 4y - 3z = 2$
$3x + 3y - 2z = 7.$

16. $2x + z = 5$
$x + y = 3$
$z - y = 1.$

17. $2x + y - 3z = 12$
$x + 3y - 4z = 6$
$x - 2y + z = 4.$

18. $2w + 3x + 4y + z = 1$
$w + 2y - z = 3$
$6x + 4y + z = 1$
$2w - 2z = 5.$

---

[3]The arbitrary variable(s) in a given general solution may be changed without solving the entire system again. For example, the general solution of

$$e_1: \ x + y + z = 6$$
$$e_2: \ 4x - 2y + z = -9$$
$$e_3: \ 3x - y + z = -4$$

can be written as

$$x \text{ arbitrary}$$
$$e_4: y = x + 5$$
$$e_5: z = 1 - 2x.$$

If we wish to make y arbitrary in place of x, from e_4 we find that

$$x = y - 5.$$

Substituting this expression for x into, say, e_1, we obtain z. Thus,

$$z = 6 - (y - 5) - y = 11 - 2y.$$

The general solution then becomes

$$y \text{ arbitrary}$$
$$x = y - 5$$
$$z = 11 - 2y.$$

19.
$$w + x - 2y + 2z = -4$$
$$w \quad\quad - y + z = -1$$
$$2w - 3x + y - z = 7$$
$$3w - x - 2y + 2z = 0.$$

20.
$$w - x - y \quad\quad = -3$$
$$2x + y + z = -1$$
$$w - x \quad\quad + z = 0$$
$$w \quad\quad - y + z = 3.$$

21.
$$w + z = 0$$
$$x + z = 1$$
$$y - w = -3$$
$$w + 2x + y = 3.$$

22.
$$2w - 5x + 3y - z = -10$$
$$w - 3x - y + z = 4$$
$$w - 5x + y + z = -2$$
$$-8x + 2y + 3z = -3.$$

23.
$$w - 2x + y - z = 1$$
$$4w + x - 2y - z = -8$$
$$w - 5x + 3y - 2z = 5$$
$$4w - 5x + 2y - 3z = 0.$$

Solve each of the following systems by the tableau procedure:

24.
$$2x + 3y = 9$$
$$3x + 2y = 1.$$

25.
$$5x + 2y = 9$$
$$2x - y = 0.$$

26.
$$2x + y \quad\quad = 3$$
$$4x - y + 2z = 17$$
$$y + z = 3.$$

27.
$$2x - 4y + z = 8$$
$$x \quad\quad - z = 1$$
$$y + z = 2.$$

Solve each of the following by the method of determinants:

28. Problem 24.

29. Problem 25.

30. Problem 26.

31. Problem 27.

2.19 NUMBERS OF EQUATIONS AND VARIABLES NOT EQUAL

Cases where the *number of equations exceeds the number of variables* are not common in applications. When such systems do arise, elimination often leads to a contradiction, showing that the system has no solution. As an example, consider the system

$$e_1: 2x + y = 7$$
$$e_2: x - 3y = 7$$
$$e_3: 2x + 4y = 10.$$

We find

$$e_4: 7y = -7 \quad\quad -2e_2 + e_1$$
$$e_5: 10y = -4 \quad\quad -2e_2 + e_3.$$

The last two statements are contradictory because y cannot be -1 and $-4/10$ simultaneously, so the system has no solutions.

It is helpful to think of e_1, e_2, and e_3 in the foregoing as three *constraints* on the two variables, x and y. When the number of constraints exceeds the number of variables, it is not likely that the system has a solution, and the greater the number of constraints which must be satisfied, the less is the likelihood that a solution exists. Of course, we know from our discussion of geometry at the beginning of the chapter that such a system can have a single solution or an unlimited number of solutions. Thus, four planes (four constraints on three variables) can have a single point in common or a line (an unlimited number of points) in common. Examples of zero, one, and an unlimited number of solutions may be found in the next problem set.

Systems having *fewer equations than variables* arise often in practice as we shall see later in the chapter. Such underconstrained systems have either no solutions, or an unlimited number of solutions. The latter case is most common in applications and is of interest because the solution, in arbitrary variable form, provides alternative ways of satisfying the constraints, and some alternatives may be superior to others. The geometry of an undercon- strained system is illustrated by two planes (two equations in three variables) which either are parallel (no solutions) or intersect in a line (unlimited num- ber of solutions). As an example of an underconstrained system which has an unlimited number of solutions, consider the 3 by 4 system

$$e_1: \quad w + \quad x + y + 2z = 9$$
$$e_2: \quad 2w - \quad x + y - \quad z = 4$$
$$e_3: \quad w + 2x - y - \quad z = 2.$$

We have:

$$e_4: \quad 3w + 2y + \quad z = 13 \qquad e_1 + e_2$$
$$e_5: \quad 5w + \quad y - 3z = 10 \qquad 2e_2 + e_3$$
$$e_6: \quad -7w \qquad + 7z = -7 \qquad e_4 - 2e_5$$
$$\qquad\qquad w = z + 1 \qquad e_6 \div 7 \text{ and solved for } w.$$

Resubstitution yields the general solution

$$z \text{ arbitrary}$$
$$e_7: w = z + 1$$
$$e_8: x = 3 - z$$
$$e_9: y = 5 - 2z.$$

2.20 NONNEGATIVITY CONSTRAINTS

In applied problems, variables generally are not permitted to have nega- tive values. If we impose this nonnegativity constraint on the foregoing example, it is clear from e_7 that w will not be negative if z is not negative. However, e_8 shows that z must be less than or equal to 3 if x is to be non- negative, and (from e_9) less than or equal to $5/2$ if y is to be nonnegative. It follows that z must be in the interval 0 to $5/2$, inclusive, if all variables are to be nonnegative.

We have seen that nonnegativity has restricted the interval of permissible values for the variable z. To find corresponding restrictions for another variable, we state the solution with that variable arbitrary and repeat the analysis just discussed. For example, if we make y arbitrary we find from e_9 that $z = (5 - y)/2$. Substituting this z into e_7 and e_8 yields $w = (7 - y)/2$

and $x = (y + 1)/2$, respectively. Analysis of the last two sentences shows the maximum value permitted for y is 5.

||||||||||||||||||||||||||||||||||||||||||||||||||||||||||||||||||||||||||||||||||||||||||||||||||||||||||||||||||||||||||||||||||||

Exercise. Take w as the arbitrary variable. Using e_7, e_8, and e_9, express x, y, and z in terms of w and state the permissible range of values for w if all variables are to be nonnegative. Answer: $x = 4 - w$; $y = 7 - 2w$; $z = w - 1$. The permissible interval for w is 1 to 7/2, inclusive.

||||||||||||||||||||||||||||||||||||||||||||||||||||||||||||||||||||||||||||||||||||||||||||||||||||||||||||||||||||||||||||||||||||

Similar analysis shows that with x arbitrary, $w = 4 - x$, $y = 2x - 1$, and $z = 3 - x$. We see that x must be at least $1/2$ if y is not to be negative and at most 3 if z is not to be negative. The permissible interval for x is $1/2$ to 3, inclusive.

When a solution has more than one variable arbitrary, the range of permissible values for one variable usually is not a fixed interval but, rather, depends upon the values assigned to the other variables. We shall consider such cases when we study linear programming in later chapters.

2.21 PROBLEM SET 4

Solve the following systems, which have more equations than variables:

1. $x - 3y + z = 1$
   $x + y - z = 3$
   $2x - 4y + z = 3$
   $x - 3y + z = 10.$

2. $x - 3y + z = 1$
   $x + y - z = 3$
   $2x - 4y + z = 3$
   $4x - 6y + z = 7.$

3. $2x - y + 3z = 5$
   $3x + 2y = 1$
   $x - 3y + 6z = 8$
   $2x + 3z = 4.$

4. $2x - y = 3$
   $x + 2y = 4$
   $3x - 4y = 2$
   $-x + 5y = 2.$

5. $x + y = 2$
   $x - y = 1$
   $-x + 3y = 5.$

6. $x + 3y + z = 6$
   $-x + y + z = 2$
   $3x + y - z = 2$
   $x + y + z = 6.$

7. $x + 3y + z = 6$
   $-x + y + z = 2$
   $3x + y - z = 2$
   $2x + 4y + z = 8.$

8. $w + x - 3y = 4$
   $w - 2x + 3z = 1$
   $w - x - y + 2z = 2$
   $2w - 3x - y + 5z = 3.$

Solve the following systems, which have fewer equations than variables:

9. $4x - 2y + 6z = 6$
   $-2x + y - 3z = -3.$

10. $x + 3y + z = 6$
    $-x + y + z = 2.$

11. $2x + y + z = 4$
    $6x + 3y + 3z = 10.$

12. $2x + y + z = 4$
    $6x + 3y + 4z = 15.$

13. $w + 2x - y + z = 5$
$w - 5x + 4y - 5z = 6$
$3w - x + 2y - 3z = 2.$

14. $w + x + z = 4$
$w - 3x + y = 1$
$2x + y + z = -1.$

15. $w + 2x - y + z = 4$
$-w + 3x + z = 2$
$2w - x - y = 2.$

16. $w - 2x + y = 1$
$w + 3x + 2z = 7$
$x + y + z = 3.$

17. $w + x - 2y = 0$
$w - x + 2z = 0.$

18. See the answer for Problem 10. If nonnegativity is imposed, what are the permissible intervals for x, y, and z?

19. See the answer for Problem 14. What are the permissible intervals for the variables if nonnegativity is imposed?

20. See the answer for Problem 16. What are the permissible intervals for the variables if nonnegativity is imposed?

2.22 APPLICATIONS–3

The manner in which the topics considered in this chapter arise in applications is illustrated by examples in this section and the two sections which follow.

Example 1. A buyer wishes to combine quality Y gasoline at $0.50 per gallon with quality Z gasoline at $0.60 per gallon and obtain 2000 gallons of gasoline worth $0.52 per gallon. How many gallons of each should he buy?

Letting y and z be the respective amounts of each quality, we require that $y + z = 2000$ gallons. The value of these amounts of gasoline is $0.50y + 0.60z$, and this must equal $1040 which is the value of 2000 gallons at $0.52 per gallon. Hence,

$$e_1: \quad y + z = 2000$$
$$e_2: \quad 0.5y + 0.6z = 1040.$$

Proceeding,

$$e_3: \quad 0.1z = 40 \qquad -0.5e_1 + e_2$$

so that $z = 400$ gallons of quality Z and $y = 1600$ gallons of quality Y. We note that only one set of values for x and y satisfies the constraints of this problem.

Example 2. Suppose that *in addition to the constraints of Example 1,* the buyer wants a 95 octane mixture and that Y is 90 octane and Z is 100 octane. This leads to the constraint $90y + 100z = 95(2000)$ or, dividing by 100, $0.9y + z = 1900$. What now should be the amounts of Y and Z?

It should be clear at once that no mixture will satisfy all constraints because equal amounts of 90 and 100 octane must be mixed to obtain 95 octane, and equal numbers for y and z will not satisfy e_1 and e_2. If we proceed to write the equations

$$\begin{aligned} e_1: \quad & y + \quad z = 2000 \\ e_2: \quad & 0.5y + 0.6z = 1040 \\ e_3: \quad & 0.9y + \quad z = 1900 \end{aligned}$$

we find

$$\begin{aligned} e_4: \quad & 0.1z = \quad 40 \qquad -0.5e_1 + e_2 \\ e_5: \quad & 0.1z = 100 \qquad -0.9e_1 + e_3. \end{aligned}$$

The last two equations are a contradiction because z cannot be both 400 and 1000. We have here a case of more equations than variables and as we stated earlier in the chapter, such systems often have no solution.

Example 3. Suppose the buyer of Example 1 also may combine quality X at \$0.40 per gallon in his mixture. What choice of components does he now have for his mixture?

We now have two equations in three variables, as follows:

$$\begin{aligned} e_1: \quad & x + \quad y + \quad z = 2000 \\ e_2: \quad & 0.4x + 0.5y + 0.6z = 1040. \end{aligned}$$

Proceeding,

$$e_3: \quad 0.1y + 0.2z = 240 \qquad -0.4e_1 + e_2$$

so that

$$y + 2z = 2400 \quad \text{or} \quad y = 2400 - 2z.$$

Then, from e_1 we find

$$x = z - 400.$$

From the last two statements we see that if x and y are to be nonnegative, z must be at least 400 and at most 1200. The mixture choices are

$$z \text{ arbitrary in the interval 400 to 1200, inclusive.}$$
$$x = z - 400$$
$$y = 2400 - 2z.$$

In this example, the system had fewer equations than variables. As we know, if such a system has a solution, it has an unlimited number of solutions. Of course, the nonnegativity constraints can do away with some or all of the solutions.

Exercise. Express the answer to Example 3 with x arbitrary, stating the permissible interval for x. Answer: $y = 1600 - 2x$, $z = x + 400$; x arbitrary in the interval 0 to 800 gallons, inclusive.

In passing we note that the highest quality gasoline probably is Z, so if the buyer wishes to maximize quality (octane) he should include the maximum amount of Z (1200 gallons) in his mixture. The other components are 800 gallons of X and 0 gallons of Y.

2.23 APPLICATIONS–4: INTRODUCTION TO OPTIMIZATION

When a system has many solutions, some solution or solutions may be better than others because they optimize an objective such as achieving minimum cost. Two examples follow:

Example 1. A buyer wishes to mix quality X gasoline (80 octane at $0.40 per gallon) with Y (90 octane at $0.50 per gallon) and Z (100 octane at $0.65 per gallon) to obtain 2000 gallons of 85 octane gasoline at minimum cost. What amounts of each quality should be used?

Here we have two constraints, one on the total amount of gasoline and the second on the octane rating of this total. The first constraint is simply $x + y + z = 2000$. The second is $80x + 90y + 100z = 85(2000)$ or, dividing by 100, $0.8x + 0.9y + z = 1700$. Hence,

$$e_1: \quad x + \quad y + z = 2000$$
$$e_2: \quad 0.8x + 0.9y + z = 1700.$$

We have

$$e_3: \quad 0.2x + 0.1y = 300 \qquad e_1 \quad e_2$$

so that

$$y = 3000 - 2x$$

and from e_1

$$z = x - 1000,$$

so that x is arbitrary in the interval 1000 to 1500, inclusive.

Now let us recall that our system consists of two planes which intersect in a line in space. We can move in one direction or the other on that line by increasing or decreasing the value of our arbitrary variable, x. Next consider our cost expression

$$\text{Cost} = 0.4x + 0.5y + 0.65z.$$

If we move in one way on our solution line in space and cost decreases, it will keep on decreasing as we continue in that direction. It follows that *we should go as far as possible* in that direction; that is, to the end point of the permissible interval for x. Consequently, to find the minimum we need only calculate cost at the endpoints, $x = 1000$ and $x = 1500$, then select the lower of the two.[4] Using $x = 1000$, our solution shows $y = 3000 - 2x = 1000$ and $z = x - 1000 = 0$. We have

$$\text{Cost (at } x = 1000) = 0.4(1000) + 0.5(1000) + 0.65(0) = \$900.$$

Similarly, at the upper end, $x = 1500$, $y = 0$, and $z = 500$.

$$\text{Cost (at } x = 1500) = 0.4(1500) + 0.5(0) + 0.65(500) = \$925.$$

The minimum-cost solution is to mix 1000 gallons of X with 1000 gallons of Y, and no Z. The minimum cost is $900.

Exercise. Express the foregoing solution with z arbitrary and state the permissible interval for z. Using the interval limits, determine the minimum-cost mixture. Answer: $x = z + 1000$; $y = 1000 - 2z$; z arbitrary in the inclusive interval 0 to 500 gallons. The cost at $z = 0$ is $900 and at $z = 500$ is $925. Thus, as in the foregoing, the minimum is $900 at $x = 1000$, $y = 1000$, and $z = 0$.

Example 2. A mixture is to be made containing x pounds of food A, y pounds of food B, and z pounds of food C. The total weight of the mixture is to be five pounds. Food A contains 500 units of vitamin per pound, B and C contain 200 and 100 units of vitamin per pound, respectively. The five-pound mixture is to contain a total of 1500 units of vitamin. Foods A, B, and C contain, respectively, 300, 600, and 700 calories per pound, and the five-pound mixture is to contain a total of 2500 calories. The problem is to determine how many pounds each of foods A, B, and C must be in a five-pound mix if the mix is to contain the required 1500 units of vitamin and the required 2500 calories.

The data of the problem can be summarized as shown in Table 2–1.

TABLE 2–1

Food	Number of Pounds	Units of Vitamin per Pound	Calories per Pound
A x		500	300
B . . . . y		200	600
C . . . . z		100	700

The conditions of the problem require that

$$\begin{aligned} e_1: \quad & x + y + z = 5 \\ e_2: \quad & 500x + 200y + 100z = 1500 \\ e_3: \quad & 300x + 600y + 700z = 2500. \end{aligned}$$

---

[4]The two costs may be the same, in which case all permissible values of the arbitrary variable will lead to the same cost.

The first equation says that the number of pounds of A plus the number of pounds of B plus the number of pounds of C, which is $x + y + z$, must equal five. The second equation says that 1500 units of vitamin are to be contained in the mixture of x pounds of A at 500 units per pound, plus y pounds of B at 200 units per pounds, plus z pounds of C at 100 units per pound. A similar description applies to the third equation.

The solution proceeds in the usual manner:

$$
\begin{array}{llll}
e_4: & 300y + 400z = & 1000 & 500e_1 - e_2 \\
e_5: & -300y - 400z = & -1000 & 300e_1 - e_3 \\
e_6: & 0 + 0 = & 0. & e_4 + e_5
\end{array}
$$

The system does not have a unique solution. Letting z be arbitrary:

$$e_7: \quad y = \frac{10 - 4z}{3} \quad \text{from } e_5$$

$$e_8: \quad x = \frac{5 + z}{3} \quad e_7 \text{ into } e_1$$

$$z \text{ arbitrary.}$$

The general solution shows that an unlimited number of different mixtures can be made which satisfy the stated conditions. For example, if we use one pound of C ($z = 1$), then the general solution says we must use

$$y = \frac{10 - 4}{3} = 2 \text{ pounds of B}$$

$$x = \frac{5 + 1}{3} = 2 \text{ pounds of A.}$$

The reader may verify that this (2, 2, 1) mixture will weigh five pounds, and will contain 1500 units of vitamin and 2500 calories.

Applying the nonnegativity constraint to

$$e_7: \quad y = \frac{10 - 4z}{3},$$

we see that y will be negative if $4z$ is larger than 10, that is, if z is larger than 10/4. If z is restricted to the interval 0 to 2.5, inclusive, none of the variables will be negative. Our final solution is

$$z \text{ arbitrary in the interval}$$
$$0 \text{ to 2.5, inclusive}$$

$$y = \frac{10 - 4z}{3}$$

$$x = \frac{5 + z}{3}.$$

||||||||||||||||||||||||||||||||||||||||||||||||||||||||||||||||||||||||||||||||||||||||||||||||||||||||||||||||||||

Exercise. If the costs per pound of A, B, and C are, respectively, $2, $3, and $1, what is the composition and cost of the minimum-cost mixture? Answer: Mix 2.5 pounds of A and 2.5 pounds of C; the minimum cost will be $7.50.

||||||||||||||||||||||||||||||||||||||||||||||||||||||||||||||||||||||||||||||||||||||||||||||||||||||||||||||||||||

2.24 APPLICATIONS–5: TWO-PRODUCT SUPPLY AND DEMAND ANALYSIS

Readers planning to work through this section may wish to review the one-product discussion of Section 2.10. Here we consider the case of two products whose supply and demand expressions are interrelated. As an example consider the following expressions in which p_1 and q_1 are the price of product #1 and the quantity demanded of product #1, respectively, and similarly for the price and quantity demanded for product #2.

<table>
<tr><th></th><th>Demand</th><th>Supply</th></tr>
<tr><td>Product #1</td><td>$p_1 = 2000 - 3q_1 - 2q_2$</td><td>$p_1 = 100 + 2q_1 + q_2$</td></tr>
<tr><td>Product #2</td><td>$p_2 = 2800 - q_1 - 4q_2$</td><td>$p_2 = 200 + 3q_1 + 2q_2.$</td></tr>
</table>

To achieve equilibrium, the two price expressions for each product must be equal. Hence,

for product #1: $2000 - 3q_1 - 2q_2 = 100 + 2q_1 + q_2$
for product #2: $2800 - q_1 - 4q_2 = 200 + 3q_1 + 2q_2.$

Rearranging the last two expressions, we have

$$e_1: 5q_1 + 3q_2 = 1900$$
$$e_2: 4q_1 + 6q_2 = 2600.$$

We obtain the equilibrium quantities supplied and demanded by solving the last pair of equations in the usual manner:

$$e_3: 18q_2 = 5400 \qquad 5e_2 - 4e_1.$$

Hence, $q_2 = 5400/18 = 300$. Then from e_2 we find $q_1 = 200$. Using these quantities in the original supply (or demand) expressions, we find $p_1 = \$800$ and $p_2 = \$1400$. The equilibrium prices and quantities are

$$E: \begin{array}{ll} p_1 = \$800 & p_2 = \$1400 \\ q_1 = 200 \text{ units} & q_2 = 300 \text{ units.} \end{array}$$

Now suppose consumer demand for product #1 rises by 90 units at any price level. From Section 2.10, we know that this results in an upward shift of 270 in the demand expression for product #1; that is, $|(-3)(90)| = 270$, where -3 is the coefficient of q_1 in its demand equation. Similarly, there is an upward shift of 90 for product #2. The new demand expressions then are

product #1: $p_1 = 2270 - 3q_1 - 2q_2$
product #2: $p_2 = 2890 - q_1 - 4q_2.$

Equating these new demand expressions with the corresponding original supply expressions leads to

$$5q_1 + 3q_2 = 2170$$
$$4q_1 + 6q_2 = 2690.$$

Solving the last pair, we find $q_1 = 275$ and $q_2 = 265$, and using these values in the new demand or the old supply expressions yields $p_1 = \$915, p_2 = \$1555$. The new equilibrium is

$$E': \begin{array}{ll} p_1 = \$915 & p_2 = \$1555 \\ q_1 = 275 \text{ units} & q_2 = 265 \text{ units.} \end{array}$$

Exercise. Solve the foregoing pair of equations, then carry out the substitution necessary to verify the values at E'.

Comparing E' with E, we find that the increased demand for product #1 raised the equilibrium quantity of #1 from 200 to 275 units, and raised the equilibrium price from \$800 to \$915. The equilibrium price for product #2 rose from \$1400 to \$1555, but the quantity declined from 300 to 265 units.

If we think of product #1 as the standard model of an appliance and #2 as the deluxe model, we can see that consumer response to rising prices of appliances (both prices up) might well be to increase their demand for the standard model and decrease demand for the deluxe model, as was the outcome in this example. The outcome of supply and demand shifts depends on the nature of the products involved and relates directly to the constants in the mathematical expressions for supply and demand. Readers interested in pursuing this matter should refer to discussions of *complementary, competing* (or *substitute*), and *independent* products in economics textbooks.

2.25 PROBLEM SET 5

1. A mixture of pellets is to be made containing x type-A pellets, y type-B pellets, and z type-C pellets. Cost, weight, and volume data for each type of pellet are shown in the table.

Pellet Type	Number of Pellets	Cost per Pellet in Cents	Weight Units per Pellet	Volume Units per Pellet
A x		2	1	4
B . . . . .y		3	2	2
C . . . . .z		1	4	3

Is it possible to make a mixture of these pellets at a cost of 85 cents if the mixture is to have 120 weight units and 130 volume units? If so, how many of each type of pellet should be in the mixture?

2. Replace the volume data in the table for Problem 1 with the numbers 1.6, 2.8, and 3.6 for A, B, and C, respectively. Leaving the other data and the conditions of the problem unchanged, set up and solve the resultant system of equations.

3. *a)* Foods A, B, and C are to be combined to make a seven-pound mixture which will contain 5000 calories. Caloric content per pound of A, B, and C are, respectively, 500, 1000, and 1500. Find the general solution, showing permissible numbers of pounds of A, B, and C which can be used in the mixture.

   b) If the costs per pound of A, B, and C are, respectively, $1, $3, and $4, what is the composition and cost of the minimum-cost mixture?

4. In the table, w, x, y, and z represent tons of a certain type of steel shipped from factories to warehouses. For example, x is the number of tons shipped from factory one to warehouse two, that is, from F_1 to W_2. The numbers 40 and 60 are the total numbers of tons available at the factories, and the numbers 30 and 70 are the total tons required by the warehouses.

	W_1	W_2	
F_1	w	x	40
F_2	y	z	60
	30	70	

Note that the *marginal* total $30 + 70$ equals the *marginal* total $40 + 60$. Prove that only one of w, x, y, and z may be chosen arbitrarily by writing the general solution for w, y, and z with x arbitrary.

5. See Problem 4. Write all the equations which must be satisfied, then write the general solution of this system of equations.

	W_1	W_2	W_3	
F_1	w	x	y	90
F_2	u	v	z	60
	30	45	75	

6. Transistors are of three types, type P, type Q, and type R. A grade A box of transistors contains 5 P's, 3 Q's, and 1 R. A grade B box contains 2 P's, 3 Q's, and 4 R's. How many boxes of each grade can be filled if the total numbers of P, Q, and R available are 80, 75, and 70, respectively, and all transistors are used?

7. See Problem 6. A grade C box contains 5 P's and 1 Q; a grade D box contains

3 P's and 1 Q; a grade E box contains 1 P and 4 Q's. How many boxes of grades C, D, and E can be filled *exactly* using 44 type P and 22 type Q transistors?

8. See Problems 6 and 7. The numbers of transistors, by type, contained in boxes of grades F, G, and H are shown in the table. Find the numbers of boxes (x, y, and z) which could be completely filled using all available transistors.

| Grade | Number of Boxes | Number of Transistors Required per Box, by Type | | |
		P	Q	R
F............x		1	2	2
G............y		2	7	1
H............z		1	3	1
Total available		12	34	14

9. Solve Problem 8 if the requirement for type Q in a box of grade G is reduced from 7 to 6.

10. Solve Problem 8 if the requirement for type R in grade F is increased from 2 to 3.

11. *a*) The table shows the numbers of hours required in each of two departments to make a unit of various products named A, B, and C. For example, product B requires 1 hour of time in Department I and 3 hours in Department II.

	A	*B*	*C*
	Hours Required per Unit of Product		
Department I 1		1	9
Department II......1		3	7

Find the numbers of units of A, B, and C which could be made if Department I has 75 hours available and Department II has 65 hours available.

b) If profits per unit of A, B, and C are, respectively, $20, $30, and $40, what is the maximum profit and the composition of the maximum-profit combination of outputs?

12. Solve Problem 11*a*) and *b*) if both departments have 40 hours available.

13. The demand and supply expressions for products #1 and #2 are:

Demand / Supply

#1: $p_1 = 1000 - 5q_1 - 4q_2$ $p_1 = 90 + 2q_1 + 3q_2$
#2: $p_2 = 900 - 2q_1 - 5q_2$ $p_2 = 120 + q_1 + 4q_2$.

a) Find the prices and quantities at equilibrium.

b) If demand for #1 is 42 units higher at any price level, find the prices and quantities at the new equilibrium.

14. The demand and supply expressions for products #1 and #2 are:

$$\begin{array}{lll} & \textit{Demand} & \textit{Supply} \\ \#1\text{:} & p_1 = 1500 - 4q_1 - 3q_2 & p_1 = 400 + 3q_1 + q_2 \\ \#2\text{:} & p_2 = 700 - q_1 - 2q_2 & p_2 = 200 + q_1 + q_2. \end{array}$$

a) Find the prices and quantities at equilibrium.

b) If the demand for #1 is 52 units lower at any price level, find prices and quantities at the new equilibrium.

2.26 REVIEW PROBLEMS

Solve each of the following systems by the elimination method:

1. $x + 10y = 25$
    $3x - 7y = 1.$

2. $x + 2y - 3z = 11$
    $3x + 2y + z = 1$
    $2x + y - 5z = 11.$

3. $x + y - 2z = 1$
    $x - 2y + z = 7.$

4. $x + y = 6$
    $2x + y = 10$
    $x - y = 1.$

5. $3x + y + z = 4$
    $x - 2y - 3z = 0$
    $2x + 3y + 4z = 6.$

6. $2x + z = 4$
    $3y + 2z = 6$
    $4x - 3y = 2.$

7. $x + y + z = 1$
    $2x - y + 3z = 5$
    $x + 5y + 2z = 4$
    $3x - 2y + 2z = 0.$

8. $w + x - 2z = 0$
    $w - x - 2y = 0$
    $w + 2x + y - 3z = 0.$

9. $x - y = 1$
    $2x + y - 3z = 2$
    $x + y - 2z = 1.$

10. $w + x + y + z = 0$
    $w - x + y - z = 0$
    $2w + x + 2y + z = 0$
    $3w + 2x + 3y + 2z = 0.$

11. The table shows, for example, that one pound of food A contains one ounce of nutrient P and two ounces of nutrient Q. Similar relations are shown for foods B and C.

Food	Ounces of Nutrient per Pound	
	Nutrient P	Nutrient Q
A.	1	2
B	2	3
C	3	2

If a mixture of x pounds of A, y pounds of B, and z pounds of C is to contain exactly 6 ounces of P and 10 ounces of Q, find the permissible values for x, y, and z.

a) Write the solution with x arbitrary, then with y arbitrary, and then with z arbitrary.

b) What are the largest and the smallest permissible amounts of A, B, and C?

12. See Problem 11. If the per pound costs of A, B, and C were 5 cents, 8 cents, and 3 cents, respectively, the cost of a mixture would be

$$M = 5x + 8y + 3z.$$

Find the composition and cost of the minimum-cost mixture.

13. See Problem 11. Suppose that the per pound costs of A, B, and C are 10 cents, 5 cents, and 3 cents, respectively. Using the results of Problem 11, find the cost and composition of the minimum-cost mixture.

14. The table shows, for example, that a pound of food A contains 2 ounces of nutrient P, 3 ounces of Q, and 1 ounce of R. Similar relations are shown for foods B and C. If a mixture of x pounds of A, y pounds of B, and z pounds of C is to contain exactly 9 ounces of P, 13 ounces of Q, and 4 ounces of R, find the permissible values for x, y, and z.

	Ounces per Pound of:		
Food	Nutrient P	Nutrient Q	Nutrient R
A 2		3	1
B 1		2	1
C 1		1	0

Write the solution in arbitrary variable form, and state the permissible ranges of values for the variables.

15. See Problem 14. If the per pound costs are 20, 10, and 5 cents for A, B, and C, respectively, find the cost and composition of the minimum-cost mixture.

16. We wish to mix x gallons of X quality gasoline with y gallons of Y and z gallons of Z to obtain 1000 gallons of 90 octane gasoline. Costs per gallon are $0.50, $0.55, and $0.65, respectively, and octane ratings are 84, 92, and 100, respectively.
   a) Write the solution in arbitrary-variable form.
   b) What is the composition and cost of the minimum-cost mixture?

17. The table shows the number of hours required to make one unit of various products (A, B, and C) in each of two departments (I and II). For example, it takes one hour of Department I time and three hours of Department II time to make one unit of product C.

	Hours to Make One Unit of:		
Department	Product A	Product B	Product C
I 1		1	1
II 2		. 1	3

a) Find the numbers of units, x, y, and z of products A, B, and C which can be made if exactly 8 hours of Department I time and 14 hours of Department II time are to be utilized.

b) If the maximum possible number of units of C are made, how many hours of each department's time will be spent on making C?

c) Is it possible to utilize the hours exactly if no units of B are made? Explain.

18. The table shows, for example, that one unit of product A consumes two hours of time in Department II. Similar relations are shown for other products and departments.

	Hours to Make One Unit of:			
Department	Product A	Product B	Product C	Product D
I 1	2	1	2
II 2	1	0	1
III 0	1	3	2

Find the numbers of complete units, w, x, y, and z of products A, B, C, and D, respectively, which can be made if exactly 12 hours of each department's time is to be utilized. (Fractional units are not permitted in the solution).

Solve each of the following by the method of determinants:

19. $3x - 2y = 7$
$2x + y = 14.$

20. $5x + 0.6y = 0.5$
$0.2x + 0.1y = 0.4.$

21. $2x + 2y - z = 1$
$3x + z = 5$
$4y - 2z = 2.$

22. $x + y = 0$
$y + z = 3$
$x + z = 1.$

Solve each of the following by the tableau procedure:

23. Problem 19.

24. Problem 20.

25. Problem 21.

26. Problem 22.

27. The demand and supply expressions for products #1 and #2 are

Demand	Supply
#1: $p_1 = 1700 - 3q_1 - q_2$	$p_1 = 100 + 2q_1 + q_2$
#2: $p_2 = 1650 - q_1 - 2q_2$	$p_2 = 50 + q_1 + 2q_2.$

a) What will be the prices and quantities at equilibrium?

b) If the demand for #2 rises 30 at any price level, what will be the prices and quantities at the new equilibrium?

28. The demand and supply expressions for products #1 and #2 are

Demand	Supply
#1: $p_1 = 2300 - 30q_1 - 10q_2$	$p_1 = 180 + 5q_1 + 2q_2$
#2: $p_2 = 2000 - 10q_1 - 20q_2$	$p_2 = 120 + q_1 + 4q_2.$

a) What will be the prices and quantities at equilibrium?

b) If the demand for #1 is 70.8 higher at any price level, what will be the prices and quantities at the new equilibrium?

3

Systems of Linear Inequalities

3.1 INTRODUCTION

MATHEMATICAL STATEMENT of the conditions imposed by a problem often requires the use of inequalities rather than equalities. For instance, if we let x represent the number of hours a plant operates per day, x does not necessarily equal 24, but x must be less than or equal to 24. Moreover, a plant cannot operate a negative number of hours, so x must be greater than or equal to zero. Other examples of conditions leading to inequalities come easily to mind. The volume of fluid stored in a tank must be greater than or equal to zero, but less than or equal to the volume of the tank; the amount of product made is restricted by plant capacity and raw material availability. In this chapter, we shall discuss linear inequalities from an algebraic point of view and offer geometrical interpretations of systems involving one or two variables.

3.2 DEFINITIONS AND FUNDAMENTAL PROPERTIES

The symbols of inequality are $<$ and $>$. The first, $<$, means *is less than;* the second, $>$, means *is greater than.* Thus $3 > 2$ is the true statement that 3 is greater than 2, and $5 < 9$ is the true statement that 5 is less than 9. It may be of use to note that the symbols $>$ and $<$ point toward the smaller quantity. The statement $a < b$ is read "a is less than b." If we wish to state that "a is less than or equal to b," we combine the inequality sign with the

93

equality sign and write $a \leq b$. Further practice with the symbols is afforded next:

> $a \geq b$ means a is greater than or equal to b.
> $a \leq 3$ means a is less than or equal to 3.
> $b > 0$ means b is greater than zero; that is, b is positive.
> $c < 0$ means c is negative.
> $a \geq 0$ means a is not negative.

One of the fundamental properties of the real number system is the property of *order*. We say that the real numbers are ordered in the sense that either any two numbers are equal, or one is greater than the other. Thus, given any two numbers, a and b, one and only one of the following must be true:

$$a < b$$
$$a = b$$
$$a > b.$$

To say that a first number is greater than a second implies that a positive number must be added to the second number to make the sum equal the first number. For example, $5 > 2$ implies that a positive number must be added to 2 to make a sum equal to 5. In general:

> $a > b$ implies that $a = b + c$, where $c > 0$.
> $a < b$ implies that $a + c = b$, where $c > 0$.

The last statement that "a is less than b" implies that a positive number ($c > 0$) must be added to a to make the sum equal b.

It is a consequence of the order property of the real numbers that $-5 < -2$ and $0 > -1$. That is, inasmuch as the positive number 3 must be added to -5 to make the sum equal -2, -5 is less than -2. Again, inasmuch as the positive number 1 must be added to -1 to make the sum 0, 0 is greater than -1.

Exercise.   What argument leads to the conclusion that -10 is greater than -20? (See the immediately preceding paragraph).

The ordering of numbers can be remembered easily by reference to Figure 3–1. If the number a is to the left of the number b, then $a < b$. If the number a is to the right of the number b, then $a > b$. We see that $-1 < 0$, $0 < 1$, $-2 > -3$, $2 > -1$.

The direction in which an inequality symbol points is referred to as its *sense*, the particular use of the word being in phrases which describe inequalities as being of the *same sense* or of *opposite sense*. Thus, $x < 3$ and $x < -1$ are of the same sense, and $x \leq 5$ and $x \geq 8$ are of opposite sense.

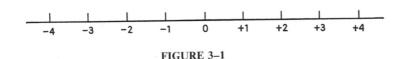

FIGURE 3–1

3.3 FUNDAMENTAL OPERATIONS ON INEQUALITIES

If both sides of an inequality are multiplied or divided by the same negative number, the sense of the inequality must be changed. For example, the true statement $2 < 3$ can be multiplied by, say, -4 to yield $-8 > -12$. Again, if both sides of the true statement $-4 < -2$ are divided by -2, we obtain $2 > 1$.

It is clear that the same number may be added to or subtracted from both sides of an inequality, and that both sides of an inequality may be multiplied or divided by the same *positive* number, without changing the sense. For example, if we add 2 to both sides of $-7 < -5$ we obtain $-5 < -3$. Subtracting 1 from both sides of the last inequality yields $-6 < -4$. Multiplying both sides of the last inequality by 5 yields $-30 < -20$. Dividing both sides of the last inequality by 10 yields $-3 < -2$.

3.4 SOLVING SINGLE INEQUALITIES

A single inequality in one variable is solved in the usual algebraic manner, except that the sense must be changed if the inequality is multiplied or divided by a negative number. Starting with $2x + 3 < 7$ we may subtract 3 from both sides to obtain $2x < 4$, and then divide both sides by 2 to yield the solution $x < 2$. Thus the original inequality is satisfied by any number less than 2. For example, if we select a number less than 2, say 1.9, and substitute it into the original inequality, we obtain the true statement $6.8 < 7$. As another example, starting with $3 - 2x \geq 5$ we may subtract 3 from both sides to obtain $-2x \geq 2$, and then divide both sides by -2, changing the sense to yield $x \leq -1$.

Exercise. Show that the solution of $3 - 2x \leq 1$ is $x \geq 1$.

A single equation in two-space has a straight line as its geometric representation. A single *inequality,* on the other hand, is represented by all points on one side of a line, the line itself being included if the equality and inequality signs appear together. If we think of a straight line as dividing a plane in half, we may say that the solutions of an inequality in two-space consist of all points in a *half space.* Consider the inequality

$$2x + 3y \leq 6.$$

If we solve for x in the usual manner, we find

$$x \leq \frac{6 - 3y}{2}.$$

The points which satisfy the *equality* part of the statement lie *on* a line. The points which satisfy the *inequality* part of the statement must lie to the *left* of the line because the inequality reads "x less than." If we had solved for y, the outcome would have been

$$y \leq \frac{6 - 2x}{3}$$

which says that the points satisfying the inequality lie *below* the line. Figure 3–2 is the graph of the inequality. It consists of the line

$$2x + 3y = 6$$

and the half space below (or to the left of) the line.[1]

The shaded area in Figure 3–2 can be described either as "below" the line or as "left" of the line. Similarly, the unshaded space on the other side can be described as "above" the line or "right" of the line. If a line slopes in the other direction, the descriptions would be below or right, above or left.

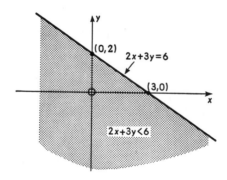

FIGURE 3–2

It is easy to determine which side of a line is required by an inequality. Simply solve the inequality (actually or mentally) for one of the variables, changing the sense if need be to secure a positive coefficient for that variable. Clearly, if the solution reads $x >$, the space is to the right; if it reads $x <$, the space is to the left; if it reads $y >$, the space is above; if it reads $y <$, the space is below. For example, if the inequality $3x - 5y < 7$ is solved for x it will be of the form $x <$; if solved for y (changing sense), it will be of the form $y >$. Hence, the space in question may be referred to as either to the left of the line or above the line.

Exercise. Specify in two ways the side of the line which contains the points satisfying $y - 3x > -3$. Answer: Left or above.

As a special case in two-space, $x > -1$ is satisfied by all points to the right of the vertical line $x = -1$. Similarly, the inequality $y < 4$ is satisfied by all points below the horizontal line $y = 4$.

3.5 PROBLEM SET 1

Mark (T) for true or (F) for false.
1. ( ) $-3 > -2$.
2. ( ) $a > b$ means a is greater than b.
3. ( ) If $4 < c$, then c could not equal 5.
4. ( ) $a \leq 0$ means a must be positive.
5. ( ) $a \geq 0$ means a must be positive.
6. ( ) If $a < b$, then $a = b + c$ for some positive number c.
7. ( ) If $a > b$, then b is less than a.
8. ( ) If $-a < 2$, then $a < -2$.
9. ( ) If $a - 2 > -5$, then $a > -3$.
10. ( ) If $x \leq y$, then $2x \leq 2y$.
11. ( ) The solution of $2 - 2x > -6$ is $x < 4$.

Specify in two ways the side of the line which contains the points satisfying
12. $3x - 7y < 6$.   13. $y - 2x > -7$.
14. $3x + 5y > 16$.   15. $-x - 2y > -1$.
16. $2x + 9y < 3$.

Solve for x:
17. $3x - 4 \leq 5$.   18. $5 - 2x > 8$.   19. $4 \leq 3 + 2x$.

Solve for x, and make a geometric sketch of the solution:
20. $3y - 2x \leq 4$.   21. $x + 2y > -6$.   22. $y - x < -3$.

Graph the following in two-space:
23. $x < 0$.   24. $y > -3$.   25. $x < 5$.

3.6 DOUBLE INEQUALITIES

The system of inequalities

$$x > 4$$
$$x \leq 9$$

requires that x be greater than 4 but less than or equal to 9. Hence, x is in an interval with the smaller number, 4, at the left and the larger number, 9, at the right. This may be written as

$$4 < x \leq 9.$$

Again, the statement

$$0 \leq x \leq 10$$

says that x is greater than or equal to 0 but less than or equal to 10.

Exercise.   Write the statement which specifies that y is less than 0 but greater than or equal to -3. Answer: $-3 \leq y < 0$.

Two inequalities are said to be *inconsistent* if both cannot be true at the same time, and one of a pair is *redundant* if it is true automatically when the other is true. Thus, if company policy states that workers may not work more than 10 hours a day ($x \leq 10$) and the union contract specifies that workers must not work more than 8 hours a day ($x \leq 8$), company policy is redundant.

Exercise.   How would the system $x < 5$ and $x > 8$ be described? Why? Answer: The inequalities are inconsistent because if x is less than 5, it cannot be greater than 8.

Double inequalities are used in describing the space between two lines. For example, the inequalities

$$i_1: y \geq 2x - 1$$
$$i_2: y \leq \quad x + 2$$

are satisfied by points in the shaded space shown in Figure 3–3. The lines, which are labeled e_1 and e_2 because they are the graphs of the *equality* parts of the statements, are found in the usual manner to intersect at $(3, 5)$. Hence, for any point in space I, the x coordinate must be less than or equal to 3.

Exercise.   Suppose that we let x have the permissible value 0. What y values are permissible if $x = 0$? (See Figure 3–3). Answer: $-1 \leq y \leq 2$.

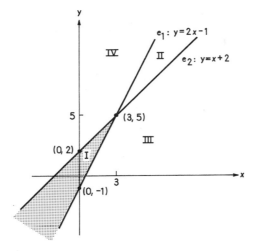

FIGURE 3-3

The exercise shows that for a given permissible value of x, the permissible values of y are those which are greater than or equal to $2x - 1$ but less than or equal to $x + 2$. That is, the solution space is completely described by

$$x \leq 3; \; 2x - 1 \leq y \leq x + 2.$$

This last result says, for example, that if $x = -1$, then both i_1 and i_2 will be true for any y value in the closed interval

$$-3 \leq y \leq 1.$$

Exercise. Describe the solution space II in the unshaded acute angle of Figure 3-3. Answer: $x \geq 3; \; x + 2 \leq y \leq 2x - 1$.

Next let us describe the solution space III in the lower obtuse angle of Figure 3-3. This would be the solution space for the inequality system

$$i_1: \; y \leq 2x - 1$$
$$i_2: \; y \leq x + 2.$$

Every point in III must be on or below *both* lines. To the left of (3, 5) this means that points must be on or below e_1 (that is, $y \leq 2x - 1$), and to the right of (3, 5) points must be on or below e_2 (that is, $y \leq x + 2$). Hence, we have

$$x \leq 3; \; y \leq 2x - 1$$
$$x \geq 3; \; y \leq x + 2.$$

‖‖

    Exercise. *a*) What inequalities have IV as their solution space?
b) Describe IV. Ánswer: *a*) $i_1 : y \geq x + 2, i_2 : y \geq 2x - 1. b) x \leq 3,$
$y \geq x + 2; x \geq 3, y \geq 2x - 1.$

‖‖

Note in our description of III in Figure 3–3 that to the right of (3, 5)

$$i_1: y \leq 2x - 1$$

is *redundant* because it holds automatically if

$$i_2: y \leq x + 2$$

is true.

‖‖

    Exercise. Which inequality is redundant in the description of the
part of IV to the right of (3, 5) in Figure 3–3? Answer: $y \geq x + 2.$

‖‖

3.7 NONNEGATIVITY CONSTRAINTS

    In the work to come, we shall deal often with quantities and prices of
goods, and other variables which cannot take on negative values. In the case
of two variables, x and y, this restriction is expressed by writing

$$x \geq 0$$
$$y \geq 0$$

along with the other inequalities at hand. Nonnegativity means that only
points in the first quadrant, including the axes, are under consideration.
    The system

$$x \geq 0$$
$$y \geq 0$$
$$x \leq 4$$
$$y \leq 2$$

has as its solutions the points on the boundary and inside the rectangle,
Figure 3–4.

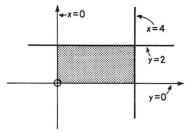

FIGURE 3–4

The algebraic solution of 2 by 2 systems of inequalities with nonnegativity constraints is facilitated by graphing the solution space and determining from the graph what algebraic manipulations of the inequalities will lead to the correct general solution. Consider the system

$$
\begin{aligned}
i_1: &\qquad x \geq 0 \\
i_2: &\qquad y \geq 0 \\
i_3: &\quad x + 2y \leq 8 \\
i_4: &\quad 7x + 4y \geq 28.
\end{aligned}
$$

The solution space for the system is shown as I on Figure 3–5. This space

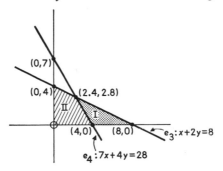

FIGURE 3–5

can be described by saying that if y is in the interval from 0 to 2.8, inclusive, x must be in the interval from e_4 to e_3. Thus the general solution is

$$
0 \leq y \leq 2.8
$$

$$
\frac{28 - 4y}{7} \leq x \leq 8 - 2y.
$$

Exercise. If we describe the solution space I in Figure 3–5, by letting x run over the values from 2.4 to 8, two statements will be required. The first is

$$
2.4 \leq x \leq 4
$$

$$
\frac{28 - 7x}{4} \leq y \leq \frac{8 - x}{2}.
$$

The second will be of the form

$$
\begin{aligned}
a \leq x \leq b \\
c \leq y \leq d.
\end{aligned}
$$

Determine from Figure 3 – 5 the quantities that should appear in the places indicated as a, b, c, and d in the form just written. Answer: a is 4, b is 8, c is 0, d is $\dfrac{8 - x}{2}$.

Consider next the system

$$i_1: \qquad x \geq 0$$
$$i_2: \qquad y \geq 0$$
$$i_3: \quad x + 2y \leq 8$$
$$i_4: 7x + 4y \leq 28.$$

The solution space for this system is shown as II on Figure 3–5. We see that for x in the interval 0 to 2.4, inclusive, y must be in the interval from 0 to line e_3, inclusive. However, if x is from 2.4 up to and including 4, the interval for y is from 0 to line e_4, inclusive. Hence, the general solution may be written:

$$\text{If } 0 \leq x \leq 2.4, \text{ then } 0 \leq y \leq \frac{8 - x}{2}.$$

$$\text{If } 2.4 \leq x \leq 4, \text{ then } 0 \leq y \leq \frac{28 - 7x}{4}.$$

||||||||||||||||||||||||||||||||||||||||||||||||||||||||||||||||||||||||||||||||||||||||||||||||||||||||||||||||||||||||||

Exercise. What quantities should appear in the places designated as a, b, c, d, e, and f if the following inequalities are to describe the solution space II in Figure 3–5?

$$\text{If } 0 \leq y \leq 2.8, \text{ then } a \leq x \leq b.$$
$$\text{If } c \leq y \leq d, \text{ then } e \leq x \leq f.$$

Answer: a is 0, b is $\dfrac{28 - 4y}{7}$, c is 2.8, d is 4, e is 0, f is $8 - 2y$.

||||||||||||||||||||||||||||||||||||||||||||||||||||||||||||||||||||||||||||||||||||||||||||||||||||||||||||||||||||||||||

3.8 APPLICATIONS

Example 1. The United Parcel Service in Massachusetts accepts a parcel for delivery only if its length plus girth (distance around) does not exceed 108 inches.

a) We want to ship fishing poles in cylindrical containers of radius r. What is the restriction on the length of the poles in terms of the radius of the cylindrical container, and what restrictions apply to the radius?

Here, the girth, g, is $2\pi r$, the circumference of a circle, and using L for the length, we have

$$L + 2\pi r \leq 108$$
$$L \leq 108 - 2\pi r$$

where, of course, r must be greater than 0 and less than its value when $L = 0$, $108/2\pi$, which is 17.188 plus a bit more; that is,

$$0 < r \leq 17.188$$

having rounded the right-hand limit *down* to three decimals to insure that the statement is true.

b) From *a)*, what lengths of poles can be shipped in containers having a 2-inch radius?

Substitution shows that

$$L \leq 108 - 2(3.1416)2$$

or

$$L \leq 95.433.$$

|||||||||||||||||||||||||||||||||||||||||||||||||||||||||||||||||||||||||||||||||||||||||||||||||||||||||||||||||||||||||||||||||||||||

Exercise. Cubical boxes are to be shipped. What is the restriction on *x*, the dimension of the side of the cube? Answer: $0 < x \leq$ 21.6 inches.

|||||||||||||||||||||||||||||||||||||||||||||||||||||||||||||||||||||||||||||||||||||||||||||||||||||||||||||||||||||||||||||||||||||||

Example 2. We plan to invest *x* dollars at 7 percent and *y* dollars at *i* percent, and we require that the combined investment yield at least 9 percent. Express the restriction on *i* in terms of *x* and *y* and state the restrictions on *x* and *y*.

We have that 7 percent of *x* plus *i* percent of *y* must be equal to or greater than 9% of $(x + y)$. Hence,

$$0.07x + 0.01iy \geq 0.09(x + y)$$

$$0.01iy \geq 0.02x + 0.09y$$

$$iy \geq 2x + 9y$$

$$i \geq \frac{2x + 9y}{y}.$$

Both *x* and *y* must be nonnegative. Additionally, *x* can be zero, but *y* cannot be zero. Hence,

$$x \geq 0$$
$$y > 0.$$

|||||||||||||||||||||||||||||||||||||||||||||||||||||||||||||||||||||||||||||||||||||||||||||||||||||||||||||||||||||||||||||||||||||||

Exercise. If $x = \$5000$ and $y = \$4000$, what is the minimum interest rate required on the $4000? Answer: 11.5 percent.

|||||||||||||||||||||||||||||||||||||||||||||||||||||||||||||||||||||||||||||||||||||||||||||||||||||||||||||||||||||||||||||||||||||||

Example 3. A mixture is to be made containing *x* units of food A and *y* units of food B. The weight and amount of nutrient per unit of each food are shown in Table 3–1. If the mixture is to contain not less than 13 ounces of nutrient, and it is to weigh not more than 50 ounces, what combinations (numbers of units) of the foods are permissible? To obtain the necessary

Table 3–1

Food	Number of Units in Mixture	Ounces of Nutrient per Unit	Ounces of Weight per Unit
A x		0.4	2
B . . . . y		0.9	3

inequalities, we find the number of ounces of nutrient in a mixture containing x units of A and y units of B, which is $0.4x + 0.9y$. According to the stated requirement, this sum must be 13 or more. Hence, we write

$$0.4x + 0.9y \geq 13.$$

Similarly, the weight constraint is found to be

$$2x + 3y \leq 50.$$

Taking the last two inequalities together with the nonnegativity constraints, we have the system

$$
\begin{aligned}
i_1: & \quad x \geq 0 \\
i_2: & \quad y \geq 0 \\
i_3: & \quad 0.4x + 0.9y \geq 13 \\
i_4: & \quad 2.0x + 3.0y \leq 50.
\end{aligned}
$$

The solution space for the system is shown in Figure 3–6. We see that x may take on values from 0 to 10, inclusive. For any x in this interval the value of

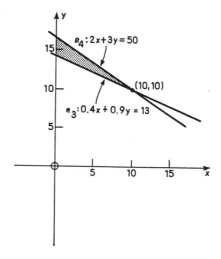

FIGURE 3–6

y must be selected in the interval from line e_3 to line e_4, inclusive. Hence, the general algebraic solution is

$$0 \le x \le 10$$

$$\frac{13 - 0.4x}{0.9} \le y \le \frac{50 - 2.0x}{3.0}.$$

3.9 PROBLEM SET 2

Make a graph showing the solution space:

1. $x \ge 0$
   $y \ge 0$
   $y \ge -1$
   $x \le 4.$

2. $x \ge 0$
   $y \ge 0$
   $x \le 3$
   $x \le y.$

3. $x \ge 0$
   $y \ge 0$
   $2x + y \ge 6$
   $x + 4y \le 8.$

4. $x \le 3$
   $y \le 1$
   $x + 3y \ge 3.$

5. $x \ge 0$
   $y \ge 0$
   $7x + 4y \le 28$
   $x + 2y \le 8.$

Find solutions, if any exist:

6. $x \ge 0$
   $y \ge 0$
   $2x + 3y \le 12$
   $x - 2y \ge 2.$

7. $x \ge 0$
   $y \ge 0$
   $4x + y \le 12$
   $x + 5y \ge 8.$

8. $x \ge 0$
   $y \ge 0$
   $4x + y \ge 12$
   $x + 5y \le 8.$

9. $x \ge 0$
   $y \ge 0$
   $4x + y \ge 12$
   $x + 5y \ge 8.$

10. A parcel service requires that length plus girth of parcels not exceed 108 inches. Containers with rectangular cross section (girth) of width *w*, height *h*, and length *L* are to be shipped.
    a) If the width is to be 16 inches, state the restriction on the height in terms of the length, and state the restrictions on the length.
    b) If the rectangular cross section is a square of side *x*, state the restriction on *x* in terms of *L* and state the restriction on *L*.

11. A mixture is to be made containing *x* units of food A and *y* units of food B. Food A weighs 3 ounces per unit; food B, 5 ounces per unit. Food A contains 0.5 ounces of nutrient per unit; food B, 1 ounce of nutrient per unit. The mixture is to have at least 8 ounces of nutrient, and the total weight is not to exceed 45 ounces. Find the general solution, showing all permissible combinations of the two foods.

12. Solve Problem 11 if the mixture is not to exceed 40 ounces in total weight.

13. Type A and type B items require the same amount of storage space per unit. Type A costs $3 per unit, B costs $8 per unit. Sufficient storage space is available for 500 units (total), and $2400 is available to spend on the items. Write the general solution, showing permissible combinations of items which may be purchased and stored without exceeding total space and money restrictions.

14. The table shows, for example, that it takes one hour in Department I and two hours in Department II to make one unit of product A. Department I has up to five hours available and II has up to seven hours available for making products A and B. Letting x be the number of units of A and y be the number of units of B, *a*) Write the general solution showing permissible combinations of units A and B which can be made with the time available. *b*) What specific combinations are possible if only complete units of product are allowed?

	Hours Required to Make One Unit of:	
Department	Product A	Product B
I 1		1
II 2		1

15. It takes two hours to make a unit of A and three hours to make a unit of B. The number of units of B must not be more than twice the number of units of A. If up to 36 hours are available to make A and B, find the permissible combinations of numbers of units of A and B.

16. A costs $2 per pound and B costs $5 per pound. If we mix x pounds of A with y pounds of B, the mixture contains $x + y$ pounds. The proportion of A in the mixture therefore is the ratio $x/(x + y)$. Mixtures are to be made at a total cost not exceeding $286. In these mixtures, the proportion of A is to be not less than 0.2 (20 percent) and not greater than 0.8 (80 percent). Write the general solution showing permissible combinations of A and B.

17. The table shows, for example, that Department III has 190 hours available and that one unit of A requires 6 hours of Department III time. Write the general solution showing the combinations of A and B which can be made in the time available.

		Hours Required to Make One Unit of:	
Department	Available Hours	Product A	Product B
I 120		4	3
II 40		1	2
III 190		6	7

18. (This problem is a new interpretation of the numbers in Problem 17.) Each unit of A made contributes $6 to overhead and profit and each unit of B contributes $7. Write the general solution showing permissible combinations that can be made in the available time if total contribution to overhead and profit is to be at least $190.

Department	Available Hours	Hours Required to Make One Unit of:	
		Product A	Product B
I	120	4	3
II	40	1	2

3.10 REVIEW PROBLEMS

Specify in two ways the side of the line which contains the points satisfying the following inequalities. (See Problem Set 1.)

1. $x - y < 5$.
2. $x + y > 3$.
3. $2x - 3y < -5$.
4. $-2x + 5y > 2$.
5. $4x - 3y > -5$.

Solve for x.

6. $2 - 3x \leq 6$.
7. $2x + 7 \geq 4$.
8. $7 - 4x \leq 0$.
9. $3 \geq 5 - 2x$.
10. $0 \leq 2x + 3$.

Write the solution, if any exists, for each of the following systems:

11. $x < 4$
    $x > 6$.
12. $x > -2$
    $x > 2$.
13. $x > 2$
    $x < 6$.

Make a graph showing the solution space for each of the following, assuming in each system that the nonnegativity constraints $x \geq 0$ and $y \geq 0$ prevail, and write the algebraic statement of the solution space:

14. $x - 2y \leq 0$
    $x + y \leq 2$.
15. $x + y \geq 2$
    $x + 2y \leq 4$.
16. $x - 2y \leq 0$
    $x + 2y \leq 4$.
17. $x - 2y \geq 0$
    $x + 2y \geq 4$
    $x \geq 3$.
18. $x + y \geq 2$
    $x - 2y \leq 0$
    $x + 2y \leq 4$.
19. $x + y \geq 2$
    $x - 2y \geq 0$
    $x \leq 3$
    $x + 2y \leq 4$.

20. *a*) We plan to invest x dollars at 6 percent and y dollars at i percent and require that the combined investment yield at least 8 percent. Express the restriction on i in terms of x and y, and state the restrictions on x and y.
    b) If $x = \$2000$ and $y = \$1000$, what is the minimum interest rate that must be obtained on the $1000?

21. A mixture of foods A and B is to weigh not more than 5 pounds and must contain at least 24 ounces of nutrient. Food A contains 6 ounces of nutrient per pound and B contains 4 ounces of nutrient per pound. Write the general solution showing the permissible mixtures of the two foods.

22. See Problem 21. If, in addition to the stated constraints, the mixture must contain at least three pounds of food A, write the solution showing the permissible mixtures of the foods.

23. See Problem 21. If, in addition to the stated constraints, it is required that the mixture must not cost more than 21 cents, and if A costs 6 cents per pound and B costs 2 cents per pound, write the general solution showing the permissible mixtures of the two foods.

24. A type A batch of a product contains one unit of substance P and three units of substance Q. A type B batch contains three units of P and four units of Q. If 70 units of P and 110 units of Q are available, write the solution showing permissible combinations of the two types of batches.

25. See Problem 24. Write the solution if a type A batch contributes $2 to profit and a type B batch contributes $1 to profit, and total profit is to be at least $65.

26. If, in addition to the conditions and facts stated in Problems 24 and 25, it is required that at least 5 percent of total output is to be of type B, write the solution showing the permissible combinations of the two types of batches.

27. The table shows, for example, that it takes two hours in I, one hour in II, and one hour in III to make one unit of product A. If the hours available in I, II, and III are, respectively, 12, 7, and 15, write the solution showing the permissible combinations of the two products.

Department	Hours Required to Make One Unit of:	
	Product A	Product B
I 2		1
II 1		1
III 1		3

4

Introduction to Linear Programming

4.1 INTRODUCTION

IN THE REAL world, we are often confronted with a number of ways of accomplishing a certain objective, some ways being better in a certain sense than others. For example, there are many different combinations of foods which will provide a satisfactory diet, but some combinations are more costly than others, and we may be interested in finding the *minimum* cost of providing dietary requirements. Again, there are many combinations of products a plant can manufacture, and we may be interested in finding the combination which leads to *maximum* profit.

The variables in real-world situations are subject to restrictions which we shall call *constraints*. In the first place, it is required in most instances that the variables not take on negative values. Furthermore, certain combinations of the variables are not permissible. For example, a product mix which requires a plant to operate more than 24 hours a day obviously is not permissible, nor is it permissible to schedule output at levels exceeding capacity.

Let us suppose that the relevant variables in a certain situation are x and y, and that the constraints can be expressed as a system of inequalities in these variables. Then suppose we have an expression called the *objective function,* which shows how a quantity θ (theta) is to be computed for given values of x and y. The problem is to find the maximum (or minimum) value of θ, subject to the constraints set forth in the system of inequalities. If the constraints and the objective function are linear, we have a problem in *linear programming.*

To be more specific, suppose that a textile mill buys unfinished cloth and uses 10 processes to convert it into 12 styles of finished material. The styles require varying amounts of time in each of the processes, and there is a capacity limitation on the time available for each process, making 10 capacity constraints. Further, a certain quantity (at least) of each style must be produced to satisfy customer demand, making 12 demand constraints. Finally, of course, negative amounts of finished cloth are not permissible in a solution, adding 12 nonnegativity constraints, for a total of $10 + 12 + 12 = 34$ linear inequality constraints.

By deducting raw material and processing cost from selling price, the profit contribution per yard produced for each of the 12 styles can be computed, and these profit contributions may be combined to form a linear objective function containing as its variables the unknown number of yards of each of the 12 styles to be produced. The problem is to determine the set of values for these variables which will maximize total profit, and not violate any of the 34 constraints. This set of values may then be used to establish the production schedule.

The situation just described involved 34 constraints on 12 variables (a 34 by 12 system), but it is by no means a large system compared to many encountered in practice. We need not be concerned about problem size, however, because large systems can be solved in a matter of seconds, or perhaps a few minutes, by entering the coefficients of the variables into one of the many computers equipped to solve such problems. In this chapter we present the fundamentals of linear programming by use of small systems, and in the next chapter we shall develop a procedure for solving larger problems.

4.2 TWO-VARIABLE CASE: AN INTRODUCTORY EXAMPLE

Table 4–1 shows that products A and B are made in Departments I and II. To make one unit of A requires 1 hour in each department. One unit of B requires 1 hour in I and two hours in II. Department I has four hours of time available, and II has six hours available. Each unit of A contributes $1 to profit, and each unit of B contributes $0.50 to profit. The problem is

TABLE 4–1

			Hours Required per Unit in:	
Product	Number of Units Made	Profit per Unit	Department I (4 hours available)	Department II (6 hours available)
A . . . . . . *x*		$1.00	1	1
B . . . . . . *y*		0.50	1	2

to determine the maximum profit that can be achieved, keeping in mind the four- and six-hour time limitations in the departments. The profit achieved from making x units of A and y units of B is

$$\text{profit} = \theta = x + 0.5y.$$

To make x units of A and y units of B will require $x + y$ hours in I and $x + 2y$ hours in II. The time limitations therefore are

$$x + y \leq 4$$
$$x + 2y \leq 6.$$

The conditions in the foregoing, taken together with nonnegativity constraints on x and y lead to the following statement of the problem:
Find θ_{max}, subject to:

$$
\begin{aligned}
&i_1: &x &\geq 0\\
&i_2: &y &\geq 0\\
&i_3: x + &y &\leq 4\\
&i_4: x + 2&y &\leq 6\\
&\theta = &x + 0.5&y.
\end{aligned}
$$

Clearly, we can find nonnegative values of x and y (that is, values satisfying i_1 and i_2) which also satisfy i_3 and i_4. The obvious values are $x = 0$ and $y = 0$. However, if we substitute $(0, 0)$ into the objective function, we find that the profit is 0. We must make some product to achieve a profit. If we make one unit of A and one unit of B ($x = 1$ and $y = 1$) we find θ is 1.5, and all inequalities are satisfied.

||||||||||||||||||||||||||||||||||||||||||||||||||||||||||||||||||||||||||||||||||||||||||||||||||||||||||||

Exercise. Is it possible to make two units of each product? If so, what profit will be achieved? Answer: Yes, $x = 2, y = 2$ satisfy all constraints in the problem. The profit will be $3.

||||||||||||||||||||||||||||||||||||||||||||||||||||||||||||||||||||||||||||||||||||||||||||||||||||||||||||

||||||||||||||||||||||||||||||||||||||||||||||||||||||||||||||||||||||||||||||||||||||||||||||||||||||||||||

Exercise. Is it possible to make three units of each product? Answer: No; $x = 3, y = 3$ satisfies neither i_3 nor i_4.

||||||||||||||||||||||||||||||||||||||||||||||||||||||||||||||||||||||||||||||||||||||||||||||||||||||||||||

Our goal is not simply to list all the various profits that might be achieved but, rather, to find the *maximum* profit that can be achieved. Consider Figure 4–1.

The solution space for the inequalities i_1 through i_4 is shown on Figure 4–1 as the space bounded by lines whose intersections (which we shall call

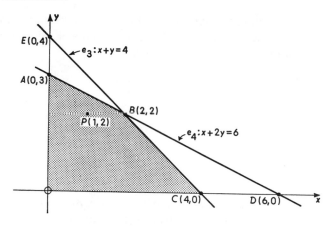

FIGURE 4–1

corners) are at *O, A, B,* and *C.* Each point in the solution space has a pair of coordinates (x, y) which, when substituted into the objective function, yield the value of θ associated with that point. For example, the point $P(1, 2)$ is in the solution space. At *P,* we find:

$$\theta = 1 + 0.5(2) = 2.$$

Again, point $A(0, 3)$ is in the solution space. At *A:*

$$\theta = 0 + 0.5(3) = 1.5.$$

Clearly, the value of θ varies from point to point in the solution space. But observe that for a given x, we obtain the largest value of θ, for this given x, by making y as large as possible, that is, by going all the way to the boundary of the solution space. We see, therefore, that the point which will maximize θ must lie on the boundary of the solution space.

Consider the boundary segment from $A(0, 3)$ to $B(2, 2)$:

$$\text{At } A, \theta = 1.5$$
$$\text{At } B, \theta = 3.$$

This means that as we go from *A* toward *B* on the boundary *AB,* θ is increasing. Clearly, therefore, in seeking the maximum θ, we should go as far as possible in this direction, that is, to corner *B.*

Consider next the boundary segment from $B(2, 2)$ to $C(4, 0)$:

$$\text{At } B, \theta = 3$$
$$\text{At } C, \theta = 4.$$

Following the above reasoning, we move along the boundary to corner *C,* and then consider the segment from *C* to *O.* Obviously, θ decreases along this boundary. Whichever direction we move from *C,* toward *O* or toward *B,* we find θ decreases. Hence, the maximum value of θ occurs at $C(4, 0)$, and

$$\theta_{max} = 4.$$

The solution of our problem is to make four units of A and zero units of B. This schedule will provide a profit of \$4, the maximum profit possible under the conditions of the problem.

‖‖

Exercise. Suppose that $\theta = x + 1.5y$ is a *cost* function and that the inequalities i_3 and i_4 of the foregoing example were of opposite sense to that shown. What is the solution of the following linear programming problem?

Find θ_{min}, subject to:

$$\begin{array}{ll} i_1: & x \geq 0 \\ i_2: & y \geq 0 \\ i_3: & x + y \geq 4 \\ i_4: & x + 2y \geq 6. \end{array}$$

Answer: The solution space is in the obtuse angle *EBD* of Figure 4–1, in the first quadrant. The minimum value of θ will occur on the *inner* boundary of the space at one of the corners *E, B,* or *D*. Checking the coordinates of these corners in the objective function shows that θ is minimal at point *B*. The minimal value is $\theta = 5$ at $(2, 2)$.

‖‖

According to the foregoing discussion, it appears that the maximum, or minimum, value of the objective function will occur at one of the corners of the solution space. An exception to this rule is the special case where θ is constant along an entire segment of the boundary, and the value of θ on this segment is the maximum or minimum. This special case presents no difficulty here, however, because it remains true that the maximum or minimum will be found by evaluating the objective function at the corners of the solution space.

4.3 A GENERAL METHOD OF SOLUTION

The maximum or minimum value of an objective function occurs at a corner of the solution space of the system of constraints. It is possible, therefore, to obtain the optimum θ by the following procedure:

1. Find all the corners of the solution space by treating the inequalities as equations and solving for points of intersection.
2. Substitute the coordinates of all corners into the constraint inequalities, discarding corners which do not satisfy all constraints.
3. Evaluate θ at the remaining corners, and select the maximum (minimum) value of θ.

4.4 NUMBER OF CORNERS

Step 1 of the last section, as applied to an *m* by 2 system (that is, a system having two variables), requires that all possible *pairs* of equations (obtained by replacing inequality signs by equality signs) be solved to find the point of intersection; that is, all 2 by 2 systems which can be set up from the constraints must be solved. If the system had *m* constraints and 3 variables, determination of corners would require the solution of all 3 by 3 systems which could be set up. Similarly, the attack on an *m* by 4 system would start with the setting up and solving of all possible 4 by 4 systems.

Recall the system at the beginning of the section entitled "Two-Variable Case: An Introductory Example." Replacing inequality by equality signs, we have four equations in two variables, a 4 by 2 system. The objective function is not counted because it has no control over the solution space. We can list six 2 by 2 systems which can be selected from the 4 by 2. Numbering the equations 1, 2, 3, and 4, these combinations are

$$
\begin{array}{ll}
1, 2 & 2, 3 \\
1, 3 & 2, 4 \\
1, 4 & 3, 4
\end{array}
$$

corresponding to the six points of intersection shown on Figure 4–1. We note in passing that in some problems the actual number of intersections is less than the possible number because of parallelism of pairs of lines.[1] A 6 by 3 system has a possible 20 corners; a 7 by 4 system has a possible 35 corners. Now, in principle, we can find the 35 corners for a 7 by 4 system, but the prospect of having to solve 35 systems of four equations in four variables makes us wonder if there might be a more efficient method for solving linear programming problems, especially in view of the fact that many of the corners will have to be discarded after they have been found because they do not satisfy the constraints. We leave for the next chapter a discussion of more efficient procedures, and confine our attention here to problems which can be solved in a reasonable length of time by elementary procedures.

4.5 GRAPHIC AID IN THE TWO-VARIABLE CASE

Step 2 of the general solution method (Section 4.3) requires that all corners be checked in the constraint inequalities, so that those not satisfying

---

[1]The number of corners possible in an *m* by *n* system can be computed as

$$
\frac{m!}{n!(m-n)!}, \; m \geq n,
$$

where, for example, $2! = (2)(1)$ and $4! = (4)(3)(2)(1)$. Thus, a 6 by 3 system has

$$
\frac{(6)(5)(4)(3)(2)(1)}{[(3)(2)(1)][(3)(2)(1)]} = 20 \text{ possible corners.}
$$

the constraints may be discarded. Only coordinates of corners satisfying the constraints may be substituted into the objective function. In the case of two variables the selection of permissible corners can be carried out by graphing equations and selecting from the graph only those corners which are on the boundary of the solution space. This procedure was carried out for the first problem of the chapter (see Figure 4–1). The only corners on the boundary of the solution space were A, B, C (and the origin, which is of no interest in a maximization problem). E and D were not considered because the graph shows they are not in the solution space, so their coordinates could not satisfy the constraints.

As another example, consider the three-part problem:

$$
\begin{aligned}
i_1: & \quad x \geq 0 \\
i_2: & \quad y \geq 0 \\
i_3: & \quad x \leq 10 \\
i_4: & \quad y \leq 6 \\
i_5: & \quad 5x + 6y \leq 60 \\
i_6: & \quad x + 2y \leq 16.
\end{aligned}
$$

Find $\theta_{\max}$ if:

$$
\begin{aligned}
1. \ \theta &= x + y \\
2. \ \theta &= 2x + 3y \\
3. \ \theta &= x + 10y.
\end{aligned}
$$

We shall solve the three parts of the problem together. Consider Figure 4–2, which shows the solution space for the system of constraints. The

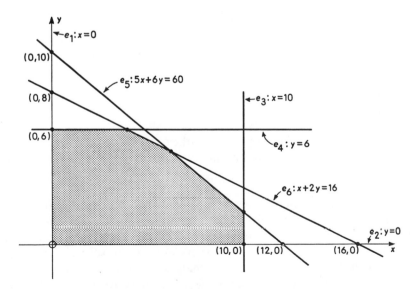

FIGURE 4–2

boundary lines are plotted in the usual manner, using intercepts. Thus, changing i_5 to an equality:

$$5x + 6y = 60$$
$$\text{If } x = 0, y = 10$$
$$\text{If } y = 0, x = 12.$$

The line connecting (0, 10) and (12, 0) is shown as e_5 on Figure 4–2. The lines obtained by plotting equations resulting from i_1 and i_2 are, of course, the coordinate axes; those obtained from i_3 and i_4 are parallel to one of the axes.

Using the symbol 1 # 4 to designate the corner where lines e_1 and e_4 intersect, and similarly for other corners, we see from Figure 4–2 that the only corners which are part of the solution space are 1 # 4, 4 # 6, 5 # 6, 3 # 5, 2 # 3, and 1 # 2. The corner 1 # 2, the origin, is of no interest. The coordinates of the remaining permissible corners are obtained in the usual manner by elimination and substitution. Thus, for 5 # 6:

$$e_5:\ 5x + 6y = 60$$
$$e_6:\ \ x + 2y = 16$$
$$e_7:\ 5x + 6y = 60 \qquad e_5$$
$$e_8:\ 5x + 10y = 80 \qquad 5e_6$$
$$e_9:\ \qquad 4y = 20 \qquad e_8 - e_7$$
$$y = 5$$
$$x = 6 \qquad \text{from } e_6.$$

The corner 5 # 6 is the point (6, 5). In the case of 4 # 6:

$$e_4:\ \qquad y = 6$$
$$e_6:\ x + 2y = 16.$$

Substituting $y = 6$ into e_6 yields $x = 4$, so this corner is (4, 6).

Exercise. Find the coordinates of the corner 3 # 5 and evaluate the three objective functions at this corner. Answer: See Table 4–2.

The corners, their coordinates, and values of the three objective functions are shown in Table 4–2.

In the last example, 13 corners appear in Figure 4–2. By charting the lines, it was possible to see that only six corners were in the solution space and were permissible points at which to evaluate the objective function. Assuming that one's graph is correct, these selected corners necessarily satisfy the constraints.

TABLE 4–2

Corner	Coordinates	Objective Function, θ		
		$x + y$	$2x + 3y$	$x + 10y$
1 # 4 . . . . . . (0, 6)		6	18	60
4 # 6 . . . . . . (4, 6)		10	26	64*
5 # 6 . . . . . . (6, 5)		11	27*	56
3 # 5 . . . . . . (10, 5/3)		35/3*	25	80/3
2 # 3 . . . . . . (10, 0)		10	20	10

*Indicates the maximum.

Example.   The minimum value of the objective function is sought in this example.

$$i_1: \qquad x \geq 0$$
$$i_2: \qquad y \geq 0$$
$$i_3: \qquad x \leq 10$$
$$i_4: \; 5x + 6y \geq 60$$
$$i_5: \; x + 2y \geq 16.$$

Find $\theta_{\min}$ if:

$$\theta = 2x + y.$$

The solution space is shown in Figure 4–3. The minimum value of θ is 10

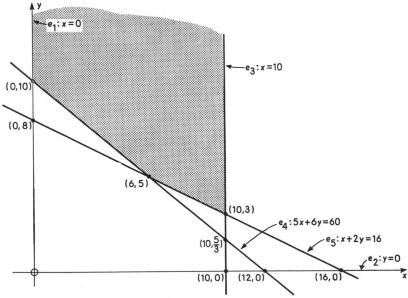

FIGURE 4–3

and occurs at the corner whose coordinates are (0, 10), as shown in Table 4–3:

TABLE 4–3

Corner	Coordinates	$\theta = 2x + y$
1 # 1	(0, 10)	10*
4 # 5	(6, 5)	17
3 # 5	(10, 3)	23

*Indicates the minimum.

Exercise. Change the senses of i_3, i_4, and i_5 in the last example and then find θ_{min} by referring to Figure 4–3. Answer: The relevant corners are at the points (10, 0), (10, 5/3), and (12, 0). The minimum θ, 20, occurs at the point (10, 0).

4.6 PROBLEM SET 1

Use the graphical procedure to eliminate nonpermissible corners; then find θ maximum or minimum, as required.

1.
$$x \geq 0$$
$$y \geq 0$$
$$4x + 3y \leq 24$$
$$x + 2y \leq 11.$$
Find θ_{max} if:
a) $\theta = x + y$
b) $\theta = x + 3y$
c) $\theta = 3x + y.$

2.
$$x \geq 0$$
$$y \geq 0$$
$$2x + y \geq 8$$
$$6x + 10y \leq 60.$$
Find θ_{min} if:
a) $\theta = 3x + 2y$
b) $\theta = 10x + y.$

3.
$$x \geq 0$$
$$y \geq 0$$
$$0.15x + 0.10y \geq 15$$
$$x + y \leq 120$$
$$0.40x \leq 32.$$
Find θ_{min} if:
$$\theta = 0.20x + 0.15y.$$

4.
$$x \geq 0$$
$$y \geq 0$$
$$x \leq 9$$
$$y \leq 6$$
$$8x + 15y \leq 120$$
$$11x + 10y \leq 110.$$
Find θ_{max} if:
$$\theta = 3x + y.$$

5.
$$x \geq 0$$
$$y \geq 0$$
$$x \leq 4$$
$$y \leq 4$$
$$3x + y \leq 9$$
$$x + y \leq 5.$$
Find θ_{max} if:
a) $\theta = 0.3x + 0.5y$
b) $\theta = 2x + y$
c) $\theta = 5x + y.$

6.
$$x \geq 0$$
$$y \geq 0$$
$$y \leq x$$
$$x + y \leq 2$$
$$x + 3y \leq 3.$$
Find θ_{max} if:
$$\theta = x + 2y.$$

7. How would each of the corners in a 6 by 3 system be found?
8. Describe the type of problem considered in the subject known as linear programming.

4.7 APPLICATIONS–1

The introductory example of Section 4.2 serves as a first application. As a second, consider a manufacturer who makes a canned food by mixing x pounds of food X with y pounds of food Y. The proportion of X in the mixture must be at least 0.2, but not more than 0.3. Additionally, for reasons of machine-time economy, batches of at least 100 pounds of mix must be made. If the costs are \$0.25 and \$0.30 per pound for X and Y, respectively, what is the minimum-cost mixture?

The proportion of X in the mixture is $x/(x + y)$. It follows that

$$\frac{x}{x + y} \geq 0.2, \qquad \frac{x}{x + y} \leq 0.3, \quad \text{and} \quad x + y \geq 100.$$

Rearranging the first two of the expressions, we find

$$i_1: \ 0.8x - 0.2y \geq \quad 0$$
$$i_2: \ 0.7x - 0.3y \leq \quad 0$$
$$i_3: \qquad x + \quad y \geq 100.$$

Plotting the lines corresponding to the equality statements, we find the solution space as shown in the shaded area of Figure 4–4. The corners we seek are 1 ⧣ 3 and 2 ⧣ 3.

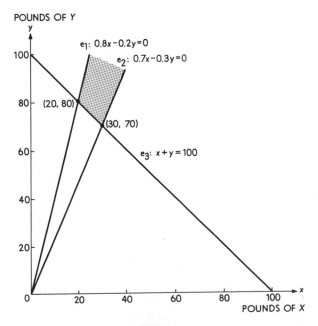

POUNDS OF Y

e_1: 0.8x−0.2y=0
e_2: 0.7x−0.3y=0
(20, 80)
(30, 70)
e_3: x+y=100

POUNDS OF X

FIGURE 4–4

Exercise. Verify that these corners are, respectively, (20, 80) and (30, 70).

The objective function is

$$\theta = 0.25x + 0.30y.$$

At (20, 80), θ = $29 and at (30, 70), θ = $28.50. The latter point yields the desired minimum.

Exercise. Find the minimum if the costs per pound of X and Y change to $0.40 and $0.45, respectively. Answer: The minimum is $43.50 at (30, 70).

4.8 PROBLEM SET 2

1. Products A and B require various times for operations on machines M1, M2, and M3. During a certain production period, M1 has 4000 minutes of time available, M2 has 6000 minutes, and M3 has 3000 minutes. Machine time per unit of A is one minute on M1, two minutes on M2, and no time on M3. Machine time per unit of B is one minute on each of the three machines. Expected profit contributions are 60 cents per unit of A made, and 75 cents per unit of B. If production during the period is limited to products A and B, how many units of each should be made to maximize profit?

2. Parts A and B are to be produced and placed in a warehouse whose capacity is 36,000 feet. Part A requires 4 feet of space per unit, B requires 3 feet. In total, 60,000 machine-hours are available for making the parts; A requires 4 machine-hours per unit, B requires 10 hours. In total, 24,000 man-hours are available for painting the parts; A requires 2 man-hours per unit for painting, B requires 3 man-hours. Profit contributions are expected to be $5 per unit of A and $3 per unit of B. How many units of each should be made to maximize profit?

3. A company makes types A and B watchbands. Type A contributes 80 cents per band to profit, B contributes 60 cents. One type A band requires twice as much time to make as a type B band. If all bands were of type B, the company could make 2500 per day; however, because of material shortages, total output (both types) cannot exceed 1600 bands per day; and for the same reason the output of type A cannot exceed 800 per day, nor can the output of B exceed 1400 per day. How many of each type should be made per day to maximize profit?

4. To make one unit of product A requires three minutes in Department I and one minute in II. One unit of B requires four minutes in I and two minutes in II. Profit contributions are $5 per unit of A made, and $8 per unit of B. However, the number of units of A made must be at least as large as the number of units of B. Find the number of units of A and B which should be made to maximize

profit if Department I and Department II have 150 and 60 minutes available, respectively. What is the maximum profit?

5. Suppose the profit contribution of A (see Problem 4) fell from $5 to $2 per unit. Other factors being the same, what combination would yield the maximum profit? What is this maximum profit?

6. (See Problem 4.) If, instead of requiring that the number of units of A must be at least as large as the number of units of B, it is required that at least 25 units of A be made, what then is the maximum achievable profit, and what numbers of units yield this profit?

7. (See Problem 4.) If, instead of requiring that the number of units of A must be at least as large as the number of units of B, it is required that the number of units of A must be at least three times the number of units of B, the Department II constraint is no longer binding; that is, the Department II time limitation does not affect the problem. Why is this so?

8. A quart of A costs $2 and contains 2.5 ounces of Zap. A quart of B costs $3 and contains 4 ounces of Zap. A mixture not to exceed 10 quarts, and having at least 36.4 ounces of Zap is to be made. Find the minimum-cost mixture.

9. What is the solution for Problem 8 if B is to constitute at least three-fourths of the total mixture?

4.9 THREE-VARIABLE CASE

If the reader plans to study the next chapter, which develops an efficient procedure for solving problems involving three or more variables, he may wish to omit the discussions of three-variable cases in Sections 4.9–4.12.

In principle, one can represent a system in three variables in graphical form, but the plotting is too complex to serve as a practical method for eliminating nonpermissible corners. In the example which follows, we shall find the coordinates of *all* the corners, check each in *all* of the inequalities to eliminate nonpermissible corners, and then evaluate the objective function at the remaining permissible corners.

$$
\begin{aligned}
i_1: & \quad x \geq 0 \\
i_2: & \quad y \geq 0 \\
i_3: & \quad z \geq 0 \\
i_4: & \quad 0.5x + y + 0.2z \geq 600 \\
i_5: & \quad 2.0x + 3y + z \leq 2000 \\
& \quad \theta = x + 4y + 3z.
\end{aligned}
$$

We have a 5 by 3 system. The number of corners possible is 10, as listed in Table 4–4. We shall illustrate the determination and checking of a few of these corners.

$$
\textit{Corner 1 \# 4 \# 5}
$$

$$
\begin{aligned}
e_1: & \quad x = 0 \\
e_4: & \quad 0.5x + y + 0.2z = 600 \\
e_5: & \quad 2.0x + 3y + z = 2000.
\end{aligned}
$$

Substituting $x = 0$ into e_4 and e_5, we have

$$e_6: \quad y + 0.2z = 600$$
$$e_7: \quad 3y + \quad z = 2000$$
$$e_8: \quad \quad 0.4z = 200 \qquad e_7 - 3e_6$$
$$\quad \quad z = 500.$$

From e_6, with $z = 500$, we find $y = 500$, so the desired corner is $(0, 500, 500)$. These coordinates satisfy all inequalities, i_1 through i_5.

$$\text{Corner } 1 \,\#\, 2 \,\#\, 4$$
$$e_1: \quad \quad \quad x = 0$$
$$e_2: \quad \quad \quad y = 0$$
$$e_4: 0.5x + y + 0.2z = 600.$$

Substitution of $x = y = 0$ into e_4 yields $z = 3000$. However, $(0, 0, 3000)$ does not satisfy i_5, and so is not a permissible corner at which to evaluate the objective function.

Exercise. Find and check the corner $2 \,\#\, 4 \,\#\, 5$. Answer: $x = 2000, y = 0, z = -2000$ does not check because z is negative, contrary to contraint i_3.

TABLE 4–4

Corner	Coordinates	Check	$\theta = x + 4y + 3z$
1 # 2 # 3	$(0, 0, 0)$	No	
1 # 2 # 4	$(0, 0, 3000)$	No	
1 # 2 # 5	$(0, 0, 2000)$	No	
1 # 3 # 4	$(0, 600, 0)$	Yes	2400
1 # 3 # 5	$(0, 2000/3, 0)$	Yes	8000/3
1 # 4 # 5	$(0, 500, 500)$	Yes	3500
2 # 3 # 4	$(1200, 0, 0)$	No	
2 # 3 # 5	$(1000, 0, 0)$	No	
2 # 4 # 5	$(2000, 0, -2000)$	No	
3 # 4 # 5	$(400, 400, 0)$	Yes	2000*

*Indicates the minimum.

Table 4–4 shows the outcome of the determination and checking of all 10 corners. The desired minimum is shown in the table to be 2000 and occurs when $x = 400$, $y = 400$, and $z = 0$.

4.10 PROBLEM SET 3

1. Find θ_{max}, and write the coordinates of the point at which the maximum occurs if:

$$\theta = 0.5x + y + 0.2z$$
$$x \geq \quad 0$$
$$y \geq \quad 0$$
$$z \geq \quad 0$$
$$x + 4y + 3z \leq 1800$$
$$2x + 3y + \quad z \leq 2000.$$

2. Find θ_{min} if: $\theta = x + 2y + 3z$

$$x \geq \quad 0$$
$$y \geq \quad 0$$
$$z \geq \quad 0$$
$$x + 2y + 3z \geq 1200$$
$$2x + \quad y + \quad z \geq \quad 600.$$

3. Find θ_{max} if: $\theta = 1.5x + 2.5y + 2.0z$

$$x \geq 0$$
$$y \geq 0$$
$$z \geq 0$$
$$x + 1.5y \qquad \leq 4.0$$
$$y + 2.5z \leq 5.0$$
$$1.5x + \quad y + 2.0z \leq 7.5.$$

4.11 APPLICATIONS–2: THREE VARIABLES

As we have seen, information leading to a linear programming problem may be presented in a word description of the nature of the circumstances involved and the restrictions which apply to the variables in the problem. In such a situation the first step is to extract the relevant information and set up the system of constraints.

Example. A diet (see Table 4–5) is to contain at least 10 ounces of nutrient P, 12 ounces of nutrient R, and 20 ounces of nutrient S. These nutrients are to be obtained from some combination of foods A, B, and C.

TABLE 4–5

Food	No. of Pounds	Cost per Pound	Ounces of Nutrient per Pound		
			P	R	S
A x		$0.04	4	3	0
B . . . . y		0.07	1	2	4
C . . . . z		0.05	0	1	5

Each pound of A costs 4 cents and has 4 ounces of P, 3 ounces of R, and 0 ounces of S. Each pound of B costs 7 cents and has 1 ounce of P, 2 ounces of R, and 4 ounces of S. Each pound of C costs 5 cents and has 0 ounces of P, 1 ounce of R, and 5 ounces of S. How many pounds of each food should be purchased if the stated dietary requirements are to be fulfilled at minimum cost? The objective function is the sum of the costs of x pounds of A at 4 cents per pound, y pounds of B at 7 cents per pound, and z pounds of C at 5 cents per pound; that is:

$$\theta = 4x + 7y + 5z.$$

The contraints are formulated in a similar manner. Our problem becomes one of minimizing θ subject to:

$$
\begin{aligned}
i_1: & & x &\geq 0 \\
i_2: & & y &\geq 0 \\
i_3: & & z &\geq 0 \\
i_4: & 4x + y & &\geq 10 \\
i_5: & 3x + 2y + z &\geq 12 \\
i_6: & 4y + 5z &\geq 20.
\end{aligned}
$$

This system has 20 possible corners whose coordinates are shown in Table 4–6, together with the evaluation of θ at permissible corners. The minimum cost is 92/3 cents. The combination of foods leading to this mimimum cost is 8/3 pounds of A, 0 pounds of B, and 4 pounds of C.

TABLE 4–6

Corner	Coordinates	Check	$\theta = 4x + 7y + 5z$
1 # 2 # 3	(0, 0, 0)	No	
1 # 2 # 4	(Impossible)		
1 # 2 # 5	(0, 0, 12)	No	
1 # 2 # 6	(0, 0, 4)	No	
1 # 3 # 4	(0, 10, 0)	Yes	70
1 # 3 # 5	(0, 6, 0)	No	
1 # 3 # 6	(0, 5, 0)	No	
1 # 4 # 5	(0, 10, −8)	No	
1 # 4 # 6	(0, 10, −4)	No	
1 # 5 # 6	(0, 20/3, −4/3)	No	
2 # 3 # 4	(5/2, 0, 0)	No	
2 # 3 # 5	(4, 0, 0)	No	
2 # 3 # 6	(Impossible)		
2 # 4 # 5	(5/2, 0, 9/2)	Yes	32.5
2 # 4 # 6	(5/2, 0, 4)	No	
2 # 5 # 6	(8/3, 0, 4)	Yes	92/3*
3 # 4 # 5	(8/5, 18/5, 0)	No	
3 # 4 # 6	(5/4, 5, 0)	Yes	40
3 # 5 # 6	(2/3, 5, 0)	No	
4 # 5 # 6	(20/9, 10/9, 28/9)	Yes	290/9

*Indicates the minimum.

4.12 PROBLEM SET 4

1. A wholesaler has 9600 feet of space available, and $5000 he can spend to buy merchandise of types A, B, and C. Type A costs $4 per unit and requires 4 feet of storage space in the warehouse. B costs $10 per unit and requires 8 feet of

space. C costs $5 per unit and requires 6 feet of space. Only 500 units of type A are available to the wholesaler. Assuming that the wholesaler expects to make a profit of $1 on each unit of A he buys and stocks, $3 per unit on B, and $2 per unit on C, how many units of each should he buy and stock in order to maximize his profit?

2. A diet is to contain at least 10 ounces of nutrient P, 12 ounces of R, and 20 ounces of S. These nutrients are to be obtained from foods A, B, and C. Each pound of A costs 3 cents and contains 3 ounces of P, 2 ounces of R, and 0 ounces of S. Each pound of B costs 6 cents and contains 1 ounce of P, 3 ounces of R, and 3 ounces of S. Each pound of C costs 7 cents and contains 0 ounces of P, 1 ounce of R, and 5 ounces of S. How many pounds of each food should be purchased if the stated dietary requirements are to be met at minimum cost?

3. Products A, B, and C are sold door to door. A costs $3 per unit, takes 10 minutes to sell (on the average), and costs $0.50 to deliver to a customer. B costs $5, takes 15 minutes to sell, and is left with the customer at the time of sale. C costs $4, takes 12 minutes to sell, and costs $1 to deliver. During any week, a salesman is allowed to draw up to $500 worth of A, B, and C (at cost) and he is allowed delivery expenses not to exceed $75. If a salesman's selling time is not expected to exceed 30 hours (1800 minutes) in a week, and if the salesman's profit (net after all expenses) is $1 each on a unit of A or B and $2 on a unit of C, what combination of sales of A, B, and C will lead to maximum profit, and what is this maximum profit?

4.13 REVIEW PROBLEMS

Solve the following two-variable linear programming problems:

1.
$$x \geq 0$$
$$y \geq 0$$
$$x + 2y \leq 40$$
$$3x + y \leq 45.$$
a) Find θ_{max} if $\theta = 2x + 3y$
b) Find θ_{max} if $\theta = 4x + y$.

2.
$$x \geq 0$$
$$y \geq 0$$
$$x + 2y \geq 40$$
$$3x + y \geq 45.$$
Find θ_{min} if $\theta = 4x + y$.

3.
$$x \geq 0$$
$$y \geq 0$$
$$x - y \geq -1$$
$$3x + 2y \leq 17$$
$$x + 4y \geq 9.$$
Find θ_{max} if $\theta = 2x + y$.

4.
$$x \geq 0$$
$$y \geq 0$$
$$x - y \geq -1$$
$$3x + 2y \geq 17$$
$$x + 4y \geq 9.$$
Find θ_{min} if $\theta = 2x + y$.

5.
$$x \geq 0$$
$$y \geq 0$$
$$x \geq 2$$
$$x + y \geq 7$$
$$3x + 4y \geq 24.$$
Find θ_{min} if $\theta = 3x + 2y$.

6. One pound of food A costs $1 and contains 2 ounces of nutrient I and 4 ounces of nutrient II. A pound of B costs $2 and contains 3 ounces of I and 1 ounce of II. A mixture is to contain at least 90 ounces of I and 80 ounces of II. Find the minimum-cost mixture.

7. (See Problem 6.) Suppose the mixture must weigh not more than 40 pounds. What will be the minimum-cost mixture now?

8. (See Problem 6.) What will be the minimum-cost mixture if at least 80 percent of the mixture must be food B?

9. To make one unit of product A requires three minutes each in Departments I and II. A unit of B requires two minutes in I and four minutes in II. A unit of either product contributes $1 to profit. If Departments I and II have 900 and 1200 minutes available, respectively, for making A and B, find the numbers of each which should be made to maximize profit, and find what the maximum profit is.

10. (See Problem 9.) Solve the problem if the number of units of A must be at least as great as the number of units of B.

Solve the following three-variable problems:

11.
$$x \geq 0$$
$$y \geq 0$$
$$z \geq 0$$
$$x + 3y + 4z \leq 30$$
$$x + 5y + 2z \leq 40.$$
Find θ_{max} if $\theta = x + 4y + 3z$.

12.
$$x \geq 0$$
$$y \geq 0$$
$$z \geq 0$$
$$4x + 8y + z \leq 52$$
$$8x + 28y + 3z \leq 168.$$
Find θ_{max} if $\theta = 3x + 9y + z$.

13.
$$x \geq 0$$
$$y \geq 0$$
$$z \geq 0$$
$$x + 3y + 3z \leq 50$$
$$x + 4y + 2z \leq 60$$
$$z \leq 10.$$
Find θ_{max} if $\theta = x + 4y + 2z$.

14.
$$x \geq 0$$
$$y \geq 0$$
$$z \geq 0$$
$$x + 2y + 7z \leq 21$$
$$5x + 17y + 28z \leq 140$$
$$x + 9y + 10z \leq 66.$$
Find θ_{max} if $\theta = x + 3y + 7z$.

15.
$$x \geq 0$$
$$y \geq 0$$
$$z \geq 0$$
$$x + 2y + 3z \leq 14$$
$$\tfrac{1}{2}x + 2y + \tfrac{1}{2}z \leq 10$$
$$x + 5y + 4z \leq 26.$$
Find θ_{max} if $\theta = x + 3y + 3z$.

16.
$$x \geq 0$$
$$y \geq 0$$
$$z \geq 0$$
$$x + 2y + 3z \leq 25$$
$$x + 3y + 2z \leq 30.$$
Find θ_{max} if $\theta = x + 3y + 2z$.

17. A man sells and installs products A, B, and C. The table shows, for example, that it takes three hours to sell a unit of B, four hours to install it, and net profit per unit is $40.

Product	Number of Units	Selling Hours per Unit	Installation Hours per Unit	Profit per Unit
A	x	1	1	$10
B	y	3	4	40
C	z	2	1	10

During a 38-hour week, the man allots no more than 18 hours to selling and no more than 20 hours to installation. Find the combination of numbers of units of A, B, and C which would yield maximum profit.

18. The table shows, for example, that two man-hours are needed to sell a unit of B, three man-hours to deliver it, and three man-hours to install it. Profit per unit of B is $60.

Product	Number of Units	Selling Hours per Unit	Delivery Hours per Unit	Installation Hours per Unit	Profit per Unit
A x	1	1	2	$20
B . . . . y	2	3	3	60
C . . . . z	3	2	5	40

If 220 man-hours are available, of which not more than 50 are to be used for selling, not more than 60 for delivery, and not more than 110 for installation, find the combination of A, B, and C which yields maximum profit and state what this maximum is.

19. The table shows, for example, that in making a unit of B, three minutes are required for stamping, 13 minutes for forming, and five minutes for painting, and each unit of B contributes $4 to profit.

Product	Number of Units	Minutes per Unit for Stamping	Minutes per Unit for Forming	Minutes per Unit for Painting	Profit per Unit
A x	1	3	1	$1
B . . . . y	3	13	5	4
C . . . . z	2	2	5	2

Minutes available for stamping, forming, and painting are 40, 144, and 70, respectively. Find the combination of numbers of units of A, B, and C which lead to maximum profit. (Note: Assume that fractional units are permissible and that a fractional unit contributes its fraction to profit.)

5

Linear Programming (continued)

5.1 INTRODUCTION

THE NUMBER of corners in a linear programming problem of realistic scope is so large that the arithmetic procedures of the last chapter become inadequate for obtaining a solution. In this chapter, we shall describe a systematic procedure which concentrates on permissible corners and does so in a manner such that each successive corner investigated is a step in the direction toward the optimum. We start with a discussion of the equation manipulations which we need later to carry us from corner to corner.

5.2 MANIPULATING ARBITRARY SOLUTIONS

If a system of linear equations has an unlimited number of solutions, and these solutions are written in a general form with, say, x and y expressed in terms of the arbitrary variable z, the general form can be manipulated so that x becomes the arbitrary variable, or so that y becomes the arbitrary variable. The manipulation involves only straightforward substitution in equations. However, we shall see that this substitution can be systematized and carried out without actually writing all the details.

5.3 ONE VARIABLE ARBITRARY

It may be verified quickly that the general solution of the system

$$2x + y + z = 12$$
$$x + y + 2z = 18$$

can be written in the form

$$z \text{ arbitrary}$$
$$e_1: x = -6 + z$$
$$e_2: y = 24 - 3z.$$

If we wish to change the solution, making y arbitrary instead of z, we find

$$y \text{ arbitrary}$$

$$e_3: z = 8 - \frac{1}{3}y \qquad \text{from } e_2$$

$$e_4: x = 2 - \frac{1}{3}y \qquad e_3 \text{ into } e_1.$$

The constants which appear in the second arbitrary solution obviously are related to the constants in the first arbitrary solution. We want to find out in a general way what this relationship is. To this end, we designate the constants of a first arbitrary solution as a, b, c, and d, carry out the substitution procedures, and observe the forms in which the constants appear in the second arbitrary solution. Suppose the first arbitrary solution is

$$z \text{ arbitrary}$$
$$e_1: y = a + bz$$
$$e_2: x = c + dz.$$

If we wish to make y arbitrary instead of z, we find

$$e_3: z = -\frac{a}{b} + \frac{1}{b}y \qquad \text{from } e_1$$

$$x = c + d\left(-\frac{a}{b} + \frac{1}{b}y\right) \qquad e_3 \text{ into } e_2$$

$$= c - \frac{ad}{b} + \frac{d}{b}y$$

$$e_4: x = \frac{cb - ad}{b} + \frac{d}{b}y.$$

To show the relationships between the constants in the two general solutions, the following tableau form is adopted for writing equations:

$$z$$
$$e_1: y \qquad a \quad b*$$
$$e_2: x \qquad c \quad d.$$

The tableau form omits the equality symbols and places the variable z at the head of the column so that attention may be directed toward the constants a, b, c, and d. The appearance of z as a column head means that the solution has z arbitrary, and the row variables, x and y, are expressed in terms of z. The asterisk (*) next to b means that we propose to rewrite the solution with *y* arbitrary instead of *z*. The element with the asterisk is called the *pivot* element, and is the element at the intersection of the row and column which specify the variables to be interchanged, that is, in this case, the intersection of the *y* row and the *z* column.

The solution with *z* arbitrary and the derived solution with *y* arbitrary are shown for comparison in the following tableaus:

$$
\begin{array}{ccc}
 & z & \\
y & a \quad b^* & \\
x & c \quad d. &
\end{array}
\qquad\qquad
\begin{array}{ccc}
 & & y \\
z & -\dfrac{a}{b} & \dfrac{1}{b} \\[2mm]
x & \dfrac{cb-ad}{b} & \dfrac{d}{b}.
\end{array}
$$

The relationship between the tableaus is stated in the form of four rules which tell how to obtain the numbers in the second tableau from those of the first:

1. Replace the pivot element by its reciprocal. Thus, b^* in the first tableau goes to $1/b$ in the second.
2. Replace the elements in the pivot column by the quotient

$$\frac{\text{Element}}{\text{Pivot}}.$$

Thus, *d* in the first tableau becomes d/b in the second.
3. Replace the elements of the pivot row by

$$-\frac{\text{Element}}{\text{Pivot}}.$$

Thus, *a* in the first tableau goes to $-a/b$ in the second.
4. Replace elements not in pivot row or pivot column by their pivot cross product. The pivot cross product for an element is

$$\frac{(\text{Element})(\text{pivot}) - \text{product of opposite diagonal elements}}{\text{Pivot}}.$$

In the present case, *c* is in neither pivot row nor column. Its pivot cross product is

$$\frac{cb - ad}{b}$$

which is shown in the appropriate location in the second tableau.

Comparison shows these rules will carry the first tableau into the second. We shall next illustrate the procedure with the numerical constants of the first example of the chapter. The first arbitrary solution was

$$
\begin{aligned}
x &= -6 + z \\
y &= 24 - 3z.
\end{aligned}
$$

In tableau form:

$$
\begin{array}{ccc}
 & & z \\
x & -6 & 1 \\
y & 24 & -3*.
\end{array}
$$

The asterisk shows that we plan to make y arbitrary instead of z. Interchanging y and z, the form of the next tableau will be

$$
\begin{array}{ccc}
 & & y \\
x & ? & ? \\
z & ? & ?.
\end{array}
$$

Following the rules, we have:
1. Replace pivot element, -3, by $1/-3$.
2. Replace pivot column element, 1, by $1/-3$.
3. Replace pivot row element, 24, by $-(24/-3) = 8$.
4. Replace -6 by its pivot cross product

$$
\frac{(-6)(-3) - (24)(1)}{-3} = 2.
$$

The second tableau is

$$
\begin{array}{ccc}
 & & y \\
x & 2 & -\dfrac{1}{3} \\
\\
z & 8 & -\dfrac{1}{3}.
\end{array}
$$

The equation reproduction of the last tableau is

$$
y \text{ arbitrary}
$$

$$
x = 2 - \frac{1}{3}y
$$

$$
z = 8 - \frac{1}{3}y.
$$

As another example, we shall show that

$$
\begin{array}{ccccc}
 & & z & & \\
u & 5 & -2 & & \\
v & 4 & -1*
\end{array}
\qquad \text{goes to} \qquad
\begin{array}{ccc}
 & & v \\
u & -3 & 2 \\
z & 4 & -1.
\end{array}
$$

That is, the solution

$$
\begin{array}{l}
z \text{ arbitrary} \\
u = 5 - 2z \\
v = 4 - z
\end{array}
\qquad \text{goes to} \qquad
\begin{array}{l}
v \text{ arbitrary} \\
u = -3 + 2v \\
z = 4 - v.
\end{array}
$$

The details of the change in tableaus are:

1. Pivot, -1, goes to reciprocal, $1/-1 = -1$.
2. Pivot column element -2 goes to

$$\frac{\text{Element}}{\text{Pivot}} = \frac{-2}{-1} = 2.$$

3. Pivot row element 4 goes to

$$-\frac{\text{Element}}{\text{Pivot}} = -\frac{4}{-1} = 4.$$

4. Element not in pivot row or column, 5, goes to

$$\text{Pivot cross product} = \frac{(5)(-1) - 4(-2)}{-1} = -3.$$

As a final example before the problem set, the reader may verify that

	y	
w	3	7*
x	5	6

goes to

	w	
y	$-\frac{3}{7}$	$\frac{1}{7}$
x	$\frac{17}{7}$	$\frac{6}{7}$.

That is, in equation form:

$$w = 3 + 7y \qquad\qquad y = -\frac{3}{7} + \frac{1}{7}w$$

$$x = 5 + 6y \quad \text{goes to} \quad x = \frac{17}{7} + \frac{6}{7}w$$

$$y \text{ arbitrary} \qquad\qquad w \text{ arbitrary.}$$

5.4 PROBLEM SET 1

Mark (T) for true or (F) for false:

1. (   ) The column in which the pivot element occurs must have a variable at its head.
2. (   ) Elements in the pivot row of the first tableau are replaced by element over pivot.
3. (   ) Elements in the pivot column of the first tableau are replaced by the negative of element over pivot.
4. (   ) The pivot cross product is computed only for elements not in the pivot row or the pivot column.
5. (   ) The replacement for p in $\begin{array}{cc} p & 2^* \\ 3 & 6 \end{array}$ would be $-\frac{p}{2}$.
6. (   ) The replacement for p in $\begin{array}{cc} 3 & 6^* \\ 2 & p \end{array}$ would be $\frac{p}{6}$.

7. ( ) The replacement for p in $\begin{smallmatrix} 3 & 6 \\ 2 & p* \end{smallmatrix}$ would be $-\dfrac{1}{p}$.

8. ( ) The replacement for p in $\begin{smallmatrix} p & 6* \\ 3 & 2 \end{smallmatrix}$ would be $\dfrac{2p-18}{6}$.

9. ( ) The replacement for 4 in $\begin{smallmatrix} 4 & 2* \\ -1 & 1 \end{smallmatrix}$ would be 3.

10. ( ) The replacement for 4 in $\begin{smallmatrix} 3 & -2* \\ 4 & 1 \end{smallmatrix}$ would be $\dfrac{11}{2}$.

Write the numbers which would appear in the second tableau:

11. $\begin{smallmatrix} 2 & 1 \\ 6 & 4* \end{smallmatrix}$.

12. $\begin{smallmatrix} 3 & 4* \\ 1 & 2 \end{smallmatrix}$.

13. $\begin{smallmatrix} 5 & -1* \\ -3 & 2 \end{smallmatrix}$.

14. $\begin{smallmatrix} -\frac{1}{2} & 2 \\ -5 & 3* \end{smallmatrix}$.

15. $\begin{smallmatrix} \frac{1}{4} & -\frac{1}{3}* \\ -2 & 1 \end{smallmatrix}$.

Convert the equations into first tableau form; then find the second tableau. Finally, write the equations specified by the second tableau.

16. $x = 3 - 2y$
    $w = 5 - 3y*$.

17. $v = 10 + 4z*$
    $y = 5 - z$.

18. $y = 8 - 2x$
    $w = 2 - x*$.

19. $u = 5 + v*$
    $w = 1 - v$.

20. $x = 3 + 2z$
    $y = 4 - 3z*$.

5.5 MORE THAN ONE VARIABLE ARBITRARY

The pivot rules developed in the last section apply without alteration when the number of variables is increased. Consider the system

$$w, z \text{ arbitrary}$$
$$x = 3 + 2w + 4z$$
$$y = 5 + w - 3z.$$

To make x arbitrary instead of z, we start with the tableau

		w	z
x	3	2	4*
y	5	1	-3

making the intersection of the x row and z column the pivot. To obtain the second tableau:

1. Pivot element 4 is replaced by reciprocal, 1/4.
2. Pivot column element -3 is replaced by

$$\frac{\text{Element}}{\text{Pivot}} = \frac{-3}{4}.$$

3. Pivot row elements 3 and 2 are replaced by the negative of

$$\frac{\text{Element}}{\text{Pivot}}, \quad -\frac{3}{4} \quad \text{and} \quad -\frac{1}{2}, \text{ respectively.}$$

4. Elements 5 and 1 are replaced by their pivot cross products.

$$5 \text{ is replaced by } \frac{(5)(4) - (3)(-3)}{4} = \frac{29}{4}$$

$$1 \text{ is replaced by } \frac{(1)(4) - (2)(-3)}{4} = \frac{5}{2}.$$

The second tableau is

	w	x
z	$-\dfrac{3}{4}$ $-\dfrac{1}{2}$	$\dfrac{1}{4}$
y	$\dfrac{29}{4}$ $\dfrac{5}{2}$	$-\dfrac{3}{4}$

which represents the equation system

$$w, x \text{ arbitrary}$$

$$z = -\frac{3}{4} - \frac{1}{2}w + \frac{1}{4}x$$

$$y = \frac{29}{4} + \frac{5}{2}w - \frac{3}{4}x.$$

As a more extensive example, let us start with the following tableau representation of a system with five variables arbitrary:

	v	w	x	y	z
r	2	4	1	-3	1	3
s	1	3	-3	5^*	-2	-3
t	3	1	2	6	4	2.

The asterisk shows we intend to make s arbitrary in place of x. The form of the second tableau, with pivot row and column entries, is:

	v	w	s	y	z
r	?	?	?	$-\dfrac{3}{5}$	?	?
x	$-\dfrac{1}{5}$	$-\dfrac{3}{5}$	$\dfrac{3}{5}$	$\dfrac{1}{5}$	$\dfrac{2}{5}$	$\dfrac{3}{5}$
t	?	?	?	$\dfrac{6}{5}$	?	?.

All the elements indicated by a question mark are obtained as pivot cross products. Thus the first three elements in the first row of the initial tableau, 2, 4, and 1, are replaced as follows:

$$2 \text{ goes to} \qquad \frac{(2)(5) - (1)(-3)}{5} = \frac{13}{5}$$

$$4 \text{ goes to} \qquad \frac{(4)(5) - (3)(-3)}{5} = \frac{29}{5}$$

$$1 \text{ goes to} \qquad \frac{(1)(5) - (-3)(-3)}{5} = -\frac{4}{5}.$$

Similar calculations lead to the following completed tableau:

	v	w	s	y	z
r	$\frac{13}{5}$	$\frac{29}{5}$	$-\frac{4}{5}$	$-\frac{3}{5}$	$-\frac{1}{5}$	$\frac{6}{5}$
x	$-\frac{1}{5}$	$-\frac{3}{5}$	$\frac{3}{5}$	$\frac{1}{5}$	$\frac{2}{5}$	$\frac{3}{5}$
t	$\frac{9}{5}$	$-\frac{13}{5}$	$\frac{28}{5}$	$\frac{6}{5}$	$\frac{32}{5}$	$\frac{28}{5}$.

5.6 PROBLEM SET 2

1. Given:

	y	z	
w	1	2	1
x	5	3	-2.

a) Make w arbitrary in place of y.
b) Make x arbitrary in place of y.
c) Make w arbitrary in place of z.
d) Make x arbitrary in place of z.

2. Given:

	w	x	y	
u	3	2	1	2
v	5	-1	2	-3.

a) Make u arbitrary in place of x.
b) Make v arbitrary in place of w.
c) Make u arbitrary in place of w.
d) Make u arbitrary in place of y.
e) Make v arbitrary in place of x.
f) Make v arbitrary in place of y.

3. Given:

	v	w	x	y
r	4	-2	4	0	6
s	6	2	-2^*	4	0
t	8	0	-6	2	$-4.$

a) Write the second tableau.
b) Write the equation systems corresponding to the given tableau and the second tableau.

4. Given:

$$w, x, \text{ and } y \text{ arbitrary}$$
$$u = 4 + 3w + 2x + y$$
$$v = 2 - w + 3x - 2y.$$

a) Write the tableau for the given system of equations.
b) Construct a second tableau which makes v arbitrary in place of x, and write the equation system corresponding to the second tableau.

5.7 INTRODUCING THE OBJECTIVE FUNCTION

If we have a system of equations and an objective function, θ, and the system has a general solution in terms of arbitrary variables, θ can also be expressed in terms of the same arbitrary variables, as shown in the next example:

$$x = 5 + y$$
$$z = 1 + 3y$$
$$\theta = 2x + y + 3z.$$

By substitution, θ is expressed in terms of the arbitrary variable y; thus:

$$\theta = 2(5 + y) + y + 3(1 + 3y)$$
$$\theta = 13 + 12y.$$

The objective function can be manipulated in tableau form along with the rest of the equations. The tableau for the present example, with y arbitrary, is

		y
θ	13	12
x	5	1^*
z	1	3.

The asterisk indicates that we are going to make x arbitrary in place of y. The usual procedures lead to the second tableau:

		x
θ	-47	12
y	-5	1
z	-14	3

which corresponds to the equations

$$\begin{aligned} \theta &= -47 + 12x \\ y &= -5 + x \\ z &= -14 + 3x. \end{aligned}$$

5.8 SLACK VARIABLES

Systems discussed thus far in this chapter have been systems of equalities. The programming problems we wish to discuss involve systems of inequalities. In this section, we shall see how to convert a system of inequalities into a system of equalities by means of *slack* variables.

Consider the inequality

$$x + 3y \le 10$$

where we require that both x and y be nonnegative; that is, the variables must be equal to or greater than zero. We define the slack variable, r, by the statement

$$x + 3y + r = 10$$

r also being nonnegative; that is, r takes up the slack between $(x + 3y)$ and 10. The original inequality is satisfied by various number pairs, one of which is

$$\begin{aligned} x &= 1 \\ y &= 2. \end{aligned}$$

For this pair the inequality is the true statement $7 \le 10$. The slack in this instance is 3; that is, if the number pair is placed into the equality $x + 3y + r = 10$ the value of r will be 3.

Again, the inequality is satisfied by

$$\begin{aligned} x &= 1 \\ y &= 3 \end{aligned}$$

because substitution leads to the true statement $10 \le 10$. In this instance, however, there is no slack, and r would take the value zero.

We shall work with slack variables in programming problems. The manner in which the problems will be set up is illustrated next. Suppose that we have to maximize

$$\theta = 2x + 3y$$

subject to the inequality constraints

$$\begin{aligned} x + 3y &\le 10 \\ 2x + y &\le 6. \end{aligned}$$

The inequalities will be converted to equalities by means of slack variables. The appearance of the problem will then be

$$\theta = 0 + 2x + 3y$$
$$x + 3y + r = 10$$
$$2x + y + s = 6.$$

All variables (including slack variables) are to be nonnegative. We shall solve the equations for the slack variables, then make a tableau. Thus:

$$\theta = 0 + 2x + 3y$$
$$r = 10 - x - 3y$$
$$s = 6 - 2x - y$$

leads to the tableau

		x	y
θ	0	2	3
r	10	-1	-3
s	6	-2	-1.

The solution of the programming problem will be found by manipulating tableaus in the usual manner, following a set of rules (stated in Section 5.12) for determining which element is to be the pivot.

5.9 PROBLEM SET 3

Construct the tableau corresponding to each of the following programming situations:

1. $\theta = 2x + y$, subject to:
$$3x + 2y \le 20$$
$$x + y \le 12.$$

2. $\theta = x + 2y + z$, subject to:
$$x + y + z \le 8$$
$$2x + 3y + z \le 12$$
$$x + 2y + 3z \le 10.$$

3. $\theta = 3w + 2.8x + y + 2z$, subject to:
$$2w + x + 2y + 3z \le 150$$
$$2x + y + 2z \le 100$$
$$3w + x + 4y \le 200.$$

5.10 MECHANICS OF THE SIMPLEX PROCEDURE

Linear programming problems can be solved by a procedure called the *simplex* method. The word simplex is a mathematical term used in *n*-space geometry and is not a synonym for *simple*. The important attribute of the simplex method is that it provides a step-by-step solution of linear programming problems in which only permissible corners are investigated and each step brings us closer to the optimal value of the objective function. In this section, we shall work through some examples of the simplex method in a purely mechanical fashion. The logic of the procedure will be discussed under the heading, "Sources of Rules and Interpretation."

The steps in the simplex solution lead to a series of tableaus. At any step, we can read the current solution (not necessarily optimal) from the tableau. We can determine also whether the solution is optimal; if it is not, we select the proper pivot and construct the next tableau, continuing in this manner until the optimum is reached.

5.11 READING THE SOLUTION; IDENTIFYING THE OPTIMUM

The solution at any step is read from the tableau by letting all arbitrary variables be zero, and assigning the numbers in the first column (the constants) as values of the corresponding nonarbitrary variables. For example, suppose that we have the tableau

		r	z	y
θ	10	-1	2	1
x	8	-2	2	2
s	12	2	-6^*	1
t	6	4	-1	3.

Disregard the pivot element for the moment. In the tableau the solution is

$$\theta = 10$$
$$x = 8$$
$$s = 12$$
$$t = 6$$
$$r = z = y = 0.$$

A solution is optimal if the numbers in the θ row *under the arbitrary variables* are all zero or negative. The specified numbers in the last tableau are $-1, 2$, and 1, so the solution is not optimal. A pivot must be selected and the next tableau formed.

5.12 SELECTION OF THE PIVOT

The pivot column is the arbitrary variable column which has the largest positive number in the θ row. In the last tableau the pivot will be in the z column.

Now, divide each *negative* element of the pivot column into the corresponding constant in the first column. Select as the pivot row that row for which the quotient is smallest in absolute value (that is, smallest disregarding sign). In the present case the quotients are

$$s \text{ row: } \frac{12}{-6} = -2$$

$$t \text{ row: } \frac{6}{-1} = -6.$$

The former of these two numbers, -2, is the smaller in absolute value, so we select the s row as the pivot row. As the foregoing tableau shows, the pivot element is -6^* at the intersection of the s row and the z column. The next step would be to form a new tableau in the usual manner, making s arbitrary in place of z. Rather than proceed further with this example, we shall start a new one at the beginning and carry it all the way through. The example is arranged as a set of exercises so that the reader can obtain practice in the methodology. When the reader comes to an exercise, he should cover the following material (which contains the answer to the exercise) and do the exercise on a scratch pad before proceeding. Consider the following linear programming problem:

$$\text{Maximize } \theta = x + y$$
$$\begin{aligned} \text{subject to: } 3x + 2y &\le 34 \\ x + 2y &\le 18 \\ x + 6y &\le 48 \\ x &\le 10. \end{aligned}$$

Exercise.   Introduce the slack variables r, s, t, and u and convert the given system into a system of equalities.

The system of equalities is found to be

$$\begin{aligned} 3x + 2y + r &= 34 \\ x + 2y + s &= 18 \\ x + 6y + t &= 48 \\ x + u &= 10. \end{aligned}$$

Exercise.   Solve each of the last equations for the slack variable and write the resultant equations and the objective function in terms of x, y, and z.

The set of equations is found to be

$$\begin{aligned} \theta &= 0 + x + y \\ r &= 34 - 3x - 2y \\ s &= 18 - x - 2y \\ t &= 48 - x - 6y \\ u &= 10 - x + 0y. \end{aligned}$$

Exercise.   Construct the first tableau.

The first tableau is found to be

		x	y
θ	0	1	1
r	34	−3	−2
s	18	−1	−2
t	48	−1	−6
u	10	−1	0.

The solution here is not optimal because the θ row has a positive one under both x and y. We may choose either the x or y column as the pivot column. (The rule is to choose the column having the largest positive number in the θ row.)

Exercise.   Take x as the pivot column, then find the pivot row.

To find the pivot row, we form the following quotients for the negative numbers in the x column:

$$r \text{ row: } \frac{34}{-3}$$

$$s \text{ row: } \frac{18}{-1}$$

$$t \text{ row: } \frac{48}{-1}$$

$$u \text{ row: } \frac{10}{-1} \cdot$$

The smallest quotient, disregarding sign, is $10/-1$, so we take u as the pivot row. The pivot element is the -1 at the intersection of the u row and x column.

Exercise.   Place an asterisk at the intersection of the u row and x column of the tableau. Manipulate the tableau in the usual manner to make u arbitrary in place of x and write the second tableau.

The second tableau is found to be

		u	y
θ	10	−1	1
r	4	3	−2
s	8	1	−2
t	38	1	−6
x	10	−1	0.

Exercise.　Write the solution exhibited by the second tableau.

The solution at this point is

$$\theta = 10$$
$$r = 4$$
$$s = 8$$
$$t = 38$$
$$x = 10$$
$$u = y = 0.$$

Exercise.　Is the present solution optimal?

The solution is not optimal because the θ row has a positive element under an arbitrary variable, namely, the one under y. The y column becomes the new pivot column.

Exercise.　Take y as the pivot column and find the pivot row.

To find the pivot row we form the quotients

$$r \text{ row: } \frac{4}{-2}$$

$$s \text{ row: } \frac{8}{-2}$$

$$t \text{ row: } \frac{38}{-6}$$

$$x \text{ row: None (why?)}$$

In absolute value the r row has the smallest quotient, and so becomes the pivot row. The pivot element is the -2 at the intersection of the r row and y column.

Exercise.　Place an asterisk next to the pivot element, -2, of the last tableau, then calculate and write the third tableau.

The third tableau is found to be

		u	r
θ	12	$\frac{1}{2}$	$-\frac{1}{2}$
y	2	$\frac{3}{2}$	$-\frac{1}{2}$
s	4	-2	1
t	26	-8	3
x	10	-1	0.

Exercise. Write the solution exhibited by the third tableau. Is this solution optimal?

The current solution from the last tableau is

$$\theta = 12$$
$$y = 2$$
$$s = 4$$
$$t = 26$$
$$x = 10$$
$$u = r = 0.$$

The solution is not optimal because the arbitrary variable column u has a positive number, $1/2$, in the θ row.

Exercise. Find the pivot element of the third tableau, then carry out the manipulations and write the fourth tableau.

The pivot is found to be the -2 at the intersection of the s row and u column. The fourth tableau is

		s	r
θ	13	$-\frac{1}{4}$	$-\frac{1}{4}$
y	5	$-\frac{3}{4}$	$\frac{1}{4}$
u	3	$-\frac{1}{2}$	$\frac{1}{2}$
t	10	4	-1
x	8	$\frac{1}{2}$	$-\frac{1}{2}.$

In the last tableau, all elements in the θ row under the variables are negative. The present solution is optimal. It is

$$\theta_{max} = 13$$
$$x = 8$$
$$y = 5$$
$$u = 2$$
$$t = 10$$
$$s = r = 0.$$

The answer to the original problem is that the maximum value of θ is 13, and it occurs when x is 8 and y is 5. The number pair (8, 5) satisfies all the original inequalities and maximizes θ. The significance of the complete solution can be seen by referring to the original inequality system:

At (8, 5), $3x + 2y \leq 34$ becomes $24 + 10 \leq 34$.

There is no slack in the last statement, so the slack variable r is zero, as shown in the complete solution. Similar demonstration shows why slack variable s also is zero.

On the other hand:

At (8, 5), $x + 6y \leq 48$ becomes $8 + 30 \leq 48$

showing a slack of 10 in this inequality. This is the meaning of $t = 10$ in the complete solution. Similar demonstration shows why $u = 2$ in the complete solution.

Observe the increase in θ from tableau to tableau in the foregoing example. At the start, θ was zero. It increased successively to 10, to 12, and finally to the maximum, 13. The progress of the solution can be seen graphically from Figure 5–1, which shows the solution space for

$$3x + 2y \leq 34$$
$$x + 2y \leq 18$$
$$x + 6y \leq 48$$
$$x \leq 10$$
$$x \geq 0$$
$$y \geq 0.$$

Observing the corners of the solution space, we see that the simplex method started with the corner (0, 0). The next tableau moved us to the corner (10, 0), the next to (10, 2), and the final tableau to (8, 5), where θ reaches its maximum. Figure 5–1 shows that the simplex progression goes from one permissible, or *feasible,* corner to an adjacent feasible corner, avoiding corners whose coordinates do not satisfy all of the constraints.

Before turning to the problem set, we review the simplex method with an example which yields the optimum after one step:

Maximize $\theta = 2w + x + z$
subject to: $w + 2x + 2z \leq 20$
$2w + 3x + z \leq 18.$

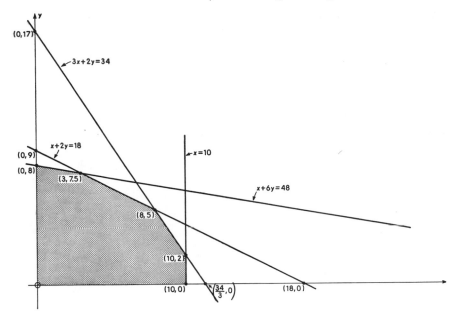

FIGURE 5–1

The usual nonnegativity restrictions apply to all variables. Introducing and solving for slack variables, we have

$$\theta = \qquad 2w + \ x + \ z$$
$$r = 20 - \quad w - 2x - 2z$$
$$s = 18 - 2w - 3x - \quad z.$$

The first tableau is

		w	x	z
θ	0	2	1	1
r	20	-1	-2	-2
s	18	-2^*	-3	$-1.$

The pivot column is w because it has the largest positive number in the θ row. The absolute value of $18/-2$ is smaller than that of $20/-1$, so s is the pivot row. The next tableau is

		s	x	z
θ	18	-1	-2	0
r	11	$\frac{1}{2}$	$-\frac{1}{2}$	$-\frac{3}{2}$
w	9	$-\frac{1}{2}$	$-\frac{3}{2}$	$-\frac{1}{2}.$

No element in the θ row under a variable is positive, so the solution is optimal. It is $\theta_{\max} = 18$.

5.13 PROBLEM SET 4

In each of the following, find the maximum value of θ by the simplex method, and state the values of the variates which make θ maximum. It is to be assumed that all variables must be zero or positive.

1. $\theta = x + 3y + 2z$
   $x + 2y + 3z \leq 50$
   $x + 3y + 5z \leq 60.$

2. $\theta = x + 4y + 3z$
   $x + 3y + 4z \leq 30$
   $x + 5y + 2z \leq 40.$

3. $\theta = x + 4y + 6z$
   $x + 3y + 6z \leq 48$
   $x + 6y + 3z \leq 90$
   $x + 9y + 10z \leq 137.$

4. $\theta = x + 6y + z$
   $x + 5y + 2z \leq 30$
   $x + 7y \leq 40$
   $2x + y + 3z \leq 70.$

5. $\theta = 3x + 7y + 6z$
   $x + y + z \leq 4$
   $x + y \leq 3.$

6. $\theta = 1.5x + 2.5y + 2z$
   $1.5x + y + z \leq 7.5$
   $x + y \leq 4.0$
   $y + 2.5z \leq 5.0.$

7. $\theta = 3x + 5y + 3z$
   $2x + 3y + 6z \leq 50$
   $3x + 4y + z \leq 40$
   $3x + 5y + 2z \leq 20.$

8. $\theta = 2x + 3y + z$
   $2x + y + 3z \leq 10$
   $x + 3y + 2z \leq 20.$

9. $\theta = x + 2y + z$
   $2x + y + 3z \leq 12$
   $x + 2y \leq 6$
   $2x + z \leq 4.$

10. $\theta = x + y + z$
    $x + 2y + z \leq 12$
    $2x + y + z \leq 20$
    $x + y + 3z \leq 15.$

5.14 SOURCES OF RULES AND INTERPRETATION

The solution space for the programming problem

$$\text{Maximize } \theta = 2x + 3y$$
$$\text{subject to: } 3x + 2y \leq 12$$
$$7x + 2y \leq 20$$
$$x \geq 0$$
$$y \geq 0$$

is shown in Figure 5–2. Introducing the nonnegative slack variables r and s, the first two inequalities become

$$3x + 2y + r = 12$$
$$7x + 2y + s = 20.$$

At the corner $(0, 0)$ of the solution space (Figure 5–2), we have $x = 0, y = 0,$ $r = 12, s = 20$. Similarly, at the other corners, we have:

At $\left(\frac{20}{7}, 0\right)$: $x = \frac{20}{7}$ $y = 0$ $r = \frac{24}{7}$ $s = 0$

At $(2, 3)$: $x = 2$ $y = 3$ $r = 0$ $s = 0$

At $(0, 6)$: $x = 0$ $y = 6$ $r = 0$ $s = 8.$

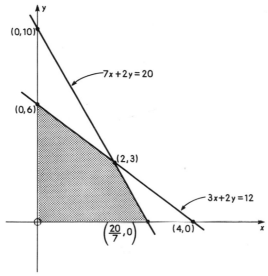

FIGURE 5–2

Each of the corners just referred to is the solution of a 4 by 4 system of equations selected from the following set of six equations:

$$x = 0$$
$$y = 0$$
$$r = 0$$
$$s = 0$$
$$3x + 2y + r = 12$$
$$7x + 2y + s = 20.$$

Any 4 by 4 system selected from the foregoing set obviously must contain at least two of the first four, so that any corner must have at least two coordinates equal to 0. Moreover, the only corners that could have two nonzero coordinates would be those coming from 4 by 4 systems which included the last two equations; that is, the equations arising from the constraints (other than nonnegativity) of the problem.

Exercise. A linear programming problem has four constraints (other than nonnegativity) on three variables. How many variables, including slacks, will arise in the solution of the problem? What will be the dimensions of the systems whose solutions will be corners of the solution space? What will be the maximum number of nonzero coordinates at a corner? Answer: There will be seven variables: three original plus four slack. Each corner will be the solution of a 7 by 7 system. The maximum number of nonzero coordinates at a corner is the same as the number of constraints; that is, four.

We have illustrated a general principle of linear programming; namely, excluding nonnegativity requirements, the maximum number of nonzero coordinates at a corner of the solution space is the same as the number of constraints. It is this principle which leads us to set up a tableau with the same number of rows as there are constraints (not counting the θ row), and then read off a solution by letting the row variables equal the constants in the first column and assigning the value zero to the arbitrary variables at the column heads. In the present case:

		x	y
θ	0	2	3
r	12	-3	-2*
s	20	-7	-2

the two positive-valued variables are

$$r = 12$$
$$s = 20.$$

Setting $x = y = 0$ completes the statement of the solution at this point. We say this solution is *feasible* because it satisfies all the constraints. We say further that it is *basic* because it is a corner of the solution space or, what amounts to the same thing, a basic solution is one in which the number of coordinates having nonzero values does not exceed the number of constraints, again not counting nonnegativity constraints.

The solution at this stage is not optimal (θ is zero). We wish to increase

$$\theta = 2x + 3y$$

by increasing either x or y from their present value, which is zero. Inasmuch as y has the larger coefficient in the objective function, we decide to increase θ by increasing y. This observation leads to the rule stated earlier for selecting the pivot column; namely, select the arbitrary variable column which has the largest positive number in the θ row.

We cannot, however, increase y without paying attention to the effect of the increase upon the constraints; in particular, no variable is permitted to be negative. Observe the equations presently at hand:

$$r = 12 - 3x - 2y$$
$$s = 20 - 7x - 2y.$$

We are interested in increasing y, leaving x at zero. Thinking of x as being zero, we see from

$$r = 12 - 2y$$

that the largest value y may assume is $12/2$ if r is to remain positive. Again, in

$$s = 20 - 2y,$$

the largest value y may assume if s is to remain positive is $20/2$. Of the two numbers, $12/2 = 6$ is the smaller and therefore is controlling. This controlling number is derived from the equation

$$r = 12 - 3x - 2y$$

so we select the r row as the pivot row; and in the next tableau, it will be seen that y will be increased to the maximum, 6.

The preceding analysis is summarized by the rule stated earlier for selecting the pivot row; namely, for each negative element in the pivot column, form the quotient

$$\frac{\text{Constant}}{\text{Element}}$$

and select as pivot the row for which this quotient is smallest in absolute value. Applied to the last tableau, the quotients are $12/-2$ and $20/-2$, and the smaller is $12/-2$, corresponding to the r row.

The rules for selecting the pivot ensure that θ will increase from step to step and that all variables will remain nonnegative. The next tableau is

		x	r
θ	18	$-\frac{5}{2}$	$-\frac{3}{2}$
y	6	$-\frac{3}{2}$	$-\frac{1}{2}$
s	8	-4	$1.$

The solution here is

$$\theta = 18$$
$$y = 6$$
$$s = 8$$
$$x = r = 0.$$

We now have r as arbitrary in place of y. By assigning zero to x and to r, y goes to 6. The solution is a basic feasible solution. Moreover, it is optimal; for at this stage, we may read the objective function from the tableau as

$$\theta = 18 - \frac{5}{2}x - \frac{3}{2}r.$$

Because of the negative coefficients, any increase in x or r from their present value, zero, will result in a *decrease* in θ. This observation is the source of the rule which states that the optimum has been reached when all elements in the θ row are negative (or, of course, zero).

To examine our example from an *interpretive* viewpoint, let us suppose our business is painting and drying tricolor and monocolor vehicles. Each day we plan to complete x tricolor and y monocolor vehicles. Painting time, drying time, and net profit per vehicle are shown in Table 5–1, together with the maximum number of hours available per day for painting and drying.

TABLE 5–1

| | Number | Hours per Vehicle to | | Net Profit per Vehicle |
Vehicle	per Day	Paint	Dry	in $ hundreds
Tricolor x		7	3	2
Monocolor . . . . y		2	2	3
Total hours available . . .		20	12	

We see from Table 5–1 that it takes $7x$ hours to paint x tricolors and $2y$ hours to paint y monocolors. The total $7x + 2y$ must not exceed 20, so $7x + 2y \leq 20$. Similar algebraic statements lead to the system we introduced at the beginning of the section; namely,

$$\theta = 2x + 3y \quad \text{(profit objective function)}$$
$$3x + 2y \leq 12 \quad \text{(drying time constraint)}$$
$$7x + 2y \leq 20 \quad \text{(painting time constraint)}.$$

Introducing the slacks, we have

$$\overset{.}{\theta} = 2x + 3y$$
$$3x + 2y + r = 12$$
$$7x + 2y + s = 20$$

which leads to the initial tableau

$$
\begin{array}{c c c c}
 & & x & y \\
\theta & 0 & 2 & 3 \\
r & 12 & -3 & -2 \\
s & 20 & -7 & -2.
\end{array}
$$

The solution here, $r = 12$, $s = 20$, $x = y = 0$, means that we have 12 drying hours and 20 painting hours available if we do not paint or dry any vehicles. The tableau shows profit $\theta = 0$ at this point.

The next (optimal) tableau is

$$
\begin{array}{c c c c}
 & & x & r \\
\theta & 18 & -\dfrac{5}{2} & -\dfrac{3}{2} \\
y & 6 & -\dfrac{3}{2} & -\dfrac{1}{2} \\
s & 8 & -4 & 1.
\end{array}
$$

(1)

The solution, $y = 6$, $s = 8$, $x = r = 0$, $\theta = 18$, says that if we paint and dry $y = 6$ monocolor cars, we will use up all the drying time (12 hours), leaving zero slack drying time ($r = 0$). Consequently, we can not handle any tricolors ($x = 0$). Further, we consume only 12 hours of painting time from the 20 hours available, so the slack (unused) painting time is $s = 8$. Finally, the profit from $y = 6$ monocolors is $\theta = 6(3) = 18$ (hundred dollars).

5.15 SENSITIVITY ANALYSIS

In the last example, the relatively low profit on tricolors made it undesirable to do any tricolor work at all. Now, if the profit per tricolor were increased, there would come a point where it would be advantageous to do tricolor work and at that point the present solution would not be optimal. Let us find out how *sensitive* the optimal solution is to changes in the per unit profit in tricolors. In particular, we ask the question, How much can the tricolor unit profit change if the present solution is to remain optimal? If we change the unit profit by c dollars per vehicle, we change the profit on x vehicles by cx dollars, so that the θ row of the immediately foregoing tableau, (1), becomes

$$\begin{array}{cccc} & & x & r \\ \theta' & 18 & \left(-\dfrac{5}{2} + c\right) & -\dfrac{3}{2}. \end{array}$$

Now, if the solution is to be optimal, the coefficients under x and r in the θ' row must not be positive. Hence,

$$-\frac{5}{2} + c \leq 0, \text{ so } c \leq \frac{5}{2}.$$

Thus, the profit per tricolor could be changed by any amount not exceeding 2.5 hundred dollars and it still would not pay to do tricolor jobs; that is, the present solution with $x = 0$ would remain in effect. Clearly, c may have any negative value, which means simply that if per unit profit is decreased in any amount we certainly would not do tricolor jobs.

It follows from the foregoing that for a variable which is zero in a maximal solution, we may change that variable's coefficient in the objective function by any amount not exceeding the absolute value of its θ row entry in the optimal solution and not alter the solution.

Exercise. Suppose that the maximization of $\theta = 5x + 4y + 2z$ leads to the following optimal tableau:

		r	s	z
θ	70	-1	-2	-3
x	10	-3	2	-1
y	5	$\frac{7}{2}$	-3	0
t	2	0	1	-2.

In the objective function, the profit per unit of Z is $2. How large could this unit profit be and not affect the optimal solution? Answer: The unit profit could change by any amount up to and including $3, so the solution would not change if the profit per unit of Z were as high as $2 + $3 = $5.

||||||||||||||||||||||||||||||||||||||||||||||||||||||||||||||||||||||||||||||||||||||||||||||||||||||||||||||||||||||||||||||||||||||||||||||||

Next, let us consider a variable *which is not zero* in the maximal solution. If the profit per monocolor vehicle changes by c dollars, then the total profit for y units changes by cy so that the new profit is cy greater than the old. Thus, $\theta' = \theta + cy$. We proceed to add c times the y row of the optimal tableau (1) to the θ row to obtain

$$
\begin{array}{ccc}
 & x & r \\
\theta' \quad 18 + 6c & -\dfrac{5}{2} - \dfrac{3c}{2} & -\dfrac{3}{2} - \dfrac{c}{2}.
\end{array}
$$

The entries under x and r must not be positive if the solution is to be an optimal one. Hence,

$$
-\frac{5}{2} - \frac{3c}{2} \leq 0 \qquad \text{and} \qquad -\frac{3}{2} - \frac{c}{2} \leq 0.
$$

These lead to

$$
c \geq -\frac{5}{3} \qquad \text{and} \qquad c \geq -3.
$$

The second inequality is redundant, so we have $c \geq -5/3$ as our condition. Thus, we would still have an optimal solution if we decreased the unit mono-color profit by any amount not exceeding $5/3$ hundred dollars, or increased it by any amount. However, since we are considering a variable which is not zero in the solution, any change in the profit per unit will affect the total profit as indicated in the θ' row of the foregoing. Thus, a change of c hundred dollars changes the optimum profit from $\theta = 18$ to $\theta' = 18 + 6c$ hundred dollars.

For practice, let us return to the last exercise and find how large the profit per unit of Y could be if the solution $x = 10, y = 5, z = 0$ is to remain optimal. If we change unit profit of Y by c dollars, θ changes by cy to $\theta' = \theta + cy$. Thus, we have

$$
\begin{array}{cccc}
 & r & s & z \\
\theta' \quad 70 + 5c & -1 + (\tfrac{1}{2})c & -2 - 3c & -3 + 0c.
\end{array}
$$

The entries under r, s, and z must not be positive. Hence,

$$
-1 + (\tfrac{1}{2})c \leq 0 \qquad -2 - 3c \leq 0 \qquad -3 \leq 0
$$

which lead to

$$
c \leq \frac{2}{7} \qquad c \geq -\frac{2}{3}.
$$

We have the double inequality

$$-\frac{2}{3} \le c \le \frac{2}{7}.$$

Hence, we could increase profit per unit of Y by 2/7 from its present 4 in the objective function, and $30/7 is the highest profit per unit which would result in the same optimal values for x, y, and z. If we made the increase, maximum profit would change from the original $70 (the maximum in the tableau) to $70 + 5c = 70 + 5(2/7) = \$500/7$.

We note also that the same optimal x, y, and z remain if we decrease unit profit of Y by the maximum, 2/3, to $10/3, at which time maximum profit would be $70 - 5(10/3) = \$160/3$.

Exercise. Find the limits on the amount of change, c, in the unit profit of X if the same x, y, and z are to be optimal. See previous exercise. Answer: $-\frac{1}{3} \le c \le 1$.

Finally, we observe that each of the entries in the tableau (1) has a meaning. To see this, we may write the equation form of the tableau as

$$\theta = 18 - \frac{5}{2}x - \frac{3}{2}r$$

$$y = 6 - \frac{3}{2}x - \frac{1}{2}r$$

$$s = 8 - 4x + 1r$$

and state that $\theta = 18$, $y = 6$, $s = 8$, $x = 0$, $r = 0$ is the optimum feasible solution. Now suppose that we ask what will happen if we let x change from the optimal value zero to 1, holding r at zero. We then have

$$\theta = 18 - \frac{5}{2} = \frac{31}{2}$$

$$y = 6 - \frac{3}{2} = \frac{9}{2}$$

$$s = 8 - 4 = 4.$$

Now, $\theta = 31/2$ at $x = 1$, $y = 9/2$, $s = 4$, $r = 0$ is a feasible solution, although it is not optimal. Two facts emerge from this arithmetic. First we see that the number $-5/2$ in the θ row under the column variable x shows how *sensitive* θ is to a unit change in x. Specifically, an increase of 1 in x causes a decrease of 5/2 in θ. Secondly, the number $-3/2$ in the y row under the column variable x shows the *exchange rate* between y and x if a feasible solution is to be maintained. Specifically, if x is increased by 1, y must decrease by 3/2 from 6, and s must decrease by 4 from 8.

With reference to our earlier car painting and drying illustration (see Tableau 1), if we increase the number of tricolor jobs, x, from its optimal value, 0, to 2, profit will change by $(-5/2)(2) = -5$ to $18 - 5 = \$13$ hundred, the number of monocolor jobs possible changes by $(-3/2)2 = -3$ to $6 - 3 = 3$, and slack painting time, s, drops to zero. We now have $s = r = 0$, so facilities are fully utilized, but profit is $5 hundred below the maximum.

Exercise. What would be the effect of scheduling two hours slack drying time, r? See tableau (1). Answer: We could then handle only five monocolor jobs, profit would drop to $15 hundred, and slack painting time would increase to 10 hours.

5.16 PROBLEM SET 5

The number of original variables and the number of constraints (not counting nonnegativity) are given in Problems 1 through 5.

Answer the following for each of the problems:
a) How many variables will be involved in the problem, counting slacks?
b) What will be the dimensions of the systems whose solutions will be corners of the solution space?
c) What will be the maximum number of nonzero coordinates at a corner?
d) What will be the minimum number of zero coordinates in a basic solution?

1. 3 constraints, 4 variables.
2. 4 constraints, 3 variables.
3. 5 constraints, 3 variables.
4. 7 constraints, 11 variables.
5. 6 constraints, 3 variables.

6. Consider the following tableau encountered in the solution of a linear programming problem:

		x	r
θ	20	2	-1
y	10	2	1
s	4	-1	$-2.$

Without recourse to rules, answer the following questions:
a) What is the equation for the objective function at this stage?
b) What values do x and r have at this stage of the solution?
c) Why does it not make sense to consider the r column as the pivot column?
d) If we manipulate the tableau to get x in place of s, how much does each unit of increase in x (from 0) add to the objective function?
e) What limits the amount we may increase x? What is the limiting amount of increase?
f) Taking (d) and (e) into consideration, how much will θ increase if x and s are interchanged in the usual manner?
g) Having chosen x as the pivot column, we observe the y row in the pivot column is $+2$. Why do we disregard this element when selecting a pivot row?

More generally, what is the significance of a positive row element in a pivot column?

7. Assume the following is a maximal tableau for a problem whose objective function is $\theta = 2x + 6y + 4z$.

		z	r	s
θ	150	-2	-3	-4
y	20	-2	1	-8
x	15	3	1	-2
t	5	-1	-4	$-1.$

a) The profit now is $4 per unit of Z. How much could it change without affecting the optimal solution? What is its largest possible value?

b) By how much could the profit per unit of X change and leave the same optimal values for x, y, and z? What would be the expression for the new profit, θ'?

c) By how much could the profit per unit of Y change and leave the same optimal values for x, y, and z? What would be the expression for the new profit, θ'?

d) What would be the effect of introducing 2 hours of slack, s?

5.17 MINIMIZATION

Linear programming problems can involve all combinations of $<$, $>$, and $=$ in their constraints. It is not difficult to adapt all problems to a standard form so that the simplex procedure we have developed can be applied. We shall not illustrate all variations here, but we shall consider one important variation, that of problems in minimization. The reader may have noted that all problems to this point in the chapter have been maximization with $\leq$ constraints. In this section we shall show one way to handle minimization with $\geq$ constraints, and then use the solution in the following section to illustrate the important dual theorem. This theorem shows a second, and more generally useful, way to solve problems in minimization. We shall not pause for exercises and illustrations in this section. Our purpose, as noted, is to find a minimization solution so that we can see by comparison how the dual theorem provides the same information.

Suppose $\theta = 2x + 3y$, and we wish to find θ_{min} subject to

$$x \geq 0$$
$$y \geq 0$$
$$3x + 2y \geq 12$$
$$7x + 2y \geq 20.$$

We want to apply the methodology already developed in this chapter to solve the problem. Therefore, we must convert the given problem into a maximization problem. We note first that to minimize θ is the same as to maximize $-\theta$. Hence, we shall seek to maximize $-2x - 3y$, which is $-\theta$. Second, it is clear that we must *subtract* positive slacks to convert the inequalities to equalities. This conversion leads to

$$3x + 2y - r = 12 \quad \text{or} \quad r = -12 + 3x + 2y$$
$$7x + 2y - s = 20 \quad \text{or} \quad s = -20 + 7x + 2y.$$

Taking these equations together with that for $-\theta$, construct the tableau

		x	y
$-\theta$	0	-2	-3
r	-12	3	2
s	-20	7	2.

The solution implied by the tableau is not feasible because r and s are negative. Our problem is to obtain a tableau with positive numbers in the second and third rows of the first column to serve as a starting point for the simplex methodology. A way of doing this is suggested by observing Figure 5–2 where the solution space for the current illustration is in the obtuse angle opposite the shaded area. Note that there are two x intercepts, one of which is a corner of the solution space. At either x intercept, y is zero. It follows that a tableau with a feasible solution can be obtained by leaving the y column as it is (that is, y remains at zero) and interchanging x with either r or s. To determine which row is the pivot row, we must try each element of the x column as a pivot element to learn which pivot results in positive numbers in the second and third rows of the first column. A quick check shows that 3 will do the job, and 7 will not. Hence, we take the 3 at the intersection of the r row and x column as pivot and construct the next tableau:

		r	y
$-\theta$	-8	$-\frac{2}{3}$	$-\frac{5}{3}$
x	4	$\frac{1}{3}$	$-\frac{2}{3}$
s	8	$\frac{7}{3}$	$-\frac{8}{3}.$

(2)

The solution exhibited in the new tableau is feasible. Moreover, it is the optimum solution because the numbers in the θ row under r and y are negative, indicating that further manipulation will not lead to a reduction in $-\theta$. The maximum value of $-\theta$ is seen to be -8; hence, the minimum value of $+\theta$ is 8, and this is the solution for the original problem.

If the feasible solution obtained by the first manipulation had not been the optimum solution, new tableaus would have been constructed following the regular rules for tableau manipulation until the optimum was reached.

Now let us learn how to convert the minimization problem of this section to a maximization problem by applying the *dual theorem.*

5.18 THE DUAL THEOREM

If we have a minimization problem (the *primal*), the dual theorem of linear programming shows how we may transform the primal to a problem in maximization (the *dual*) in a manner such that the solution of one problem contains the same information as the solution of the other. In particular, the maximum of the dual equals the minimum of the primal. Alternatively, if the primal problem is one of maximization, the dual is one of minimization, and the solutions of both contain the same information. The proof of the dual theorem is beyond the scope of this text, and we shall limit our discussion here to a demonstration of the use of the dual in solving the minimization problem discussed in Section 5.17.

The problem of Section 5.17 was to minimize $\theta = 2x + 3y$ subject to non-negativity of variables and to two linear constraints. Let us write the primal problem as

Primal

$r: 3x + 2y \geq 12$

$s: 7x + 2y \geq 20$

$2x + 3y = \theta,$

where r and s at the left specify the names for the slack variables. The dual problem is written as follows:

Dual

$x: \quad 3r + \quad 7s \leq 2$

$y: \quad 2r + \quad 2s \leq 3$

$12r + 20s = \theta.$

Observe that the first row of the dual was constructed from the coefficients of x in the first column of the primal, with the sense of the inequality changed from $\geq$ to $\leq$. The remaining rows of the dual were obtained in the same manner from the corresponding columns in the primal. We now apply the standard technique to maximize θ in the dual, taking x and y as slacks. The initial tableau becomes

		r	s
θ	0	12	20
x	2	-3	-7^*
y	3	-2	$-2.$

Taking -7 as the pivot, the second tableau is found to be

		r	x
θ	$\frac{40}{7}$	$\frac{24}{7}$	$-\frac{20}{7}$
s	$\frac{2}{7}$	$-\frac{3^*}{7}$	$-\frac{1}{7}$
y	$\frac{17}{7}$	$-\frac{8}{7}$	$\frac{2}{7}.$

The new pivot is $-3/7$. The third tableau is found to be

	s	x	
θ	8	-8	-4
r	$\dfrac{2}{3}$	$-\dfrac{7}{3}$	$-\dfrac{1}{3}$
y	$\dfrac{5}{3}$	$\dfrac{8}{3}$	$\dfrac{2}{3}.$

$$(3)$$

The solution is optimal, and the maximum of the dual is 8. Now let us compare (3) with (2), the solution found for this problem in Section 5.17:

	r	y	
$-\theta$	-8	$-\dfrac{2}{3}$	$-\dfrac{5}{3}$
x	4	$\dfrac{1}{3}$	$-\dfrac{2}{3}$
s	8	$\dfrac{7}{3}$	$-\dfrac{8}{3}.$

$$(2)$$

From (2) we concluded in Section 5.17 that 8 was the minimum value of the original (primal) problem. We see, then, that the minimum of the primal is the same as the maximum of the dual in (3). Moreover, note in (2) that the minimum occurs at $x = 4, s = 8, r = y = 0$. To get this information from the dual solution (3) we must use the negative of the entries in the θ *row* for s and x, and zero for the values of the variables in the first *column*. It may be verified that, excluding the numbers in the first row and column, the number at the intersection of any row and column in the dual solution (3) is the negative of the number at the corresponding intersection of the primal solution (2), so the two solutions contain the same information. Consequently, the simplex procedure of this chapter can be used for both maximization problems and, by using the dual, for minimization problems.

5.19 PROBLEM SET 6

Convert each of the primal problems into its dual and write the *first* tableau for the dual. It is not necessary to carry out iterations.

1. Given $\theta = 5x + 3y$, find $\theta_{\min}$
   subject to

   $$4x + 2y \geq 20$$
   $$6x + y \geq 10.$$

2. Given $\theta = 4x + 2y$, find $\theta_{\min}$
   subject to

   $$3x + 2y \geq 10$$
   $$x + 2y \geq 20.$$

3. Given $\theta = x + y + z$, find $\theta_{\min}$
   subject to

   $$3x + 3y + z \geq 2$$
   $$2x + y \geq 4$$
   $$4x + 2y + 2z \geq 6.$$

4. Given $\theta = 2x + 3y + z$, find $\theta_{\min}$
   subject to

   $$x + y + z \geq 10$$
   $$2x + 3y + z \geq 15$$
   $$3x + y + 2z \geq 20.$$

If the maximum of the dual has the following (partial) tableau, what is the minimum of the primal, and at what point does the minimum occur?

5.

	x	y
θ	10	-3	-2
r	4		
s	1		

6.

	x	t	z
θ	50	-5	-1	-7
r	4			
y	1			
s	3			

Solve by the dual procedure.

7. A quart of A costs $2 and contains 2.5 ounces of Zap. A quart of B costs $3 and contains 4 ounces of Zap. A mxiture not to exceed 10 quarts, and having at least 36.4 ounces of Zap is to be made. Find the minimum-cost mixture.

8. A diet is to contain at least 10 ounces of nutrient P, 12 ounces of nutrient R, and 20 ounces of nutrient S. These nutrients are to be obtained from some combination of foods A, B, and C. Each pound of A costs 4 cents and has 4 ounces of P, 3 ounces of R, and 0 ounces of S. Each pound of B costs 7 cents and has 1 ounce of P, 2 ounces of R, and 4 ounces of S. Each pound of C costs 5 cents and has 0 ounces of P, 1 ounce of R, and 5 ounces of S. Find the minimum-cost mixture and state its composition.

5.20 COMMENT ON ELECTRONIC COMPUTERS

The reader who has worked his way through the details of the simplex method may question whether it is superior to the longhand method of finding and checking all possible corners, as discussed in an earlier chapter. The answer depends upon the size of the problem at hand. Textbook illustrative problems with a few variables and a few constraints can be solved by longhand methods quite easily. Real-world problems, on the other hand, often have many variables and a large number of constraints. The particular advantage of the simplex method in these real-world problems is that simplex procedures can be carried out on electronic computers.

If we had a system with 15 constraints on 10 variables, a formula of the earlier chapter says that the number of corners that might have to be investigated in a longhand attack would be

$$\frac{15!}{5!10!} = \frac{(15)(14)(13)(12)(11)}{(5)(4)(3)(2)(1)} = 3003$$

and each corner would be found by solving a 10 by 10 system. Clearly, it would be impractical to tackle such a problem longhand.

The electronic computer, on the other hand, when supplied with proper instructions for the simplex procedure and the data of the problem, provides answers in a very short time. The 15 by 10 problem just mentioned could be solved in seconds on a large computer. Indeed, the availability of electronic computers has had much to do with the rapid development of linear programming techniques and their application to some very complex problems in industry.

5.21 REVIEW PROBLEMS

Solve by the simplex procedure. Nonnegativity constraints prevail (but are not written) in each problem.

1. Given $\theta = x + 6y + 10z$, find θ_{max} subject to

$$x + 5y + 10z \leq 150$$
$$3x + 20y + 25z < 500$$
$$x + 15y + 15z \leq 300.$$

2. Given $\theta = x + 6y + 3z$, find θ_{max} subject to

$$x + 5y + 4z \leq 39$$
$$2x + 13y + 5z \leq 90$$
$$x + 2y + 3z \leq 40.$$

3. Given $\theta = 2x + 3y + z$, find θ_{max} subject to

$$3x + 2y + z \leq 8$$
$$2x + 3y + z \leq 10$$
$$5x + 3y + 2z \leq 17.$$

4. Given $\theta = x + 4y + 7z$, find θ_{max} subject to

$$x + 3y + 7z \leq 70$$
$$x + 5y + 5z \leq 100$$
$$x + 10y + 8z \leq 190.$$

5. Given $\theta = 3x + 2y + z$, find θ_{max} subject to

$$3x + y + z \leq 35$$
$$2x + 10y + 3z \leq 140$$
$$4x + 4y + z \leq 50.$$

6. Given $\theta = x + 10y + 3z$, find θ_{max} subject to

$$x + 9y + 3z \leq 240$$
$$3x + 35y + z \leq 800$$
$$x + 12y + 5z \leq 300.$$

7. Given $\theta = 2x + y + 2z$, find θ_{max} subject to

$$x + y + 2z \leq 8$$
$$2x + y + z \leq 10$$
$$3x + y + 3z \leq 15.$$

8. Given $\theta = 3x + y + 7z$, find θ_{max} subject to

$$2x + y + 7z \leq 21$$
$$17x + 5y + 28z \leq 140$$
$$9x + y + 10z \leq 66.$$

9. The following is the optimal tableau derived while maximizing a problem for which $\theta = 2x + y + 2z$. If the current values of x, y, and z are to remain optimal, by how much the coefficients of the variables in the objective function change?

What is the new θ' for the maximum increase in each case?

a) Variable x.   b) Variable y.   c) Variable z.

		t	s	r
θ	11	$-\frac{1}{3}$	$-\frac{1}{3}$	$-\frac{1}{3}$
z	1	$-\frac{1}{3}$	$\frac{2}{3}$	$-\frac{1}{3}$
y	3	1	-1	-1
x	3	$-\frac{1}{3}$	$-\frac{1}{3}$	$\frac{2}{3}$.

d) What is the maximum permitted increase in the slack t, which is zero in the present solution?

e) See d). What would be the effect of this increase?

Solve PROBLEMS 10–12 by the dual procedure:

10. Given $\theta = 2x + 3y + z$, find the *minimum* value of θ subject to

$$x + y + z \geq 10$$
$$2x + 3y + z \geq 15$$
$$3x + y + 2z \geq 20.$$

11. Given $\theta = 2x + y + z$, find $\theta_{\min}$ subject to

$$2x + 3y + 2z \geq 10$$
$$x + 4y + 3z \geq 20.$$

12. Quality X gasoline costs $0.40 per gallon and has an octane rating of 70. The per gallon costs and octane ratings are $0.50 and 90 for Y, $0.60 and 100 for Z. We wish to buy at least 1000 gallons and mix them, requiring that the mixture have an octane rating of at least 85. Find the minimum cost.

6

Compact Notation:
Vectors, Matrices,
and Summation

6.1 INTRODUCTION

BREVITY IN mathematical statements is achieved through the use of symbols. Thus the brief expression $247 \div 793$ takes the place of what would be a very lengthy statement if one chose to write a complete description of the steps involved in the long division. The price paid for brevity, of course, is the effort spent in learning the meaning of the symbol.

In this chapter, we shall learn the symbols for matrices and vectors and apply them in the statement and solution of input-output problems and other problems involving linear systems. Then we shall introduce the summation symbol and show its application in linear systems and in statistics.

6.2 MATRICES AND VECTORS

Numerical data arranged in a form which we shall come to call a matrix are very common in everyday life. For example, suppose that a company has six gasoline stations, three in region #1 and three in region #2. January sales volume, in thousands of gallons, is shown for each station in Table 6–1.

We note that sales of Station #1 in Region #2 were 15 thousand gallons. The position of the *entry* (also called *element*) 15 is at the intersection of the first row and second column, and we may symbolize the entry as a_{12} (read "a sub one two")[1] where the first subscript refers to the row and the second to

---

[1] A comma is required to avoid ambiguity if the number of rows or columns exceeds nine. For example, an entry in the 12th row 3rd column would be $a_{12,3}$.

TABLE 6–1
Tiger Oil Company
Sales in Thousands of Gallons
January

Station	Region #1	Region #2
#1	10	15
#2	12	18
#3	8	12

the column. To avoid having to specify which of a pair of subscripts specifies the row and which the column, we shall *always* use the row, column order convention.

Exercise. In Table 6–1, *a*) what are a_{11} and a_{32}? *b*) Write the symbol for the entry 8. Answer: *a*) 10 and 12. *b*) a_{31}.

If we keep in mind that rows are stations and columns are regions, we can omit the stub and caption of the table and write the *matrix:*

Sales Matrix by Station by Region

$$\begin{pmatrix} 10 & 15 \\ 12 & 18 \\ 8 & 12 \end{pmatrix}. \qquad [1]$$

Definition: A *matrix* is a rectangular array of numbers. Matrices are enclosed in grouping symbols such as parentheses or brackets.

If we use *m* by *n* to mean a matrix of *m* rows by *n* columns, *m* by *n* is called the *order* of the matrix or, more descriptively, the *shape* of the matrix. Thus, a 5 by 4 matrix has the shape of a rectangle 5 (rows) down by 4 (columns) across.

Definition: A 1 by *n* matrix, such as the 1 by 3 matrix (4 7 6), is called a *row vector,* and an *n* by 1 matrix, such as the 2 by 1 matrix

$$\begin{pmatrix} 5 \\ 7 \end{pmatrix},$$

is called a *column vector.* Entries in row and column vectors often are referred to as the *components* of the vectors. The matrix [1] may be thought of as consisting of two column vectors or three row vectors.

Exercise. Write the first row vector in the matrix, [1]. Answer: (10 15)

Next, suppose Tiger Oil Company sells one grade of gasoline but, due to transportation costs, prices vary in the two regions as shown in Table 6–2.

TABLE 6–2
Tiger Oil Company
Regional Price Variation

Region	Price in Cents per Gallon
#1	55
#2	60

We can represent the price matrix as a column or a row vector. Thus,

Price Vector by Region

$$\begin{pmatrix} 55 \\ 60 \end{pmatrix} \quad \text{or} \quad (55 \quad 60). \qquad [2]$$

6.3 PRODUCT OF A NUMBER AND A MATRIX

In matrix algebra, an ordinary number is called a *scalar*. To multiply a matrix by a scalar, we multiply each entry in the matrix by the scalar. For example

$$12\begin{pmatrix} 1 & 3 \\ 4 & 2 \end{pmatrix} = \begin{pmatrix} 12 & 36 \\ 48 & 24 \end{pmatrix}.$$

To see the origin of the word scalar, think of the entries 1, 3, 4, 2 in the left matrix as being in feet. Scaling these to inches requires that each entry be multiplied by 12. Similarly, Tiger Oil Company may wish to set February sales quotas which are 10% higher than actual January sales. The quotas will then be 1.1 times the January sales matrix, [1]. Thus,

[1]	[3]
January Sales	February Quota

$$1.1 \begin{pmatrix} 10 & 15 \\ 12 & 18 \\ 8 & 12 \end{pmatrix} \qquad \begin{pmatrix} 11 & 16.5 \\ 13.2 & 19.8 \\ 8.8 & 13.2 \end{pmatrix}.$$

Exercise. Tiger Oil Company's price vector is (55 60) in cents per gallon. What would be the scaling factor and the resultant price vector if prices are to be in dollars per 1000 gallons? Answer: 55 cents per gallon is $550 per 1000 gallons, so the scaling factor is 10. 10 (55 60) = (550 600).

6.4 ADDITION AND SUBTRACTION OF MATRICES

Matrices are added or subtracted by adding or subtracting *corresponding* entries. Since entries must correspond, the matrices involved must have the same shape. We cannot compute

$$\begin{pmatrix} 3 & 1 & 5 \\ 2 & 4 & 7 \end{pmatrix} + \begin{pmatrix} 5 & 2 \\ -3 & 6 \end{pmatrix}$$

because the first is 2 by 3 and the second is 2 by 2. However,

$$\begin{pmatrix} 3 & 1 \\ 2 & 4 \end{pmatrix} + \begin{pmatrix} 5 & -6 \\ 2 & 0 \end{pmatrix} = \begin{pmatrix} 8 & -5 \\ 4 & 4 \end{pmatrix}$$

and

$$\begin{pmatrix} 5 & 8 \\ -1 & 3 \end{pmatrix} - 2\begin{pmatrix} 1 & 0 \\ 3 & -2 \end{pmatrix} = \begin{pmatrix} 3 & 8 \\ -7 & 7 \end{pmatrix}.$$

Exercise. Compute

$$3\begin{pmatrix} 4 & -1 \\ 2 & 5 \end{pmatrix} - 2\begin{pmatrix} 1 & 2 \\ -3 & 4 \end{pmatrix}.$$ Answer: $\begin{pmatrix} 10 & -7 \\ 12 & 7 \end{pmatrix}$.

Returning to the affairs of Tiger Oil Company, recall its February sales quota matrix, [3]. This is the middle matrix in the following. The left is the new matrix, [4], of *actual* February sales, and the difference is matrix [5] which shows the deviation of actual February sales from quota.

[4]	[3]	[5]
February Sales	February Quota	Deviation from Quota

$$\begin{pmatrix} 12 & 15.5 \\ 13 & 20 \\ 8.5 & 12.9 \end{pmatrix} - \begin{pmatrix} 11 & 16.5 \\ 13.2 & 19.8 \\ 8.8 & 13.2 \end{pmatrix} = \begin{pmatrix} 1 & -1 \\ -0.2 & 0.2 \\ -0.3 & -0.3 \end{pmatrix}$$

From [5] we see, for example, that Station #1, Region #2 sales were 1 thousand gallons below quota.

Exercise. In [5], interpret the entry a_{21}. Answer: Station #2, Region #1 sales were 0.2 thousand gallons below quota.

6.5 MULTIPLICATION OF MATRICES

Matrix multiplication is a specialized form of multiplication devised for very practical reasons. Consider the following:

$$(3 \quad 2)\binom{5}{4} = 23.$$

We obtain the product, 23, as the sum of 3(5) + 2(4). To fix this in mind, we may place our left index finger on the 3 and our right index finger on the 5. Multiply to obtain 15. Then move the left finger *across* to the 2 and the right finger *down* to the 4, multiply to obtain 8, then add this to 15 to get 23. A product obtained in this manner is called the *inner product*. Observe that the inner product of a row vector by a column vector is an ordinary number, a scalar. As another example,

$$(2 \quad -3 \quad 1 \quad 0)\begin{pmatrix} 4 \\ -2 \\ 3 \\ -1 \end{pmatrix} = 8 + 6 + 3 + 0 = 17.$$

Exercise. Evaluate

$$(3 \quad -1)\binom{4}{2}. \qquad \text{Answer: 10.}$$

It is clear that if the "left finger across, right finger down" technique is to be defined there must be as many numbers across as there are down. That is, the first matrix must have as many columns as the second has rows. In our last example we had a 1 by 4 multiplied by a 4 by 1. If we indicate this as (1 by 4) (4 by 1), observe that the two middle numbers are the same, 4. When the shapes of two matrices are indicated in this manner and the middle numbers are the same, the two are said to be *conformable for multiplication.* Alternatively, we may say that two matrices are conformable for multiplication if the number of columns in the first is the same as the number of rows in the second. In the next example, we multiply a 2 by 2 and a 2 by 3.

$$\begin{pmatrix} 1 & 2 \\ 3 & 4 \end{pmatrix}\begin{pmatrix} 5 & 6 & 7 \\ 8 & 9 & 10 \end{pmatrix} = \begin{pmatrix} 21 & 24 & 27 \\ 47 & 54 & 61 \end{pmatrix}.$$

The calculations leading to the rightmost matrix are:

First row: $1(5) + 2(8) = 21$
$1(6) + 2(9) = 24$
$1(7) + 2(10) = 27.$

Second row: $3(5) + 4(8) = 47$
$3(6) + 4(9) = 54$
$3(7) + 4(10) = 61.$

Exercise. Compute the following product:

$$\begin{pmatrix} 1 & 2 & 3 \\ 4 & 5 & 6 \end{pmatrix} \begin{pmatrix} 0 & 1 \\ 2 & 3 \\ 4 & 5 \end{pmatrix}. \quad \text{Answer: } \begin{pmatrix} 16 & 22 \\ 34 & 49 \end{pmatrix}.$$

As an example of the applied meaning of matrix multiplication, let us multiply the gallons sold matrix (entries in thousands of gallons) by the price per thousand gallons vector for the Tiger Oil Company.

Station	Region #1	#2	Region	Dollars per Thousand Gallons
#1	10	15	#1	550
#2	12	18	#2	600
#3	8	12		

The product is

Station	Total Dollar Volume of Sales, Both Regions	
#1	10(550) + 15(600)	$14,500
#2	12(550) + 18(600) =	$17,400
#3	8(550) + 12(600)	$11,600

Thus, if Mr. Jones owns both #1 stations, his dollar volume of sales is 10 thousand gallons at $550 per thousand in Region #1, plus 15 thousand gallons at $600 per thousand in Region #2 for a total of $14,500 in both regions. Similarly, the owner of the #2 stations grosses $17,400 and the owner of the #3 stations grosses $11,600.

Matrix addition (subtraction) has the associative and commutative properties of ordinary algebra. Matrix multiplication has the associative property, but *the commutative property does not hold in general for matrix multiplication.* For example,

$$\begin{pmatrix} 1 & 2 \\ 3 & 4 \end{pmatrix} \begin{pmatrix} 5 & 6 \\ 7 & 8 \end{pmatrix} = \begin{pmatrix} 19 & 22 \\ 43 & 50 \end{pmatrix},$$

but if we interchange the factor matrices we find

$$\begin{pmatrix} 5 & 6 \\ 7 & 8 \end{pmatrix} \begin{pmatrix} 1 & 2 \\ 3 & 4 \end{pmatrix} = \begin{pmatrix} 23 & 34 \\ 31 & 46 \end{pmatrix}.$$

Thus, if we *premultiply* by

$$\begin{pmatrix} 1 & 2 \\ 3 & 4 \end{pmatrix}$$

we get one result, but if we *postmultiply* we get a different result.

Exercise. What do we obtain if we

a) premultiply $\begin{pmatrix}1\\2\end{pmatrix}$ by (3 4),

b) if we postmultiply $\begin{pmatrix}1\\2\end{pmatrix}$ by (3 4)?

Answer: *a*) 11.   *b*)

$$\begin{pmatrix}3&4\\6&8\end{pmatrix}.$$

We *can* design matrices which commute with each other in multiplication. For example,

$$\begin{pmatrix}1&0\\0&1\end{pmatrix}\begin{pmatrix}2&3\\4&5\end{pmatrix}=\begin{pmatrix}2&3\\4&5\end{pmatrix}\begin{pmatrix}1&0\\0&1\end{pmatrix}=\begin{pmatrix}2&3\\4&5\end{pmatrix}.$$

However, as we have seen, the commutative property does not hold in general.

We conclude this section with the suggestion that the reader keep in mind that an *m* by *n* matrix can be postmultiplied by an *n* by *p* matrix (they are conformable), and the result will be an *m* by *p* matrix. Thus, we can multiply a 5 by 8 and an 8 by 11, and the product matrix will be 5 by 11.

Exercise. *a*) What will be the shape of the product matrix if a 5 by 7 is postmultiplied by a 7 by 6? *b*) What happens if we premultiply the 5 by 7 by the 7 by 6? Answer: *a*) 5 by 6. *b*) The matrices are not conformable for multiplication.

6.6 PROBLEM SET 1

Perform the following operations:

1. (2 3 4) + (1 −2 3).

2. $\begin{pmatrix}1\\-3\end{pmatrix}+\begin{pmatrix}-3\\5\end{pmatrix}.$

3. $\begin{pmatrix}2\\4\end{pmatrix}+\begin{pmatrix}5\\7\end{pmatrix}+\begin{pmatrix}3\\2\end{pmatrix}.$

4. (6 −1 2) − (3 −2 4) + (5 −1 6).

5. $3\begin{pmatrix}7\\2\end{pmatrix}.$

6. −2(5 −9 3).

7. $(2\ 7)\begin{pmatrix}3\\5\end{pmatrix}.$

8. $(1\ -3\ 6\ 2)\begin{pmatrix}2\\1\\2\\-3\end{pmatrix}.$

9. $\begin{pmatrix} 3 & 1 & 2 \\ 1 & 4 & 1 \end{pmatrix} + \begin{pmatrix} 1 & -5 & -2 \\ -3 & 2 & 4 \end{pmatrix}.$

10. $\begin{pmatrix} 4 & 7 \\ 1 & 3 \end{pmatrix} + \begin{pmatrix} 1 & -2 \\ 6 & -3 \end{pmatrix} - \begin{pmatrix} 2 & -4 \\ 5 & 7 \end{pmatrix}.$

11. $4\begin{pmatrix} 1 & -3 & 2 \\ 5 & 1 & -3 \end{pmatrix} - 3\begin{pmatrix} 2 & 5 & -3 \\ 1 & 2 & -1 \end{pmatrix}.$

12. $\begin{pmatrix} 1 & 2 \\ 3 & 5 \end{pmatrix}\begin{pmatrix} 3 & 0 & 5 \\ 1 & -2 & 0 \end{pmatrix}.$

13. $(5\ 6\ 7)\begin{pmatrix} 1 & 2 \\ 0 & 3 \\ 3 & 1 \end{pmatrix}.$

14. $\begin{pmatrix} 3 & 0 \\ -5 & 4 \end{pmatrix}\begin{pmatrix} 1 & 6 \\ 2 & 0 \end{pmatrix}.$

15. $\begin{pmatrix} 2 & 1 & 1 & 0 \\ 1 & 3 & 0 & 2 \\ -1 & -2 & 1 & 4 \end{pmatrix}\begin{pmatrix} 5 & 6 \\ 1 & 1 \\ 2 & 3 \\ 0 & -1 \end{pmatrix}.$

Note: In Problems 16 and 17, the matrix with ones on the diagonal and zeros elsewhere is called the *unit* or *identity* matrix. Observe that multiplication of a given matrix by the unit matrix yields the given matrix.

16. $\begin{pmatrix} 1 & 0 & 0 \\ 0 & 1 & 0 \\ 0 & 0 & 1 \end{pmatrix}\begin{pmatrix} 1 & 4 \\ 2 & 5 \\ 3 & 6 \end{pmatrix}.$

17. $\begin{pmatrix} 1 & 4 \\ 2 & 5 \\ 3 & 6 \end{pmatrix}\begin{pmatrix} 1 & 0 & 0 \\ 0 & 1 & 0 \\ 0 & 0 & 1 \end{pmatrix}.$

18. Interest at the rates 0.06, 0.07, and 0.08 is earned on respective investments of $3000, $2000, and $4000. *a*) Express the total amount of interest earned as the product of a row vector by a column vector. *b*) Compute the total interest by matrix multiplication.

19. Two canned meat spreads, Regular and Superior, are made by grinding beef, pork, and lamb together. The numbers of pounds of each meat in a 15 pound batch of each brand are as follows:

Brand	Pounds of		
	Beef	Pork	Lamb
Superior 8		2	5
Regular. 4		8	3

a) Suppose we wish to make 10 batches of Superior and 20 of Regular. Multiply the meat matrix in the table and the batch vector (10 20) and interpret the result.

b) Suppose that the per pound prices of beef, pork, and lamb are $1.50, $0.80, and $1.00, respectively. Multiply the price vector and the meat matrix and interpret the results.

6.7 MATRIX SYMBOLS

We shall use a bold face capital letter for a matrix. Thus, A is a matrix. Readers may use a wavy underscore, $\underset{\sim}{A}$, to designate a matrix. Similarly, bold face small letters, such as b, will be used to designate a vector. Subscripted small letters will be used to symbolize the entries in a matrix. Thus, a_{23} is the entry at the intersection of the 2nd row and 3rd column. The 3 by 4 matrix A will then mean

$$A = \begin{pmatrix} a_{11} & a_{12} & a_{13} & a_{14} \\ a_{21} & a_{22} & a_{23} & a_{24} \\ a_{31} & a_{32} & a_{33} & a_{34} \end{pmatrix}.$$

We shall refer to A as the *compact* form and to the righthand expression as the *expanded matrix form*.

Exercise. Write the expanded matrix form of the 2 by 3 matrix B.
Answer:

$$B = \begin{pmatrix} b_{11} & b_{12} & b_{13} \\ b_{21} & b_{22} & b_{23} \end{pmatrix}.$$

The expanded form of the 1 by 4 vector c is

$$c = (c_1 \quad c_2 \quad c_3 \quad c_4)$$

and similarly for a column vector.

6.8 LINEAR EQUATIONS IN MATRIX FORM

To secure the advantage of matrix representation of equations, we shall use the same letter with varying subscripts to designate different variables. That is, instead of using x, y, and z as three variables, we shall use x_1, x_2, and x_3. Consider the linear equation

$$3x_1 + 2x_2 = 7$$

and observe that the technique of matrix multiplication makes it possible to express the equation as

$$(3 \quad 2)\begin{pmatrix} x_1 \\ x_2 \end{pmatrix} = 7.$$

Exercise. Write the linear equation specified by

$$(5 \quad 1 \quad 4)\begin{pmatrix} x_1 \\ x_2 \\ x_3 \end{pmatrix} = 10.$$

Answer: $5x_1 + x_2 + 4x_3 = 10.$

Similarly, the equation

$$a_1 x_1 + a_2 x_2 + a_3 x_3 = b$$

is

$$(a_1 \quad a_2 \quad a_3)\begin{pmatrix} x_1 \\ x_2 \\ x_3 \end{pmatrix} = b$$

and the general linear equation in n variables,

$$a_1 x_1 + a_2 x_2 + a_3 x_3 + \cdots + a_n x_n = b$$

becomes

$$(a_1 \quad a_2 \quad a_3 \ldots a_n)\begin{pmatrix} x_1 \\ x_2 \\ x_3 \\ \vdots \\ x_n \end{pmatrix} = b.$$

We can now see the advantage of compact form. Namely, all the information in the expanded vector expression just written can be summarized by stating simply that the equation is

$$ax = b$$

and that a is 1 by n. We do not have to state the shape of x because conformability requires that this vector be n by 1.

Exercise. If p is 1 by 2 and we have $py = q$, a) what is the shape of y? b) Write the expanded vector form of the equation. c) Write the usual algebraic form. Answer: a) y must be 2 by 1.

b) $(p_1 \quad p_2)\begin{pmatrix} y_1 \\ y_2 \end{pmatrix} = q.$

c) $p_1 y_1 + p_2 y_2 = q.$

||||||||||||||||||||||||||||||||||||||||||||||||||||||||||||||||||||||||||||||||||||||||||||||||||||||||||||||

Exercise. If a customer buys u_1 units of product #1, u_2 units of #2, and so on to u_g units of product #g at unit prices of, respectively, $p_1, p_2, \ldots, p_g$, write the expression for the total cost, t, in usual algebraic form and in compact form. What are the shapes of ***p*** and ***u***? Answer: $p_1 u_1 + p_2 u_2 + \cdots + p_g u_g = t$; ***pu*** $= t$; ***p*** is 1 by g, and ***u*** is g by 1.

||||||||||||||||||||||||||||||||||||||||||||||||||||||||||||||||||||||||||||||||||||||||||||||||||||||||||||||

It is now easy to express any system of linear equalities or inequalities in matrix form. For example,

$$3x_1 + 2x_2 + x_3 = 5$$
$$2x_1 + x_2 - x_3 = 4$$

becomes

$$\begin{pmatrix} 3 & 2 & 1 \\ 2 & 1 & -1 \end{pmatrix}\begin{pmatrix} x_1 \\ x_2 \\ x_3 \end{pmatrix} = \begin{pmatrix} 5 \\ 4 \end{pmatrix}.$$

The general 2 by 3 system (2 equations, 3 variables) is

$$a_{11}x_1 + a_{12}x_2 + a_{13}x_3 = b_1$$
$$a_{21}x_1 + a_{22}x_2 + a_{23}x_3 = b_2,$$

where now we need double subscripts to specify the row and column position of a coefficient. The foregoing system can be described in compact form as

$$Ax = b, \text{ where } A \text{ is 2 by 3.}$$

Note again that if A is 2 by 3, the vector x must be 3 by 1 and the product vector, b, is 2 by 1.

||||||||||||||||||||||||||||||||||||||||||||||||||||||||||||||||||||||||||||||||||||||||||||||||||||||||||||||

Exercise. If $cy \le d$ represents a 3 by 2 system of $\le$ inequalities, write a) the expanded matrix form and b) the usual algebraic form of the system.

Answer: a)
$$\begin{pmatrix} c_{11} & c_{12} \\ c_{21} & c_{22} \\ c_{31} & c_{32} \end{pmatrix}\begin{pmatrix} y_1 \\ y_2 \end{pmatrix} \le \begin{pmatrix} d_1 \\ d_2 \\ d_3 \end{pmatrix}.$$

b) $c_{11}y_1 + c_{12}y_2 \le d_1$
$c_{21}y_1 + c_{22}y_2 \le d_2$
$c_{31}y_1 + c_{32}y_2 \le d_3.$

||||||||||||||||||||||||||||||||||||||||||||||||||||||||||||||||||||||||||||||||||||||||||||||||||||||||||||||

The general m by n system,

$$a_{11}x_1 + a_{12}x_2 + a_{13}x_3 + \cdots + a_{1n}x_n = b_1$$
$$a_{21}x_1 + a_{22}x_2 + a_{23}x_3 + \cdots + a_{2n}x_n = b_2$$
$$\vdots$$
$$a_{m1}x_1 + a_{m2}x_2 + a_{m3}x_3 + \cdots + a_{mn}x_n = b_m$$

can be specified simply as

$$Ax = b \qquad \text{where } A \text{ is } m \text{ by } n.$$

Checking the shapes again, we see that if A is m by n, the vector x is n by 1, and the product vector b is m by 1.

Exercise. In the p by q system $Cy = d$, what are the shapes of C, y, and d? Answer: C is p by q, y is q by 1, and d is p by 1.

Finally, let us state the linear programming problem of the last chapter in compact form. We have m constraints on n variables, an m by n system. Further, all n variables must be greater than or equal to zero. Using $\mathbf{0}$ to mean a vector or matrix whose entries are all zeros, the statement $x \geq \mathbf{0}$ gives the nonnegativity requirement. The objective function, $\theta = c_1 x_1 + c_2 x_2 + \cdots + c_n x_n$ can be written compactly as cx. The objective is to maximize cx subject to the m by n system of constraints $Ax \leq b$. The problem may be stated as

> Maximize cx
> Subject to $Ax \leq b$ where A is m by n
> and $x \geq \mathbf{0}$.

Analyzing the shapes of the matrices, A is m by n, so x is n by 1 and b is m by 1. Since in cx the vector c premultiplies x, c is 1 by n.

Exercise. Given that the components of c are 1, 3, and 2; the components of b are 10, 20, and 30; x is an appropriate vector, and

$$A = \begin{pmatrix} 5 & 3 & 4 \\ 2 & 4 & 2 \\ 1 & 1 & 5 \end{pmatrix}.$$

Write the linear programming problem

> Maximize cx
> subject to: $Ax \leq b$
> $x \geq \mathbf{0}$

first in expanded vector form, then in ordinary algebraic form.

Answer: The expanded vector form is

$$\text{Maximize} \quad (1 \quad 3 \quad 2)\begin{pmatrix} x_1 \\ x_2 \\ x_3 \end{pmatrix}$$

$$\text{subject to:} \quad \begin{pmatrix} 5 & 3 & 4 \\ 2 & 4 & 2 \\ 1 & 1 & 5 \end{pmatrix}\begin{pmatrix} x_1 \\ x_2 \\ x_3 \end{pmatrix} \leq \begin{pmatrix} 10 \\ 20 \\ 30 \end{pmatrix}$$

$$\begin{pmatrix} x_1 \\ x_2 \\ x_3 \end{pmatrix} \geq \begin{pmatrix} 0 \\ 0 \\ 0 \end{pmatrix}.$$

The ordinary algebraic form is

$$\begin{aligned}
\text{Maximize} \quad & x_1 + 3x_2 + 2x_3 \\
\text{subject to:} \quad & 5x_1 + 3x_2 + 4x_3 \leq 10 \\
& 2x_1 + 4x_2 + 2x_3 \leq 20 \\
& x_1 + x_2 + 5x_3 \leq 30 \\
& x_1 \geq 0 \\
& x_2 \geq 0 \\
& x_3 \geq 0.
\end{aligned}$$

It is often useful in the solution of linear programming problems to convert an inequality constraint into an equality by adding a new nonnegative variable (called a *slack* variable and designated by s) to the constraint. For example, $2x_1 + 3x_2 \leq 15$ might be converted to $2x_1 + 3x_2 + s_1 = 15$ by addition of the slack variable s_1, where, of course, $s_1 \geq 0$. Consider the following linear programming problem which contains two constraints (other than nonnegativity) and, therefore, employs two slack variables to convert the constraints into equalities.

$$\begin{aligned}
\text{Maximize} \quad & 3x_1 + 2x_2 \\
\text{subject to:} \quad & 3x_1 + 4x_2 + s_1 = 20 \\
& x_1 + 2x_2 + s_2 = 12 \\
& x_1 \geq 0 \\
& x_2 \geq 0 \\
& s_1 \geq 0 \\
& s_2 \geq 0.
\end{aligned}$$

The reader may verify that the problem at hand can be written in expanded matrix form as:

$$\text{Maximize} \quad (3 \quad 2)\begin{pmatrix} x_1 \\ x_2 \end{pmatrix}$$

$$\text{subject to:} \quad \begin{pmatrix} 3 & 4 \\ 1 & 2 \end{pmatrix}\begin{pmatrix} x_1 \\ x_2 \end{pmatrix} + \begin{pmatrix} s_1 \\ s_2 \end{pmatrix} = \begin{pmatrix} 20 \\ 12 \end{pmatrix}$$

$$\begin{pmatrix} x_1 \\ x_2 \end{pmatrix} \geq \begin{pmatrix} 0 \\ 0 \end{pmatrix}$$

$$\begin{pmatrix} s_1 \\ s_2 \end{pmatrix} \geq \begin{pmatrix} 0 \\ 0 \end{pmatrix}.$$

In compact matrix notation the last is of the form:

$$\text{Maximize} \quad cx$$
$$\text{subject to:} \quad Ax + s = b$$
$$x \geq 0$$
$$s \geq 0.$$

6.9 PROBLEM SET 2

Write the following in expanded matrix form:

1. $2x_1 + 3x_2 = 5$
   $x_1 + 2x_2 = 3.$

2. $x_1 + 2x_2 + y_1 = 10$
   $2x_1 + 3x_2 + y_2 = 12.$

3. $3x_1 + x_3 - x_4 \leq 5$
   $2x_1 + x_2 - 5x_4 \leq 10$
   $x_2 + 3x_3 + x_4 \leq 8.$

Write the following in usual algebraic form:

4. $\begin{pmatrix} 3 & 1 & 2 \\ 1 & 4 & 1 \end{pmatrix}\begin{pmatrix} x_1 \\ x_2 \\ x_3 \end{pmatrix} = \begin{pmatrix} 5 \\ 4 \end{pmatrix}.$

5. $\begin{pmatrix} 2 & 3 \\ 4 & 6 \\ 1 & 7 \end{pmatrix}\begin{pmatrix} x_1 \\ x_2 \end{pmatrix} = \begin{pmatrix} 5 \\ 10 \\ 6 \end{pmatrix}.$

6. $(x_1 \quad x_2)\begin{pmatrix} 3 & 5 \\ 1 & 4 \end{pmatrix} = \begin{pmatrix} 2 \\ 6 \end{pmatrix}.$

7. $\begin{pmatrix} p_{11} & p_{12} & p_{13} & p_{14} \\ p_{21} & p_{22} & p_{23} & p_{24} \\ p_{31} & p_{32} & p_{33} & p_{34} \end{pmatrix}x = q,$ if x and q are appropriate vectors.

8. $\begin{pmatrix} 2 & 1 & 5 \\ 4 & 6 & 2 \end{pmatrix}\begin{pmatrix} x_1 \\ x_2 \\ x_3 \end{pmatrix} + \begin{pmatrix} y_1 \\ y_2 \end{pmatrix} = \begin{pmatrix} 10 \\ 5 \end{pmatrix}.$

9. Maximize $(3 \quad 2 \quad 4)\begin{pmatrix} x_1 \\ x_2 \\ x_3 \end{pmatrix}$ subject to $\begin{pmatrix} 2 & 1 & 4 \\ 3 & 5 & 2 \end{pmatrix}\begin{pmatrix} x_1 \\ x_2 \\ x_3 \end{pmatrix} \leq \begin{pmatrix} 10 \\ 15 \end{pmatrix}$

and $x \geq 0$.

10. Maximize cx
    subject to: $Ax + s = b$
    $$x \geq 0$$
    $$s > 0$$

    where

    $$b = \begin{pmatrix} 12 \\ 15 \\ 10 \\ 20 \end{pmatrix}$$

    $c = (2 \ 5 \ 4 \ 3)$,

    $$A = \begin{pmatrix} 1 & 3 & 2 & 0 \\ 0 & 1 & 3 & 2 \\ 5 & 3 & 0 & 2 \\ 1 & 4 & 6 & 0 \end{pmatrix}$$

    and s and x are appropriate vectors.

11. The coefficient matrix of a system of equalities, A, is p by q. The variables, y, and the constants, g, are vectors with the proper number of components.
    a) What are the proper shapes for y and for g?
    b) Write the system in compact matrix notation.

6.10 APPLICATIONS—1: MARKOV CHAINS

To set the stage for this section, suppose that a restaurant chain notes that 30 percent of the dinners it sells each week are beef dinners, and 70 percent are other dinners. The chain manager has a special arrangement for volume buying of beef at relatively low prices, and would like to raise the proportion of beef dinners sold. He carries out a promotional campaign to increase beef sales and collects the information shown in Table 6–3.

TABLE 6–3
Transition Proportions
One Week to the Next Week

One Week	Next Week	
	Beef	*Other*
Beef	0.8	0.2
Other	0.6	0.4

The matrix of proportions is called the *transition matrix*. The number 0.8 means that 80 percent of those buying beef dinners one week buy beef dinners again the next week. Similarly, 20 percent of those buying beef one week buy other dinners the next week.

Exercise.   Interpret the second row of Table 6–3. Answer: 60 percent of those buying other dinners one week change to beef dinners the next week, and the remaining 40 percent buy other dinners again the next week.

The state of affairs at the beginning of the section was 30 percent beef and 70 percent other. We shall call the vector (0.3 0.7) the *state* vector. Writing the state vector to the left of the transition matrix, we have

$$
(0.3 \text{ beef } 0.7 \text{ other}) \quad \begin{array}{c} \\ \text{Beef} \\ \text{Other} \end{array} \begin{array}{cc} \text{Beef} & \text{Other} \\ \begin{pmatrix} 0.8 & 0.2 \\ 0.6 & 0.4 \end{pmatrix}. \end{array}
$$

If we wish to find what proportion will buy beef after a one week transition, we note that 80 percent of the 30 percent who bought beef one week will buy it the next and an additional 60 percent of the 70 percent who bought other dinners one week will buy beef the next week. The sum is 0.3(0.8) + 0.7(0.6) = 0.66 and we found this by the usual inner product matrix multiplication procedure. In the same manner 0.3(0.2) + 0.7(0.4) = 0.34 is the proportion buying other dinners next week, and the new state vector is (0.66 beef 0.34 other). In brief,

$$
(0.3 \ 0.7)\begin{pmatrix} 0.8 & 0.2 \\ 0.6 & 0.4 \end{pmatrix} = (0.66 \ 0.34).
$$

Now suppose the promotion activity is maintained and the transition matrix remains constant from week to week. Then the state vector for week #1, (0.66 0.34) can be used as a premultiplier of the transition matrix to obtain the state vector for week #2.

Exercise.   What proportions will buy beef and other in week #2?

Answer: $(0.66 \ 0.34)\begin{pmatrix} 0.8 & 0.2 \\ 0.6 & 0.4 \end{pmatrix} = (0.732 \ 0.268)$

which is 73.2 percent beef and 26.8 percent other dinners.

If we continue the chain, we may examine the successive state vectors as shown in Table 6–4.

TABLE 6–4
Successive State Vectors

Week	Beef	Other
Beginning	(0.30	0.70)
#1	(0.66	0.34)
#2	(0.732	0.268)
#3	(0.7464	0.2536)
#4	(0.74928	0.25072)
#5	(0.749856	0.250144)

Observe that the components of each state vector sum to 1, as must be the case because they are proportions of a whole. Also, for the same reason, the rows of any transition matrix must each sum to 1. Note also that the state vector appears to be approaching (0.75 0.25). What would happen if we used this state vector as a multiplier of the transition matrix? We would obtain

$$(0.75 \quad 0.25)\begin{pmatrix} 0.8 & 0.2 \\ 0.6 & 0.4 \end{pmatrix} = (0.75 \quad 0.25).$$

We see that the transition of (0.75 0.25) leads to the same state, (0.75 0.25), and we call this vector the *steady state*. The actual calculation of successive state vectors will never yield exactly (0.75 0.25), so it is a matter of importance to learn how to find the steady state by a method other than tabulating successive state vectors and guessing the steady state from the sequence of results. In our problem, let us call the steady state $(v_1 \quad v_2)$. Then it must be true that

$$(v_1 \quad v_2)\begin{pmatrix} 0.8 & 0.2 \\ 0.6 & 0.4 \end{pmatrix} = (v_1 \quad v_2),$$

where, of course, $v_1 + v_2 = 1$. The matrix multiplication yields

$$e_1\!: \ 0.8v_1 + 0.6v_2 = v_1 \quad \text{or} \quad -0.2v_1 + 0.6v_2 = 0$$
$$e_2\!: \ 0.2v_1 + 0.4v_2 = v_2 \qquad\qquad 0.2v_1 - 0.6v_2 = 0.$$

Clearly, e_1 and e_2 are the same equation, so we need only one of them. Taking e_1 with $v_1 + v_2 = 1$ we have

$$e_1\!: \ -0.2v_1 + 0.6v_2 = 0$$
$$e_3\!: \qquad v_1 + \quad v_2 = 1$$
$$e_4\!: \qquad\qquad 0.8v_2 = 0.2 \qquad 0.2e_3 + e_1.$$

From the last statement, $v_2 = 0.25$ and from e_3, v_1 is 0.75, so we have the steady state vector (0.75 0.25) which we obtained by guessing from the sequence in Table 6–4.

Exercise. Find the steady state for the transition matrix

$$\begin{pmatrix} 0.5 & 0.5 \\ 1 & 0 \end{pmatrix}.$$

Answer: (2/3 1/3).

Next, let us suppose that once a customer buys beef he is so satisfied that he will buy beef the next time. This results in a 1 in the transition matrix. Thus,

$$(0.3 \text{ beef} \quad 0.7 \text{ other}) \qquad \begin{matrix} \text{Beef} \\ \text{Other} \end{matrix} \begin{pmatrix} 1.0 & 0.0 \\ 0.6 & 0.4 \end{pmatrix}.$$

with headers Beef Other above the matrix.

Inasmuch as 60 percent of those buying other meals change to beef, then continue to buy beef, it is reasonable to expect that ultimately all customers will buy beef and the steady state will be (1 0). In this situation, beef has absorbed all the business, and the chain leading to this steady state is called an *absorbing Markov chain*.

Exercise. Verify that the steady state for the last written beef-other matrix is (1 0).

We have used 2 by 2 transition matrices to illustrate this section. Clearly, the methodology can be applied to *n* by *n* matrices. All that would change is the complexity of the calculations, so we shall not pursue this topic further. We should mention that Markov was a mathematician whose name has been given to processes of the type discussed in this section; namely, processes in which the future state is completely determined by the present state and not at all by the way in which the present state arose. Finally, we should call attention to the fact that the transition matrices in our discussion have been constant as we changed from state to state, and this fact can be made explicit by referring to the processes as *stationary* Markov processes.

6.11 PROBLEM SET 3

1. Eager and Beaver Companies each have 50 percent of the market for a product. Because of a promotion campaign, buyers are switching between Eager and Beaver according to the following transition matrix.

$$\begin{array}{cc} & \text{Eager} \quad \text{Beaver} \\ \begin{array}{c} \text{Eager} \\ \text{Beaver} \end{array} & \begin{pmatrix} 0.6 & 0.4 \\ 0.5 & 0.5 \end{pmatrix}. \end{array}$$

a) What do the numbers 0.6 and 0.4 mean?

b) What will be the market shares after the first and second transitions?

c) What are the steady state market shares?

2. At a point in time 95 percent of the population were spenders of copper pennies and 5 percent were savers. Because of the increasing value of pennies, only 30 percent of the spenders remain spenders, and 10 percent of the savers become spenders.

a) What will be the (spender saver) state vector after one transition?

b) What is the steady state vector?

3. At a point in time, 1 percent of the population use a drug and 99 percent do not. In a year, $\frac{1}{10}$ of one percent of non-users become users, but all users remain users.

a) What will be the percent of users and non-users after one transition?

b) What is the steady state?

4. Carry out the multiplication and interpret the result.

$$\begin{array}{cc} \text{State} & \begin{array}{cc} \#1 & \#2 \end{array} \\ \begin{array}{cc} \#1 & \#2 \end{array} & \begin{array}{c} \#1 \\ \#2 \end{array} \begin{pmatrix} 1 & 0 \\ 0 & 1 \end{pmatrix}. \\ \begin{array}{cc} (a & b) \end{array} & \end{array}$$

5. The following chain is cyclical, meaning that it returns periodically to the same state. Write the successive state vectors until the initial one, (0.6 0.3 0.1) reappears.

$$(0.6 \ 0.3 \ 0.1) \begin{pmatrix} 0 & 0 & 1 \\ 1 & 0 & 0 \\ 0 & 1 & 0 \end{pmatrix}.$$

6.12 SQUARE MATRICES

An *n* by *n* matrix has the same number of rows and columns and is called a square matrix. We may specify a square matrix simply by stating *n*. Thus, a 3 by 3 matrix is a square matrix of order 3. In the remainder of this chapter, all matrices except vectors will be square matrices.

6.13 UNIT MATRIX

A *unit* (or identity) matrix is a square matrix whose diagonal elements from upper left to lower right are each 1, and whose other elements all are zeros. Thus,

$$I = \begin{pmatrix} 1 & 0 & 0 \\ 0 & 1 & 0 \\ 0 & 0 & 1 \end{pmatrix}$$

is a unit matrix of order 3.

The unit matrix of order 2 is

$$I = \begin{pmatrix} 1 & 0 \\ 0 & 1 \end{pmatrix}.$$

The most important property of the unit matrix is illustrated by the statements

$$AI = A \quad \text{and} \quad IA = A.$$

That is, the product of any given matrix and the unit matrix is the given matrix itself. In other words, the unit matrix behaves like the number 1 in ordinary arithmetic where we say

$$(a)(1) = (1)(a) = a.$$

The reader may verify the unity property of I by carrying out the following multiplication in which the first written matrix, I, when multiplied by the second matrix, yields the second matrix as the product.

$$\begin{pmatrix} 1 & 0 & 0 \\ 0 & 1 & 0 \\ 0 & 0 & 1 \end{pmatrix} \begin{pmatrix} 2 & 3 & 4 \\ -1 & -2 & 0 \\ 5 & 2 & -3 \end{pmatrix} = \begin{pmatrix} 2 & 3 & 4 \\ -1 & -2 & 0 \\ 5 & 2 & -3 \end{pmatrix}.$$

Exercise.   Given

$$A - \begin{pmatrix} a & b \\ c & d \end{pmatrix}$$

verify that $AI = A$ and that $IA = A$.

6.14 ROW OPERATIONS

Row operations on a matrix consist either of multiplying (dividing) a row by a nonzero constant, adding a multiple of one row to another row, or interchanging two rows. For example, given the matrix

$$A = \begin{pmatrix} 1 & 2 \\ 3 & 9 \end{pmatrix}$$

we may multiply the first row by -2 to give another matrix

$$\begin{pmatrix} -2 & -4 \\ 3 & 9 \end{pmatrix}.$$

Or, dividing the second row of A by 3 leads to the matrix

$$\begin{pmatrix} 1 & 2 \\ 1 & 3 \end{pmatrix}.$$

If we multiply the first row of A by -3 and add the result to the second row (leaving the first row unchanged), we have

$$\begin{pmatrix} 1 & 2 \\ 0 & 3 \end{pmatrix}.$$

Exercise.   What row operation performed on

$$\begin{pmatrix} 2 & 5 \\ 6 & 13 \end{pmatrix}$$

will lead to another matrix which will have a 0 in place of the 6, and what will the new matrix be? Answer: Multiply the first row of the given matrix by -3 and add to the second row, obtaining

$$\begin{pmatrix} 2 & 5 \\ 0 & -2 \end{pmatrix}.$$

If we have a statement of matrix equality, such as

$$AB = C$$

and we perform the same row operations on A and C (not on B and C) we are led to new statements of equality. For example,

$$\begin{pmatrix} 1 & 2 \\ 0 & 3 \end{pmatrix} \begin{pmatrix} 2 & -4 \\ -1 & 5 \end{pmatrix} = \begin{pmatrix} 0 & 6 \\ -3 & 15 \end{pmatrix}$$

is a true statement. If we multiply the first row of the leftmost matrix by 3 and the first row of the rightmost matrix by 3 we obtain

$$\begin{pmatrix} 3 & 6 \\ 0 & 3 \end{pmatrix} \begin{pmatrix} 2 & -4 \\ -1 & 5 \end{pmatrix} = \begin{pmatrix} 0 & 18 \\ -3 & 15 \end{pmatrix}$$

which the reader may verify is also a true statement. Or, starting over, if we take the leftmost and rightmost matrices and add twice the first row to the second row, we obtain

$$\begin{pmatrix} 1 & 2 \\ 2 & 7 \end{pmatrix} \begin{pmatrix} 2 & -4 \\ -1 & 5 \end{pmatrix} = \begin{pmatrix} 0 & 6 \\ -3 & 27 \end{pmatrix}$$

which computation will prove to be a true statement.

Exercise.   Starting with the original matrices in the foregoing, select the leftmost and rightmost matrices and multiply the second row by -2 and add to the first. Write the new statement and verify that it is an equality.

6.15 INVERSE OF A MATRIX

The inverse of a square matrix A, if one exists, is another matrix, written as A^{-1}, such that the product of the two is the unit matrix. Thus,

$$AA^{-1} = I \quad \text{and} \quad A^{-1}A = I.$$

The symbol A^{-1} is read as "A inverse." The -1 superscript is *not* an exponent in the usual algebraic sense. We can verify that two matrices are inverses by carrying out the multiplication specified in the definition. Thus, the two leftmost matrices in the following are inverses because their product is the unit matrix.

$$\begin{pmatrix} 8 & 5 \\ 3 & 2 \end{pmatrix} \begin{pmatrix} 2 & -5 \\ -3 & 8 \end{pmatrix} = \begin{pmatrix} 1 & 0 \\ 0 & 1 \end{pmatrix}.$$

Exercise. Verify that the following matrices are inverses of each other.

$$\begin{pmatrix} 13 & 3 \\ 4 & 1 \end{pmatrix} \begin{pmatrix} 1 & -3 \\ -4 & 13 \end{pmatrix}.$$

Answer: Multiplication of the two matrices leads to the unit matrix

$$\begin{pmatrix} 1 & 0 \\ 0 & 1 \end{pmatrix}.$$

We shall see later that not every matrix has an inverse. However, if the inverse of A does exist, it is unique: that is, a matrix which has an inverse has exactly one inverse. The uniqueness of the inverse may not seem obvious, so let us see what would happen if in addition to A^{-1}, A had another inverse, call if B. Then, according to the definition of the inverse, the product AB would have to be I; that is,

$$AB = I.$$

Now if we multiply both sides of this expression by A^{-1} we have

$$A^{-1}AB = A^{-1}I.$$

Noting that on the left $A^{-1}A$ can be replaced by I, and then IB can be replaced by B, and on the right $A^{-1}I$ can be replaced by A^{-1}, we see that

$$B = A^{-1}.$$

So the supposedly different inverse, B, turns out to be the original inverse, A^{-1}. We shall make use of the uniqueness of the inverse a little later when we will ask what the expression (?) must be to make the following a true statement.

$$I = (?)A.$$

The answer will be that (?) must be A^{-1}.

We are now ready to attack the main objective of this section of our study which is to compute the inverse of a given matrix. After we have learned the computational procedures, we shall discover that they are directly applicable to the solution of n by n systems of linear equations.

Briefly, the inverse of a given matrix can be found by writing the given matrix at the left, and the corresponding unit matrix next to it, at the right. Then select and carry out row operations which will convert the given matrix into the unit matrix, and *apply the same operations to the matrix at the right.* When the left (given) matrix becomes the unit matrix, the matrix on the right will be the desired inverse. To illustrate, let us find the inverse of the matrix

$$\begin{pmatrix} 3 & 2 \\ 1 & 1 \end{pmatrix}.$$

We start by writing the given matrix to the left and the unit matrix to the right; thus,

$$\begin{pmatrix} 3 & 2 \\ 1 & 1 \end{pmatrix} \begin{pmatrix} 1 & 0 \\ 0 & 1 \end{pmatrix}.$$

To change the left matrix into the unit matrix will require several steps (which we perform on both matrices). We start by getting a 1 in the upper left corner. Divide the first row above by 3 to obtain

$$\begin{pmatrix} 1 & \frac{2}{3} \\ 1 & 1 \end{pmatrix} \begin{pmatrix} \frac{1}{3} & 0 \\ 0 & 1 \end{pmatrix}.$$

Next, get a 0 in the 2nd row, 1st column of the left matrix by multiplying the first row above by -1 and adding to the second row. We have

$$\begin{pmatrix} 1 & \frac{2}{3} \\ 0 & \frac{1}{3} \end{pmatrix} \begin{pmatrix} \frac{1}{3} & 0 \\ -\frac{1}{3} & 1 \end{pmatrix}.$$

Multiply the second row, just above, by 3 to obtain a 1 in the 2nd row, 2nd column of the left matrix. Thus,

$$\begin{pmatrix} 1 & \frac{2}{3} \\ 0 & 1 \end{pmatrix} \begin{pmatrix} \frac{1}{3} & 0 \\ -1 & 3 \end{pmatrix}.$$

Finally, multiply the second row, just above, by $-2/3$ and add to the first to obtain a zero in the 1st row, second column of the left matrix. This gives

$$\begin{pmatrix} 1 & 0 \\ 0 & 1 \end{pmatrix} \begin{pmatrix} 1 & -2 \\ -1 & 3 \end{pmatrix}.$$

We now have the identity matrix at the left, so the inverse of

$$\begin{pmatrix} 3 & 2 \\ 1 & 1 \end{pmatrix}$$

is the matrix at the right; namely,

$$\begin{pmatrix} 1 & -2 \\ -1 & 3 \end{pmatrix}.$$

Exercise. Apply the procedure of the last example and find the inverse of the matrix

$$A = \begin{pmatrix} 7 & 3 \\ 2 & 1 \end{pmatrix}.$$

Answer: The inverse is the matrix

$$A^{-1} = \begin{pmatrix} 1 & -3 \\ -2 & 7 \end{pmatrix}.$$

We next illustrate the computation of the inverse of a matrix of order 3, the left matrix in the following:

$$\begin{pmatrix} 2 & 3 & 1 \\ 1 & 4 & 2 \\ 5 & 6 & 4 \end{pmatrix} \begin{pmatrix} 1 & 0 & 0 \\ 0 & 1 & 0 \\ 0 & 0 & 1 \end{pmatrix}.$$

Divide the first row above by 2 to obtain

$$\begin{pmatrix} 1 & \frac{3}{2} & \frac{1}{2} \\ 1 & 4 & 2 \\ 5 & 6 & 4 \end{pmatrix} \begin{pmatrix} \frac{1}{2} & 0 & 0 \\ 0 & 1 & 0 \\ 0 & 0 & 1 \end{pmatrix}.$$

Multiply the first row above by -1 and add to the second row to obtain

$$\begin{pmatrix} 1 & \frac{3}{2} & \frac{1}{2} \\ 0 & \frac{5}{2} & \frac{3}{2} \\ 5 & 6 & 4 \end{pmatrix} \begin{pmatrix} \frac{1}{2} & 0 & 0 \\ -\frac{1}{2} & 1 & 0 \\ 0 & 0 & 1 \end{pmatrix}.$$

Multiply the first row above by -5 and add to the third row to obtain

$$\begin{pmatrix} 1 & \frac{3}{2} & \frac{1}{2} \\ 0 & \frac{5}{2} & \frac{3}{2} \\ 0 & -\frac{3}{2} & \frac{3}{2} \end{pmatrix} \begin{pmatrix} \frac{1}{2} & 0 & 0 \\ -\frac{1}{2} & 1 & 0 \\ -\frac{5}{2} & 0 & 1 \end{pmatrix}.$$

The tactics we have followed are these: first, get a 1 in the first column, first row, then use combinations of this row with each of the other rows to get zeros in the first columns of these rows. We now repeat these tactics, starting by getting a 1 in the second column of the second row, then using this row to get zeros in the second columns of the other rows.

Divide the second row by 5/2 to obtain the new second row

$$0 \quad 1 \quad \frac{3}{5} \quad \bigg| \quad -\frac{1}{5} \quad \frac{2}{5} \quad 0.$$

If we multiply this new second row by $-3/2$ and add to the first row, then multiply the new second row by $3/2$ and add to the third row, we obtain the new matrices

$$\begin{pmatrix} 1 & 0 & -\frac{2}{5} \\ 0 & 1 & \frac{3}{5} \\ 0 & 0 & \frac{12}{5} \end{pmatrix} \begin{pmatrix} \frac{4}{5} & -\frac{3}{5} & 0 \\ -\frac{1}{5} & \frac{2}{5} & 0 \\ -\frac{14}{5} & \frac{3}{5} & 1 \end{pmatrix}.$$

Finally, we repeat the tactics by getting a 1 in the third row of the third column and then using this row to get zeros in the third column of the other rows. The new third row, obtained by dividing the previous third row by 12/5, is

$$0 \quad 0 \quad 1 \quad \bigg| \quad -\frac{7}{6} \quad \frac{1}{4} \quad \frac{5}{12}.$$

If we multiply this new third row by 2/5 and add to the first row, then multiply the new third row by $-3/5$ and add to the second row we find

$$\begin{pmatrix} 1 & 0 & 0 \\ 0 & 1 & 0 \\ 0 & 0 & 1 \end{pmatrix} \begin{pmatrix} \frac{1}{3} & -\frac{1}{2} & \frac{1}{6} \\ \frac{1}{2} & \frac{1}{4} & -\frac{1}{4} \\ -\frac{7}{6} & \frac{1}{4} & \frac{5}{12} \end{pmatrix}$$

which has the matrix *I* to the left and the desired inverse to the right.

Exercise.   Multiply the original matrix of the last illustration by the inverse and verify that the product is I.

Inversion of large matrices is a task best left for electronic computers. However, we should practice the procedure long enough to understand that even though the arithmetic is tedious, the basic methodology is not complicated. To see why the method works, consider the true statement

$$A = IA.$$

Suppose that we carry out row operations on the left of the equal sign to change the left matrix to I, and maintain the equality by applying the same operations to the first matrix on the right of the equal sign (which starts out as I). The end result will be I on the left and something times A on the right; thus,

$$I = (?)A.$$

We showed earlier that the inverse of a matrix is unique. Hence, there is only one appropriate entry for (?) in the foregoing. It is A^{-1}. It follows that if we write a given matrix A with I to its right, then change the left to I by row operations which are applied also to the right, the end result will be A^{-1} on the right.

Not every matrix has an inverse. For example, consider the matrix

$$\begin{pmatrix} 1 & 1 \\ 2 & 2 \end{pmatrix}.$$

If we set this matrix up with the unit matrix to the right, we have

$$\begin{pmatrix} 1 & 1 \\ 2 & 2 \end{pmatrix} \begin{pmatrix} 1 & 0 \\ 0 & 1 \end{pmatrix}.$$

The indicated row operation is to multiply the first row by -2 and add to the second row. We find

$$\begin{pmatrix} 1 & 1 \\ 0 & 0 \end{pmatrix} \begin{pmatrix} 1 & 0 \\ -2 & 1 \end{pmatrix}.$$

No row operations for the left matrix can be found which will provide a 1 in the lower left corner and a 0 in the upper right. The given matrix has no inverse and is said to be *singular*. In the matrix context, the word singular does not mean one or single but, rather, connotes that the matrix is peculiar in that it has no inverse.

Before turning to the problem set, we note that steps may be required to set up and maintain the inversion problem in the form required by the tactics discussed thus far. Consider the problem of inverting

$$\begin{pmatrix} 0 & 1 \\ 2 & 3 \end{pmatrix}.$$

Writing this matrix with the unit matrix, we have,

$$\begin{pmatrix} 0 & 1 \\ 2 & 3 \end{pmatrix} \begin{pmatrix} 1 & 0 \\ 0 & 1 \end{pmatrix}.$$

We may put this into the desired form with a 1 at the upper left by multiplying the second row by 1/2 and adding to the first. We find,

$$\begin{pmatrix} 1 & \frac{5}{2} \\ 2 & 3 \end{pmatrix} \begin{pmatrix} 1 & \frac{1}{2} \\ 0 & 1 \end{pmatrix}.$$

The usual tactics may now be applied. See Problem 1 of the following problem set.

6.16 PROBLEM SET 4

Find the inverse of each of the following (if an inverse exists).

1. $\begin{pmatrix} 0 & 1 \\ 2 & 3 \end{pmatrix}$.

2. $\begin{pmatrix} 9 & 4 \\ 2 & 1 \end{pmatrix}$.

3. $\begin{pmatrix} 2 & 2 \\ 3 & 5 \end{pmatrix}$.

4. $\begin{pmatrix} 2 & 2 \\ 6 & 6 \end{pmatrix}$.

5. $\begin{pmatrix} 1 & -1 \\ -1 & 2 \end{pmatrix}$.

6. $\begin{pmatrix} 1 & 3 \\ 2 & 0 \end{pmatrix}$.

7. $\begin{pmatrix} 7 & 3 \\ 2 & 1 \end{pmatrix}$.

8. $\begin{pmatrix} 3 & 3 \\ 2 & 2 \end{pmatrix}$.

9. $\begin{pmatrix} 2 & 8 & -11 \\ -1 & -5 & 7 \\ 1 & 2 & -3 \end{pmatrix}$.

10. $\begin{pmatrix} 1 & -1 & 0 \\ 2 & 1 & 3 \\ 3 & 0 & 3 \end{pmatrix}$.

11. $\begin{pmatrix} 4 & -18 & -3 \\ 0 & 1 & 0 \\ -5 & 24 & 4 \end{pmatrix}$.

12. $\begin{pmatrix} 2 & 2 & 3 \\ 0 & 1 & 1 \\ 4 & 0 & 3 \end{pmatrix}$.

13. $\begin{pmatrix} 1 & 1 & 1 \\ 1 & 1 & 1 \\ 2 & 2 & 2 \end{pmatrix}$.

14. $\begin{pmatrix} 0 & -1 & 1 \\ -1 & 1 & 2 \\ 1 & 0 & -2 \end{pmatrix}$.

6.17 APPLICATIONS–2: MATRIX SOLUTION OF LINEAR SYSTEMS

In this section we shall show how the solution of an n by n system of linear equations is accomplished using matrix inversion. First, consider the 2 by 2 system

$$2x_1 + 3x_2 = 17$$
$$x_1 + 2x_2 = 10.$$

The system can be written in expanded matrix form as

$$\begin{pmatrix} 2 & 3 \\ 1 & 2 \end{pmatrix}\begin{pmatrix} x_1 \\ x_2 \end{pmatrix} = \begin{pmatrix} 17 \\ 10 \end{pmatrix}.$$

If we multiply both sides of the last equation by the inverse of the coefficient matrix, which is

$$\begin{pmatrix} 2 & 3 \\ 1 & 2 \end{pmatrix}^{-1}$$

we have

$$\begin{pmatrix} 2 & 3 \\ 1 & 2 \end{pmatrix}^{-1}\begin{pmatrix} 2 & 3 \\ 1 & 2 \end{pmatrix}\begin{pmatrix} x_1 \\ x_2 \end{pmatrix} = \begin{pmatrix} 2 & 3 \\ 1 & 2 \end{pmatrix}^{-1}\begin{pmatrix} 17 \\ 10 \end{pmatrix}.$$

Inasmuch as the product of a matrix and its inverse is the unit matrix, we may rewrite the last equation as

$$\begin{pmatrix} 1 & 0 \\ 0 & 1 \end{pmatrix}\begin{pmatrix} x_1 \\ x_2 \end{pmatrix} = \begin{pmatrix} 2 & 3 \\ 1 & 2 \end{pmatrix}^{-1}\begin{pmatrix} 17 \\ 10 \end{pmatrix}.$$

The left member now is the matrix

$$\begin{pmatrix} x_1 \\ x_2 \end{pmatrix}$$

so we may write

$$\begin{pmatrix} x_1 \\ x_2 \end{pmatrix} = \begin{pmatrix} 2 & 3 \\ 1 & 2 \end{pmatrix}^{-1}\begin{pmatrix} 17 \\ 10 \end{pmatrix}.$$

The last equation shows that x_1 and x_2 (the elements of the solution vector) can be computed as soon as the inverse of the coefficient matrix is at hand. The reader may verify that the desired inverse is

$$\begin{pmatrix} 2 & 3 \\ 1 & 2 \end{pmatrix}^{-1} = \begin{pmatrix} 2 & -3 \\ -1 & 2 \end{pmatrix}.$$

Hence the solution of the original system is given by

$$\begin{pmatrix} x_1 \\ x_2 \end{pmatrix} = \begin{pmatrix} 2 & -3 \\ -1 & 2 \end{pmatrix}\begin{pmatrix} 17 \\ 10 \end{pmatrix}$$

that is, after computing the right member,

$$\begin{pmatrix} x_1 \\ x_2 \end{pmatrix} = \begin{pmatrix} 4 \\ 3 \end{pmatrix}$$

so that $x_1 = 4$ and $x_2 = 3$.

In general, if A is the matrix of coefficients and b is the column vector of constants, the matrix expression for an n by n system of linear equations is

$$Ax = b.$$

If both sides of the last equation are multiplied by the inverse of the co-efficient matrix, A^{-1}, we have

$$A^{-1}Ax = A^{-1}b$$

from which it follows that

$$x = A^{-1}b.$$

The key to the solution of the system is A^{-1}, the inverse of the coefficient matrix. Once A^{-1} has been found, the solution vector, x, can be computed by premultiplying b, the vector of constants, by A^{-1}. Consider the system

$$\begin{aligned} 2x_1 + 2x_2 + 3x_3 &= 3 \\ x_2 + x_3 &= 2 \\ x_1 + x_2 + x_3 &= 4. \end{aligned}$$

The coefficient matrix is

$$A = \begin{pmatrix} 2 & 2 & 3 \\ 0 & 1 & 1 \\ 1 & 1 & 1 \end{pmatrix}$$

and the reader may verify that the inverse of this matrix is

$$A^{-1} = \begin{pmatrix} 0 & -1 & 1 \\ -1 & 1 & 2 \\ 1 & 0 & -2 \end{pmatrix}.$$

The vector of constants is

$$b = \begin{pmatrix} 3 \\ 2 \\ 4 \end{pmatrix}.$$

The solution vector, x, is found as the product $A^{-1}b$ which is

$$\begin{pmatrix} x_1 \\ x_2 \\ x_3 \end{pmatrix} = A^{-1}b = \begin{pmatrix} 0 & -1 & 1 \\ -1 & 1 & 2 \\ 1 & 0 & -2 \end{pmatrix}\begin{pmatrix} 3 \\ 2 \\ 4 \end{pmatrix} = \begin{pmatrix} 2 \\ 7 \\ -5 \end{pmatrix}.$$

We find that $x_1 = 2$, $x_2 = 7$, and $x_3 = -5$. The reader may verify that these values satisfy all three original equations. It is worth pointing out that once A^{-1} has been computed, we have in effect solved *all* linear systems which have A as the coefficient matrix. In the case of the present system, for example, if we were to change the constants on the right from 3, 2, and

4 to, say, 0, 1, and 2, respectively, the solution of the system would be the foregoing A^{-1} multiplied by the new vector of constants; that is,

$$\begin{pmatrix} x_1 \\ x_2 \\ x_3 \end{pmatrix} = \begin{pmatrix} 0 & -1 & 1 \\ -1 & 1 & 2 \\ 1 & 0 & -2 \end{pmatrix} \begin{pmatrix} 0 \\ 1 \\ 2 \end{pmatrix} = \begin{pmatrix} 1 \\ 5 \\ -4 \end{pmatrix}$$

so that $x_1 = 1$, $x_2 = 5$, $x_3 = -4$ would be the solution of the system

$$2x_1 + 2x_2 + 3x_3 = 0$$
$$x_2 + x_3 = 1$$
$$x_1 + x_2 + x_3 = 2.$$

Exercise. If the system

$$7x_1 + 3x_2 = 5$$
$$2x_1 + x_2 = 7$$

is expressed in matrix form as $Ax = b$, what are A, x, and b; what is A^{-1}? Answer: A is the coefficient matrix

$$\begin{pmatrix} 7 & 3 \\ 2 & 1 \end{pmatrix},$$

x is the solution vector

$$\begin{pmatrix} x_1 \\ x_2 \end{pmatrix},$$

b is the vector of constants

$$\begin{pmatrix} 5 \\ 7 \end{pmatrix},$$

and the inverse of the coefficient matrix is found by computation to be

$$A^{-1} = \begin{pmatrix} 1 & -3 \\ -2 & 7 \end{pmatrix}.$$

Exercise. The solution of the foregoing exercise is $x = A^{-1}b$. Substitute the proper numbers into $A^{-1}b$ and compute the solution vector, x. Answer:

$$x = \begin{pmatrix} x_1 \\ x_2 \end{pmatrix} = \begin{pmatrix} 1 & -3 \\ -2 & 7 \end{pmatrix} \begin{pmatrix} 5 \\ 7 \end{pmatrix} = \begin{pmatrix} -16 \\ 39 \end{pmatrix}.$$

As an application of the material we have just studied, suppose a company makes liquid products X_1, X_2, and X_3 which contain different amounts of additives A_1, A_2, and A_3 per gallon as shown in Table 6–5.

TABLE 6–5

Liquid	Gallons Made	Pounds of Additive per Gallon		
		A_1	A_2	A_3
X_1	x_1	1	1	1
X_2	x_2	1	2	1
X_3	x_3	2	0	1
Available additive, end of week		a_1	a_2	a_3

The additives deteriorate if not used within a week, so each Saturday the company schedules production of $(x_1 \ x_2 \ x_3)$ gallons of $(X_1 \ X_2 \ X_3)$ to use up the additives on hand. These amounts vary from week to week and are shown as $(a_1 \ a_2 \ a_3)$ in Table 6–5. If we schedule $(x_1 \ x_2 \ x_3)$ gallons of the liquids, the pounds of additive A_1 used will be $x_1(1) + x_2(1) + x_3(2)$, and this should equal a_1, the amount available. Hence, $x_1 + x_2 + 2x_3 = a_1$. Combining this condition with the conditions on additives A_2 and A_3 leads us to the system

$$
\begin{aligned}
x_1 + x_2 + 2x_3 &= a_1 \\
x_1 + 2x_2 \quad\;\; &= a_2 \\
x_1 + x_2 + x_3 &= a_3.
\end{aligned}
$$

The coefficient matrix is

$$
A = \begin{pmatrix} 1 & 1 & 2 \\ 1 & 2 & 0 \\ 1 & 1 & 1 \end{pmatrix}
$$

and it may be verified that the inverse is

$$
A^{-1} = \begin{pmatrix} -2 & -1 & 4 \\ 1 & 1 & -2 \\ 1 & 0 & -1 \end{pmatrix}.
$$

Hence, the solution vector is

$$
\begin{pmatrix} x_1 \\ x_2 \\ x_3 \end{pmatrix} = \begin{pmatrix} -2 & -1 & 4 \\ 1 & 1 & -2 \\ 1 & 0 & -1 \end{pmatrix} \begin{pmatrix} a_1 \\ a_2 \\ a_3 \end{pmatrix}.
$$

Suppose that on a given Saturday the amounts of additives available are

$a_1 = 20$ pounds of A_1, $a_2 = 30$ pounds of A_2, and $a_3 = 20$ pounds of A_3. To use up these additives, the production schedule should be

$$\begin{pmatrix} x_1 \\ x_2 \\ x_3 \end{pmatrix} = \begin{pmatrix} -2 & -1 & 4 \\ 1 & 1 & -2 \\ 1 & 0 & -1 \end{pmatrix} \begin{pmatrix} 20 \\ 30 \\ 20 \end{pmatrix} = \begin{pmatrix} 10 \\ 10 \\ 0 \end{pmatrix},$$

which is 10 gallons of X_1, 10 gallons of X_2, and no X_3.

Exercise. If the additives available at a week's end are $(a_1 \ a_2 \ a_3)$ $= (80 \ 100 \ 70)$, what should the production schedule be? Answer: $(x_1 \ x_2 \ x_3) = (20 \ 40 \ 10)$.

Of course, it may not be possible to schedule production to use all the additives available on a given Saturday. For example, if $(a_1 \ a_2 \ a_3) = (40 \ 60 \ 50)$, the solution vector is $(x_1 \ x_2 \ x_3) = (60 \ 0 \ -10)$, and the value $x_3 = -10$ is not possible. In such a case we might choose to maximize either the total amount of liquid made or the total amount of additive used, depending upon cost considerations, and we would apply the linear programming techniques of Chapters 4 and 5 to determine the optimum production schedule. In this context, it should be noted that the manipulations carried out in the solution of linear systems in earlier chapters are operations on matrices. In this book, however, we have chosen to make our major study of such problems (Chapters 1–5) not depend upon an ability to use matrix algebra, and we shall not recast these problems in matrix form.

Finally, it should be repeated that matrix inversion is defined only for square matrices. If the reader wishes to work with m by n systems where $m \neq n$, or if he encounters a singular square matrix (one with no inverse), he should refer to Chapter 2 to find the solution for the system, if one exists.

6.18 PROBLEM SET 5

1. Consider the system

$$8x_1 + 5x_2 = 2$$
$$3x_1 + 2x_2 = 1.$$

a) Relating the system to $Ax = b$, what is A?, x?, b?
b) Compute A^{-1}.
c) Write $x = A^{-1}b$ in expanded matrix form.
d) Compute the solution from part (c).
e) What would be the solution vector if the elements in the vector of constants, 2 and 1, were changed to each of the following:
   (1) 1, 0.   (2) 0, 1.   (3) 1, 1.   (4) 3, 4.   (5) −3, 1.

2. Answer parts (*a*) through (*d*) of Problem 1 for the system

$$4x_1 + 3x_2 = 2$$
$$9x_1 + 7x_2 = 3.$$

e) What would be the solution vector if the elements in the vector of constants, 2 and 3, were changed to each of the following:

(1) 1, 0.   (2) 0, 1.   (3) 1, 1.   (4) 2, 1.   (5) −1, 2.

3. Answer parts (*a*) through (*d*) of Problem 1 for the system

$$6x_1 + 8x_2 = 3$$
$$2x_1 + 3x_2 = 1.$$

e) What would be the solution vector if the elements in the vector of constants, 3 and 1, were changed to each of the following:

(1) 1, 1.   (2) 0, 1.   (3) 1, 0.   (4) 2, 3.   (5) −1, 1.

4. Given the system

$$8x_1 - 7x_2 = b_1$$
$$-5x_1 + 5x_2 = b_2$$

find the missing elements to complete the following equation

$$\begin{pmatrix} x_1 \\ x_2 \end{pmatrix} = \begin{pmatrix} & \\ & \end{pmatrix} \begin{pmatrix} b_1 \\ b_2 \end{pmatrix}.$$

5. Answer parts (*a*) through (*d*) of Problem 1 for the system

$$3x_1 \qquad + 5x_3 = 3$$
$$2x_1 + 2x_2 + 5x_3 = 7$$
$$x_2 + x_3 = 2.$$

e) What would be the solution vector if the elements in the vector of constants, 3, 7, and 2, were changed to each of the following:

(1) 1, 2, 1.   (2) 2, 3, 4.   (3) 1, 0, −1.   (4) 1, 6, 2.   (5) 2, 10, 1.

6. Answer parts (*a*) through (*d*) of Problem 1 for the system

$$7x_1 + 3x_2 \qquad = 1$$
$$3x_2 + 5x_3 = 2$$
$$x_1 + x_2 + x_3 = 3.$$

e) What would be the solution vector if the elements in the vector of constants were changed from 1, 2, and 3 to each of the following:

(1) 1, −1, 1.   (2) 2, 3, 0.   (3) −1, 2, −2.   (4) 10, −1, 1.
(5) 0, −1, 0.

7. Given the system

$$2x_1 + 2x_2 + 3x_3 = b_1$$
$$x_2 + x_3 = b_2$$
$$4x_1 \qquad + 3x_3 = b_3$$

find the missing elements to complete the following equation:

$$\begin{pmatrix} x_1 \\ x_2 \\ x_3 \end{pmatrix} = \begin{pmatrix} & \\ & \\ & \end{pmatrix} \begin{pmatrix} b_1 \\ b_2 \\ b_3 \end{pmatrix}.$$

8. Given the system

$$x_1 + 2x_2 + x_3 = b_1$$
$$2x_1 + x_2 + x_3 = b_2$$
$$3x_1 \qquad + 2x_3 = b_3.$$

a) Verify by multiplication that the inverse of the coefficient matrix is

$$A^{-1} = \begin{pmatrix} -\frac{2}{3} & \frac{4}{3} & -\frac{1}{3} \\ \frac{1}{3} & \frac{1}{3} & -\frac{1}{3} \\ 1 & -2 & 1 \end{pmatrix}.$$

b) Compute the solution vector if the b's are, respectively,
(1) 3, 0, 3.   (2) 6, 3, 0.

9. Given the system

$$x_1 + x_2 + x_3 + x_4 = b_1$$
$$x_1 + 2x_2 + 2x_3 + 2x_4 = b_2$$
$$x_1 + 2x_2 + 3x_3 + 3x_4 = b_3$$
$$x_1 + 2x_2 + 3x_3 + 4x_4 = b_4.$$

a) Verify by multiplication that the inverse of the coefficient matrix is

$$\begin{pmatrix} 2 & -1 & 0 & 0 \\ -1 & 2 & -1 & 0 \\ 0 & -1 & 2 & -1 \\ 0 & 0 & -1 & 1 \end{pmatrix}.$$

b) What is the solution vector if the b's are, respectively,
(1) 1, 1, 1, 1.   (2) 1, 0, 1, 0.

10. The table shows that we plan to make x_1 units of product X_1. The machine hours used in making a unit of X_1 are 1 hour on machine M_1 and 1 hour on M_3. H_1 is the total number of hours available on M_1. Other entries in the table have corresponding interpretations.

| | | Machine Hours per Unit on Machine | | |
Product	Units Made	M_1	M_2	M_3
X_1	x_1	1	0	1
X_2	x_2	0	1	2
X_3	x_3	2	3	0
Total hours available		H_1	H_2	H_3

a) Set up the system of equations which must be solved if all available machine hours are to be used in making $(x_1\ x_2\ x_3)$ units of the products.
b) Set up the coefficient matrix and find its inverse. What numbers of units can be made if the numbers of available hours, $(H_1\ H_2\ H_3)$, are:
c) (160 80 200)?   d) (400 400 400)?   e) (320 192 256)?

6.19 APPLICATIONS–3: INPUT-OUTPUT ANALYSIS

Input-output analysis is concerned with an equilibrium model in which the industries of an economy produce just enough output to satisfy inter-industry requirements and nonindustry (final) demand. The problem is to determine the industry outputs required at various levels of final demand, given the relationships for interindustry requirements. To develop the model in a simple fashion, we assume that an economy has only two industries, No. 1 and No. 2, whose outputs are measured in dollars and are designated as x_1 and x_2, respectively. Final demands for the outputs of industries No. 1 and No. 2 will be symbolized by d_1 and d_2 dollars, respectively. In addition to supplying output to satisfy final demand, each industry must satisfy inter-industry requirements for its output. That is, No. 1 must provide enough output to satisfy its requirement for its own output and to satisfy the industry No. 2 requirement for the output of No. 1. Similarly, No. 2 must satisfy its requirement for its own product and the industry No. 1 requirement for the output of industry No. 2.

We start with interindustry requirements. Suppose that each dollar of industry No. 1 output requires \$0.40 of its own output and \$0.10 of industry No. 2 output. Similarly, each dollar of industry No. 2 output requires \$0.50 of industry No. 1 output and \$0.20 of industry No. 2 output. These inter-industry requirements are summarized in the *technological matrix* in the following manner:

Technological Matrix

	User No. 1	User No. 2
Producer No. 1	0.4	0.5
Producer No. 2	0.1	0.2

The first column under *user* states that each dollar of industry No. 1 output requires \$0.4 of industry No. 1 output and \$0.1 of industry No. 2 output.

Exercise. What is the meaning of the second *user* column of the foregoing technological matrix? Answer: Each dollar of industry No. 2 output requires \$0.5 of industry No. 1 output and \$0.2 of industry No. 2 output.

Recalling that total outputs of No. 1 and No. 2 are x_1 and x_2, respectively, it follows that industry No. 1 must produce $0.4x_1$ to satisfy its own requirement and $0.5x_2$ to satisfy the industry No. 2's requirement for the output of industry No. 1. Thus, for interindustry requirements, No. 1 must produce $0.4x_1 + 0.5x_2$.

Exercise. How much must industry No. 2 produce to satisfy interindustry requirements? Answer: $0.1x_1 + 0.2x_2$.

In addition to satisfying interindustry requirements, output must also be sufficient to fill final demands, d_1 and d_2. Suppose that final demands for outputs of No. 1 and No. 2 are $90 and $200, respectively. If equilibrium is to be achieved, industry No. 1 must produce $0.4x_1 + 0.5x_2$ to meet interindustry demand, plus 90 to satisfy final demand. That is, its output, x_1, must be such that

$$x_1 = 0.4x_1 + 0.5x_2 + 90.$$

Exercise. What must x_2 be to meet interindustry demand and final demand? Answer: $x_2 = 0.1x_1 + 0.2x_2 + 200$.

We now have the linear system

$$e_1: x_1 = 0.4x_1 + 0.5x_2 + 90$$
$$e_2: x_2 = 0.1x_1 + 0.2x_2 + 200$$

or

$$e_1: 0.6x_1 - 0.5x_2 = 90$$
$$e_2: -0.1x_1 + 0.8x_2 = 200.$$

Exercise. Find the values of x_1 and x_2 which satisfy the foregoing system by the elimination procedure. Answer: $x_1 = 400$, $x_2 = 300$.

To interpret the answer in the exercise, let us substitute numbers into the original equation, e_1. We find

$$400 = 160 + 150 + 90$$

This means that industry No. 1 produces $400 of output, uses $160 of this itself, supplies industry No. 2 with $150, and satisfies final demand of $90.

Exercise.   What is the disposition of the output of industry No. 2? Answer: Industry No. 2 output is $300. It uses $60 of this itself, supplies $40 to industry No. 1, and satisfies final demand of $200.

In the circumstances we have described, all requirements are met, and all output is used. We have solved the input-output problem for the given levels of final demand. But suppose that the levels of final demand change, what then will be the required outputs, given that the technological matrix (that is, the interindustry demand relationships) is constant? We could, of course, repeat the solution for each set of demands, but it is more efficient to employ the matrix solution discussed in the previous section of this chapter. First, let us display the relevant information in the following manner:

Producer	User		Final Demand	Total Output
	No. 1	No. 2		
No. 1. . . . . . . .a_{11}		a_{12}	d_1	x_1
No. 2. . . . . . . .a_{21}		a_{22}	d_2	x_2

The *user* columns constitute the technological matrix, *A*. Thus

$$A = \begin{pmatrix} a_{11} & a_{12} \\ a_{21} & a_{22} \end{pmatrix}.$$

The next column is the (final) demand vector, *D*. Thus,

$$D = \begin{pmatrix} d_1 \\ d_2 \end{pmatrix}$$

and the last column is the total output vector, *X*. Thus,

$$X = \begin{pmatrix} x_1 \\ x_2 \end{pmatrix}.$$

In the technological matrix, a_{ij} represents the amount of industry *i*'s output required in the output of $1 by industry *j*. Thus, for example, if we had a 9 by 9 matrix, a_{37} would represent the amount of industry No. 3 output required in the output of $1 by industry No. 7.

Exercise.   Interpret a_{21}. Answer: This is the amount of industry No. 2 product required in the output of $1 by industry No. 1.

Following our introductory example, we may write

$$x_1 = a_{11}x_1 + a_{12}x_2 + d_1$$
$$x_2 = a_{21}x_1 + a_{22}x_2 + d_2.$$

The last may be rearranged to give

$$(1 - a_{11})x_1 - a_{12}x_2 = d_1$$
$$-a_{21}x_1 + (1 - a_{22})x_2 = d_2.$$

Changing now to matrix-vector form, we have

$$\begin{pmatrix} 1 - a_{11} & -a_{12} \\ -a_{21} & 1 - a_{22} \end{pmatrix} \begin{pmatrix} x_1 \\ x_2 \end{pmatrix} = \begin{pmatrix} d_1 \\ d_2 \end{pmatrix}.$$

In the last equation, the leftmost matrix is the same as $I - A$, where I is the identity matrix. That is,

$$\begin{pmatrix} 1 & 0 \\ 0 & 1 \end{pmatrix} - \begin{pmatrix} a_{11} & a_{12} \\ a_{21} & a_{22} \end{pmatrix} = \begin{pmatrix} 1 - a_{11} & -a_{12} \\ -a_{21} & 1 - a_{22} \end{pmatrix}.$$

We may therefore write the matrix-vector form compactly as

$$(I - A)X = D.$$

We can now solve for the solution (output) vector, X, by pre-multiplying both sides by the inverse, $(I - A)^{-1}$ to obtain

$$X = (I - A)^{-1}D.$$

The last statement shows that the core mathematical problem in input-output computations is the calculation of the inverse, $(I - A)^{-1}$, given the technological matrix, A. Let us demonstrate by repeating the example with which this section started. The relevant data are:

Producer	User No. 1	No. 2	Final Demand	Total Output
No. 1.	0.4	0.5	90	x_1
No. 2.	0.1	0.2	200	x_2

The technological matrix is

$$A = \begin{pmatrix} 0.4 & 0.5 \\ 0.1 & 0.2 \end{pmatrix}$$

or, in fractional form,

$$A = \begin{pmatrix} \dfrac{2}{5} & \dfrac{1}{2} \\[2ex] \dfrac{1}{10} & \dfrac{1}{5} \end{pmatrix}.$$

Subtracting A from the identity matrix yields

$$\begin{pmatrix} 1 & 0 \\ 0 & 1 \end{pmatrix} - \begin{pmatrix} \frac{2}{5} & \frac{1}{2} \\ \frac{1}{10} & \frac{1}{5} \end{pmatrix} = \begin{pmatrix} \frac{3}{5} & -\frac{1}{2} \\ -\frac{1}{10} & \frac{4}{5} \end{pmatrix}.$$

We proceed in the manner of the last section to find $(I - A)^{-1}$.

$$\begin{pmatrix} \frac{3}{5} & -\frac{1}{2} \\ -\frac{1}{10} & \frac{4}{5} \end{pmatrix} \begin{pmatrix} 1 & 0 \\ 0 & 1 \end{pmatrix}.$$

Divide row 1 by 3/5:

$$\begin{pmatrix} 1 & -\frac{5}{6} \\ -\frac{1}{10} & \frac{4}{5} \end{pmatrix} \begin{pmatrix} \frac{5}{3} & 0 \\ 0 & 1 \end{pmatrix}.$$

Multiply row 1 by 1/10, add to row 2

$$\begin{pmatrix} 1 & -\frac{5}{6} \\ 0 & \frac{43}{60} \end{pmatrix} \begin{pmatrix} \frac{5}{3} & 0 \\ \frac{1}{6} & 1 \end{pmatrix}.$$

Divide row 2 by 43/60:

$$\begin{pmatrix} 1 & -\frac{5}{6} \\ 0 & 1 \end{pmatrix} \begin{pmatrix} \frac{5}{3} & 0 \\ \frac{10}{43} & \frac{60}{43} \end{pmatrix}.$$

Multiply row 2 by 5/6, add to row 1:

$$\begin{pmatrix} 1 & 0 \\ 0 & 1 \end{pmatrix} \begin{pmatrix} \frac{80}{43} & \frac{50}{43} \\ \frac{10}{43} & \frac{60}{43} \end{pmatrix}.$$

Hence we have

$$(I - A)^{-1} = \begin{pmatrix} \frac{80}{43} & \frac{50}{43} \\ \frac{10}{43} & \frac{60}{43} \end{pmatrix}$$

or, after factoring,

$$(I - A)^{-1} = \frac{10}{43} \begin{pmatrix} 8 & 5 \\ 1 & 6 \end{pmatrix}.$$

The solution vector, $X = (I - A)^{-1}D$ then is

$$X = \frac{10}{43}\begin{pmatrix} 8 & 5 \\ 1 & 6 \end{pmatrix}\begin{pmatrix} 90 \\ 200 \end{pmatrix}$$

so that

$$x_1 = \frac{10}{43}(720 + 1000) = 400$$

$$x_2 = \frac{10}{43}(90 + 1200) = 300$$

and we have the same solution found earlier in the chapter. The matrix solution is more efficient, however, because with it we may now find quickly the solution vector for varying final demand levels. For example, if final demand changes to 215 for industry No. 1 and 430 for No. 2, we have

$$X = \frac{10}{43}\begin{pmatrix} 8 & 5 \\ 1 & 6 \end{pmatrix}\begin{pmatrix} 215 \\ 430 \end{pmatrix}.$$

||||||||||||||||||||||||||||||||||||||||||||||||||||||||||||||||||||||||||||||||||||||||||||||||||||||||||||||||||||||||||||||||||||||||||||||||||||

Exercise. Compute the elements of the output vector for the demand levels just stated. Answer: $x_1 = 900$, $x_2 = 650$.

||||||||||||||||||||||||||||||||||||||||||||||||||||||||||||||||||||||||||||||||||||||||||||||||||||||||||||||||||||||||||||||||||||||||||||||||||||

Matrix inversion with a 3 by 3 matrix is a tiresome and error-prone calculation when carried out by hand and we shall not add arithmetical complications by enlarging our two-industry economy. The methodology is perfectly general, however, and applies to economies with hundreds or even thousands of industries. Input-output analysis is a thriving offshoot of mathematical economics which has been nurtured by the computational power of modern large-scale computers.

Before turning to the Problem Set, let us review our two-industry input-output analysis methodology and present a final example. We have as given the values for the technological matrix, A, and the values for final demand vector, D. We proceed as follows:

1. Compute $I - A$; that is, subtract the technological matrix from the identity matrix.
2. Compute the inverse, $(I - A)^{-1}$.
3. Write the output (solution) vector as $X = (I - A)^{-1}D$.
4. Compute the components, x_1 and x_2, of the output vector X by matrix multiplication.

Suppose that final demands for the outputs of industries No. 1 and No. 2 are 330 and 550, respectively. One dollar of industry No. 1 output requires $\frac{1}{6}$ dollars of its own output and $\frac{1}{3}$ dollars of industry No. 2 output. Finally, $1 of industry No. 2 output requires $\frac{1}{4}$ dollars of its own output and $\frac{1}{2}$ dollars of industry No. 1 output. The given information is tabulated next.

	User		Final	Total
Producer	No. 1	No. 2	Demand	Output
No. 1......	$\frac{1}{6}$	$\frac{1}{2}$	330	x_1
No. 2......	$\frac{1}{3}$	$\frac{1}{4}$	550	x_2

The technological matrix is

$$A = \begin{pmatrix} \frac{1}{6} & \frac{1}{2} \\ \frac{1}{3} & \frac{1}{4} \end{pmatrix}.$$

Subtracting A from the identity matrix yields

$$(I - A) = \begin{pmatrix} 1-\frac{1}{6} & 0-\frac{1}{2} \\ 0-\frac{1}{3} & 1-\frac{1}{4} \end{pmatrix} = \begin{pmatrix} \frac{5}{6} & -\frac{1}{2} \\ -\frac{1}{3} & \frac{3}{4} \end{pmatrix}.$$

We omit the steps in the calculation of the inverse of $(I - A)$. The exercise following verifies that this is

$$(I - A)^{-1} = \frac{1}{11}\begin{pmatrix} 18 & 12 \\ 8 & 20 \end{pmatrix}.$$

Exercise. Verify that the foregoing is the correct inverse by showing that $(I - A)(I - A)^{-1} = I$. (We omit showing the steps but pause to note that it is wise to check an inverse by this procedure).

Finally, the output vector is

$$X = \frac{1}{11}\begin{pmatrix} 18 & 12 \\ 8 & 20 \end{pmatrix}\begin{pmatrix} 330 \\ 550 \end{pmatrix} = \frac{110}{11}\begin{pmatrix} 18 & 12 \\ 8 & 20 \end{pmatrix}\begin{pmatrix} 3 \\ 5 \end{pmatrix}.$$

Hence,

$$x_1 = 10(54 + 60) = 1140, \quad \text{and} \quad x_2 = 10(24 + 100) = 1240.$$

6.20 PROBLEM SET 6

1. Given the following:

| | User | | | |
| | No. 1 | No. 2 | Final Demand | Total Output |
Producer				
No. 1.......	$\frac{1}{5}$	$\frac{3}{10}$	d_1	x_1
No. 2.......	$\frac{3}{5}$	$\frac{1}{10}$	d_2	x_2

a) What is the technological matrix, A?
b) What does the first column of A mean?
c) What is $(I - A)$?
d) What is $(I - A)^{-1}$?
e) Write the solution equation in matrix-vector form, using the elements found for $(I - A)^{-1}$.
f) Compute the elements x_1 and x_2 of the output vector if final demand is 270 for industry No. 1 and 405 for No. 2.
g) Compute the elements of the output vector if final demand is 540 for industry No. 1 and 810 for No. 2.

2. Each dollar of industry No. 1 output requires $\frac{1}{10}$ dollars of its own output and $\frac{3}{10}$ dollars of industry No. 2 output. Each dollar of industry No. 2 output requires $\frac{3}{5}$ dollars of industry No. 1 output and $\frac{1}{5}$ dollars of its own output. Let d_1, d_2 be final demands and x_1, x_2 total outputs.
a) Compute the matrix which completes the equation

$$\begin{pmatrix} x_1 \\ x_2 \end{pmatrix} = \begin{pmatrix} & \end{pmatrix}\begin{pmatrix} d_1 \\ d_2 \end{pmatrix}.$$

b) Find the elements of the output vector if $d_1 = 81$ and $d_2 = 135$.
c) Find the elements of the output vector if the demands in (b) double.
d) Find the elements of the output vector if $d_1 = 216$ and $d_2 = 171$.

3. Each dollar of industry No. 1 output requires $\frac{1}{6}$ dollars of its own output and $\frac{3}{5}$ dollars of industry No. 2 output. Each dollar of industry No. 2 output requires $\frac{1}{4}$ dollars of its own output and $\frac{3}{4}$ dollars of industry No. 1 output. Find the outputs of industry No. 1 and industry No. 2 if demands for final products are as follows: (Be careful in setting up the technological matrix).
a) $d_1 = 210$ and $d_2 = 420$.
b) $d_1 = 420$ and $d_2 = 630$.

4. Suppose that we have three industries whose total outputs are to be x_1, x_2, and x_3, respectively. Final demands are d_1, d_2, and d_3.
a) What is the meaning of a_{ij} in the general case?
b) What is the meaning of a_{13} in the three-industry case?
c) Write the elements of the technological matrix for the three industries.
d) Write the expanded matrix-vector expression for the output vector. That is, complete the equation

$$\begin{pmatrix} x_1 \\ x_2 \\ x_3 \end{pmatrix} = \begin{pmatrix} & \end{pmatrix}\begin{pmatrix} & \end{pmatrix}.$$

6.21 SUMMATION SYMBOL

The letter Σ (sigma) is the mathematical symbol for summation. The expression

$$\sum_{j=1}^{3} j$$

is read as "sigma j, j going from 1 to 3" and means to insert 1 for j, then 2 for j, then 3 for j, and sum the results. Thus:

$$\sum_{j=1}^{3} j = 1 + 2 + 3 = 6.$$

Similarly:

$$\sum_{j=2}^{5} j = 2 + 3 + 4 + 5 = 14$$

$$\sum_{j=1}^{4} 2j = 2(1) + 2(2) + 2(3) + 2(4) = 20$$

$$\sum_{j=1}^{3} (j - 1) = (1 - 1) + (2 - 1) + (3 - 1) = 3$$

$$\sum_{j=2}^{3} j^3 = 2^3 + 3^3 = 8 + 27 = 35.$$

We shall have frequent occasion for indicating the sum of n terms, without specifying a particular value for n. Consider, for example, the sum of the first n integers. In expanded form, we would indicate this sum by

$$1 + 2 + 3 + \cdots + n$$

where the three dots are read "and so on" and mean that the unwritten terms of the series are formed according to the same rule which applies to the first written terms; that is, in the present case, each number is formed by adding one to the preceding number. In compact summation notation the sum of the first n integers is

$$\sum_{j=1}^{n} j$$

because, by definition, this symbol expands to

$$\sum_{j=1}^{n} j = 1 + 2 + 3 + \cdots + n.$$

6.22 PROBLEM SET 7

Find the numerical values of

1. $\sum_{p=4}^{7} p.$

2. $\sum_{q=1}^{n} q$, if n is 5.

3. $\sum_{u=6}^{9} u^2.$

4. $\sum_{j=1}^{5} 3j.$

5. $\sum_{p=1}^{3} p^3.$

Express in summation notation:

6. The sum of the first q integers.
7. The sum of the squares of the first n integers.
8. $1 + 8 + 27 + 64.$
9. $(1 + 2) + (2 + 2) + (3 + 2) + (4 + 2).$
10. $3(1) + 3(2) + 3(3) + 3(4) + 3(5).$

6.23 SUMMATION ON INDICES: LINEAR EQUATIONS

As a first step toward expressing equations in compact summation form, we abandon the practice of using different letters for different variables and constants. We choose one letter and tag it with different subscripts to indicate different variables, or choose one letter and tag it with subscripts to indicate different constants. For example, x_1, x_2, x_3 can be used to represent three different variables, and c_1, c_2, c_3 to represent three different constants.

Now, consider the symbol

$$\sum_{j=1}^{3} x_j.$$

As before, the statement means to replace j first by 1, then by 2, then by 3, and sum the results. We see that

$$\sum_{j=1}^{3} x_j = x_1 + x_2 + x_3.$$

Similarly:

$$\sum_{j=1}^{5} a_j = a_1 + a_2 + a_3 + a_4 + a_5$$

$$\sum_{j=1}^{n} x_j = x_1 + x_2 + x_3 + \cdots + x_n.$$

It is clear that

$$\sum_{j=1}^{3} a_j x_j = a_1 x_1 + a_2 x_2 + a_3 x_3$$

$$\sum_{j=1}^{n} a_j x_j = a_1 x_1 + a_2 x_2 + \cdots + a_n x_n.$$

Next, consider the expression

$$\sum_{j=1}^{2} a_j x_j = b$$

where the a_j and b are constants. In expanded form, the expression becomes

$$a_1 x_1 + a_2 x_2 = b$$

which is a linear equation in the two variables, x_1 and x_2. It is clear that any linear equation can be written in compact summation notation as

$$\sum_{j=1}^{n} a_j x_j = b$$

because any linear equation is of the form to which the last expression expands; namely:

$$a_1 x_1 + a_2 x_2 + \cdots + a_n x_n = b.$$

For example, the linear equation

$$5x_1 + 2x_2 + 4x_3 = 6$$

is of the stated form with

$$n = 3$$
$$a_1 = 5$$
$$a_2 = 2$$
$$a_3 = a_n = 4$$
$$b = 6.$$

6.24 PROBLEM SET 8

Write the expanded form of:

1. $\sum_{j=1}^{5} y_j.$

2. $\sum_{j=1}^{3} c_j x_j.$

3. $\sum_{j=1}^{n} b_j y_j.$

4. $\sum_{j=1}^{5} p_j x_j = 10.$

5. $\sum_{j=1}^{n} a_j x_j = c.$

Express in compact summation form:

6. $x_1 + x_2 + x_3 + x_4$.
7. $a_1x_1 + a_2x_2 + a_3x_3 + a_4x_4 + a_5x_5 + a_6x_6 = b$.
8. $c_1x_1 + c_2x_2 + \cdots + c_9x_9$.
9. $a_1x_1 + a_2x_2 + \cdots + a_qx_q$.
10. Identify n, b, and each of the a's in

$$\sum_{j=1}^{n} a_j x_j = b$$

with a number in the equation

$$x_1 + 2x_3 + 5x_4 = 7.$$

6.25 SUMMATION FORM FOR SYSTEMS

We have seen how to express a single linear equation in terms of the summation symbol by tagging variables with a subscript. To extend the notation to systems of equations, another subscript is required to identify the different equations of the system. Consider the following expression:

$$\sum_{j=1}^{3} a_{ij}x_j = b_i \qquad i = 1, 2.$$

The symbolism means to insert $i = 1$ first, obtaining

$$\sum_{j=1}^{3} a_{1j}x_j = b_1.$$

The insert $i = 2$ in the original summation, obtaining

$$\sum_{j=1}^{3} a_{2j}x_j = b_2.$$

The expression with $i = 1$ expands to

$$\sum_{j=1}^{3} a_{1j}x_j = a_{11}x_1 + a_{12}x_2 + a_{13}x_3 = b_1.$$

The expression with $i = 2$ expands in the same manner. Thus the original expression

$$\sum_{j=1}^{3} a_{ij}x_j = b_i \qquad i = 1, 2$$

means the following system of two linear equations in three variables:

$$a_{11}x_1 + a_{12}x_2 + a_{13}x_3 = b_1$$
$$a_{21}x_1 + a_{22}x_2 + a_{23}x_3 = b_2.$$

The first number in each double subscript names the equation; the second names the variable. Thus, a_{21} is in the second equation, and it is the coefficient of variable number one, that is, x_1. Similarly, a_{49} would be in the fourth equation, and would be the coefficient of x_9 in that equation.

Exercise.   Write the system of equations specified by

$$\sum_{j=1}^{2} c_{ij} x_j = b_i \qquad i = 1, 2, 3.$$

Answer: The system has three equations, each of the form $c_{i1} x_1 + c_{i2} x_2 = b_i$. The three equations can be generated by letting i equal 1, then 2, then 3.

Consider the linear system

$$a_{11} x_1 + a_{12} x_2 + a_{13} x_3 + a_{14} x_4 = b_1$$
$$a_{21} x_1 + a_{22} x_2 + a_{23} x_3 + a_{24} x_4 = b_2$$
$$a_{31} x_1 + a_{32} x_2 + a_{33} x_3 + a_{34} x_4 = b_3.$$

Note that the beginning subscript on all a's in an equation and the subscript on b are the same as the equation number. For example, the first equation has ones in the positions just described. If we let i represent the equation number, it follows in the present case that i must take on the values 1, 2, and 3 to generate the three equations. Next, letting j be the number indicating the variable, it follows that each equation is a summation with j going from 1 to 4, inclusive. The foregoing system of equations, therefore, is represented by

$$\sum_{j=1}^{4} a_{ij} x_j = b_i \qquad i = 1, 2, 3.$$

The first equation of the system is generated from the last expression by letting i be 1, and then constructing the sum for j going from 1 through 4. The other two equations are generated in the same manner after letting i be 2, then 3.

Exercise.   How would a system of three equations in five variables be symbolized in sigma notation? Answer: The symbols would be the same as those written in the immediately foregoing expression, except the upper limit on j would be 5 rather than 4.

We are now ready to symbolize any system of linear equations. Suppose that we have m linear equations in n variables, an m by n system. To indicate this system in expanded form, we would write

$$a_{11}x_1 + a_{12}x_2 + \cdots + a_{1n}x_n = b_1$$
$$a_{21}x_1 + a_{22}x_2 + \cdots + a_{2n}x_n = b_2$$
$$\vdots \qquad \vdots \qquad\qquad \vdots \qquad \vdots$$
$$a_{m1}x_1 + a_{m2}x_2 + \cdots + a_{mn}x_n = b_m.$$

In compact summation form the system is simply

$$\sum_{j=1}^{n} a_{ij}x_j = b_i \qquad i = 1, 2, \ldots, m.$$

6.26 LINEAR PROGRAMMING PROBLEMS IN SUMMATION NOTATION

The linear programming problems of the last chapter consisted of a system of linear inequalities with nonnegativity constraints, and an objective function which was to be maximized. The nonnegativity constraints can be expressed as

$$x_j \geq 0 \text{ for all } j.$$

The objective function is of the form

$$\theta = c_1x_1 + c_2x_2 + \cdots + c_nx_n$$

where the letter c is adopted to represent the constants in this function. In symbols:

$$\theta = \sum_{j=1}^{n} c_jx_j.$$

The system of linear inequalities can be expressed in the same manner as a system of equalities. Hence the compact summation form of the linear programming problems of the last chapter is

$$\text{Maximize } \sum_{j=1}^{n} c_jx_j$$

$$\text{subject to: } \sum_{j=1}^{n} a_{ij}x_j \leq b_i \qquad i = 1, 2, \ldots, m$$

$$x_j \geq 0 \qquad \text{for all } j.$$

6.27 PROBLEM SET 9

1. Write in expanded form:

$$\sum_{j=1}^{4} a_{ij}x_j = b_i \qquad i = 1, 2.$$

2. Write in expanded form:

$$\text{Maximize } \sum_{j=1}^{3} c_j x_j$$

$$\text{subject to: } \sum_{j=1}^{3} a_{ij}x_j \leq b_i \qquad i = 1, 2, \ldots, 4$$

$$x_j \geq 0 \qquad \text{for all } j.$$

3. A system has p linear equations in q variables.
   a) Write the system in expanded form, using ... where necessary.
   b) Write the system in compact summation form.

4. A linear programming maximization problem has p "less than or equal" constraints on q variables.
   a) Write the statement of the problem in expanded form, using ... where necessary.
   b) Write the problem in compact summation form.

5. Express the following in compact summation notation:

$$a_{11}x_1 + a_{12}x_2 + \cdots + a_{18}x_8 = b_1$$
$$a_{21}x_1 + a_{22}x_2 + \cdots + a_{28}x_8 = b_2$$
$$\vdots \qquad \vdots \qquad \qquad \vdots \qquad \vdots$$
$$a_{61}x_1 + a_{62}x_2 + \cdots + a_{68}x_8 = b_6.$$

6. In the system

$$2x_1 + 3x_2 + 4x_4 = 20$$
$$x_1 + 9x_3 + 8x_4 = 50$$
$$5x_1 + 7x_3 + 6x_4 = 100.$$

   a) What is m? What is n?
   b) What symbol would represent the following constants which appear in the system: 2, 3, 8, 9, 7, 6?
   c) What constants in the system correspond to a_{13}, b_1, a_{31}, a_{22}, a_{32}?

7. We buy x_1 units of item number one at d_1 dollars per unit, x_2 units of item two at d_2 dollars per unit, and so on. Write the expression for the total cost of 10 different items:
   a) In expanded form.
   b) In compact summation notation.

8. Repeat parts (a) and (b) of Problem 7 if we buy p different items.

6.28 APPLICATIONS OF Σ IN STATISTICS

Statistical formulas which describe the computations to be performed on data make extensive use of the summation operation and its symbol. Suppose that we have tabulated sales data for eight weeks, as follows:

Week i	Sales in $ Thousands S_i
1	10
2	12
3	8
4	9
5	14
6	11
7	12
8	12

Total sales for the eight weeks would be

$$\sum_{i=1}^{8} S_i = \$88 \text{ thousand.}$$

Often, the index of summation is not specified and in such cases we simply assume that the summation runs over all the data. Thus, $\Sigma\, S$ means the sum of weekly sales for as many weeks (n) as are entered in the table. Viewed in this tabular light, $\Sigma\, S$ is an instruction to sum the data numbers whose name is S. If the values of S are in a column, the instruction $\Sigma\, S$ means to sum the column headed S.

The arithmetic (pronounced arith-met'ic) average is the ordinary average found by dividing the sum of the data items by the number of items. We find average weekly sales to be

$$\frac{\text{Sum of weekly sales}}{\text{Number of weeks}} = \frac{88}{8} = \$11 \text{ thousand.}$$

In general, if we have n items of sales data, S, we may denote the arithmetic average as $\overline{S}$ (read S bar) and compute it as

$$\overline{S} = \frac{\sum_{i=1}^{n} S_i}{n} \quad \text{or, more briefly,} \quad \overline{S} = \frac{\sum S}{n}.$$

Exercise. If your grades on four examinations are 70, 85, 95, 80 and g represents grades, what formula denotes your average grade? What is n? Compute this average.

Answer: $\overline{g} = (\Sigma\, g)/n$; $n = 4$; $\overline{g} = 330/4 = 82.5$.

Squares and products of variables play an important role in statistics. Common forms are:

$$\Sigma\, x^2 = \text{The sum of the squares of the values of } x.$$
$$(\Sigma\, x)^2 = \text{The square of the sum of the } x\text{'s.}$$
$$\Sigma\, xy = \text{The sum of the } xy \text{ products.}$$
$$(\Sigma\, x)(\Sigma\, y) = \Sigma\, x\, \Sigma\, y = \text{The product of the sum of the } x\text{'s times the sum of the } y\text{'s.}$$

To illustrate, suppose that x represents sales of a company during a given month and y represents expenses during the same month. Table 6–6 shows sales-expense data for four months to the left of the double divider. To the right of the double divider we have computed extensions headed x^2, y^2, and xy. The column headed x^2 contains the squares of the respective values of x, similarly for the column headed y^2. The xy column contains the xy products for each data pair. The number of data pairs (which is the number of rows in the table) is the number n and here $n = 4$.

TABLE 6–6

x	y	x^2	y^2	xy
5	3	25	9	15
10	5	100	25	50
6	4	36	16	24
8	4	64	16	32
29	16	225	66	121

We see that $\Sigma\, x^2 = 225$, $\Sigma\, y^2 = 66$, and $\Sigma\, xy = 121$. Note carefully, however, that

$$(\Sigma\, x)^2 = (29)^2 = 841.$$

Exercise. Find $(\Sigma\, y)^2$ and $\Sigma\, y\, \Sigma\, x$ from Table 6–1.
Answer: $(\Sigma\, y)^2 = (16)^2 = 256$, $\Sigma\, x\, \Sigma\, y = (29)(16) = 464$.

Now let us show the utility of the summation notation in describing statistical calculations. First, we plot the sales-expense data of Table 6–6 as points. See Figure 6–1. It is clear that the points do not all lie on the same straight line, but we could sketch a line freehand which would come close to the points, and such a line would describe the approximate relationship between expenses and sales and serve as a *model* from which expenses at various sales levels could be estimated. However, if we use a freehand sketch, each of us might obtain a somewhat different line and debate as to whether

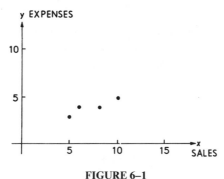

FIGURE 6–1

one line is a better *fit* of the data than another. We appeal to statistical theory to settle the issue. According to this theory, the line which best fits the points (according to the least squares criterion) has an equation of the form

$$y = mx + b$$

where the slope, *m*, and the intercept, *b*, are computed from the data by means of the formulas

$$m = \frac{n \Sigma\, xy - \Sigma\, x \Sigma\, y}{n \Sigma\, x^2 - (\Sigma\, x)^2},$$

$$b = \frac{\Sigma\, y - m \Sigma\, x}{n}.$$

From Table 6–6, we have that

$$\begin{aligned} n &= 4 & \Sigma\, xy &= 121 \\ \Sigma\, x &= 29 & \Sigma\, x \Sigma\, y &= 464 \\ \Sigma\, y &= 16 & \Sigma\, x^2 &= 225 \\ & & (\Sigma\, x)^2 &= 841. \end{aligned}$$

Substituting, we find the slope of the best fitting line to be

$$\frac{4(121) - (29)(16)}{4(225) - (29)^2} = \frac{484 - 464}{900 - 841} = \frac{20}{59} = 0.34.$$

With this value for *m*, we find the *y*-intercept to be

$$b = \frac{16 - (0.34)(29)}{4} = 1.54.$$

The desired equation then is

$$y = 0.34x + 1.54.$$

The last example illustrates the fact that an understanding of the meaning of the summation symbol makes the techniques of statistical analysis available to the non-statistician.

Exercise. If we let $x = 0$ and $x = 10$ in the last equation, the corresponding values of y are 1.54 and 4.94, respectively. On Figure 6–1, plot lightly, with pencil, the points (0, 1.54) and (10, 4.94) and verify that the line joining them does indeed come close to the points.

6.29 PROPERTIES OF THE Σ OPERATION

An elementary, but important, property of summation is illustrated by

$$\sum_{i=1}^{3} 5 = 5 + 5 + 5 = 15.$$

We see here that the quantity being summed is the *constant*, 5, when $i = 1$, $i = 2$, and $i = 3$. We could write more directly

$$\sum_{i=1}^{3} 5 = 3(5) = 15.$$

Exercise. Compute the value of $\sum_{i=1}^{8} 2$. Answer: $8(2) = 16$.

In general, if c is any *constant*, then

$$\sum_{i=1}^{n} c = nc.$$

The word *constant* here means that the expression is independent of the index of summation.

Exercise. What is $\sum_{j=1}^{m} p$? Answer: mp.

The point of the last exercise is that the expression p is constant with respect to the index of summation.

A second property of the summation operation is that it may be applied term by term to an expression.

To illustrate,

$$\sum_{i=1}^{3}(x_i + y_i) = x_1 + y_1 + x_2 + y_2 + x_3 + y_3$$

$$= x_1 + x_2 + x_3 + y_1 + y_2 + y_3$$

$$\sum_{i=1}^{3}(x_i + y_i) = \sum_{i=1}^{3}x_i + \sum_{i=1}^{3}y_i.$$

The property obviously applies whatever the value of n; that is,

$$\sum_{i=1}^{n}(x_i + y_i - z_i) = \sum_{i=1}^{n}x_i + \sum_{i=1}^{n}y_i - \sum_{i=1}^{n}z_i.$$

Finally, a constant factor of the quantity being summed can be taken outside the summation symbol. Thus,

$$\Sigma\, 3x_i = 3\,\Sigma\, x_i \text{ and, in general, } \Sigma\, cx_i = c\,\Sigma\, x_i$$

where c is a constant with respect to the index of summation.

Summarizing the properties just discussed, we have

1. $\displaystyle\sum_{i=1}^{n}cx_i = c\sum_{i=1}^{n}x_i;$ or $\Sigma\, cx_i = c\,\Sigma\, x_i.$

2. $\displaystyle\sum_{i=1}^{n}c = nc;$ or $\Sigma\, c = nc,$ assuming c is constant with respect to the index which goes from 1 to n.

3. $\displaystyle\sum_{i=1}^{n}(x_i + y_i - z_i) = \sum_{i=1}^{n}x_i + \sum_{i=1}^{n}y_i - \sum_{i=1}^{n}z_i,$

or

$$\Sigma\,(x_i + y_i - z_i) = \Sigma\, x_i + \Sigma\, y_i - \Sigma\, z_i.$$

For practice with these properties, let us compute

$$\Sigma\,(4x + 2)$$

for the following data: $x = 1, 2,$ and 3. We may write

$$\Sigma\,(4x + 2) = \Sigma\, 4x + \Sigma\, 2 = 4\,\Sigma\, x + 3(2)$$

where, in the last term, the factor 3 is the value of n for the three data items. Now, $\Sigma\, x$ is $1 + 2 + 3 = 6$, so

$$4\,\Sigma\, x + 3(2) = 4(6) + 6 = 30.$$

Alternatively, we could have performed the calculation as shown in **Table 6–7**,

TABLE 6–7

x	$4x + 2$
1	6
2	10
3	14
	30

which shows that $\Sigma\,(4x + 2) = 30$.

Exercise.   Given the data y: 5, 7, 2, 9, compute $\Sigma\,(2y - 1)$ by the two procedures just illustrated. Answer: 42.

Continuing, let us find an alternate expression for

$$\Sigma(x - \bar{x})^2.$$

The expression implies that we have a set of n values for the variable x, and we have computed the average of these n values, $\bar{x}$. Hence, the values of x vary, but $\bar{x}$ is a constant. Squaring the expression leads to

$$\Sigma(x - \bar{x})^2 = \Sigma(x^2 - 2x\bar{x} + \bar{x}^2) = \Sigma\,x^2 - \Sigma\,2x\bar{x} + \Sigma\,\bar{x}^2.$$

Inasmuch as $\bar{x}$ is a constant, we may write the last as

$$\Sigma\,x^2 - 2\bar{x}\Sigma\,x + n\bar{x}^2.$$

Substituting from the definition of $\bar{x}$,

$$\bar{x} = \frac{\Sigma\,x}{n},$$

we find

$$\Sigma x^2 - 2\,\frac{\Sigma x}{n}\left(\Sigma x\right) + n\frac{\left(\Sigma x\right)^2}{n^2} = \Sigma x^2 - \frac{\left(\Sigma x\right)^2}{n}.$$

To illustrate what has just been proved, consider the data $x = 4, 6, 11$. Here, $\bar{x} = 21/3 = 7$, and $n = 3$. Direct calculation of $\Sigma(x - \bar{x})^2$ is shown in Table 6–8.

TABLE 6–8

x	$(x - \bar{x})$	$(x - \bar{x})^2$
4	-3	9
6	-1	1
11	4	16
$\bar{x} = 7$		26

We find that by direct calculation, $\Sigma(x - \bar{x})^2 = 26$. The alternate computation requires only Σx^2 and $(\Sigma x)^2$, as shown next.

x	x^2
4	16
6	36
11	121
21	173

$$\sum x^2 - \frac{\left(\sum x\right)^2}{n} = 173 - \frac{(21)^2}{3} = 173 - 147 = 26.$$

Both the direct and the alternate method of computation lead to the same result, 26. In this example, as in others which arise in statistics, the advantage of the alternate method of computation is that it is simpler than the direct calculation.

As an alternative way of providing practice in the use of Σ properties, let us specify that

$$\sum_{i=1}^{n}(2x_i - 3)$$

is to be expressed in terms of n and $\bar{x}$. Expanding, we have $\Sigma(2x_i - 3)$ $= \Sigma(2x_i) - \Sigma(3) = 2\Sigma x_i - 3n = 2n\bar{x} - 3n$, as specified. Observe the substitution of $n\bar{x}$ for Σx_i to obtain the result in the specified form. As another example, let us express

$$\sum_{i=1}^{n} x_i(y_i - \bar{y})$$

in terms of $\Sigma x_i y_i$, n, Σx_i, and Σy_i. Expanding, we find

$$\sum x_i y_i - \sum y_i \bar{y} = \sum x_i y_i - \bar{y}\sum x_i = \sum x_i y_i - \frac{\left(\sum y_i\right)\left(\sum x_i\right)}{n}$$

as specified.

Exercise. The average of the x_i's is $\bar{x}$. If a is added to each x_i, the new average will be $\Sigma(x_i + a)/n$. Express the new average in terms of the old average and a. Interpret the outcome. Answer: The new average is $\bar{x} + a$, which means that if the constant a is added to each x_i, the average will be increased by a.

6.30 THE VARIANCE AND STANDARD DEVIATION

Conventionally in statistics, $\bar{x}$ represents the average of a *sample* of items from an overall group called the *population*. The symbol μ (mu) is used to represent the average of the population. Thus, if 10 students in a class of 30 averaged 75 percent on a test and the class average was 80 percent, we would use $\bar{x} = 75$ for the sample of 10 students, but $\mu = 80$ percent for the population of 30 students. Similarly, the measures we shall discuss in this section are represented by s^2 and s in samples and σ and σ^2 for populations, where σ is the small Greek letter sigma. Our reason for using σ and μ in this section is that we plan to develop in the next section an important result concerning the variation of the numbers in a *population* about the population average.

The *variance* of a population of n data items, symbolized by σ^2, is defined as

$$\sigma^2 = \frac{\sum_{i=1}^{n} (x_i - \mu)^2}{n}.$$

The standard deviation of the population, σ, is the square root of the variance. To use this formula for a given set of x_i's, we must first find the average, $\mu = \Sigma x_i / n$, in the usual manner as shown at the foot of the first column of the next table.

x_i	$x_i - \mu$	$(x_i - \mu)^2$
4	-5	25
13	4	16
10	1	1
$\Sigma x_i = 27$		$42 = \Sigma(x_i - \mu)^2$
$\mu = 27/3 = 9.$		

Then, in the second column, we compute the deviation of each x_i from the average, $\mu = 9$. Squaring the second column entries leads to the third column entries. The variance is the average of the third column,

$$\sigma^2 = \frac{42}{3} = 14.$$

The standard deviation is $\sigma = \sqrt{14} = 3.74$.

Exercise. Compute the variance and standard deviation of the two-item population 4, 10. Answer: Variance, σ^2, is 9; standard deviation, σ, is 3.

The standard deviation is a measure of variability. Admittedly, this measure has little intuitive appeal. However, it has important practical applications which we shall see in the next section, and in a later chapter on probability.

6.31 TCHEBYSHEFF'S THEOREM

To set the stage for this section, suppose that a college student learns that the 30 students who graduated last year with a major in a certain subject area are receiving annual salaries averaging $25 thousand, with a standard deviation of $3 thousand. By itself, the average salary seems quite impressive, but the student realizes that very different groups of numbers could lead to this average. For example, the majority of salaries could be considerably below $25 thousand, with a relatively small group of very high salaries pulling the average up; or the salaries may spread more or less evenly over a wide interval; or salaries may cluster very closely around $25 thousand. Clearly, these circumstances have different implications to the student considering the subject area as a major, and it would be helpful to know whether or not the individual salaries cluster closely about the average. The standard deviation is a measure of variability around the average, and our objective is to show how to use the standard deviation to make specific statements, such as, "at least 75 percent of all the salaries are in the interval $19 to $31 thousand," based upon the given average and standard deviation. The method is:

a) Arbitrarily pick any number greater than or equal to 1 and call this number k. For example, suppose $k = 2$.

b) Form the interval $\mu \pm k$(standard deviation). In our example, this is $25 \pm 2($3) thousand, or $19 to $31 thousand.

c) Compute the fraction $1 - 1/k^2$. For our example, $1 - 1/2^2 = 3/4$.

d) Make the statement, "At least $1 - 1/k^2$ of the numbers in the population are in the interval $\mu \pm k$(standard deviation)." Our result is "At least 3/4 of the salaries are in the interval $19 to $31 thousand."

If we had started with the multiple $k = 3$ in step a), the interval would be b) $25 \pm 3($3) or $16 to $34 thousand. In c) we would find $1 - 1/k^2 = 1 - 1/3^2 = 1 - 1/9$ or 8/9. The statement d) would be "At least 8/9 of the numbers in the population are in the interval $16 to $34 thousand."

‖‖

Exercise. If we start with the multiplier $k = 2.5$ in a), what would be the resulting statement in d)? Answer: "At least 0.84 or 84 percent of the salaries are in the interval $17,500 to $32,500."

‖‖

The average and the standard deviation play important roles in a world where summarization is needed to aid in comprehension of large masses of data. The procedure we have just illustrated is important because it applies to any population of numerical data. The procedure is justified by the theorem we shall now prove.

Tchebysheff's Theorem: At least $1 - 1/k^2$ of the data items in a population are in the inclusive interval which extends k standard deviations on either side of the population mean, μ.

We start the proof by writing the definition of the variance,

$$\sigma^2 = \frac{\sum\limits_{i=1}^{n}(x_i - \mu)^2}{n} \tag{1}$$

Now if we pick an interval extending $k\sigma$ on either side of the average, μ, a data item either is in this interval or it is outside the interval. Note in particular that for an item outside the interval, the deviation $(x_i - \mu)$ is larger absolutely than $k\sigma$, so for such an item, $(x_i - \mu)^2 > (k\sigma)^2$. Now suppose the number of items inside the interval is m, so the remaining $n - m$ are outside. Let us separate the two groups by writing (1) in two parts. Thus,

$$\sigma^2 = \frac{\sum\limits_{i=1}^{m}(x_i - \mu)^2 + \sum\limits_{i=m+1}^{n}(x_i - \mu)^2}{n} \tag{2}$$

Observe that all the individual terms in the sums in the numerator are squares, so none can be negative. Consequently, if we discard the first sum of squares in the numerator, we will have a smaller value on the right (or the same value if the discarded amount is zero). It follows that

$$\sigma^2 \geq \frac{\sum\limits_{i=m+1}^{n}(x_i - \mu)^2}{n} \tag{3}$$

Now remember that there are $n - m$ terms in the numerator sum in (3), and for each of the x_i,

$$(x_i - \mu)^2 > (k\sigma)^2$$

because we are using only items which are more than $k\sigma$ from μ. Consequently, if we replace each $(x_i - \mu)^2$ in (3) by $(k\sigma)^2$ we are using smaller values at the right, and it remains true that

$$\sigma^2 \geq \frac{\sum\limits_{i=m+1}^{n}(k\sigma)^2}{n}.$$

Inasmuch as $(k\sigma)^2$ is a constant, and we are summing $n - m$ of these constants, we have

$$\sigma^2 \geq \frac{(n - m)(k\sigma)^2}{n}.$$

Cancelling the σ^2, and writing $(n - m)/n$ as $1 - m/n$, we have

$$1 \geq \left(1 - \frac{m}{n}\right)k^2 \quad \text{or} \quad \frac{1}{k^2} \geq 1 - \frac{m}{n},$$

from which

$$\frac{m}{n} \geq 1 - \frac{1}{k^2}.$$

Now, since m is the number of items in the interval and n is the total number of items, m/n is the proportion of items in the interval. Hence, the last statement says that the proportion in the inclusive interval $\mu \pm k\sigma$ is at least (greater than or equal to) $1 - 1/k^2$, which proves the theorem.

It is often stated that an average is misleading because it is not *typical* or *representative* as illustrated by the comment, Sure, the average salary is $25,000, but that does not mean much to me because it is not representative. What is implied here, at least in part, is that salaries vary so greatly that the average is not a reliable indicator for an individual. The moral of this section is that the standard deviation provides a means of assessing representativeness, and the theorem just proved shows a way of using the standard deviation to this end which applies to any population of data.

To close this section let us practice the use of the theorem by stating an example in a form slightly different from that used in the introductory examples.

Example. If the average and standard deviation of monthly salaries paid in a certain job category are $1200 and $100, respectively, what can we say about the proportion of workers earning between $800 and $1600, inclusive? Here we note that the interval $\mu \pm k\sigma$ is $1200 \pm 4(100)$, so $k = 4$. Hence, $1 - 1/k^2 = 1 - 1/16 = 15/16$ or 0.9375, so that at least 93.75 percent of all workers earn between $800 and $1600, inclusive.

Exercise. From the data of the foregoing example, what can we say about the proportion of workers earning between $1000 and $1400, inclusive? *Answer:* At least 75 percent have salaries in the interval.

6.32 PROBLEM SET 10

1. Given the values $x = 2, 3,$ and $7,$ compute the following:
   a) $\Sigma x.$ b) $\Sigma x^2.$ c) $\bar{x}.$ d) $\Sigma(x - \bar{x})^2.$
   e) The square of the average of the x's.
   f) The average of the squares of the x's.
2. Given the values $y = 5, 7, 12,$ compute the following:
   a) $\Sigma y.$ b) $\Sigma y^2.$ c) $\bar{y}.$ d) $\Sigma(y - \bar{y})^2.$
   e) The square of the average of the y's.
   f) The average of the squares of the y's.
3. Compute the following from the tabulated values of x and y:
   a) $\Sigma xy.$ b) $\Sigma x \Sigma y.$ c) $\Sigma x^2.$
   d) $(\Sigma x)^2.$ e) $\Sigma(x - \bar{x})(y - \bar{y}).$

x	y
4	7
5	8
9	10
2	3

4. Compute the following from the tabulated values of x and y.
   a) Σxy.
   b) $\Sigma x \Sigma y$.
   c) Σx^2.
   d) $(\Sigma x)^2$.
   e) $\Sigma(x - \bar{x})(y - \bar{y})$.

x	y
1	2
3	4
5	6

5. a) Plot the following points on a graph.

x	y
3	1
4	3
5	6
6	6
7	9

   b) Find the equation of the best fitting line, $y = mx + b$, where

$$m = \frac{n \Sigma xy - \Sigma x \Sigma y}{n \Sigma x^2 - (\Sigma x)^2}, \quad \text{and} \quad b = \frac{\Sigma y - m \Sigma x}{n}.$$

   c) Sketch the line found in (b) on the graph in (a).

6. Carry out the instructions in 5(a), 5(b), and 5(c) for the data

x	y
3	4
12	16
6	8
9	12

7. The *coefficient of linear correlation* is a measure of the degree of relationship between two variables. It is designated as r, and its value is always in the interval -1 to $+1$, inclusive. Values near -1 or $+1$ indicate high correlation, and values near zero indicate low correlation. The coefficient is computed from the formula

$$r = \frac{n \Sigma xy - \Sigma x \Sigma y}{\sqrt{[n \Sigma x^2 - (\Sigma x)^2][n \Sigma y^2 - (\Sigma y)^2]}}$$

The sign of the square root is $+$. The slope, m, of the best fitting line is positive when r is positive, and negative when r is negative. That is, r is positive when y tends to increase as x increases and negative when y tends to decrease as x increases.

If the observed points, (x, y), all lie on a straight line rising to the right (m positive), r will be 1. Verify this statement by computing r for the following points, all of which lie on the line $y = 2x + 5$.

x	y
0	5
2	9
3	11

8. (See Problem 7). Compute the correlation coefficient from the following observed values of x and y.

x	y
-2	0
-1	0
0	1
1	1
2	3

Compute the average, variance and standard deviation.

9. x_i: 19, 23, 17, 21, 20.
10. x_i: 10, 15, 6, 12, 7, 10.
11. x_i: 8, 9, 11, 1, 2, 3, 4, 8, 8.
12. Given the values $x = 3, 4, 8$,
   a) Show that $\Sigma(x - \bar{x}) = 0$.
   b) Prove that for *any* group of data, the sum of the deviations of the data items from their average is 0. That is, prove

$$\sum_{i=1}^{n}(x_i - \bar{x}) = 0.$$

13. a) Prove that

$$\sum_{i=1}^{n}(x_i - 5)^2 = \sum_{i=1}^{n}x_i^2 - 10n\bar{x} + 25n.$$

   b) Illustrate (a) with the data $x = 5, 4, 9$.
14. a) If $\bar{x}$ is the average of the x_i, express the following in terms of $\bar{x}$, n, and b, where b is a constant.

$$\frac{\sum_{i=1}^{n}(x_i - b)}{n}.$$

   b) If we have 1000 numbers whose average is 60 and we subtract 7 from each of these numbers, what will be average of the resultant group of 1000 numbers?
15. Express in terms of Σx_i^2, $\bar{x}$, and n:

$$\sum_{i=1}^{n}(x_i + 5)(x_i - 3).$$

16. A professor reports that the average grade on a test was 72 percent, with a standard deviation of 3 percentage points. Using the multiple $k = 5$ in the Tchebysheff procedure, what statement can be made about the class grades?
17. The average of the straight-time hourly wages of a group of workers is $5 with a standard deviation of $0.40. What statement can be made about the proportion of workers who earn between $4 and $6 per hour, inclusive?

6.33 REVIEW PROBLEMS

Perform the following operations:

1. $(1\ 4\ 6) + (3\ -1\ 0) - 5(3\ 5\ -4)$.

2. $\begin{pmatrix} 8 \\ 2 \end{pmatrix} - 3\begin{pmatrix} 4 \\ 6 \end{pmatrix} + 2\begin{pmatrix} 1 \\ 5 \end{pmatrix}$.

3. If **p** and **q** are two-component column vectors, express the compact vector statement $\mathbf{p}x + \mathbf{q}y = \mathbf{0}$
   a) In expanded vector form.
   b) In usual algebraic form.

4. If **x**, **y**, and **r** are three-component row vectors, express the compact vector statement $\mathbf{x} + \mathbf{y} = \mathbf{r}$
   a) In expanded vector form.
   b) In usual algebraic form.

5. If **x** and **0** are four-component column vectors, express the compact vector statement $\mathbf{x} \geq \mathbf{0}$
   a) In expanded vector form.
   b) In usual algebraic form.

6. $\begin{pmatrix} 3 & 2 & 0 \\ 1 & 5 & 4 \end{pmatrix} + \begin{pmatrix} 1 & -3 & 6 \\ 2 & 4 & 5 \end{pmatrix} - 2\begin{pmatrix} 1 & 1 & 3 \\ 4 & 0 & -1 \end{pmatrix}$.

7. $\begin{pmatrix} 3 & -2 \\ 1 & 7 \end{pmatrix} - \begin{pmatrix} -3 & 4 \\ 6 & 0 \end{pmatrix} - 3\begin{pmatrix} 1 & 2 \\ 1 & 3 \end{pmatrix} + 5\begin{pmatrix} 1 & -1 \\ -2 & 2 \end{pmatrix}$.

8. $(1, 0, 2)\begin{pmatrix} 2 & -1 & 3 & 5 \\ 0 & 2 & -2 & 1 \\ 1 & 4 & 2 & 3 \end{pmatrix}$.

9. $\begin{pmatrix} 1 & 3 & 2 & 2 \\ 2 & 0 & 1 & -1 \\ 0 & 1 & 2 & 3 \end{pmatrix}\begin{pmatrix} 3 & 4 \\ 2 & 0 \\ -1 & 5 \end{pmatrix}$.

10. $\begin{pmatrix} 2 & 3 \\ 1 & -1 \end{pmatrix}\begin{pmatrix} 1 & 3 & 5 \\ 2 & 4 & 6 \end{pmatrix}$.

11. $\begin{pmatrix} 4 & 1 & -1 \\ 2 & 0 & 3 \\ 0 & -2 & 4 \end{pmatrix}\begin{pmatrix} 0 & 1 & -1 \\ 1 & -1 & 0 \\ -1 & 0 & 1 \end{pmatrix}$.

12. The numbers of trees, bushes, and shrubs used in landscaping small, medium, and large lots are shown as the requirements matrix in the table. Costs per unit are shown at the left.

		Requirements Matrix Number Needed to Landscape a		
Unit Cost	Item	Small Lot	Medium Lot	Large Lot
$30	Red maple	0	1	2
20	Hard maple	1	1	1
20	Yew	2	4	5
40	Arborvitae.	0	2	2
50	Spruce	0	2	3
20	Rhododendron	2	3	6
10	Laurel	2	2	4
20	Azalea	2	5	8

a) A contractor orders plantings for six small, 10 medium, and five large lots. Multiply the planting vector and the requirements matrix and interpret the results.

b) Multiply the cost vector and the requirements matrix and interpret the results.

Write the following in expanded matrix form:

13. $\begin{aligned} x_1 + 2x_2 \qquad\quad + \quad x_4 &= 12 \\ 3x_2 - x_3 + 2x_4 &= 6 \\ 5x_1 - \quad x_2 + x_3 - \quad x_4 &= 7. \end{aligned}$

14. $\begin{aligned} x_1 + \quad 2x_2 &\le 18 \\ 2x_1 + \quad 3x_2 &\le 24 \\ -x_1 \qquad\qquad &\le 0 \\ -x_2 &\le 0. \end{aligned}$

15. $\begin{aligned} x_1 + 2x_2 + s_1 &= 10 \\ x_1 + 3x_2 + s_2 &= 15. \end{aligned}$

Write the following in usual algebraic form:

16. $\begin{pmatrix} 1 & 2 \\ 3 & 4 \\ 5 & 5 \end{pmatrix} \begin{pmatrix} x_1 \\ x_2 \end{pmatrix} = \begin{pmatrix} 7 \\ 8 \\ 9 \end{pmatrix}.$

17. $\begin{pmatrix} 3 & 0 & 1 \\ 0 & 2 & 4 \\ 2 & 3 & 0 \end{pmatrix} \begin{pmatrix} x_1 \\ x_2 \\ x_3 \end{pmatrix} \le \begin{pmatrix} 14 \\ 10 \\ 12 \end{pmatrix}.$

18. $\begin{pmatrix} 3 & 1 \\ 2 & 2 \\ 1 & 4 \end{pmatrix} \begin{pmatrix} x_1 \\ x_2 \end{pmatrix} + \begin{pmatrix} s_1 \\ s_2 \\ s_3 \end{pmatrix} = \begin{pmatrix} 20 \\ 30 \\ 40 \end{pmatrix}.$

19. Maximize $(2 \ 3) \begin{pmatrix} x_1 \\ x_2 \end{pmatrix}$ subject to $\begin{pmatrix} 1 & 1 \\ 2 & 1 \\ 1 & 2 \end{pmatrix} \begin{pmatrix} x_1 \\ x_2 \end{pmatrix} \le \begin{pmatrix} 10 \\ 15 \\ 20 \end{pmatrix}$ and $x \ge 0$.

20. Minimize $(1 \ 2 \ 3 \ 4) \begin{pmatrix} x_1 \\ x_2 \\ x_3 \\ x_4 \end{pmatrix}$ subject to $\begin{pmatrix} 1 & 0 & 2 & 3 \\ 2 & 1 & 0 & 4 \end{pmatrix} \begin{pmatrix} x_1 \\ x_2 \\ x_3 \\ x_4 \end{pmatrix} \ge \begin{pmatrix} 25 \\ 30 \end{pmatrix}$ and $x \ge 0$.

21. In a section of the country, 35 percent live in urban areas and 65 percent in rural areas at a point in time. Urban-rural movement each year is described by the transition matrix

$$\begin{array}{cc} & \begin{array}{cc} \text{Urban} & \text{Rural} \end{array} \\ \begin{array}{c} \text{Urban} \\ \text{Rural} \end{array} & \begin{pmatrix} 0.8 & 0.2 \\ 0.4 & 0.6 \end{pmatrix}. \end{array}$$

What will be the proportions in urban and rural areas after

a) One year?   b) Two years?   c) Three years?

d) What is the steady state?

22. Brand X has 25 percent of the market and other brands share the rest of the market. Because of a promotional effort, 50 percent of those buying other brands shift to Brand X each month, while 70 percent of those buying Brand X continue to buy this brand.

a) What percent of the market will Brand X have after one month?

b) What is the steady state?

Find the inverse of each of the following matrices (if an inverse exists):

23. $\begin{pmatrix} 0 & 1 \\ 1 & 0 \end{pmatrix}$.

24. $\begin{pmatrix} 3 & 1 \\ 0 & 2 \end{pmatrix}$.

25. $\begin{pmatrix} 1 & 2 \\ 3 & 3 \end{pmatrix}$.

26. $\begin{pmatrix} 1 & 1 \\ 2 & 1 \end{pmatrix}$.

27. $\begin{pmatrix} 5 & 3 \\ 2 & 2 \end{pmatrix}$.

28. $\begin{pmatrix} 1 & 2 \\ 3 & 4 \end{pmatrix}$.

29. $\begin{pmatrix} 1 & 0 & 1 \\ 0 & 1 & 2 \\ 2 & 3 & 0 \end{pmatrix}$.

30. $\begin{pmatrix} 1 & 2 & -1 \\ 1 & 0 & 1 \\ 0 & 5 & -5 \end{pmatrix}$.

31. $\begin{pmatrix} 1 & 0 & 2 \\ 0 & 1 & 3 \\ 1 & 2 & 0 \end{pmatrix}$.

32. $\begin{pmatrix} 0 & 0 & 1 \\ 0 & 1 & 0 \\ 1 & 0 & 0 \end{pmatrix}$.

33. $\begin{pmatrix} 60 & 30 & 20 \\ 30 & 20 & 15 \\ 20 & 15 & 12 \end{pmatrix}$.

34. $\begin{pmatrix} 1 & 1 & 1 \\ 2 & 1 & 1 \\ 1 & 2 & 2 \end{pmatrix}$.

35. $\begin{pmatrix} 1 & 2 & 1 \\ 2 & 1 & 1 \\ 1 & 2 & 2 \end{pmatrix}$.

36. $\begin{pmatrix} 2 & 4 & 1 \\ 3 & 1 & 2 \\ 0 & 5 & 6 \end{pmatrix}$.

37. $\begin{pmatrix} 2 & 0 & 1 \\ 3 & 1 & 3 \\ 0 & 1 & 4 \end{pmatrix}$.

38. Consider the system

$$7x_1 + 11x_2 = 3$$
$$5x_1 + 8x_2 = 5.$$

a) Relating the system to $Ax = b$, what is A?, x?, b?
b) Compute A^{-1}.
c) Write $x = A^{-1}b$ in expanded matrix form.
d) Compute the solution from part (c).
e) What would be the solution vector if the elements in the vector of constants, 3 and 5, were changed to
   (1) 1, 1.   (2) 0, 1.   (3) 1, 0.   (4) 2, −3.

39. Answer parts (a) through (d) of Problem 38 for the system

$$2x_1 + 4x_2 = 4$$
$$5x_1 + 6x_2 = 8.$$

e) What would be the solution vector if the elements in the vector of constants, 4 and 8, were changed to
   (1) 4, 12.   (2) 12, 16.   (3) 20, 40.   (4) 0, 4.

40. Answer parts (a) through (d) of Problem 38 for the system

$$3x_1 \qquad + 2x_3 = 1$$
$$5x_1 + 2x_2 + 5x_3 = 2$$
$$x_2 + x_3 = 3.$$

e) What would be the solution vector if the elements in the vector of constants, 1, 2, and 3 were changed to
   (1) 1, 1, 1.   (2) 1, 0, 1.   (3) 1, 1, 0.

41. Answer parts (a) through (d) of Problem 38 for the system

$$2x_1 + 3x_2 - 15x_3 = 3$$
$$-5x_1 - 7x_2 + 35x_3 = 2$$
$$3x_1 + 4x_2 - 21x_3 = 1.$$

e) What would be the solution vector if the elements in the vector of constants, 3, 2, and 1, were changed to

(1) 1, 1, 1.   (2) 0, 1, 1.   (3) 1, 1, 0.

42. The table shows that x_1 units of product X_1 are to be made and that a unit of X_1 requires 1 hour of time on machine M_1, 3 hours on M_2, and 2 hours on M_3. Machine M_1 has H_1 hours of available time. Similar interpretations apply to the other entries.

Product	Units To Be Made	Hours of Machine Time per Unit of Product on		
		M_1	M_2	M_3
X_1 . . . . x_1		1	3	2
X_2 . . . . x_2		2	1	3
X_3 . . . . x_3		3	2	1
Total Machine Hours Available		H_1	H_2	H_3

Find the numbers of units which can be made if the machine time availability vector $(H_1\ H_2\ H_3)$ is

a) (180 540 360).   b) (180 180 180).   c) (270 360 360).

43. Given the system

$$3x_1 + 2x_2 = b_1$$
$$4x_1 + 3x_2 = b_2$$

find the missing elements to complete the following equation:

$$\begin{pmatrix} x_1 \\ x_2 \end{pmatrix} = \begin{pmatrix} & \\ & \end{pmatrix}\begin{pmatrix} b_1 \\ b_2 \end{pmatrix}.$$

44. Given the system

$$2x_1 + x_2 + x_3 = b_1$$
$$x_1 + x_2 + x_3 = b_2$$
$$x_1 + 2x_2 + x_3 = b_3$$

find the missing elements to complete the following equation:

$$\begin{pmatrix} x_1 \\ x_2 \\ x_3 \end{pmatrix} = \begin{pmatrix} & & \\ & & \\ & & \end{pmatrix}\begin{pmatrix} b_1 \\ b_2 \\ b_3 \end{pmatrix}.$$

45. Consider the system

$$x_1 + x_2 + x_3 = 3$$
$$2x_1 + 3x_2 + 2x_3 = 5$$
$$x_1 + 2x_2 + x_3 = 2.$$

a) Show that the matrix of coefficients does not have an inverse.
b) Find the solution of the system by the methods of Chapter 2.

c) Change the constant in the first equation from 3 to 4.
   (1) Does the change in the constant have any effect upon the matrix of coefficients?
   (2) Does the change have any effect upon the solution in part (b)?
   (3) Our example illustrates what it means if the matrix of coefficients in an n by n system has no inverse. What does it mean if the coefficient matrix has no inverse?

46. a) Write the technological matrix from the following input-output data.

Producer	User		Final Demand	Total Output
	No. 1	No. 2		
No. 1	$\frac{1}{5}$	$\frac{3}{10}$	d_1	x_1
No. 2	$\frac{1}{5}$	$\frac{1}{10}$	d_2	x_2

b) What does the second column of the technological matrix mean?
c) What is $(I - A)$?   d) What is $(I - A)^{-1}$?
e) Write the equation for the output vector X in matrix-vector form using the computed elements of $(I - A)^{-1}$ and the elements x_1, x_2, d_1, d_2.
f) Compute the elements x_1 and x_2 of the output vector if final demand for industry No. 1 is 180 and for industry No. 2 is 216.
g) Compute the elements x_1 and x_2 of the output vector if final demand is 36 for industry No. 1 and 108 for industry No. 2.

47. Each dollar of industry No. 1 output requires $\frac{1}{10}$ dollars of its own output and $\frac{1}{2}$ dollars of industry No. 2 output. Each dollar of industry No. 2 output requires $\frac{3}{5}$ dollars of industry No. 1 output and $\frac{1}{5}$ dollars of its own output. Let d_1, d_2 be final demand and x_1, x_2 be total output, respectively, for industries No. 1 and No. 2.
   a) Compute the matrix which completes the following equation:

$$\begin{pmatrix} x_1 \\ x_2 \end{pmatrix} = \left(\qquad \right) \begin{pmatrix} d_1 \\ d_2 \end{pmatrix}.$$

   b) Find the elements x_1 and x_2 of the output vector if final demand is 315 for industry No. 1 and 504 for industry No. 2.
   c) Find the elements x_1 and x_2 of the output vector if final demand is 63 for industry No. 1 and 189 for No. 2.

Find the numerical values of:

48. $\sum_{n=3}^{6} (n^2 - n)$.   49. $\sum_{i=1}^{5} 2i^2$.   50. $\sum_{p=3}^{7} (2p + 1)$.

Express in summation notation:

51. The sum of the first 100 positive integers.

52. The sum of the 5th power of the first 10 positive integers.

53. The sum of the first 50 positive even integers.

54. The sum of the first 50 odd integers.

Write in expanded form:

55. $\displaystyle\sum_{j=10}^{15} c_j y_j.$ 56. $\displaystyle\sum_{j=10}^{10+n} c_j x_j.$ 57. $\displaystyle\sum_{i=m}^{m+n} a_i x_i.$

58. $\displaystyle\sum_{i=1}^{4} x_i = 25.$ 59. $\displaystyle\sum_{j=1}^{3} 2x_j = 7.$ 60. $\displaystyle\sum_{j=1}^{6} c_j x_j = 15.$

Express in compact summation form:

61. $a_1 x_1 + a_2 x_2 + a_3 x_3.$

62. $a_1 x_1 + a_2 x_2 + \cdots + a_9 x_9.$

63. $a_1 x_1 + a_2 x_2 + \cdots + a_n x_n.$

64. $a_1 x_1 + a_2 x_2 + \cdots + a_n x_n + a_{n+1} x_{n+1} + a_{n+2} x_{n+2} + \cdots + a_{n+m} x_{n+m}.$

65. Identify m, c, and each of the b's in

$$\sum_{j=1}^{m} b_i y_i = c$$

with a number in the equation

$$5y_1 + 3y_2 - y_3 + 4y_6 = 12.$$

66. Write the following system in expanded form

$$\sum_{j=1}^{3} a_{ij} x_j = b_i \qquad i = 1, 2, 3, 4.$$

67. Write in expanded form

$$\text{Maximize } \sum_{j=1}^{2} c_j x_j$$

$$\text{subject to: } \sum_{j=1}^{2} a_{ij} x_j \le b_i \qquad i = 1, 2$$

$$x_j \ge 0 \text{ for all } j.$$

68. A linear system has p equations in p variables.

    a) Write the system in expanded form, using $\cdots$ where necessary.

    b) Write the system in compact summation form.

69. A linear programming minimization problem has 15 "greater than or equal" constraints on 25 variables.

    a) Write the problem in expanded form using $\cdots$ to shorten the expressions.

    b) Write the problem in compact summation form.

70. Consider the following system and its compact representation

$$3x_1 + 2x_2 + 5x_3 = 6$$
$$x_1 + 7x_2 - x_3 = 10$$
$$4x_2 - 8x_3 = 2$$
$$6x_1 + 9x_2 = 11$$

$$\sum_{j=1}^{n} a_{ij}x_j = b_i \quad i = 1, 2, \ldots, m.$$

a) What is m? What is n?

b) What symbol would represent the following constants which appear in the system: 3, 7, 10, 9, 11?

c) What constants in the system correspond to a_{23}, a_{31}, b_4, a_{33}, a_{12}?

71. To find the average of a group of n numbers, we add up the numbers and divide by n; thus, the average of the three numbers 2, 7, 6 is

$$\frac{2 + 7 + 6}{3} = 5.$$

a) Write the expression for the average of any three numbers, x_1, x_2, and x_3 in expanded form and in compact summation form.

b) Write the expression for the average of any n numbers in expanded form (using $\cdots$) and in compact summation form.

72. To compute the weighted average of a group of numbers, each number, x_i, is multiplied by its weight, w_i; the products so obtained are summed and the sum is divided by the sum of the weights to obtain the average.

a) Given the numbers 3 and 5 with weights 2 and 8, respectively, compute the weighted average.

b) Write the compact formula for the weighted average of n numbers, where each number, x_i, is weighted by w_i.

c) Write the formula in (b) in expanded form.

73. Compute the following for the values $x = 2, 3, 7$.

a) Σx.   b) Σx^2.   c) $\bar{x}$.   d) $\Sigma (x - \bar{x})^2$.

e) Compute the square of the average of the x's.

f) Compute the average of the squares of the x's.

74. a) Plot the following points on a graph.

x	y
4	2
0	1
6	4
2	3
6	2

b) Find the equation of the best fitting line, $y = mx + b$, where

$$m = \frac{n\Sigma xy - \Sigma x \Sigma y}{n\Sigma x^2 - (\Sigma x)^2} \quad \text{and} \quad b = \frac{\Sigma y - m\Sigma x}{n}.$$

c) Sketch the line found in (b) on the graph in (a).

Find the coefficient of linear correlation for the data in (a) by evaluating the formula

$$\frac{n\Sigma xy - \Sigma x \Sigma y}{\sqrt{[n\Sigma x^2 - (\Sigma x)^2][n\Sigma y^2 - (\Sigma y)^2]}}.$$

75. The *harmonic* average (or harmonic mean) of a set of numbers is the reciprocal of the average (arithmetic) of the reciprocals of the numbers. It is used on occasion to average ratios. The formula for the harmonic mean, *h*, is

$$\frac{1}{\dfrac{\displaystyle\sum_{i=1}^{n}(1/x_1)}{n}} = \frac{n}{\displaystyle\sum_{i=1}^{n}(1/x_i)}.$$

Compute the harmonic mean of $x = 2, 3, 4$.

Find the average, variance, and standard deviation:

76. x_i: 1, 2, 7, 5, 10, 5.   77. x_i: 57, 69, 51, 63, 60.

78. If μ and σ^2 are the average and variance of the x_i and we multiply each x_i by the constant a and compute a new mean and variance,

a) Express the new mean in terms of μ and a.

b) Express the new variance and standard deviation in terms of μ and a.

c) The numbers x_i: 2, 6, 0, 4, 3 have $\mu_1 = 3$ and $\sigma_1^2 = 4$. If we triple these numbers what will be the new average, variance, and standard deviation?

79. Express

$$\sum_{i=1}^{n} x_i(x_i - \mu)$$

a) in terms of Σx_i^2, *n*, and μ;

b) in terms of Σx_i^2, Σx_i, and *n*.

7

Logarithms

7.1 INTRODUCTION

THIS CHAPTER begins with a development of the fundamentals of logarithms and then proceeds to a discussion of the application of logarithms in computation and in solving certain types of equations. Inasmuch as logarithms are exponents, the reader who feels the need might well review the discussion of exponents in Appendix 2. Exponents play a role not only in this chapter, but in every chapter to come.

7.2 DEFINITION OF A LOGARITHM

If b is any positive number and $b^L = a$, L may be called the logarithm of a to the base b. Thus, $\log_b a = L$, where $\log_b$ is the abbreviation for "logarithm to the base b." In words, the logarithm of a number to a stated base is the exponent of the power to which the base must be raised to yield the number. If we write $2^3 = 8$, then we may state that the logarithm of 8 to the base 2 is 3; thus, $\log_2 8 = 3$. Again, if we write $3^2 = 9$, it follows that $\log_3 9 = 2$.

Exercise. Write the symbols for the logarithm of 81 to the base 3. What is this logarithm? Why? Answer: The symbols are $\log_3 81$. This logarithm is 4 because $3^4 = 81$.

From a definitional point of view, any positive number can serve as a base for a system of logarithms. From a practical point of view, it makes sense to select a base which makes the resultant system of logarithms as helpful as possible. This criterion leads to the selection of *common* logarithms, which are logarithms to the base 10, for most ordinary computational purposes, and to the selection of *natural* logarithms, which are logarithms to the irrational base $e = 2.718+$, for the analytical procedures of the calculus. In this chapter, we shall consider only logarithms to the base 10.

7.3 COMMON LOGARITHMS OF NUMBERS

The common logarithm of a number is the exponent of the power to which 10 must be raised to yield that number. Thus:

$$\log_{10} 100 = 2$$

because the base, 10, must be raised to the second power to yield 100. That is, $10^2 = 100$. From the statements

$$10^3 = 1000$$
$$10^2 = 100$$
$$10^1 = 10$$
$$10^0 = 1$$
$$10^{-1} = 0.1$$
$$10^{-2} = 0.01$$
$$10^{-3} = 0.001$$

we may write the following:

$$
\begin{aligned}
\log_{10} 1000 &= 3 \\
\log_{10} 100 &= 2 \\
\log_{10} 10 &= 1 \\
\log_{10} 1 &= 0 \\
\log_{10} 0.1 &= -1 \\
\log_{10} 0.01 &= -2 \\
\log_{10} 0.001 &= -3.
\end{aligned}
$$

There is no real number which, when applied as a power of 10, will yield zero or any negative number. Consequently, we restrict our attention to logarithms of positive numbers. Moreover, we adopt the convention of letting *log* mean the logarithm to the base 10. Thus, log N is the logarithm to the base 10 of the positive number N.

The particular advantage of the base 10 is its obvious relation to the decimal system. Consider $10^{0.5}$, which is the square root of 10. By longhand calculation, we can find $10^{0.5} = 3.1623$. It follows that log $3.1623 = 0.5000$. It is true also that $10^{1.5} = (10^1)(10^{0.5}) = 31.623$. Hence: log $31.623 = 1.5000$. By similar reasoning, it follows that

$$
\begin{aligned}
\log 316.23 &= 2.5000 \\
\log 3162.3 &= 3.5000 \\
\log 31623 &= 4.5000.
\end{aligned}
$$

The point being made here is that if the logarithm of a single number to the base 10 is known, then the logarithm of every number with the same sequence of digits is known, irrespective of the position of the decimal point in the sequence.

7.4 CHARACTERISTIC AND MANTISSA

The number 31.623 is between 10 and 100 or, in exponent form, between 10^1 and 10^2. The logarithm of 31.623, consequently, is between 1 and 2 which is to say

$$\log 31.623 = 1+$$

where the part of the number indicated by the plus sign is a positive proper decimal fraction and, in the present case, is 0.5000. The integral part of the logarithm, 1, is called the *characteristic* because it is a common characteristic of the logarithms of all numbers from 10 up to 100. The positive proper decimal fraction, 0.5000, is called the *mantissa* (which means an additional part). Mantissas are found from a table such as Table I at the end of the book. The advantage of the base 10 is that a given sequence of digits requires only one entry in the table, irrespective of the position of the decimal point in the sequence.

Exercise.   The logarithm of 100 to the base 5 is between what two integers? Why? Answer: Log_5 100 is between 2 and 3 because 100 is between 5^2 and 5^3.

Two steps are required in finding the logarithm of a number: the characteristic is determined by applying an easily remembered rule, and the mantissa is found from a table. The characteristic is found by counting the number of decimal places *from* the *standard position* of the decimal point *to* the actual position, counts to the right being positive, to the left negative. The standard position of the point is immediately after the first nonzero digit in the number, reading from left to right. Thus, in 0.003 $_\wedge$ 67 the symbol $_\wedge$ shows the standard position is after the 3. To go from standard position to the actual position of the point in this number is to move three places to the left, so the characteristic of the logarithm is -3. Again, the characteristic of the logarithm of 3 $_\wedge$ 16.23 is seen to be $+2$. The reader may verify the figures in Table 7–1.

We now shall write both the characteristic and the mantissa, finding the latter in the four-place Table I at the end of the book. Consider log 394. The characteristic of this logarithm is 2. Referring to Table I, we locate the row for 39, and then move across this row until we are under the column headed 4 and find the entry 5955, which means 0.5955 because *all tabulated mantissas are positive proper decimal fractions.* The desired logarithm is 2 plus 0.5955. Hence:

$$\log 394 = 2.5955.$$

TABLE 7–1

Number	Characteristic of the Logarithm of the Number
27	1
0.00027	−4
3150	3
2.6	0
15	1
2.003	0
0.5	−1
15,000	4

Suppose the problem had been to find log 0.0394. In this case the characteristic is −2, but the mantissa is the same, +0.5955. The complete logarithm is −2 *plus* 0.5955. It would be incorrect to write this as −2.5955 because the last written number is −2 *minus* 0.5955. The error can be avoided by writing negative characteristics of logarithms at the right of the mantissa. Thus:

$$\log 0.0394 = 0.5955 - 2.$$

Some prefer to replace −2 by the equivalent 8 − 10 and write

$$\log 0.0394 = 8.5955 - 10.$$

The choice of method is of no consequence. Indeed, we may wish to use other forms, such as:

$$\log 0.0394 = 1.5955 - 3$$
$$\log 0.0394 = 2.5955 - 4.$$

For practice purposes the reader may verify the following:

$$
\begin{aligned}
\log 16 \quad &= 1.2041 \\
\log 0.00016 &= 0.2041 - 4 \\
\log 0.1 \quad &= 0.0000 - 1 \\
\log 101 \quad &= 2.0043 \\
\log 9060 \quad &= 3.9571 \\
\log 0.607 \quad &= 0.7832 - 1 \\
\log 3.47 \quad &= 0.5403 \\
\log 1 \quad &= 0.0000 \\
\log 0.444 \quad &= 0.6474 - 1 \\
\log 70,000 \quad &= 4.8451 \\
\log 0.00007 &= 5.8451 - 10 \\
\log 0.0516 \quad &= 2.7126 - 4 \\
\log 0.0516 \quad &= 1.7126 - 3 \\
\log 0.0516 \quad &= 0.7126 - 2.
\end{aligned}
$$

7.5 PROBLEM SET 1

Find the logarithms of the following numbers:

1. 34.6	2. 3.46	3. 0.346
4. 0.0346	5. 34,600	6. 0.903
7. 1.01	8. 258	9. 0.01
10. 0.0125	11. 10^3	12. 10^{-1}
13. $10^{0.786}$	14. 10^{132}	15. $10^{2.3456}$
16. 2340	17. 0.00478	18. 1
19. 9.87	20. 604	21. 0.032
22. 0.000007	23. 24,500	24. 6
25. 0.1	26. 10	27. 2.03
28. 568	29. 0.000101	30. 3.28
31. 0.0001	32. 57	33. 5
34. 16	35. 1.6	36. 1240
37. 111	38. 15.6	39. 12
40. 0.04	41. 0.139	42. 0.17
43. 1.75	44. 829.0	45. 0.0566
46. 155,000	47. 0.002	48. 0.137
49. 5.27	50. 0.00000000478	

Between what two integers would the following logarithms lie?

51. $\log_3 50$ 52. $\log_2 100$ 53. $\log_8 6$ 54. $\log_{20} 500$

Find the following logarithms:

55. $\log_2 32$ 56. $\log_{20} 400$ 57. $\log_5 625$ 58. $\log_4 2$

7.6 ANTILOGARITHMS

In base 10 terminology the expression $10^2 = 100$ could be described by stating that the logarithm of 100 is 2, or by stating that 100 is the number whose logarithm is 2. The latter statement is written

$$\text{antilog } 2 = 100$$

antilog being an abbreviation for the word *antilogarithm*. In general, antilog *c* means the number whose logarithm is *c*. If *c* is an integer, its antilogarithm can be written without referring to Table I. Thus:

$$\text{antilog } 3 = 1000$$

because the number whose logarithm is 3 is 10 to the third power, which is 1000. Again:

$$\text{antilog }\quad 0 = 1$$
$$\text{antilog } -2 = 0.01.$$

If *c* consists in part of a positive proper decimal fraction, the determination of its antilogarithm is carried out in two steps. First, the proper sequence of digits is found from Table I; then the decimal point is inserted into the sequence. Consider

$$N = \text{antilog } 1.5611.$$

N is the number whose logarithm has a mantissa of 0.5611. Referring to Table I, we can find the proper sequence of digits for N; then, in accordance with the characteristic, 1, the decimal point is inserted one place to the right of standard position in the sequence. Locating 5611 in the field (or body) of the mantissa table, we find the corresponding sequence of digits from the side and top margins to be 364. Inserting the decimal point one place to the right of standard position in the sequence yields

$$\text{antilog } 1.5611 = 36.4.$$

Similarly:

$$\text{antilog } 2.5611 = 364$$
$$\text{antilog } (0.5611 - 1) = 0.364$$
$$\text{antilog } (8.5611 - 10) = 0.0364$$
$$\text{antilog } 0.5611 = 3.64.$$

The reader may verify the following:

$$\text{antilog } (0.7024 - 1) = 0.504$$
$$\text{antilog } 2.5740 = 375$$
$$\text{antilog } 0.1931 = 1.56$$
$$\text{antilog } (8.9542 - 10) = 0.0900$$
$$\text{antilog } 3.8476 = 7040$$
$$\text{antilog } (0.9996 - 2) = 0.0999.$$

7.7 PROBLEM SET 2

Evaluate:

1. antilog 1.9042
2. antilog (0.7427 − 1)
3. antilog (8.9581 − 10)
4. antilog 0.4900
5. antilog (0.4900 − 3)
6. log 0.4900
7. antilog (2.7033 − 3)
8. antilog 4.9557
9. log 0.4620
10. antilog 0.2900
11. log 0.2900
12. antilog 2.4698
13. antilog (7.4487 − 10)
14. antilog 0.2480
15. log 248.0
16. antilog (0.9571 − 4)
17. antilog (4.8494 − 2)
18. antilog (3.7810 − 3)
19. antilog 0.9170
20. antilog (2.4886 − 5)
21. log 0.7050
22. antilog 0.7050
23. antilog 0.0043
24. antilog 1.0043
25. antilog 0.0000

7.8 INTERPOLATION

Logarithms do not behave in exact proportion; for example, log 3 is not halfway between log 2 and log 4. However, log 1.935 is very nearly halfway between log 1.93 and log 1.94. From Table I:

$$\log 1.93 = 0.2856$$
$$\log 1.94 = 0.2878.$$

If we assume proportionality and take log 1.935 as halfway between the last two numbers, we have

$$\log 1.935 = 0.2867.$$

Reference to a table of mantissas more extensive than Table I shows that

$$\log 1.935 = 0.28668$$

which, if rounded to four decimal places, would be the same as the number obtained by assuming proportionality. In general, the mantissa of the logarithm of a number having four significant digits can be estimated accurately to four decimal places by proportional interpolation in Table I. Occasionally, an error of one will occur in the fourth decimal place.

To find the mantissa of log 3274 from Table I, we find four tenths of the difference between log 327 and log 328, and add the fractional difference to log 327. Thus:

Number	Mantissa
327	5145
3274	? ⟩14
328	5159

$$\text{Difference of mantissas} = 14$$
$$0.4(14) = 5.6$$
$$\text{Round to nearest whole number} = 6$$
$$\text{Add 6 to smaller mantissa} = 6 + 5145 = 5151.$$

Finally, inserting the characteristic:

$$\log 3274 = 3.5151.$$

As a second example, we find log 0.07427:

Number	Mantissa
742	8704
7427	? ⟩6
743	8710

$$\text{Difference of mantissas} = 6$$
$$0.7(6) = 4.2$$
$$\text{Round to nearest whole number} = 4$$
$$\text{Add 4 to smaller mantissa} = 4 + 8704 = 8708.$$

Inserting the characteristic, we have:

$$\log 0.07427 = 0.8708 - 2.$$

When rounding to the nearest whole number, as shown in the foregoing illustrations, a number such as 5.5 may present itself for rounding, in which case there is no nearest whole number. Of the several rules one might advance for rounding exact halves, none is distinctly superior for estimating mantissas. We shall arbitrarily round up when the fractional part is 0.5 or greater, and round down when the fractional part is less than 0.5. As an example:

$$\log 0.8005$$

Number	Mantissa
800	9031
8005	?
801	9036

> 5

Difference of mantissas = 5
$$0.5(5) = 2.5$$
Round to = 3
Desired mantissa (?) = $9031 + 3 = 9034$.

Inserting the mantissa:

$$\log 0.8005 = 0.9034 - 1.$$

Exercise. Find the logarithm of 536.6 and the logarithm of 0.09104. Answer: 2.7297 and $0.9592 - 2$.

If, in finding an *antilogarithm*, the sought-for mantissa is not shown exactly in the table, interpolation is carried out, using the two mantissas which straddle the sought-for figure and the two corresponding number sequences from the margins of the table. As an example, we find:

$$\text{antilog } 1.5241$$

Marginal Number	Mantissa
334	5237
334?	5241 →4
335	5250

> 13

? is 4/13, rounded to nearest tenth = 0.3.
Desired marginal sequence is 3343.

Inserting the decimal point into the marginal sequence, we have:

$$\text{antilog } 1.5241 = 33.43.$$

In the last example, we sought 5241 in the mantissa table and found it to be straddled by 5237 and 5250. The corresponding marginal sequences were 334 and 335, respectively. The desired sequence is 334?, where question mark is to be determined. We found 5241 was 4/13 (or 0.3 to the nearest tenth) of the way between the smaller and larger mantissas and, on the basis of proportionality, assumed the desired marginal number was three tenths of the way between 334 and 335, that is, 334.3. All we wish to know from this last number is the four-digit sequence whose mantissa is 5241, and this sequence is 3343. Moving the point one place to the right of standard position as specified by the characteristic in antilog 1.5241 leads to the answer 33.43.

As another example, we find:

$$\text{antilog } 0.5960 - 2$$

Marginal Number	*Mantissa*
394	5955
394?	5960
395	5966

>⁵ >11

? is $5/11 = 0.454$; round to 0.5.
Desired marginal sequence is 3945.

Inserting the decimal point as specified by the -2 characteristic:

$$\text{antilog } 0.5960 - 2 = 0.03945.$$

As a final example, we find:

$$\text{antilog } 4.5481$$

Marginal Number	*Mantissa*
353	5478
353?	5481
354	5490

>³ >12

? $= 3/12 = 0.25$; round to 0.3.
Desired marginal sequence is 3533.

Inserting the decimal point as specified by the 4, we have:

$$\text{antilog } 4.5481 = 35,330.$$

In the last example the digit 3 in the fourth position was estimated by interpolation (following the earlier stated decision to round 0.5 or more up). We know nothing about the digit which should be in the fifth position; the number zero in this position is simply a filler used to locate the decimal point properly.

Exercise.   Find antilog 2.7045 and antilog (0.8708 − 3). Answer: 506.4 and 0.007427.

7.9 CHANGING TO PROPER FORM

If we are to be able to evaluate antilog c, it is necessary that c consist of a whole number and a nonnegative decimal fraction. Occasions arise where it is necessary to manipulate numbers in order to arrive at a proper form. The manipulation consists simply of adding and subtracting the same number. For example, antilog -0.6925 is not in proper form because the mantissa is negative. To achieve proper form, we add and subtract one in the following manner:

$$-0.6925 = 1 - 0.6925 - 1$$
$$= (1 - 0.6925) - 1$$
$$= 0.3075 - 1.$$

Hence:

$$\text{antilog}\ (-0.6925) = \text{antilog}\ (0.3075 - 1) = 0.203.$$

If the last problem had been to find antilog -1.6925, we would have obtained a number equal to -1.6925, but in proper form, by adding and subtracting 2. Thus:

$$\text{antilog}\ (-1.6925) = \text{antilog}\ (0.3075 - 2) = 0.0203.$$

The equivalence between the given number and the proper form of the number exists, of course, because adding and subtracting the same number is the same as adding zero.

Exercise. Find antilog (-3.2807). Answer: 0.000524.

Suppose, next, that we are asked to multiply $(0.3413 - 1)$ by 3, and then find the antilogarithm of the product. The product is:

$$3(0.3413 - 1) = 1.0239 - 3 = 0.0239 - 2.$$

The desired antilog is:

$$\text{antilog}\ (0.0239 - 2) = 0.01057.$$

On the other hand, if we are asked to divide $0.6990 - 1$ by 3, and then find the antilogarithm of the quotient, we would change $0.6990 - 1$ (by adding and subtracting 2) to $2.6990 - 3$, so that division will yield a whole number as a characteristic. Thus:

$$\text{antilog}\ \frac{0.6990 - 1}{3} = \text{antilog}\ \frac{2.6990 - 3}{3}$$

$$= \text{antilog}\ (0.8997 - 1) = 0.7938.$$

Finally, if asked to subtract 0.4879 from 0.3642 and find the antilogarithm of the difference, we would avoid a negative mantissa by changing 0.3642 to 1.3642 − 1. Thus:

$$0.3642$$
$$-0.4879$$

becomes

$$1.3642 - 1$$
$$-0.4879$$
$$0.8763 - 1.$$

The desired number is:

$$\text{antilog } (0.8763 - 1) = 0.7522.$$

7.10 PROBLEM SET 3

Evaluate:

1. log 24.65
2. log 89.85
3. log 0.03175
4. log 3272
5. log 0.004932
6. antilog 0.3762
7. antilog 0.5950
8. antilog 0.4860
9. antilog (0.9070 − 2)
10. antilog (8.7438 − 10)
11. log 0.6080
12. antilog 0.6080
13. log 1.3670
14. antilog 1.3670
15. antilog (0.3670 − 3)
16. log 1.0350
17. antilog 1.0350
18. log 347.8
19. antilog 0.6120
20. log 0.4777
21. antilog 0.4777
22. log 4.8620
23. antilog 4.8620
24. log 0.003727
25. antilog (0.9232 − 3)
26. log 5.623
27. antilog (2.8493 − 3)
28. log 10.05
29. antilog 3.7417
30. antilog (0.3742 − 3)

NOTE: Computations with logarithms often lead to situations requiring conversion of a number to proper form in the manner shown in the illustrations just before this problem set. The following 10 problems offer practice in conversion to proper form.

31. Multiply (0.3076 − 2) by 3, and find the antilogarithm of the product.
32. Multiply (0.4062 − 1) by −2, and find the antilogarithm of the product.
33. Subtract 0.8762 from 0.3866, and find the antilogarithm of the difference.
34. Subtract (0.7842 − 1) from (0.3458 − 2), and find the antilogarithm of the difference.
35. Subtract (0.8493 − 2) from (0.3859 − 1), and find the antilogarithm of the difference.

36. Subtract 1.5987 from 0.9642, and find the antilogarithm of the difference.
37. Divide (0.3784 − 1) by 3, and find the antilogarithm of the quotient.
38. Divide (2.6389 − 3) by 2, and find the antilogarithm of the quotient.
39. Divide (0.9847 − 3) by 5, and find the antilogarithm of the quotient.
40. Divide (2.3568 − 3) by 4, and find the antilogarithm of the quotient.

7.11 RULES OF LOGARITHMS

In a given system of logarithms, all logarithms are exponents of the same base and are therefore subject to the rules of exponents. In the expression

$$(10^x)(10^y) = 10^{x+y}$$

x is the logarithm of the number 10^x, y is the logarithm of the number 10^y, and $x + y$ is the logarithm of the product of the two numbers. If we let

$$a = 10^x$$
$$b = 10^y$$

then

$$ab = 10^{x+y}$$

and

$$\log a = x$$
$$\log b = y$$
$$\log ab = x + y.$$

It follows that

$$\log ab = \log a + \log b.$$

That is, the logarithm of a product is the sum of the logarithms of the factors. In similar fashion, if we let

$$a = 10^x$$
$$b = 10^y$$

then

$$a/b = 10^{x-y}$$

and

$$\log a = x$$
$$\log b = y$$
$$\log a/b = x - y.$$

It follows that

$$\log a/b = \log a - \log b.$$

That is, the logarithm of a quotient is the logarithm of the numerator minus the logarithm of the denominator.

Finally, if we let $a = 10^x$, then

$$a^b = (10^x)^b = 10^{bx}$$

and

$$\log a = x$$
$$\log a^b = bx.$$

It follows that

$$\log a^b = b(\log a).$$

That is, the logarithm of a number to a power is the exponent of the power times the logarithm of the number.

The fundamental rules for logarithms are:

$$\log ab = \log a + \log b$$
$$\log a/b = \log a - \log b$$
$$\log a^b = b(\log a).$$

As examples of these rules:

$$\log (3)(2) = \log 3 + \log 2$$
$$\log (6/3) = \log 6 - \log 3$$
$$\log 3^2 = 2 \log 3.$$

Exercise.   Given that $\log a = 0.3000$, $\log b = 2.8642$, and $\log c = 1.7642$, find $\log \dfrac{a^3 b^{1/2}}{c}$ without using a table of logarithms.

Answer: 0.5679.

The fundamental rules can be stated in another manner which specifies computational procedures by observing that if two numbers are equal, then their antilogarithms are equal. Thus, if

$$\log ab = \log a + \log b$$

then it follows that

$$\text{antilog } (\log ab) = \text{antilog } (\log a + \log b).$$

Now, of course, the antilogarithm of the logarithm of a number is the number itself. Hence:

$$ab = \text{antilog } (\log a + \log b).$$

The last expression states that the product of two numbers, ab, can be found by adding the logarithms of the factors and then finding the antilogarithm of the sum. *Observe that it would be incomplete* (therefore incorrect) *to say that numbers are multiplied by adding their logarithms.*

The computational procedure for finding a quotient is

$$a/b = \text{antilog} (\log a - \log b).$$

That is, to divide a by b, subtract $\log b$ from $\log a$, and then find the antilogarithm of the difference.

Finally:

$$a^b = \text{antilog} \, b(\log a).$$

That is, to raise a number to a power, multiply the logarithm of the number by the exponent of the power, and then find the antilogarithm of the product.

In summary, the computational rules for logarithms are:

$$ab = \text{antilog} (\log a + \log b)$$
$$a/b = \text{antilog} (\log a - \log b)$$
$$a^b = \text{antilog} (b \log a).$$

The computational rules just summarized and the fundamental rules for logarithms given earlier are not the same rules. Attention is directed toward the distinction between the two sets of rules by the italicized sentence a dozen or so lines back. The computational rule for multiplication via logarithms is not simply to add the logarithms of the factors. It is, rather, to add the logarithms of the factors and then find the antilogarithm of the sum. If we simply add the logarithms of the factors, we have the logarithm of the product of the two factors, not the product itself. In the same manner, subtraction of logarithms leads (by the fundamental rule of logarithms) to the logarithm of a quotient, whereas the computational rule for a quotient is to subtract logarithms and then find the antilogarithm of the difference.

Exercise.   Given that $\log a = 1$, $\log b = 2$, $\log c = 3$, what is
(a) $\log a/b$, (b) ac, (c) $\log bc$, (d) $\log (b/c)$, (e) $(\log c)/(\log b)$?
Answer: (a) -1, (b) 10,000, (c) 5, (d) -1, (e) 1.5.

7.12 COMPUTATIONS WITH LOGARITHMS

The following series of examples shows computations carried out by means of logarithms.

Multiply $(24)(25)$:

$$
\begin{aligned}
(24)(25) &= \text{antilog} (\log 24 + \log 25) \\
&= \text{antilog} (1.3802 + 1.3979) \\
&= \text{antilog} \, 2.7781 \\
&= 599.9.
\end{aligned}
$$

The exact product is $(24)(25) = 600$. The error in the number obtained by logarithms is a consequence of the approximate nature of a four-place table of mantissas. Such errors can be reduced by appeal to more extensive tables.

To emphasize that logarithm computations are based on the rules of exponents, let us review the last example noting that

$$\log 24 = 1.3802 \quad \text{means} \quad 24 - 10^{1.3802}$$
$$\log 25 = 1.3979 \quad \text{means} \quad 25 = 10^{1.3979}.$$

Hence,

$$(24)(25) = (10^{1.3802})(10^{1.3979}).$$

By a rule of exponents,

$$(24)(25) = 10^{1.3802 + 1.3979}$$
$$(24)(25) = 10^{2.7781}.$$

The last statement says 2.7781 is the power to which 10 must be raised to yield $(24)(25)$; that is, 2.7781 is the logarithm of $(24)(25)$, so

$$\log (24)(25) = 2.7781$$

and

$$(24)(25) = \text{antilogarithm } (2.7781)$$
$$= 599.9.$$

The reader will find it instructive to relate the logarithm computations in the next two examples to rules of exponents.

Divide 257 by 13.8:

$$\frac{257}{13.8} = \text{antilog } (\log 257 - \log 13.8)$$

$$= \text{antilog } (2.4099 - 1.1399)$$

$$= \text{antilog } 1.2700$$

$$= 18.62.$$

Raise 2.46 to the third power:

$$(2.46)^3 = \text{antilog } (3 \log 2.46)$$
$$= \text{antilog } [3(0.3909)]$$
$$= \text{antilog } 1.1727$$
$$= 14.88.$$

The horizontal statements of the last three examples provide exact mathematical expressions of the problems and their solutions. In practice, problems often are set up vertically, as shown next:

Compute $\dfrac{(2.56)(38.4)}{51.7}$:

$$\log 2.56 = 0.4082$$
$$\log 38.4 = \underline{1.5843}$$
$$1.9925$$
$$\log 51.7 = \underline{1.7135}$$
$$0.2790$$

Answer is antilog $(0.2790) = 1.901$.

Compute $\dfrac{42.3}{895}$:

$$\log 42.3 = 1.6263 = 3.6263 - 2$$
$$\log 895 = 2.9518 = \underline{2.9518}$$
$$0.6745 - 2$$

Answer is antilog $(0.6745 - 2) = 0.04726$.

In the last example, notice that the subtraction of the logarithms as first written would have led to a negative mantissa. Consequently, 2 was added to and subtracted from 1.6263 before undertaking the subtraction.

Compute $\dfrac{4.23}{0.0895}$:

$$\log 4.23 = 0.6263 \qquad = 1.6263 - 1$$
$$\log .0895 = 0.9518 - 2 = \underline{0.9518} - 2$$
$$0.6745 + 1$$

Answer is antilog $(1.6745) = 47.26$.

Compute $(0.17)^5$:

$$\log 0.17 = 0.2304 - 1$$
$$5 \log 0.17 = 1.1520 - 5$$

Answer is antilog $(0.1520 - 4) = 0.0001419$.

Compute $(32.6)^{-3}$:

$$\log 32.6 = 1.5132$$
$$-3 \log 32.6 = -4.5396 = 5 - 4.5396 - 5$$
$$= 0.4604 - 5$$

Answer is antilog $(0.4604 - 5) = 0.00002887$.

Multiplication, division, and raising to integral powers can be carried out by ordinary procedures or by use of logarithms, but logarithms are indispensable in computing roots and other fractional powers of numbers.

Compute the cube root of 37.4:

$$(37.4)^{1/3} = \text{antilog}\left[\frac{1}{3}(\log 37.4)\right]$$

$$= \text{antilog}\frac{1.5729}{3}$$

$$= \text{antilog } 0.5243$$

$$= 3.345.$$

Compute $(0.462)^{1/4}$:

$$\log 0.462 = 0.6646 - 1 = 3.6646 - 4$$

$$\frac{1}{4}\log 0.462 \qquad\qquad = 0.9162 - 1$$

Answer is antilog $(0.9162 - 1) = 0.8245$.

Compute $(226)^{-0.75}$:

$$-0.75 \log 226 = -0.75(2.3541) = -1.7656 = 2 - 1.7656 - 2$$
$$= 0.2344 - 2.$$
Answer is antilog $(0.2344 - 2) = 0.01716$.

Compute $(0.302)^{-2/3}$:

$$\log 0.302 = 0.4800 - 1$$

$$-\frac{2}{3}\log 0.302 = -\frac{2}{3}(0.4800 - 1)$$

$$= -\frac{2}{3}(-0.5200)$$

$$= 0.3467$$

Answer is antilog $0.3467 = 2.222$.

In the foregoing examples, observe the manipulations undertaken to convert numbers into a form which permits evaluation of the antilogarithm.

7.13 PROBLEM SET 4

1. Define "logarithm of a number to the base b."
2. What is the logarithm of 16 to the base 2?
3. The logarithm of 50 to the base 6 is between what two integers?
4. Why does $\log_{10} (0.001)$ equal -3?
5. Where is the standard position of the decimal point?

   Mark (T) for true or (F) for false:
6. (   ) Given that log 2 is 0.3010, then log 0.2 is -1.3010.
7. (   ) Given that antilog 0.4771 is 3, then antilog $(0.4771 - 1)$ is 0.3.
8. (   ) $\log 2 + \log 3 = \log 6$.

9. (   ) $ab = \log a + \log b$.

10. (   ) $\log 250 - \log 125 = \log 125$.

11. (   ) $\log (ab/c) = \log a + \log b - \log c$.

12. (   ) $2 \log 4 = \log 16$.

13. (   ) $\sqrt{172} = \text{antilog}\left(\dfrac{\log 172}{2}\right)$.

14. (   ) $\dfrac{\log 30}{\log 5} = \log 30 - \log 5$.

15. (   ) $\dfrac{\log 30}{\log 5} = \log (\log 30) - \log (\log 5)$.

16. (   ) antilog $(\log 2 + \log 4) = $ antilog $(\log 2) + $ antilog $(\log 4)$.

Carry out the following computations by means of logarithms:

17. $(362)(2.58)$

18. $(0.00716)(329)$

19. $(3.654)(12.96)$

20. $(28.4)(0.0512)$

21. $(0.017)(0.00389)$

22. $(53.45)(2.719)$

23. $\dfrac{527}{315}$

24. $\dfrac{2.43}{816}$

25. $\dfrac{5.247}{0.0821}$

26. $\dfrac{0.3472}{0.0516}$

27. $\dfrac{217.3}{4.296}$

28. $(2.31)^5$

29. $(0.627)^{1/4}$

30. $(0.0345)^3$

31. $\sqrt[3]{10}$

32. $(2446)^{1/2}$

33. $(0.03)^{-1/2}$

34. $(0.426)^{0.7}$

35. $\dfrac{(27.5)(3.64)}{19.1}$

36. $\dfrac{(0.317)(316.2)}{107.6}$

37. $(0.0745)^{-2/3}$

38. $(425)(0.1567)$

39. $\dfrac{(2.3)^{1/2}}{(5.6)^{2/3}}$

40. $\dfrac{3.67}{814.3}$

41. $(432)^{-4/5}$

42. $(0.045)^{-2/3}$

43. $(0.562)^{3/4}$

44. $(85)^{0.2}$

45. Find the cube root of 259.

46. Find the fourth root of 0.0856.

7.14 LOGARITHMS IN THE SOLUTIONS OF EQUATIONS

An equation of the form $a^x = b$ is called an *exponential* equation because the variable, x, appears as an exponent. The equation can be solved easily for x by taking the logarithm of both sides, then applying the rule which states that the logarithm of a number to a power is the exponent of the power times the logarithm of the number. Thus:

$$\log a^x = \log b$$

$$x \log a = \log b$$

$$x = \frac{\log b}{\log a}.$$

The equation

$$ab^x + c = d$$

is approached in the same manner. However, the logarithm of a sum cannot be taken term by term, so we must remove c from the left side if we are to use logarithms in bringing x down from its exponential position. Hence:

$$ab^x = d - c$$

$$\log (ab^x) = \log (d - c)$$

$$\log a + \log b^x = \log (d - c)$$

$$\log a + x \log b = \log (d - c)$$

$$x \log b = \log (d - c) - \log a$$

$$x = \frac{\log (d - c) - \log a}{\log b}.$$

The procedure followed in solving for a variable which is an exponent is: First, isolate the term containing the variable exponent; second, take the logarithms of both sides of the equation, and apply the rules of logarithms to make the equation linear in the variable; finally, solve the resultant linear equation in the usual manner. As another example:

$$5^{x-2} + a = b$$
$$5^{x-2} = b - a$$
$$\log (5^{x-2}) = \log (b - a)$$
$$(x - 2) \log 5 = \log (b - a).$$

The last equation is linear in x. Its solution follows:

$$x \log 5 - 2 \log 5 = \log (b - a)$$

$$x \log 5 = \log (b - a) + 2 \log 5$$

$$x = \frac{\log (b - a) + 2 \log 5}{\log 5} = \frac{\log (b - a) + \log 25}{\log 5}.$$

Inasmuch as the sum of two logarithms is the logarithm of a product, the last solution could be written as

$$x = \frac{\log [25 \, (b - a)]}{\log 5}.$$

7.15 PROBLEM SET 5

Solve for x:

1. $a^x = b$.

2. $b^x - 3 = 2d$.

3. $bc^x + d = k$.

4. $(ab)^x = c$.

5. $(a + b)^x = c$.

6. $a(b + c)^x = d$.

7. $a^x = 1$.

8. $a^{x-3} = 6$.

9. $a^x b^x = c.$ 10. $\dfrac{1}{a + bc^x} = d.$

Using the rules of logarithms, write the right members of each of the following in a different form:

11. $x = \dfrac{\log a + \log b}{\log c}.$ 12. $x = \dfrac{2(\log a + \log b)}{\log c}.$

13. $x = \dfrac{\log\left(\dfrac{c}{2b^2}\right)}{\log c}.$ 14. $x = \dfrac{2\log a - \log b}{\dfrac{1}{2}\log c}.$

7.16 ANTILOGARITHMS IN THE SOLUTIONS OF EQUATIONS

Some equations in which the variable appears with a constant as an exponent can be solved by first taking the logarithm of both sides and isolating the term involving the logarithm of the variable, and then taking the antilogarithm of both sides. As a simple example, consider $x^3 = 50$. Taking the logarithm of both sides yields

$$\log x^3 = \log 50$$

$$3 \log x = \log 50$$

$$\log x = \frac{\log 50}{3}.$$

Now, taking the antilogarithm of both sides, we have

$$\text{antilog } (\log x) = \text{antilog}\left(\frac{\log 50}{3}\right)$$

$$x = \text{antilog}\left(\frac{\log 50}{3}\right).$$

The outcome could have been predicted from the rules of logarithms. That is, if $x^3 = 50$, then x is the cube root of 50, and can be found by dividing log 50 by 3 and taking the antilogarithm of the quotient.

Consider next the equation

$$a = b(c + x)^d.$$

It is not permissible to apply the exponent d to the terms of the sum. Taking the logarithm of both sides:

$$\log a = \log [b(c + x)^d].$$

The right member is a product, and the logarithm of a product is the sum of the logarithms of the factors. Thus:

$$\log a = \log b + \log (c + x)^d.$$

Applying the rule for the logarithm of a number to a power, we may bring *d* down as shown next:

$$\log a = \log b + d \log (c + x).$$

Proceeding to isolate the term containing the variable, we have

$$\log a - \log b = d \log (c + x)$$

$$\frac{\log a - \log b}{d} - \log (c + x).$$

We cannot, of course, change the right member to the sum of the logarithms of *c* and *x*. However, taking the antilogarithm of both sides yields

$$\text{antilog}\left(\frac{\log a - \log b}{d}\right) = c + x.$$

Finally:

$$\left[\text{antilog}\left(\frac{\log a - \log b}{d}\right)\right] - c = x.$$

As a third example, we solve the equation

$$x^a = c + b.$$

The following steps are self-explanatory:

$$\log x^a = \log (c + b)$$

$$a \log x = \log (c + b)$$

$$\log x = \frac{\log (c + b)}{a}$$

$$x = \text{antilog}\left[\frac{\log (c + b)}{a}\right].$$

7.17 PROBLEM SET 6

Solve for *x:*

1. *a)* $x^5 - 20 = 0.$
   b) $x^5 - a = 0.$

2. *a)* $2x^3 - 5 = 0.$
   b) $2x^3 - a = 0.$

3. *a)* $50 = \sqrt{x} - 2.$
   b) $a = \sqrt{x} - 2.$

4. *a)* $x^3 = 30.$
   b) $x^a = b.$

5. *a)* $3x^{2.5} = 36.$
   b) $ax^b = c.$

6. *a)* $3(1 + x)^5 = 18.$
   b) $a(1 + x)^n = b.$

7. *a)* $100(1 + x)^{-10} = 85.$

   b) $a(1 + x)^{-n} = b.$

8. *a)* $2 = \dfrac{\log x}{\log 20}.$

   b) $c = \dfrac{\log x}{\log b}.$

9. *a*) $2x^5 + 6 = 50$.

   b) $ax^n + c = b$.

10. *a*) $\dfrac{1}{-3 + 2x^{0.3}} = 0.5$.

    b) $\dfrac{1}{a + bx^c} = d$.

7.18 COMPUTATIONS ARISING IN THE SOLUTIONS OF EQUATIONS

Given the problem of finding the value of x for which $3^x = 5$, we proceed to solve, finding

$$\log (3^x) = \log 5$$

$$x \log 3 - \log 5$$

$$x = \frac{\log 5}{\log 3}.$$

Read the last statement carefully. It says to divide the logarithm of 5 by the logarithm of 3. Finding the logarithms and carrying out the division long-hand, we have

$$x = \frac{0.6990}{0.4771} = 1.465.$$

The longhand division could, of course, be replaced by the following calculation:

$$x = \frac{\log 5}{\log 3} = \text{antilog } [\log (\log 5) - \log (\log 3)]$$

$$= \text{antilog } [\log 0.6990 - \log 0.4771]$$

$$= \text{antilog } [(0.8445 - 1) - (0.6786 - 1)]$$

$$= \text{antilog } 0.1659$$

$$= 1.465.$$

The point of the last two calculations is that

$$\frac{\log 5}{\log 3}$$

cannot be evaluated as $\log 5 - \log 3$. The ratio is a quotient of two logarithms, not the logarithm of a quotient.

The equation

$$w = x(3 + y)^z$$

can be manipulated to evaluate any one of the letters, given numerical values for the other three letters. The next set of computations illustrates these evaluations.

Compute w, given:

$$x = 10$$
$$y = 2$$
$$z = 1.6$$
$$w = 10(3 + 2)^{1.6} = 10(5)^{1.6}.$$

Evaluating the last expression, we find

$$\begin{array}{rl} \log 10 = & 1.0000 \\ +1.6 \log 5 = & \underline{1.1184} \\ & 2.1184. \end{array}$$

Hence,

$$w = \text{antilog } 2.1184 = 131.3.$$

Compute x, given:

$$w = 200$$
$$y = 1$$
$$z = 2$$
$$200 = x(3 + 1)^2$$
$$\frac{200}{16} = x$$
$$12.5 = x.$$

Compute z, given:

$$w = 80$$
$$x = 39.6$$
$$y = 0.5$$
$$80 = 39.6(3 + 0.5)^z$$
$$\left(\frac{80}{39.6}\right) = (3.5)^z$$
$$\log\left(\frac{80}{39.6}\right) = \log (3.5)^z$$
$$\log 80 - \log 39.6 = z \log 3.5$$
$$\frac{\log 80 - \log 39.6}{\log 3.5} = z$$
$$\frac{1.9031 - 1.5977}{0.5441} = z$$
$$\frac{0.3054}{0.5441} = z$$
$$0.5613 = z.$$

Compute y, given:

$$w = 305$$

$$x = 1.26$$

$$z = 3$$

$$305 = 1.26(3 + y)^3$$

$$\frac{305}{1.26} = (3 + y)^3$$

$$\log \frac{305}{1.26} = \log (3 + y)^3$$

$$\log 305 - \log 1.26 = 3 \log (3 + y)$$

$$\text{antilog} \left[\frac{\log 305 - \log 1.26}{3} \right] = 3 + y$$

$$\text{antilog} \left[\frac{2.4843 - 0.1004}{3} \right] = 3 + y$$

$$\text{antilog } 0.7946 = 3 + y$$

$$6.231 = 3 + y$$

$$3.231 = y.$$

The equation we have just been working with is of a form which has applications in the next chapter. The form

$$w = x \left[\frac{(2 + y)^z - 1}{y} \right]$$

also will have applications. The procedures followed in evaluating one variable, given numbers for the others, are analogous to those illustrated in the foregoing, except that the equation at hand cannot be solved for y. To understand the problem which arises, suppose we start to solve for y:

$$\frac{wy}{x} = (2 + y)^z - 1$$

$$\frac{wy}{x} + 1 = (2 + y)^z$$

$$\log \left(\frac{wy}{x} + 1 \right) = z \log (2 + y).$$

Neither side can be expanded to make y available, because logarithms of sums or differences cannot be taken term by term.

7.19 PROBLEM SET 7

Solve for x:

1. $2^x = 25$.

2. $3^{-x} = 2$.

3. $(0.5)^x = 10$.

4. $(0.2)^x = 0.6$.

5. Given the equation $p = q(5 + x)^{v}$:
   a) Compute p if $q = 2$, $x = 3$, and $y = 1.4$.
   b) Compute q if $p = 72$, $x = 1$, and $y = 2$.
   c) Compute y if $p = 4.87$, $q = 1.89$, and $x = 0.25$.

6. Given the equation $p = q\left[\dfrac{(1 + x)^{v} - 2}{x}\right]$:
   a) Compute p if $q = 10$, $x = 3$, and $y = 2$.
   b) Compute q if $p = 2$, $x = 0.2$, and $y = 10$.
   c) Compute y if $p = 1000$, $q = 10$, and $x = 0.8$.

7. Compute y:

$$y = 20\left[\frac{1 - (1 + 0.1)^{-10}}{0.1}\right].$$

8. Compute x:

$$100 = x\left[\frac{1 - (1 + 0.05)^{-5}}{0.05}\right].$$

9. Compute p:

$$5000 = 100\left[\frac{1 + (1 + 0.03)^{-p}}{0.03}\right].$$

10. Compute x:

$$6 = \frac{4.5}{0.5(2^{x} - 3)}.$$

7.20 REVIEW PROBLEMS

Determine the following logarithms:

1. $\log_5 5$	2. $\log_a a$	3. $\log_8 1$
4. $\log_b 1$	5. $\log_2 16$	6. $\log_6 36$
7. $\log_4 2$	8. $\log_{125} 5$	9. $\log_{10} 0$
10. $\log_4 \frac{1}{2}$	11. $\log_{10} (10)^{10}$	12. $\log_3 \frac{1}{9}$

Between what two integers do the following logarithms lie?

13. $\log_6 124$	14. $\log_3 1000$	15. $\log_2 0.3$
16. $\log_{20} 10$		

Evaluate the following (interpolation not needed):

17. $\log 457$	18. antilog 0.2304
19. $\log 0.00109$	20. antilog 3.9201
21. $\log 1$	22. antilog 3
23. $\log 3000$	24. antilog 0.0000
25. $\log 1.9400$	26. antilog 1.9400
27. $\log 0.0607$	28. antilog 0.0607
29. antilog (-2.5200)	30. antilog -0.2

31. antilog$\frac{1}{2}$(2.3228 − 4)   32. log 10^{-25}.

Evaluate the following (use interpolation):

33. log 144.3   34. antilog 2.6670
35. log 0.9091   36. antilog 0.9091
37. log 5.536   38. antilog (1.0022 − 3)
39. log 0.008896   40. antilog (3.9545 − 2)
41. log 25.28   42. antilog (3.5746 − 2)
43. log 0.3005   44. antilog 0.3005

45. log 1.001   46. antilog$\frac{1}{3}$(−2.40)

47. log 0.09999   48. antilog (0 − 1.3600)

Given that log a = 2.1, log b = 0.7, and log c = 0.4 − 1; compute the following without reference to tables:

49. log abc 50. $\dfrac{\log a}{\log b}$ 51. log $\sqrt{c}$

52. log$\dfrac{a}{c}$ 53. log c^2 54. log$\dfrac{ab}{c}$

55. $\dfrac{\log ab}{\log c}$ 56. log $(abc)^2$

Carry out the following computations by means of logarithms:

57. (26.3)(0.167)   58. (55.6)/(2310)
59. $\sqrt{218}$ 60. (0.0137)(0.201)
61. (1.090)/(0.00217)   62. $\sqrt[3]{0.0136}$
63. (1.236)(45.81)   64. (65.93)/(257.4)
65. $28^{-2/3}$ 66. (0.908)(0.3025)
67. (0.01121)/(0.07582)   68. $(0.2718)^{1.6}$
69. (426.3)(100.8)   70. (2.037)/(0.005056)

71. $\dfrac{1}{\sqrt[5]{2}}$

Using the rules of logarithms, write the following in a different but equivalent form:

72. log a + 2 log b.   73. 2 log a − log b.   74. $\frac{1}{3}$log c^6.

75. 2 log a + 2 log b.   76. $\dfrac{\log a}{x}$.   77. $\frac{1}{x}$log $(1 + x)$.

Solve for x:

78. $ab^x = c$.   79. $a^{-x} = 5$.   80. $(a + 1)^x = b + 2$.

81. $\dfrac{a}{1 + b^{-2x}} = 3$.   82. $2a^{1-x} = 10$.   83. $7^{-2x} − b = 0$.

84. $x^{10} − b = 0$.   85. $2(1 − x)^{-n} = b$.   86. log $x^3 = 3b$.

87. $3x^{-a} − b = 5$.   88. $\dfrac{a}{1 − 2x^{-3}} = b$.

Compute *x:*

89. $x^{10} - 1 = 0.$

90. $10^{-2x} = 100.$

91. $(0.5)^x = 2.$

92. $(0.3)^{-x} = 457.$

93. $\left(\dfrac{4}{x}\right)^{-5} = 0.26.$

94. $5 = x[1 + (1 + 0.01)^{50}].$

95. $3000 = 50\left[\dfrac{1 + (1 + 0.02)^{-x}}{.02}\right].$

96. $(1 + x) = \text{antilog } (0.0253).$

97. $(1 + x)^{-25} = 0.46.$

8

Mathematics of Finance

8.1 INTRODUCTION

SIMPLE INTEREST plays a role in many business transactions, but the calculations are elementary and will not be treated here. This chapter applies the subject matter of Chapter 7 to a number of compound interest problems which arise in financial transactions. The reader will find the forms of the equations in this chapter similar to the forms of many equations of Chapter 7. Here, however, the formulas are stated in the conventional symbols of the mathematics of finance, and the reader should set out at once to learn the meaning of each symbol as it appears.

We shall develop and apply formulas to answer questions such as the following, for varying interest rates and frequencies of compounding:

a) How much will a current deposit of $5000 grow to in ten years?
b) How much should be deposited now if the amount deposited must grow to $5000 in eight years?
c) If $500 is deposited every year, what will be the total amount in the account at the end of 15 years?
d) How much should be deposited every six months in order to accumulate $10,000 in 5 years?
e) What sum deposited now will provide an income of $1000 per year for the next 15 years?
f) If $30,000 is borrowed now to pay for a house, how much must be paid back each month to cancel the debt in 20 years?

Finally, we shall investigate the outcome of compounding interest more and more frequently and develop the formula for *instantaneous,* or *continuous,* compounding.

259

8.2 COMPOUND AMOUNT FORMULA

A person who has $500 in a bank account at 4 percent interest per year and withdraws the interest earned each year will have an annual interest income of

$$0.04(500) = \$20$$

and the balance in the account will remain at $500. If, however, he does not withdraw the interest, it is logical that interest for the second year be computed on the $520 which was in his account at the beginning of the second year. Second-year interest earnings would then be

$$0.04(520) = \$20.80.$$

The $20.80 may be thought of as interest on the original $500 deposit, plus interest on the $20 of interest earned during the first year and credited to the account at the end of the first year. The practice of computing interest on interest is called *compounding* interest. We now develop the basic mathematical formulation of compound interest.

Let P = Amount placed at interest (the principal)
i = Interest rate per period
n = Number of periods P has been at interest
S = Sum of P and interest earned after n periods
= The compound amount.

It makes no difference for our purposes here whether a period is a year, a half year, or whatever. To be specific, let us make the interest period one year.

At the end of the first year the amount of interest earned on P dollars will be iP. The sum of principal and interest at the end of the first year will be

$$P + iP = P(1 + i)^1.$$

Hence, for $n = 1$:

$$S = P(1 + i)^1.$$

During the second year, interest will be earned on $P(1 + i)$ dollars and will amount to $iP(1 + i)$. Adding this interest to the amount in the account at the beginning of the second year, we find the amount in the account at the end of the second year will be

$$S = P(1 + i) + iP(1 + i).$$

Factoring the right member, we have for $n = 2$:

$$S = P(1 + i)(1 + i) = P(1 + i)^2.$$

The reader may verify that for $n = 3$:

$$S = P(1 + i)^3.$$

In fact, the equality between n and the exponent of $(1 + i)$ is perfectly general. Hence:

Compound Amount Formula: $S = P(1 + i)^n$.

The formula as stated is solved for S. It can be solved easily for any of the other letters. The following examples suggest the types of problems which require the determination of one of the variables in the compound amount formula.

Example. If $200 is deposited at 4 percent, compounded annually, what will be the sum of the deposit plus interest (that is, the compound amount) at the end of five years?

$$S = 200(1 + 0.04)^5.$$

We calculate S as follows, using four-place logarithm tables:

$$\log 200 = 2.3010$$
$$5 \log 1.04 = \underline{0.0850}$$
$$\log S = 2.3860$$
$$S = \$243.2.$$

Exercise. How much will $100 grow to in 6 years at 5 percent interest compounded annually? Answer: $134.0.

Example. How much should be deposited now at 5 percent, compounded annually, if the account is to grow to $500 in 10 years?

In this problem the amount at the end of 10 years is $S = \$500$. The unknown is the amount to be deposited now, P. Hence:

$$500 = P(1 + 0.05)^{10}$$

$$P = \frac{500}{(1.05)^{10}}$$

$$\log 500 = 2.6990$$

$$10 \log 1.05 = \underline{0.2120}$$

$$\log P = 2.4870$$

$$P = \$306.90.$$

Recall what the $306.90 represents. It is the sum of money which, if placed at 5 percent interest *now,* would grow to $500 10 years from now. It is helpful to think of P (the $306.90) as the *present value* of the future sum S (the $500). Clearly, the present value of a future sum is less than the future sum.

Exercise.   Mr. X has an obligation to pay you $100 five years from now. You need money now and Mr. X offers to pay off his obligation under the condition that the present value of the debt be computed on the basis of interest at 5 percent compounded annually. How much should Mr. X pay you? Answer: $78.34.

Example.   How long will it take for a sum of money to double itself at 6 percent, compounded annually?

Here, it matters not what sum of money is involved. In any case the future amount, S, is to be twice the present value, P. Hence:

$$\frac{S}{P} = 2.$$

We are asked to find n:

$$S = P(1 + 0.06)^n$$

$$\frac{S}{P} = (1.06)^n$$

$$2 = (1.06)^n.$$

Solving the last equation, we find:

$$n = \frac{\log 2}{\log 1.06} = \frac{0.3010}{0.0253} = 11.9.$$

We see that money at 6 percent, compounded annually, will double in about 11.9 years; that is, the compound amount at the end of 12 years will be slightly more than twice the original deposit.

Exercise.   How long will it take for a sum of money to increase by 25 percent at compound interest of 5 percent per year? Answer: 4.57 years.

Example.   At what rate of interest, compounded annually, will a sum of money increase by 75 percent in 10 years?

This problem is similar to the last, except that the unknown is the interest rate i. The ratio of S to P will be 1.75. Hence:

$$1.75 = (1 + i)^{10}.$$

The steps in the solution are:

$$\log 1.75 = 10 \log (1 + i)$$

$$\frac{\log 1.75}{10} = \log (1 + i)$$

$$\text{antilog} \left[\frac{\log 1.75}{10} \right] = 1 + i$$

$$\text{antilog} (0.0243) = 1 + i$$

$$1.058 = 1 + i$$

$$0.058 = i.$$

The desired interest rate is 5.8 percent.

Exercise. At what rate of interest compounded annually will a sum of money double itself in 10 years? Answer: 7.2 percent.

8.3 PROBLEM SET 1

1. Solve $S = P(1 + i)^n$:
   a) For P.   b) For n.   c) For i.
2. What will be the value seven years hence of a deposit now of $75 at 4 percent compounded annually?
3. For every $100 deposited now at 5 percent compounded annually, how much will be available eight years from now?
4. If $100 is deposited now at 3 percent compounded annually, what will be the sum of principal and interest 10 years from now? How much of this sum will be interest?
5. The rate of growth of a fish population is 4 percent per year. If a lake was stocked with 100 fish 20 years ago, what should be the number of fish in the lake now?
6. The rate of growth of a fish population is 4 percent per year. How many fish should be placed in the lake now to provide a population of 1000 fish 25 years from now?
7. What is the present value of $100 due eight years hence if interest is calculated at 4 percent compounded annually?
8. How much should be deposited now for each $100 wanted 10 years from now if interest is figured at 5 percent compounded annually?
9. If output per man-hour increases by 5 percent annually and is currently 100 units per man-hour, what was output per man-hour five years ago?
10. What sum deposited now at 4 percent will provide enough money to pay a debt of $500 due seven years from now?

11. How much will $50 grow to in 20 years at 3 percent, compounded annually?
12. How long will it take for a sum of money to triple itself at 4 percent, compounded annually?
13. At what rate of interest, compounded annually, will a sum of money double itself in six years?
14. If the population of a city changed from 25,000 to 39,000 in five years, what was the average annual percent rate of increase?
15. How much money should be invested at 4 percent, compounded annually, if the account is to grow to $500 six years from now?
16. If a company's output increases by 6.5 percent per year, how many years will it take for output to rise 50 percent above its current level?
17. A man wishes to have $2500 available in a bank account when his daughter's first-year college expenses begin. How much must he deposit now at 3.5 percent, compounded annually, if the girl is to start in college six years from now?
18. If the population of an area is 28,000 and is growing at the rate of 7 percent per year, what will be the population 10 years from now?
19. If the purchasing power of a dollar were to decrease by 2 percent per year, how long would it take for the purchasing power to reach 50 cents?
[NOTE: The applicable formula is $S = P(1 - i)^n$.]

8.4 SUM AND PRESENT VALUE OF AN ORDINARY ANNUITY

Suppose that we plan to deposit $100 each year for 10 years to an account which earns 3 percent interest, compounded annually, and we want to compute the total amount which will be in the account after the last of the 10 payments. Such a series of periodic payments is called an *annuity,* and our problem is to find the (future) sum of the annuity.

On the other hand, suppose that we plan to make a lump-sum deposit *now* in an account paying 3 percent, compounded annually, and we want to find out what this present sum should be if it is to be sufficient to provide an income of $100 per year for the next 10 years. In this case, we seek the *present value* of a series of periodic payments; that is, we seek the present value of an annuity.

8.5 SUM OF AN ORDINARY ANNUITY

The symbols used in this section are

n = Number of periods
i = Interest rate per period
R = Payment (rent) per period
S = Sum of the annuity.

An *ordinary* annuity is a series of equal payments, each payment of which is made at the *end* of a period. Although the formulas for an annuity in which payments are made at the beginning of each period are easily developed from those for an ordinary annuity, we shall confine our discussion to ordinary annuities. Moreover, we shall consider only the case where the time of payment coincides with the time that interest is computed, that is, for example, when payment and interest computations both occur at the end of the year.

Consider an ordinary annuity for which

$$R = \$10 \text{ per year}$$
$$i = 4 \text{ percent, compounded annually}$$
$$n = 5 \text{ years.}$$

The first of the five payments is made at the *end* of the first year and will accumulate at interest for four years. The second payment will accumulate at interest for three years, and so on for the remaining payments, as shown in Table 8–1. As the table shows, the last payment draws no interest because the sum of the annuity is computed immediately after the last payment

TABLE 8–1

Payment Number	Amount of Payment	Number of Years at 4 Percent Interest	Compound Amount of Payment
1	$10	4	$10(1.04)^4$
2	10	3	$10(1.04)^3$
3	10	2	$10(1.04)^2$
4	10	1	$10(1.04)^1$
5	10	0	10

has been made. The desired quantity, S, is the sum of the amounts in the last column of the table. It could be computed by obvious longhand procedures. However, we wish to simplify such calculations, and to do so, we resort to a trick. The trick is to write S as the sum of the last column of the table, then write $1.04S$ by multiplying each amount in the last column of the table by 1.04, and, finally, subtract S from $1.04S$, as shown next:

$$1.04S = \qquad 10(1.04) + 10(1.04)^2 + 10(1.04)^3 + 10(1.04)^4 + 10(1.04)^5$$
$$\underline{S = \quad 10 + 10(1.04) + 10(1.04)^2 + 10(1.04)^3 + 10(1.04)^4 \qquad\qquad}$$
$$0.04S = -10 \qquad\qquad\qquad\qquad\qquad\qquad\qquad\qquad\qquad + 10(1.04)^5$$

The third line, obtained by term-by-term subtraction, is

$$0.04S = -10 + 10(1.04)^5.$$

Rearranging the last statement, we may write

$$S = 10\left[\frac{(1.04)^5 - 1}{0.04}\right].$$

Evaluation of the last expression yields $S = \$54$.

If we locate the starting numbers

$$R = 10$$
$$i = 4 \text{ percent}$$
$$n = 5$$

in the last expression and replace the numbers with the corresponding letters, we have what proves to be the formula for the sum of an ordinary annuity:

Formula for Sum of Ordinary Annuity: $S = R \left[\dfrac{(1 + i)^n - 1}{i} \right].$

This formula can be applied to evaluate any one of the letters (except i), given the values of the other letters. Some typical examples follow.

Example.   At the time of birth of a son, a man decides to deposit a certain amount of money each year in the form of an ordinary annuity. He wishes the sum of the annuity to be $8000 after the 18th payment, which sum is to provide for a part of the boy's college education. If payments accumulate at 3 percent, compounded annually, how much should each payment be?

Here, we are to find the amount of the periodic payment, which is R. Substituting in the formula:

$$8000 = R \left[\frac{(1.03)^{18} - 1}{0.03} \right]$$

$$0.03(8000) = R(0.7)$$

$$\$342.90 = R.$$

In the last calculation, $(1.03)^{18}$ was evaluated with four-place mantissas. Generally, when working with such mantissas, we write answers to four significant digits. However, the error introduced when an approximate mantissa is multiplied by a two-digit number, such as 18, makes us distrust the fourth digit in $342.90 and leads us to state the answer as approximately $343.

It is instructive to observe in the last problem that the 18 payments amount to about $6170 and provide a sum of $8000. The $1830 difference is the interest earned during the term of the annuity.

Exercise.   How much must be deposited each year at 5 percent compounded annually to accumulate $1000 10 years from now? Answer: About $79.5.

Example.   If deposits of $50 per year are made to an account paying 4 percent compounded annually, what would the amount in the account be after the 20th payment?

$$S = 50 \left[\frac{(1.04)^{20} - 1}{0.04} \right].$$

If we appeal to four-place mantissas in the computation of $(1.04)^{20}$ and round the final answer to three significant digits, the amount turns out to be approximately $1490.

Exercise. A man places $100 every year in an account for his son. How much will be in the account after five such deposits if interest is 5 percent compounded annually? Answer: $552.

Example. If a deposit of $50 per year is made to an account in the form of an ordinary annuity, and interest is 4 percent, compounded annually, how many payments will it take to accumulate $1500?

Here, we are asked to find n, given

$$S = \$1500$$
$$i = 0.04$$
$$R = 50.$$

Substituting into the formula for the sum of an annuity:

$$1500 = 50\left[\frac{(1.04)^n - 1}{0.04}\right].$$

Manipulation of the expression leads to

$$(1.04)^n = 2.2.$$

Solving for n, we find

$$n = \frac{\log 2.2}{\log 1.04} = \frac{0.3424}{0.0170} = 20.1.$$

The accumulated sum will not have quite reached $1500 after the 20th payment. See the last example.

Exercise. Approximately how many annual payments of $100 each would be required to accumulate a sum of $1000 if interest is calculated at 4 percent compounded annually? Answer: 8.6 years; that is, nine payments would be required, and the sum after the ninth payment would exceed $1000. See Problem 7 in Problem Set 2.

8.6 FORMULA FOR THE SUM OF ANY GEOMETRIC SERIES

A *geometric* series is one in which each term is a constant multiple, r, of the preceding term. If the first term is 1, we have

$$\text{Geometric Series} = 1 + r + r^2 + r^3 \text{ and so on.}$$

Observe that if we have $n = 4$ terms, the last term has 3 as its exponent. Thus, if the series has n terms, the last has $n - 1$ as its exponent. Letting S_n denote the first n terms of a geometric series, we may write

$$S_n = 1 + r + r^2 + r^3 + \cdots + r^{n-1}.$$

A concise formula for evaluating S_n can be derived if we multiply S_n by r to obtain rS_n, then subtract this from S_n to obtain $S_n - rS_n$, as follows:

$$S_n = 1 + r + r^2 + r^3 + \cdots + r^{n-1}$$

Subtract

$$rS_n = \quad r + r^2 + r^3 + \cdots + r^{n-1} + r^n$$

Yielding

$$S_n - rS_n = 1 \quad 0 \quad 0 \quad 0 \qquad\qquad 0 \quad - r^n.$$

Thus we see that

$$S_n - rS_n = 1 - r^n$$

so that

$$S_n(1 - r) = 1 - r^n$$

and

$$S_n = \frac{1 - r^n}{1 - r}.$$

To illustrate the formula, let us compute

$$S_n = 1 + \frac{1}{2} + \frac{1}{4} + \frac{1}{8} + \frac{1}{16} + \frac{1}{32} + \frac{1}{64} + \frac{1}{128}.$$

Here we have $n = 8$ terms and $r = 1/2$. Hence,

$$S_8 = \frac{1 - \left(\frac{1}{2}\right)^8}{1 - \frac{1}{2}} = \frac{1 - \frac{1}{256}}{\frac{1}{2}} = 2 - \frac{1}{128} = \frac{255}{128}.$$

Exercise.   Consider the series $1 + 2 + 4 + 8 + 16 + 32$. *a*) Find the sum by ordinary addition. *b*) What are n and r? *c*) Compute the sum by the formula. Answer: *a*) 63. *b*) $n = 6$, $r = 2$. *c*) 63.

The sum of an ordinary annuity of $1 per period for n periods is

$$S_n = 1 + (1 + i) + (1 + i)^2 + (1 + i)^3 + \cdots + (1 + i)^{n-1}.$$

This is a standard geometric series with $r = 1 + i$. Hence,

$$S_n = \frac{1 - (1 + i)^n}{1 - (1 + i)} = \frac{(1 + i)^n - 1}{i},$$

which is the formula we arrived at informally earlier in the discussion.

8.7 PROBLEM SET 2

Use logarithms where necessary or convenient in the solutions of the following:

1. If $100 is deposited each year for 10 years to an account drawing 5 percent interest, compounded annually, how much will be in the account just after the 10th deposit?

2. Find the sum of an ordinary annuity consisting of five annual $200 payments if the interest rate is 3 percent, compounded annually.

3. What should be the rent (periodic payment) for an ordinary annuity which is to accumulate to $1000 in 10 payments if interest is 5 percent, compounded annually?

4. How much should one deposit each year at 4 percent, compounded annually, if he wishes to have $500 in his account just after the fifth deposit?

5. Using the formula for the sum of an ordinary annuity, find n if

$$S = \$1000$$
$$i = 5 \text{ percent}$$
$$R = \$50.$$

6. Approximately how many annual payments of $100 each are needed to accumulate $1000 if interest is computed at 3 percent, compounded annually?

7. In the last exercise of Section 8.5, what would be the amount in the account after the ninth payment?

8. Find the sum by the formula for a geometric series and verify by ordinary addition.

   a) $1 + \frac{1}{3} + \frac{1}{9} + \frac{1}{27} + \frac{1}{81}.$

   b) $1 + 3 + 9 + 27 + 81 + 243.$

8.8 PRESENT VALUE OF AN ORDINARY ANNUITY

The present value of an ordinary annuity of $1 per period for n periods at interest rate i is the amount we would be willing to pay *now* (at the present) in order to receive $1 a year for n years. Figure 8 – 1 shows an annuity of $1

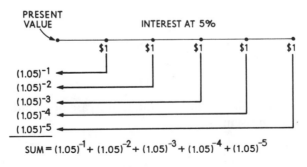

FIGURE 8–1

per year for five years at 5 percent. The first $1 will be payable one year from now, so to find its present value we compute $(1 + 0.05)^{-1}$. The second payment is two years in the future, its present value is $(1 + 0.05)^{-2}$, and so on.

The sum of the $n = 5$ present values is

$$S = (1.05)^{-1} + (1.05)^{-2} + (1.05)^{-3} + (1.05)^{-4} + (1.05)^{-5}.$$

Now let us multiply S by (1.05) and subtract S from the product to obtain $(1.05)S - S$. Thus,

$$1.05S = 1 + (1.05)^{-1} + (1.05)^{-2} + (1.05)^{-3} + (1.05)^{-4}$$

Subtract

$$\begin{array}{c} S = \\ \overline{0.05S = 1} \end{array} \quad \begin{array}{c} (1.05)^{-1} + (1.05)^{-2} + (1.05)^{-3} + (1.05)^{-4} + (1.05)^{-5} \\ \overline{0} \qquad \overline{0} \qquad \overline{0} \qquad \overline{0} \qquad - (1.05)^{-5} \end{array}.$$

Hence,

$$0.05S = 1 - (1.05)^{-5} \quad \text{and} \quad S = \frac{1 - (1.05)^{-5}}{0.05}.$$

In the last expression, $n = 5$ and $i = 0.05$. In general, we may show that the present value of \$1 per period for n periods at rate i is

$$\frac{1 - (1 + i)^{-n}}{i}$$

so that the formula for the present value, A, of \$$R$ per period for n periods at rate i is

$$A = R\left[\frac{1 - (1 + i)^{-n}}{i}\right].$$

It is instructive also to develop the last formula by applying present value calculations to the formula for the (future) sum of an annuity of \$$R$ per period. We start with the compound amount formula which states

$$S = P(1 + i)^n$$

where S is the future sum of the present value, P. Solving for P, we have

$$P = \frac{S}{(1 + i)^n} = S(1 + i)^{-n}.$$

The present value of the future sum, S, can be found by multiplying S by the *compound discount factor*

$$(1 + i)^{-n}.$$

The same holds true for any future sum. In the case of an ordinary annuity, we know the future sum is

$$R\left[\frac{(1 + i)^n - 1}{i}\right].$$

The present value of this sum, which we shall call A, is found (by applying the compound discount factor) to be

$$A = R\left[\frac{(1 + i)^n - 1}{i}\right](1 + i)^{-n}.$$

The last expression can be written in two forms, one of which has a negative exponent. Thus:

Present Value of Ordinary Annuity Formulas:

$$A = R\left[\frac{1 - (1 + i)^{-n}}{i}\right]$$

$$A = R\left[\frac{(1 + i)^n - 1}{i(1 + i)^n}\right].$$

Let us review the concept designated by the symbol A. From one viewpoint, we say that periodic payments of $\$R$ per period will accumulate to a future sum, S, and A is the present value of this sum. From another viewpoint, A is the amount which must be deposited *now* to an account (given the interest rate) if we wish to withdraw $\$R$ a year for n years from the account, at the end of which time the account is to be exhausted.

Example.   A man wishes to make a lump-sum deposit of an amount which will provide tuition payments of $2000 for his son. Four withdrawals of $2000 will be made, the first taking place one year after the account has been established; the account is to be exhausted after the fourth withdrawal. What should be the amount of the lump-sum deposit if interest is 4 percent compounded annually?

$$A = \$2000\left[\frac{(1.04)^4 - 1}{0.04(1.04)^4}\right].$$

Evaluation in the usual manner leads to

$$A = \$2000\left[\frac{0.169}{0.04676}\right] = \$7230.$$

The accuracy of the last answer is poor. If the number $(1.04)^4$ is found longhand, and calculations are carried to the nearest cent, the answer proves to be $7259.79. We shall soon see how to apply special tables to improve accuracy without appealing to longhand calculations.

Further work with the formula for the present value of a series of future payments is provided in the following examples and exercises.

Example.   If $1000 is deposited now at 4 percent, compounded annually, and withdrawals of $50 are made each year, starting at the end of the current year, approximately how many withdrawals will be made before the account is exhausted?

In this problem, $1000 is the present value of an ordinary annuity of $50 per period for an unknown number of periods. It is clear that 20 withdrawals will amount to $1000. However, the account is accumulating interest as time goes on, which will make extra withdrawals possible. The unknown number of withdrawals is n in the formula for the present value of an ordinary annuity. Substituting, we find

$$1000 = 50\left[\frac{1 - (1.04)^{-n}}{0.04}\right].$$

The last statement reduces to

$$(1.04)^{-n} = 0.2.$$

Solving the equation, we have

$$-n \log 1.04 = \log 0.2$$

$$-n = \frac{\log 0.2}{\log 1.04}$$

$$-n = \frac{0.3010 - 1}{0.0170}$$

$$n = \frac{0.6990}{0.0170} = 41.1.$$

At first glance, it appears that $1000 could not provide 41 withdrawals of $50 each. However, the account diminishes slowly at first because of the contribution of interest. At the end of the first year, for example, interest of

$$0.04(1000) = \$40$$

has been earned and added to the account. After the first $50 withdrawal the account has diminished by only $10.

Exercise.   A man stipulates that upon his death his life insurance ($50,000) shall be deposited in a bank to provide annual payments of $5000 to an heir. If the account earns interest at 5 percent compounded annually, approximately how many annual payments can be made? Answer: 14 plus payments ($n = 14.2$).

Example.   A man borrows $1500 to pay for work done on his house and agrees to repay the debt by five equal annual payments with interest at 6 percent, compounded annually. If the first payment is to be made at the end of the first year, how much should each payment be?

It is important to understand in this example that the $1500 is received *now*. It is a present value. This present value is to be equated to a series of future payments. Thus, $1500 is the present value of an ordinary annuity whose rent, R, is to be determined.

$$1500 = R\left[\frac{(1.06)^5 - 1}{0.06(1.06)^5}\right]$$

$$R = 1500\left[\frac{0.06(1.06)^5}{(1.06)^5 - 1}\right].$$

Applying four-place mantissas in evaluating the power,

$$R = \$356 \text{ (approximately)}.$$

Exercise.   See the last exercise. If the heir were to receive 10 equal annual payments from the account established by $50,000 at 5 percent, how much would each payment be, approximately? Answer: $6470, approximately.

8.9 RELATIONSHIP OF PRESENT VALUE AND FUTURE SUM OF AN ANNUITY

We have seen that an annuity is a series of equal periodic payments or withdrawals of an amount R, where R is called the rent of the annuity. Figure 8–2 shows the future sum, S, of an annuity of R per period for four

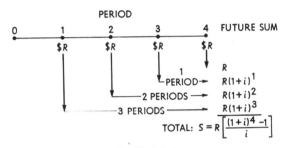

FIGURE 8–2

periods at compound interest rate i. The first deposit is made at the end of the beginning period and grows at compound interest to the value $R(1 + i)^3$. The second deposit gathers interest for two periods and grows to $R(1 + i)^2$. The third deposit grows to $R(1 + i)^1$. The last deposit enters as simply R in the sum because the sum is evaluated at the time of the last deposit, so this deposit gathers no interest. The future sum is

$$S = R + R(1 + i)^1 + R(1 + i)^2 + R(1 + i)^3$$
$$= R[1 + (1 + i)^1 + (1 + i)^2 + (1 + i)^3].$$

Thus, the future sum of an annuity is seen to be a series of compound interest calculations which could be carried out by the compound amount formula at the beginning of the chapter. However, we have derived a formula which greatly simplifies the work. According to this formula,

$$S = R[1 + (1 + i)^1 + (1 + i)^2 + (1 + i)^3] = R\left[\frac{(1 + i)^4 - 1}{i}\right].$$

The utility of the formula may be appreciated by noting that if the annuity was for 100 periods, the compound amount expression would contain 100 terms in the brackets, but the only change in the simplifying formula would be the replacement of the exponent 4 by 100.

274   *Mathematics*

Exercise. Would the sum of an annuity of $50 per period for 10 periods at 5 percent be greater than, equal to, or less than $500? Answer: Greater than, because all payments (except the last) gather interest.

Figure 8–3 shows the present value counterpart of Figure 8–2. Here we view the amounts, $R, from the present rather than in the future. The first

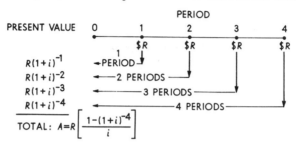

FIGURE 8–3

R is one period in the future. Its present value, $R(1 + i)^{-1}$, is less than R. The second R is two periods in the future. Its present value, $R(1 + i)^{-2}$, is less than the present value of the first R. Similarly for the third and fourth R's. The total present value of the four R's is

$$A = R(1 + i)^{-1} + R(1 + i)^{-2} + R(1 + i)^{-3} + R(1 + i)^{-4}$$
$$= R[(1 + i)^{-1} + (1 + i)^{-2} + (1 + i)^{-3} + (1 + i)^{-4}].$$

Thus, we could compute A by evaluating all the terms and adding the results. Again, however, we were able to derive a formula which greatly simplifies the work. This formula is

$$A = R\left[\frac{1 - (1 + i)^{-4}}{i}\right].$$

Exercise. If you wish to deposit now in an account paying 5 percent interest an amount sufficient to provide 10 annual payments of $1000 each, would the required deposit be greater than, equal to, or less than $10,000. Why? Answer: Less than $10,000 because the interest gathered by the account will provide part of each $1000 payment.

Problem Set 3 affords practice in computing the present value of an ordinary annuity. The relevant formulas are:

$$A = R\left[\frac{1 - (1 + i)^{-n}}{i}\right] \quad \text{or} \quad A = R\left[\frac{(1 + i)^n - 1}{i(1 + i)^n}\right].$$

8.10 PROBLEM SET 3

Use logarithms where necessary or convenient in the solutions of the following:

1. What is the present value of an ordinary annuity of $50 per year for five years if interest is 5 percent, compounded annually?

2. What sum of money deposited now at 5 percent, compounded annually, will make possible seven annual withdrawals of $100 each, the first withdrawal to be made one year hence?

3. If the present value of an ordinary annuity at 4 percent, compounded annually, is $300, approximately how many annual payments of $20 could the annuity provide?

4. A man is committed to pay $25 per year on a debt. He deposits $300 in a bank paying 4 percent, compounded annually, and plans to withdraw $25 each year to pay on the debt. Approximately how many years will the account provide funds for payment of the debt?

5. If the present value of an ordinary annuity of 10 payments is $500 and the interest rate is 5 percent, compounded annually, what is the amount of each payment?

6. A man borrows $1000 at 5 percent, compounded annually. The debt is to be repaid in five equal annual installments, the first payment to be made at the end of the first year. Find the amount of each payment.

7. What is the present value of an ordinary annuity of $50 per year for four years, interest at 2 percent, compounded annually? (Compute answer to the nearest cent; do not use logarithms.)

8.11 CONSTRUCTION OF TABLES

Up to this point, we have carried out compound interest computations step by step in order to develop familiarity with the formulas and their meanings. However, compound interest problems are so common in business transactions that tabular aids have been constructed to ease the computational burden and, at the same time, to increase accuracy.

The expression which is fundamental to all the tables is the compound amount of $1 for n periods at i percent per period, $(1 + i)^n$. Suppose we let i be 1 percent, then list $(1.01)^n$ for successive integral values of n. Thus:

$$n = 1 \qquad (1.01)^1 = 1.01$$
$$n = 2 \qquad (1.01)^2 = 1.01(1.01) = 1.0201$$
$$n = 3 \qquad (1.01)^3 = 1.01(1.0201) = 1.030301$$

and so on. We continue the successive multiplications by 1.01 and tabulate the results for as many values of n as desired. We repeat the entire process for other values of i and compile a table for as many values of i as desired. Table II at the end of the book has been computed for n running from 1 to 40 at rates of 1%, 2%, and so on to 8%. We see at a glance from this table that $1 at compound interest of 8 percent for 28 years will grow to $8.62711; under the same circumstances, $100 would grow to

$$100(8.62711) = \$862.71.$$

Exercise. How much will $1000 grow to in 20 years at 7 percent compounded annually? Answer: Using Table II, $1000(3.86968) = $3869.68.

The reciprocals of the numbers in Table II are the compound discount factor

$$\frac{1}{(1 + i)^n} = (1 + i)^{-n}.$$

Values for the last expression are given in Table III and are called the present value of $1. We see that if interest is reckoned at 4 percent, compounded annually, a debt of $1 due 15 years hence could be settled by a payment of $0.555265 now. Under the same conditions, $100 would have a present value of

$$100(0.555265) = $55.53.$$

Exercise. The population of a town is now 500 and has been growing at a rate of 5 percent per year. What was the population 10 years ago? Answer: Using Table III, $(500)(0.613913) = 307$.

As already indicated, tabular entries apply to the unit amount, $1. When an amount different from $1 is at hand, the tabular entry is multiplied by this amount. Entries for the sum of an ordinary annuity of $1 are given in Table IV. The symbol for the sum of $1 per period for n periods at i percent per period is $s_{\overline{n}|i}$, which we read as "s angle n i." The small letter s refers to the unit amount, $1. The sum, S, of an annuity of R per period is

$$S = Rs_{\overline{n}|i}.$$

According to Table IV, 40 annual deposits of $1 each in the form of an ordinary annuity at 6 percent, compounded annually, will yield the sum

$$s_{\overline{40}|6\%} = 154.76197.$$

If each deposit were $100, the sum would be

$$S = 100s_{\overline{40}|6\%}$$
$$= 100(154.76197)$$
$$= $15,476.20.$$

Exercise. How much money will be accumulated by annual deposits of $200 over a period of 25 years to an account paying 5 percent interest compounded annually. Answer: Using Table IV, $200(47.72710) = $9545.42.

We see from the formula

$$s_{\overline{n}|i} = \frac{(1 + i)^n - 1}{i}$$

that each entry in Table IV can be computed from the corresponding $(1 + i)^n$ entry in Table II by subtracting one from the latter and dividing by the interest rate.

Table V gives the present value of an ordinary annuity of $1 per period. The a in the symbol $a_{\overline{n}|i}$ designates present value, and the small letter refers to the unit value, $1. The formula

$$a_{\overline{n}|i} = \frac{1 - (1 + i)^{-n}}{i}$$

shows that each entry in Table V can be obtained from the corresponding entry in Table III by an obvious calculation. Table V shows that the present value of 25 annual payments in the form of an ordinary annuity at 4 percent, compounded annually, is

$$a_{\overline{25}|4\%} = \$15.62.$$

If each payment were $100, the present value would be

$$A = 100a_{\overline{25}|4\%} = 100(15.62208) = \$1562.21.$$

Exercise.   How much must be deposited now at 7 percent compounded annually to make possible five withdrawals of $1000, the first being made a year from now? Answer: Using Table V, $1000 (4.10020) = \$4100.20$.

The formula in the title of Table VI is

$$\frac{1}{a_{\overline{n}|i}} = \frac{i}{1 - (1 + i)^{-n}}.$$

Thus, each entry of Table VI is the reciprocal of the corresponding entry of Table V. To see the usefulness of Table VI, suppose that a man borrows $1 at 5 percent, compounded annually, and plans to pay off the debt in three equal annual payments as an ordinary annuity, and we wish to determine the amount of the payment. If we substitute into the formula for the present value of an annuity

$$A = Ra_{\overline{n}|i}$$

we have

$$1 = Ra_{\overline{3}|5\%}$$

so that

$$\frac{1}{a_{\overline{3}|5\%}} = R.$$

To evaluate this result from Table V would require the division of one by the tabular entry. Table VI makes this division unnecessary. Using the Table VI entry:

$$0.367209 = R.$$

Thus, with the annuity payback procedure of this example, $0.367209 must be repaid each period for each dollar borrowed. If $1000 had been borrowed, the payment per period would have been

$$1000 \frac{1}{a_{\overline{3}|\,5\%}} = 1000(0.367209) = \$367.21.$$

Exercise.   A man deposits $10,000 now at 6 percent compounded annually and plans to exhaust the account by making equal annual withdrawals for the next 20 years. How much will the annual withdrawal be? Answer: Using Table VI, $10,000(0.087185) = \$871.85$.

Finally, Table VII contains entries for

$$\frac{1}{s_{\overline{m}|\,i}}.$$

Table VII is easily derived from Table VI. To see how, consider the difference

$$\frac{1}{a_{\overline{m}|\,i}} - \frac{1}{s_{\overline{m}|\,i}}$$

which is

$$\frac{i}{1 - (1 + i)^{-n}} - \frac{i}{(1 + i)^n - 1}.$$

If we factor out i and combine the two expressions, we obtain

$$i \left\{ \frac{(1 + i)^n - 1 - 1 + (1 + i)^{-n}}{[1 - (1 + i)^{-n}][(1 + i)^n - 1]} \right\}.$$

Inside the braces, multiplication of the factors in the denominator leads to a product which equals the numerator, so the whole expression equals one. Hence,

$$\frac{1}{a_{\overline{m}|\,i}} - \frac{1}{s_{\overline{m}|\,i}} = i.$$

It follows that

$$\frac{1}{s_{\overline{m}|\,i}} = \frac{1}{a_{\overline{m}|\,i}} - i.$$

Hence, each entry in Table VII can be found by subtracting i from the corresponding entry in Table VI.

Table VII shows what periodic deposit will accumulate to $1. For example, to accumulate $1000 by means of 10 annual deposits at 4 percent, compounded annually, each deposit would have to be

$$R = \frac{S}{s_{\overline{10}|\,4\%}} = 1000(0.0832909) = \$83.29.$$

> **Exercise.** A man plans to accumulate $5000 to purchase a boat. He wishes to make six equal annual deposits to an account paying 5 percent compounded annually, the first deposit to be at the end of the year. How much will each deposit be? Answer: Using Table VII, $5000(0.1470175) = \$735.09$.

Up to the present, we have taken n as a number of years and i as the interest rate per year. In practice, interest often is compounded twice a year (semiannually), four times a year (quarterly), or 12 times a year (monthly). In such cases, n is interpreted as a number of periods, and the stated rate is changed to a rate per period. For example, if a problem states interest at 6 percent, compounded semiannually for 5 years, n will be 10 periods, and i will be 3 percent per period. Again, interest at 4 percent, compounded quarterly, for 5 years leads to

$$n = 20$$
$$i = 1 \text{ percent}$$

and interest at 2 percent, compounded semiannually for $3\frac{1}{2}$ years leads to

$$n = 7$$
$$i = 1 \text{ percent.}$$

The point to be kept in mind is that interest rates conventionally are stated on an annual basis and must be changed in the manner illustrated if interest is compounded more often than once a year.

> **Exercise.** If a problem states that interest is 4 percent compounded semiannually for five years, what interest rate and what number of periods would be used in solving the problem? Answer: Use 2 percent for 10 periods.

8.12 SUMMARY OF TABLES AND FORMULAS

In all expressions where i and n appear, it is assumed that i is the interest rate per period and n is the number of periods. Thus, a rate of 4 percent compounded semiannually for five years leads to $i = 2$ percent, $n = 10$ periods; a rate of 4 percent compounded quarterly for five years yields $i = 1$ percent, $n = 20$ periods.

Table II shows how much $1 will grow to in n periods. Figure 8–4 shows $1 at the left and the amount, $x, which this $1 will have grown to n periods

FIGURE 8–4

later, at the *right.* The tabular entries are the desired values of $x. The compound amount formula for computing the amount S which a principal, P, will grow to is

$$S = P(1 + i)^n.$$

Example.   How much will a deposit of $1000 now grow to in 10 years at 4 percent compounded semiannually? Answer: $1000(1.48595) = $1485.95. See Table II, 2 percent, 20 periods.

Table III shows how much must be deposited if the account is to grow to $1 in n periods. Figure 8–5 shows that the tabular entries, $x, refer to the

FIGURE 8–5

beginning of the time period, and the $1 is at the end of the time period. The compound discount formula for computing the amount P which will grow to an amount S after n periods is

$$P = S(1 + i)^{-n}.$$

P often is referred to as the present value of the future sum, S.

Example.   How much must be deposited now in an account at 8 percent compounded quarterly if the amount in the account 10 years from now is to be $1000? Answer: $1000(0.452890) = $452.89. See Table III, 2 percent, 40 periods.

Table IV shows how much will be accumulated by making equal periodic payments of $1 each. Figure 8–6 shows periodic deposits of $1 with the

FIGURE 8–6

accumulated amount, $x, at the right, immediately after the last payment. As before, $x refers to the tabulated entries. Equal periodic payments at the ends of periods constitute an ordinary annuity. The formula for the sum, S, of an ordinary annuity of R per period is

$$S = R\left[\frac{(1 + i)^n - 1}{i}\right]$$

where R is referred to as the rent.

Example. How much will be accumulated by depositing $1000 every six months for five years at 4 percent compounded semiannually? Answer: $1000(10.94972) = $10,949.72. See Table IV, 2 percent, 10 periods.

Table V shows what amount must be at hand at the beginning to make possible *n* periodic withdrawals of $1 each. Figure 8–7 shows the tabulated

FIGURE 8–7

value, $x, at the left and the $1 withdrawals made possible by $x distributed over time at the right. The tabulated values are the present value of an annuity of $1 per period. The present value, *A*, of an annuity of $R per period is

$$A = R\left[\frac{1 - (1 + i)^{-n}}{i}\right].$$

Example. What sum deposited now at 7 percent compounded annually will provide for withdrawals of $1000 every year for the next 10 years, the first withdrawal to be made a year from now? Answer: $1000(7.02358) = $7023.58. See Table V, 7 percent, 10 periods.

Table VI shows how much must be paid each period to pay off (amortize) a beginning debt of $1. Figure 8–8 shows $1 at the left will be amortized by

FIGURE 8–8

equal periodic payments of $x (the tabulated values). To amortize a beginning amount of *A* dollars would require periodic payments of $R each, where

$$R = A\left[\frac{i}{1 - (1 + i)^{-n}}\right].$$

Example. What equal semiannual payments for five years would be required to amortize a current debt of $1000 with interest at 6 percent compounded semiannually? Answer: $1000(0.117231) = $117.23. See Table VI, *i* = 3 percent, 10 periods.

Table VII shows how much must be deposited each period to accumulate $1 some time later. The account into which payments of $x per period are made is called a sinking fund. Figure 8–9 shows payments of $x leading up

FIGURE 8–9

to an accumulated $1 at the end, just after the last payment. The periodic payment R which will build a sinking fund up to $$S$ is

$$R = S\left[\frac{i}{(1 + i)^n - 1}\right].$$

Example.   What quarterly deposit to an account paying 8 percent compounded quarterly will build an account worth $1000 in five years? Answer: $1000(0.0411567) = $41.16. See Table VII, 2 percent, 20 periods.

Practice in identifying the table appropriate in a given circumstance is afforded by the following examples.

Example.   A man borrows $2000 at 12 percent, compounded monthly, to pay for an addition to his house. The debt is to be repaid in equal monthly installments over a three-year period, the first installment being due one month after receipt of the $2000. Find the amount of each payment.

Reducing the problem to an original debt of $1, we want to find the periodic payment which will cancel (be equivalent to) a present debt of $1. Table VI is indicated; n is 36 periods, and i is 1 percent per period. Thus:

$$2000(0.033214) = $66.43.$$

Table VI shows that a periodic payment of $0.033214 is required for each dollar borrowed. Consequently, a periodic payment of $66.43 is required if $2000 is borrowed.

In passing, we note that the problem at hand is an illustration of *amortization* of a debt. Amortization refers to payment of a debt due *now* by a series of payment in the future.

Example.   A company's board of directors has instructed the financial officer to accumulate $10,000 to retire bonds six years hence. The sum is to be accumulated by making equal annual payments to a bank account. Assuming interest at 7 percent, compounded annually, what should be the amount of each payment?

On a unit basis, we want to find the periodic payment which will cancel (be equivalent to) a future obligation of $1. Table VII is appropriate; n is 6, i is 7 percent. The table shows that six payments of $0.1397958 each will accumulate to $1. To accumulate $10,000, each payment must be

$$10,000(0.1397958) = $1397.96.$$

An account into which periodic payments are made in order to accumulate a future sum often is called a *sinking fund.* We may say that the objective sought in the last example can be achieved by making periodic deposits of $1397.96 into a sinking fund.

Example.   If the population of a town is currently 8000 and it is growing at a rate of 4 percent per year, what will be the population 10 years from now?

This problem is solved by straightforward application of Table II, which shows that the compound amount of one at 4 percent for 10 periods is 1.48024. Hence:

$$8000(1.48024) = $11,842.$$

Example. A house lot is offered for sale at no money down, $25 per quarter for 10 years, starting at the end of the first quarter after purchase. Assuming interest at 12 percent, compounded quarterly, what is the equivalent cash value of the lot?

This is a problem involving periodic payments. The cash value (now) is the present value of an ordinary annuity. From Table V, with n as 40 and i as 3 percent, we have

$$25(23.11477) = \$577.87.$$

Example. What sum deposited now at 8 percent, compounded annually, will provide $1000 eight years from now?

This is a problem of finding the present value of a future sum. Using Table III, we find

$$1000(0.540269) = \$540.27.$$

Example. A man wishes to buy life insurance sufficient to provide his beneficiaries an income of $3000 per year, starting a year after his death and continuing for five years. Assuming that the receipts from the insurance policy can be invested at 4 percent, approximately how much insurance should the man buy?

The time of death is not relevant in this problem. Whenever the man dies, an ordinary annuity is to be established with the proceeds of his insurance. We seek the present value of an annuity, *present* referring to the time of death. From Table V, with 5 for n and 4 percent for i, we find

$$3000(4.45182) = \$13,355.50.$$

The man should buy $13,000 or $14,000 worth of life insurance.

Example. A man has contributed $1000 per year to help care for a relative. The relative inherits money after seven contributions have been received and wishes to pay for the aid he has been given. Assuming interest at 4 percent, compounded annually, how much should the relative pay the man?

The $1000 contributions in this problem may be considered as an ordinary annuity. The sum of the annuity after seven payments is the amount of the relative's debt. From Table IV:

$$1000(7.89829) = \$7898.29.$$

Mortgage Payments. Most readers will first encounter a transaction involving a large sum of money when they buy a home. In the typical situation, a bank loans the home buyer the amount he needs for the purchase, and the home buyer amortizes his debt by making monthly payments for a period of years. Suppose, for example, that the price of the home is $40,000, and that 25 percent, or $10,000, must be paid in cash. The remaining $30,000 debt is to be amortized by equal monthly payments for 15 years with interest at 8 percent compounded monthly, so that the number of periods is $12(15) = 180$, and the monthly rate is $\frac{8}{12}$ percent, or $\frac{2}{3}$ percent. The entries in the appropriate table, VI, do not include fractional rates and large numbers of periods, so an extension, Table XI, has been included to provide practice in mortgage payment examples. With $i = 8\%$ and $n = 180$, we find from Table XI that $1 will be amortized by monthly payments of $0.009 556 5208, so the monthly payment on $30,000 will be

$$\$30,000(0.009\ 556\ 5208) = \$286.70.$$

It is interesting to note that the total amount paid by the buyer will be $180(\$286.70) = \$51,606$. Of this amount, $\$51,606 - \$30,000 = \$21,606$, will be interest.

Exercise. If the mortgage in the foregoing example were for a 10-year period, what would be the monthly payment? What total amount of interest would be paid over the period of the mortgage? Answer: $363.98 per month; total interest would be $13,677.60.

8.13 INTERPOLATION IN THE TABLES TO APPROXIMATE *i* or *n*

An extensive set of tables[1] or a high-precision computer is needed to handle accurately the wide variety of problems encountered in financial calculations. Problems in determining the interest rate, *i*, or the number of payments, *n*, given the appropriate information, may require interpolation to obtain an approximate answer even when an extensive set of tables is at hand. The next series of examples illustrates interpolation in the tables of this book.

Example. A purchaser agrees to pay $15,000 for a house. He pays $10,000 cash and $700 per year for 10 years, starting a year after purchase. Approximately what rate of interest, compounded annually, is he paying on his debt?

His debt is the current amount, $5000. The periodic payments are made to amortize this debt. If $700 per year will amortize $5000, then $0.14 per year will amortize $1; that is,

$$\frac{700}{5000} = 0.14.$$

Referring to Table VI, we glance across the row for $n = 10$, seeking a number close to 0.14. We find:

At 6 percent the entry is 0.135868
At ?   0.14
At 7 percent the entry is 0.142378.

[1]See, for example, *Financial Compound Interest and Annuity Tables*, published by the Financial Publishing Company, Boston, Mass.

Interpolating for the question mark, the interest rate is found to be about 6.6 percent.

Example. Approximately how long will it take for money to double itself at 4 percent, compounded semiannually?

For money to double means that the compound amount of $1 must be $2. Referring to Table II, we glance down the 2 percent column seeking the entry 2.00000. We find:

$$\text{At } n = 35 \text{ the entry is } 1.99989$$
$$\text{At ? } \qquad\qquad 2.00000$$
$$\text{At } n = 36 \text{ the entry is } 2.03989.$$

Thus, money will almost double after 35 periods (17½ years).

Example. If 10 annual payments of $200 each will provide a sum of $2300 after the 10th payment, what rate of interest, compounded annually, is being granted?

This problem involves the sum of an ordinary annuity. To apply Table IV, we note that if $200 accumulates to $2300, then $1 will accumulate to

$$\frac{2300}{200} = 11.5.$$

Moving across the row for $n = 10$ in Table IV, seeking the entry 11.5, we find:

$$\text{At 3 percent the entry is } 11.46388$$
$$\text{At ? } \qquad\qquad 11.50000$$
$$\text{At 4 percent the entry is } 12.00611.$$

Interpolation shows that the rate indicated by the question mark is approximately 3.1 percent.

8.14 PROBLEM SET 4

Solve the following by reference to the appropriate table at the end of the book. It is to be assumed that all periodic payments are in the form of an ordinary annuity.

1. If $100 is deposited every six months for four years at 4 percent, compounded semiannually, what will be the accumulated amount after the last deposit?
2. What periodic payment would be required to amortize a $1000 debt in eight semiannual payments if interest is 6 percent, compounded semiannually?
3. If interest is 8 percent, compounded quarterly, to what sum will a deposit of $500 grow in 10 years?
4. What annual payment to a sinking fund is required if $5000 is to be available after the ninth payment? Interest is 7 percent, compounded annually.
5. What sum of money deposited now will grow to $1000 in eight years at 4 percent, compounded semiannually?
6. How much should be deposited now at 5 percent, compounded annually, to make possible withdrawals of $500 per year for the next 10 years?
7. A man borrows $3000 to pay for construction of a garage. He is to repay the debt in equal semiannual installments over the next 10 years. What should be the amount of each payment if interest is 10 percent, compounded semiannually?

8. If it is stated that productivity of a manufacturing company, as measured by dollars of output per man-hour worked, has increased 5 percent per year and is presently $8.50 per man-hour, what was the corresponding figure for productivity 10 years ago?

9. A man deposits $200 every quarter at 8 percent, compounded quarterly. How much will he have in his account at the end of 10 years?

10. A man wishes to establish an account from which his son can draw $1500 a year for four years. If the account earns interest at 6 percent, compounded annually, how much should the man deposit in the account?

11. Taxes on a piece of property are $900 per year. If taxes increase 2 percent per year, what will be the tax on the property 40 years from now?

12. A man plans to buy a house five years from now, and he wants to accumulate $3000 for a down payment by depositing equal amounts each quarter in a bank which pays 4 percent, compounded quarterly. What should be the amount of each deposit?

13. Mr. Smith purchases a house at $30,000. He pays $10,000 cash and takes a 10-year mortgage for the remainder at 7 percent compounded monthly. What will his monthly mortgage payment be? How much interest will he pay during the 10 years?

14. Mr. Brown takes out a $30,000, 15-year mortgage, at 8 percent compounded monthly. What will his monthly mortgage payment be? How much interest will he pay during the fifteen years?

Solve the following by interpolation in the tables:

15. How long will it take for $500 to grow to $700 at 4 percent, compounded annually?

16. A business is offered for sale at $50,000 cash, or $30,000 down payment and six equal annual payments of $4240. Approximately what rate of interest, compounded annually, is being charged in the periodic payment alternative?

17. Approximately how many semiannual payments of $250 to an account drawing interest at 6 percent, compounded semiannually, will be required to accumulate $2000?

8.15 MULTISTEP PROBLEMS

The solution of a problem may require more than one step and may involve more than one interest rate, as shown by the following examples.

Example. A man borrows $1000 at 6 percent compounded semiannually which he must pay back 10 years hence. He decides to build up enough money to pay the debt when it is due by making annual deposits into an account which pays interest at 4 percent compounded annually, the first deposit to be made a year after he receives the $1000, and the last deposit to be made on the day the debt is due. How much should he deposit each year?

The problem has two parts with different rates of interest for the parts. First, according to Table II with $i = 3$ percent and $n = 20$, the amount of the debt which will be due 10 years hence is $1000(1.80611) = $1806.11. This amount is shown at the right on Figure 8–10. Second, the man must

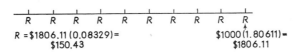

FIGURE 8–10

establish a sinking fund to build up $1806.11 by paying R into his account each year for 10 years. According to Table VII, with $i = 4$ percent and $n = 10$, the required amount is

$$R = \$1806.11(0.0832909) = \$150.43$$

which is shown in Figure 8–10.

Example. A man wishes to know what amount he should put into an account paying 5 percent compounded annually if the account is to provide a return of four annual payments of $2000 each, the first payment to be 11 years from the time of establishing the account. Figure 8–11 shows that to

FIGURE 8–11

provide for an annuity of $2000 per year for four years, starting 11 years hence, would require having $2000(3.54595) = \$7091.90 in the account at the end of 10 years (see Table V, 5 percent, four periods). Next, the present value of $7091.90 is computed to be $7091.90(0.613913) = \$4353.81 by use of Table III, 5 percent, 10 periods. This is the amount which must be placed in the original account to accomplish the desired purpose. See Figure 8–11.

We conclude with a problem whose solution we shall outline here. The details are required as the answer for Problem 1 of Problem Set 5.

A man desires to provide a semiannual income of $5000 for 20 years, starting 15½ years from now. To this end, he has deposited $25,000 in a bank account, and he has a guaranteed inheritance of $10,000 which he will receive 10 years from now and add to the account. He knows these amounts will not provide the income he desires 15 years hence, so he plans to make additional periodic deposits to his account to make up the difference. Assuming interest at 6 percent compounded semiannually, how much should he add to his account every six months, starting six months from now? The outline of the solution follows:

    a) Find the amount he will need (15 years hence) to provide $5000 every 6 months thereafter for the following 20 years.

    b) Find the amount that will be available 15 years hence from his deposits of $25,000 and $10,000.

    c) Find the deficit by subtracting (*b*) from (*a*).

    d) Find the periodic payment which will make up this deficit.

8.16 PROBLEM SET 5

1. See the problem discussed just prior to the Problem Set. Compute the values outlined in (a), (b), (c), and (d).

2. A note for $1000 is due in eight years with interest of 6 percent compounded semiannually. What is the (present) value of the note five years before it is due if this value is computed on the basis of 4 percent compounded annually?

3. A man deposits $5000 now at 4 percent compounded quarterly. He plans to let the account gather interest for 10 years, then, a year later, to start making annual withdrawals from the account, which now draws interest of 5 percent compounded annually, and to exhaust the account by five such withdrawals. How much will each withdrawal be?

4. How much should be deposited now at 6 percent compounded semiannually to make possible four annual withdrawals of $1000 each, the first occurring nine years from now. Assume that the 6 percent rate compounded semiannually holds for eight years, then changes to 5 percent compounded annually.

5. A man plans to make 20 semiannual payments to an account, at 6 percent compounded semiannually, and then use the accumulated amount to establish an ordinary annuity of $1000 per year for five years, interest at 5 percent compounded annually. What should be the amount of each semiannual payment?

6. If a man makes 12 quarterly payments of $100 each into an account earning 4 percent compounded quarterly, then stops payments and lets the accumulated amount grow at 5 percent compounded annually for 10 years, what will be the final amount in the account?

8.17 CONTINUOUS COMPOUNDING

Prior to the widespread installation of electronic computers in banks, interest typically was computed and added to savings accounts every quarter year. Investors who withdrew funds prior to the interest date lost the interest on the amount withdrawn. It was an aid to investors when banks began to move toward monthly compounding. In time, some banks began to compound daily and advertised, "interest from day of deposit to day of withdrawal." At this point, the duration of the compounding period had declined from three months (quarterly), to one month, to one day. Conceivably, the period could have progressed to hourly, every minute, every second, every 1/1000 of a second (which would mean over 31 billion compoundings per year), and so on, *ad infinitum.*

Conceptually, time moves on *continuously,* without breaks, so that a conversion period of even 1/1000 of a second is not continuous conversion. However, as we shall see, it is possible to determine mathematically the outcome of continuous conversion to any desired degree of accuracy, and the process is exceedingly simple. Some banks which had been compounding daily now have taken the ultimate step and offer interest *compounded continuously.* We shall next develop the background for continuous compounding.

At 4 percent interest, $1 grows to $1.04 in one year. At 4 percent compounded semiannually, $1 grows to $1.0404, a slightly larger amount, in one year. The more frequently interest is compounded, the greater is the amount of interest earned in a year. Table II at the end of the book shows that $1 will grow to $1.04060 in a year if 4 percent interest is compounded four times during the year. Electronic computer computations show that 10, 100, 1000, and 10,000 conversions give the respective amounts

$$1.040728$$
$$1.040802$$
$$1.040810$$
$$1.040811.$$

Because the sequence is increasing very slowly, it would appear that no matter how many conversions are made, the first six digits in the amount will be 1.04081, and this turns out to be true. Moreover, as we shall see, it is not necessary to generate the foregoing sequence in order to determine that the result of continuous compounding, accurate to six digits, is 1.04081.

Our purpose is to discover what happens to the compound amount as the number of conversions becomes larger and larger. Let us start by observing that interest at i per period on $1 leads to a compound amount of

$$(1 + i)^1$$

at the end of the period. The same rate compounded twice during the period leads to the compound amount

$$\left(1 + \frac{i}{2}\right)^2.$$

In general, compounding x times a period leads to the compound amount

$$\left(1 + \frac{i}{x}\right)^x$$

at the end of the period. We are interested in learning what happens to this last expression if x gets larger and larger and larger. It will be profitable (for a reason soon to be seen) to consider first the unlikely interest rate 100 percent; that is, $i = 1$. We see at once that $1 at interest of 100 percent per period will grow to a compound amount of

$$(1 + 1)^1 = \$2$$

in one year.

If we compound twice a year, $x = 2$, we have the compound amount

$$\left(1 + \frac{1}{2}\right)^2 = (1.5)^2 = \$2.25.$$

The following tabulation shows how the compound amount changes as x, the number of conversions, increases.

x		$\left(1 + \frac{1}{x}\right)^x$
1	$(2)^1$	$= \$2.000$
2	$(1.5)^2$	$= \$2.250$
10	$(1.1)^{10}$	$= \$2.594$
100	$(1.01)^{100}$	$= \$2.705$
1,000	$(1.001)^{1000}$	$= \$2.717$
		$\$2.71828$
		$\$2.71828\ 18284$
		$\$2.71828\ 18284\ 59045$

It appears from the foregoing table that $(1 + 1/x)^x$ will never exceed 2.72 or, using a finer statement, the expression will never exceed 2.7183.

The number sequence in the right-hand column of the foregoing tabulation presents finer and finer approximations of an important mathematical constant to which mathematicians have assigned the symbol e. We shall see in another chapter that e is the base of the Napierian, or natural, system of logarithms and plays an important role in the calculus. We can compute e to any desired number of places, but it can be proved that e is irrational, so that the decimal sequence never ends. A familiar number of the same type is π. We can write π as 3.14 or, more exactly, 3.1416 or 3.14159 26535 89793, and so on, but the sequence never ends. Values for e and π used in computations always are approximations. Common approximations are $\pi = 3.1416$ and $e = 2.7183$.

Our reason for employing $i = 1$ in the investigation of behavior of the expression $(1 + i/x)^x$ as x increased without limit was to discover the important constant, e. Further investigation, not carried out here, shows that as x grows without limit,

$$\left(1 + \frac{i}{x}\right)^x \qquad \text{becomes } e^i$$

$$\left(1 + \frac{i}{x}\right)^{-x} \qquad \text{becomes } e^{-i}$$

$$\left(1 + \frac{i}{x}\right)^{nx} \qquad \text{becomes } e^{ni}$$

$$\left(1 + \frac{i}{x}\right)^{-nx} \qquad \text{becomes } e^{-ni}.$$

The expressions in e at the right in the foregoing are referred to as compound amounts and present values with interest compounded *continuously*. Thus, assuming interest at rate i compounded continuously,

$$e^i = \text{compound amount of \$1 for 1 period}$$
$$e^{-i} = \text{present value of \$1 due 1 period hence}$$
$$e^{ni} = \text{compound amount of \$1 for } n \text{ periods}$$
$$e^{-ni} = \text{present value of \$1 due } n \text{ periods hence.}$$

At interest of i compounded continuously, $\$P$ will grow in n periods to $\$S$, where

$$S = Pe^{ni}$$

and the present value, P, of $\$S$ due n periods hence is

$$P = Se^{-ni}.$$

Formulas can be developed also for applying continuous interest to periodic payment problems, but we shall not pursue this topic here.

Before turning to examples of continuous interest calculations, attention is called to Table IX at the end of the book. Table IX provides values of e^x and e^{-x} for selected values of x.

Example.   How much will a dollar grow to in a year at 4 percent compounded continuously? Answer: We need the number $e^{0.04}$. From Table IX, with $x = 0.04$, we find $e^x = 1.0408$ so that, to four places of decimals, $1 will grow to $1.0408.

Example.   How much will $100 grow to in five years at 5 percent compounded continuously? Answer: The desired formula is

$$S = Pe^{ni}$$

with $P = \$100$, $n = 5$, and $i = 0.05$. We have

$$S = 100e^{0.25} = 100(1.2840) = \$128.40$$

where the expression $e^{0.25}$ was found to be 1.2840 from Table IX.

Example.   What is the present value of $100 due 10 years from now if money is worth 4 percent compounded continuously? Answer: We must compute

$$P = 100e^{-10(0.04)} = 100e^{-0.4}.$$

From Table IX we find $e^{-0.4} = 0.6703$; hence, $P = \$67.03$.

8.18 APPROXIMATING THE TIME FOR MONEY TO DOUBLE

Earlier in the chapter we calculated by logarithms the time it would take for money to double at a certain rate of interest. A simple and useful approximation of the time to double can be derived in the case of continuous compounding. We note that for $1 to become $2 in n years at rate i,

$$\$1(e^{ni}) = \$2.$$

Taking logarithms of both sides, we have

$$\log (e^{ni}) = \log 2$$

and applying a law of logarithms,

$$ni (\log e) = \log 2$$

so that

$$n = \frac{\log 2}{i(\log e)} = \frac{\log 2}{i \log(2.718)}$$

$$= \frac{0.3010}{i(0.4343)} = \frac{0.6931}{i}.$$

Rounding the numerator to 0.69 and then multiplying numerator and denominator by 100 yields:

$$\textbf{Approximate time to double} = \frac{69}{i\%}.$$

Thus, at 6 percent, money will double in about $\frac{6\%}{6} = 11.5$ years, or between 11 and 12 years.

‖‖‖

Exercise.   Approximately how long will it take money to double at 10 percent? Answer: Almost seven years.

‖‖‖

8.19 PROBLEM SET 6

Find the compound amount for each of the following, assuming interest is compounded continuously.

1. $50 @ 5 percent for one year.
2. $100 @ 10 percent for eight years.
3. $1 @ 50 percent for one year.
4. $100 @ 10 percent for 30 years.
5. $1000 @ 1 percent for 100 years.
6. $10 @ 8 percent for 25 years.

Find the present value for each of the following, assuming interest is compounded continuously.

7. $100 due in five years @ 6 percent.
8. $50 due in one year @ 8 percent.

9. $2.7183 due in one year @ 100 percent.

10. $100 due in 100 years @ 1 percent.

11. $1000 due in five years @ 6 percent.

12. $10 due in two years @ 50 percent.

13. Find the compound amount of $100 at 5 percent compounded annually for 40 years (Table II) and also the compound amount of $100 at 5 percent compounded continuously for 40 years. By how much does the latter exceed the former?

14. According to theory (stated but not proved in the foregoing section) $\left(1 + \frac{2}{x}\right)^x$ becomes e^2 as x grows without limit. From Table IX, or by computing (2.7183) (2.7183), we find $e^2 = 7.3891$. Compute $\left(1 + \frac{2}{x}\right)^x$ for $x = 2$ and $x = 4$, longhand. Then, using common logarithms (Table I), compute the expression for $x = 10$ and $x = 100$. By how much does $\left(1 + \frac{2}{100}\right)^{100}$ differ from e^2?

15. Approximately how long will it take money to double at
    a) 3 percent?   *b*) 5 percent?   *c*) 7 percent?

16. Derive an approximation formula for the time it will take for money to triple at *i* percent interest.

8.20 REVIEW PROBLEMS

In Problems 1–10, compute *n* by logarithms to the nearest tenth.

1. How long will it take population to triple if it grows at a compound rate of 10 percent per annum?

2. How long will it take for money to double itself at 3 percent compounded annually?

3. How long will it take sales to grow from $50,000 to $80,000 if sales grow at a compound rate of 12 percent per annum?

4. If population has been increasing 8 percent a year and now is 100,000, how long ago was population only 25,000?

5. If a fund has been growing at 6 percent per year and is now $12,000, when was its value $7500?

6. How many annual payments of $100 will $1000 at 4 percent compounded annually provide?

7. How long will it take for annual deposits of $100 to grow to $2000 at 5 percent compounded annually?

8. If a school system can service 5000 students and the student population is now 3000 and growing by 15 percent per year, when will the system's capacity be reached?

9. How many annual payments of $250 each will $5000 at 3 percent compounded annually provide?

10. How long will it take for annual deposits of $200 each to accumulate to $5000 at 2 percent compounded annually?

In Problems 11–13, find the rate to the nearest 10th of a percent by means of logarithms.

11. At what rate of interest compounded annually will $1000 grow to $1800 in 10 years?

12. At what rate of interest compounded annually will a sum of money triple itself in 25 years?

13. At what rate of growth will a population increase from 50,000 to 200,000 in 10 years?

Solve Problems 14–19 by interpolation in the tables.

14. How long will it take for money to double itself at 3 percent compounded annually?

15. At what rate of interest compounded annually will money double itself in 15 years?

16. At what rate of interest compounded annually will $10,000 provide 12 annual payments of $1000 each?

17. A man agrees to repay a current debt of $10,000 by making 20 equal annual payments of $769.10, the first payment to be made one year from now. What rate of interest is the man paying?

18. How many annual payments of $400 each will $8000 at 3 percent provide?

19. A man makes 20 annual payments of $100 each to an account, and just after the last payment he learns that the account total is $3000. What rate of interest has he received?

Problems 20–29 may be solved by use of Table II–VII. They can be used also to practice longhand calculations by means of the formulas.

20. How much should be deposited quarterly to an account paying 4 percent compounded quarterly if the account balance is to be $4000 just after the 10th deposit?

21. If a current debt of $2000 increases at 8 percent compounded semiannually, what will be the debt 20 years from now?

22. How much should be deposited now in an account paying 4 percent compounded quarterly if, starting one quarter from now and for nine more quarters thereafter, withdrawals of $500 each are to be made?

23. If a man deposits $1000 every six months to an account paying 6 percent compounded semiannually, what will be the account balance just after the seventh deposit?

24. How much should be deposited now at 5 percent compounded annually to provide $1000 20 years from now?

25. A man agrees to repay a current debt of $5000 with interest at 5 percent compounded annually, in 10 equal annual payments. Find the amount of each payment.

26. A company uses 1 million gallons of water a year at the present rate of consumption. If the company's water needs increase 5 percent every six months, what will be its rate of consumption 15 years from now?

27. How much should be deposited each year at 5 percent compounded **annually** to accumulate $10,000 just after the 11th deposit?

28. How much insurance should a man carry on his life if at his death the insurance, deposited at 4 percent compounded semiannually, is to provide his beneficiaries $1000 every six months for 15 years?

29. It has been estimated that productivity in manufacturing industries in the United States has increased at 4 percent annually. If this is so, and if the same rate continues, by what percent will productivity 25 years hence exceed current productivity?

30. A house is purchased at $50,000, with $25,000 down and the remainder to be amortized by monthly payments for 10 years. If interest is 8 percent compounded monthly, what is the amount of the monthly payment? How much interest will be paid during the period of the mortgage? (Use Table XI.)

31. A condominium apartment is purchased for $10,000 cash and $40,000 to be amortized by monthly payments for 15 years at 7 percent compounded monthly. What is the monthly payment? How much interest will be paid during the period of the mortgage? (Use Table XI.)

Problems 32–34 are multistep problems.

32. Ten years ago, a man deposited $5000 in an account paying 4 percent compounded quarterly. He now finds he can convert to an account paying 5 percent compounded annually and make equal withdrawals from this account for the next 10 years, the first withdrawal to be one year from now. Find the amount of the withdrawal.

33. How much money deposited now at 4 percent compounded annually will provide four payments of $1000 each, the first payment to be made six years from now?

34. A man wishes to build up a fund which will provide him with $1000 per year for 10 years, the first payment to be made 11 years from now To build the fund, he plans to make equal deposits every six months (20 deposits) to an account paying 6 percent compounded semiannually. Find the amount of the deposit required. (Assume that the rate changes to 5 percent annually after 10 years.)

In Problems 35–38, find the compound amount assuming that interest is compounded continuously.

35. $100 @ 4 percent for five years.   36. $1 @ 8 percent for one year.

37. $50 @ 10 percent for five years.   38. $1000 @ 3 percent for 20 years.

In Problems 39–42, find the present value assuming that interest is compounded continuously.

39. $1000 due in 10 years @ 4 percent.   40. $100 due in one year @ 20 percent.

41. $50 due in 100 years @ 2 percent.   42. $1 due in one year @ 50 percent.

43. Using the approximation formula, about how long will it take money to double at *a*) 4 percent? *b*) 12 percent?

44. Derive a formula to approximate how long it will take money to increase by 50 percent at *i* percent interest, compounded continuously.

9

Functions and Limits

9.1 INTRODUCTION

SOME NATURAL QUESTIONS arising in applied mathematics are: Is cost increasing or decreasing? Is the rate of increase in cost slowing down or speeding up? Is profit increasing and, if so, does it reach a maximum? If interest is compounded more and more frequently, is there a limit to the amount to which a deposit will grow? These are questions involving *change* and their answers depend upon the form of the mathematical expression (or model) which represents the applied situation at hand. In this chapter we shall first study the general graphical aspects of curves which relate to questions involving change. Then we shall study the particular characteristics of commonly used models. Finally, we develop the limit concept and the first rules of the subject area, differential calculus, which provide efficient methods for dealing with problems of change.

9.2 FUNCTIONAL NOTATION

The word *function* is used in mathematics in the sense of *depends upon,* as when we say the area of a circle is a function of (depends upon) its radius, or that the sum a deposit of $1000 will grow to is a function of the number of compounding periods and the interest rate per period. In the case of the circle,

$$\text{Area} = A = \pi r^2,$$

we may call attention to the fact that the *variable* upon which the area depends (the *independent variable*) is r by writing

$$A(r) = \pi r^2,$$

where $A(r)$ is read as "the A function of r," or simply, "A of r." In the case of the bank deposit of \$1000, we write

$$S(i, n) = 1000(1 + i)^n$$

where "S of i and n" is a function of the two independent variables, i and n. In this chapter we shall be concerned with functions of one variable only. We shall require that an expression must be single-valued if it is to qualify as a function in order that there be no ambiguity when we speak of the value of a function.[1] Thus, to make the expression $x^{1/2}$ (which could have the two values $+3$ and -3 when x is 9) qualify as a function, *we shall define*

$$g(x) = x^{1/2}$$

and any expression involving an even root to mean only the nonnegative value of the expression.

It seems unnecessarily complicated to change the linear expression $y = 3x + 2$ to the functional form $y(x) = 3x + 2$ because it is clear in $y = 3x + 2$ that x is the independent variable. However, we shall often encounter statements involving more than one letter, such as

$$\frac{NF}{L} + \frac{iL}{2}$$

where it is not clear which letter represents the independent variable and which letters are to be treated as constants. The functional statement

$$C(L) = \frac{NF}{L} + \frac{iL}{2}$$

makes it clear that L is the independent variable and, since only L is specified as a variable in $C(L)$, all other letters are to be treated as constants. The functional form avoids possible ambiguity, and this is one of its advantages.

Another advantage of a functional statement is that instead of having to say, Find the value of y when x is 2, we may simply say, Find $y(2)$, which means the same thing. For example, if

$$y(x) = x^2 - 3x + 10$$

then

$$y(2) = 2^2 - 3(2) + 10 = 8.$$

Exercise. If $h(y) = 3y^2 - 5$, find $h(0)$, $h(-2)$, $h(1)$. Answer: $-5, 7, -2$.

[1]See also Appendix 1, Section A1.5.

Again, if

$$f(k) = k^2 - 2kg$$

we would find

$$f(5) = 25 - 10g,$$

but if

$$h(g) = k^2 - 2kg$$

then

$$h(5) = k^2 - 10k.$$

Exercise.   If $p(q) = q^2 - r^2 + 5$ and $h(r) = q^2 - r^2 + 5$, find:
a) $p(2)$. b) $h(2)$. Answer: a) $9 - r^2$. b) $q^2 + 1$.

It is helpful to think of a function as a *rule* which states how to find the value of the function for a stated value of the independent variable. Thus, in

$$f(x) = x^2 - 5$$

the rule is that the value of the function is obtained by squaring the independent variable and subtracting 5. Following this rule, we would have $f(x + 1)$ as

$$f(x + 1) = (x + 1)^2 - 5 = x^2 + 2x + 1 - 5 = x^2 + 2x - 4.$$

We shall be interested shortly in determining how much a function *changes* when the independent variable increases by 1. Starting at any value, x, the function changes from $f(x)$ to $f(x + 1)$ and the change is $f(x + 1) - f(x)$. For the function at hand

$$f(x + 1) - f(x) = (x^2 + 2x - 4) - (x^2 - 5) = 2x + 1.$$

Thus, if we go from $x = 3$ to 4, the function increases by $2(3) + 1 = 7$, but if we go from $x = 4$ to 5, the function increases by 9.

Exercise.   a) How much does $g(x) = 3x + 2$ change as x goes from x to $x + 1$? b) Why is the answer to a) constant? Answer:
a) $g(x + 1) - g(x) = [3(x + 1) + 2] - (3x + 2) = 3$. b) $g(x) = 3x + 2$ is linear, and the change in the function for an x change of 1 is the slope of the line, which is the constant, 3.

If we have a curve such as

$$C(x) = 100 + 6x + 0.01x^2,$$

illustrated in Figure 9–1, the change in the function when x increases by 1

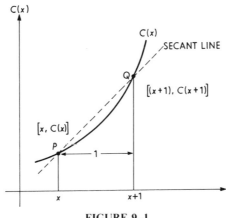

FIGURE 9–1

from x to $x + 1$ is the slope of the line (called a *secant* line) joining the two points $[x, C(x)]$ and $[(x + 1), C(x + 1)]$. This secant slope depends on the x value of interest, becoming larger (steeper secants) for larger values of x. Hence, the secant slope is a function of x which we shall call $s(x)$. We have

$$
\begin{aligned}
s(x) &= C(x + 1) - C(x) \\
&= [100 + 6(x + 1) + 0.01(x + 1)^2] - (100 + 6x + 0.01x^2) \\
&= (100 + 6x + 6 + 0.01x^2 + 0.02x + 0.01) - (100 + 6x + 0.01x^2) \\
s(x) &= 0.02x + 6.01.
\end{aligned}
$$

For example, if $x = 100$, the slope of the secant joining the points on the curve where $x = 100$ and $x = 101$ is $s(100) = 0.02(100) + 6.01 = 8.01$.

||||||||||||||||||||||||||||||||||||||||||||||||||||||||||||||||||||||||||||||||||||||||||||||||||||||||||||||||||||||

Exercise.   For the foregoing function, what is the slope of the secant line joining points on the curve where $x = 200$ and 201? Answer: 10.01.

||||||||||||||||||||||||||||||||||||||||||||||||||||||||||||||||||||||||||||||||||||||||||||||||||||||||||||||||||||||

9.3 MARGINAL COST AND REVENUE

Suppose the function we have been working with, $C(x)$, is the cost of making x units of a product. We found that as x, output, goes from 100 to 101, total cost $C(x)$ increased by $8.01. The additional or extra cost incurred in making the 101st unit is $8.01 and is called the *marginal cost* of the 101st unit. Thus, $s(x) = 0.02x + 6.01$ gives the marginal cost of the $(x + 1)$st unit.

Exercise. What is the marginal cost of the 151st unit? Answer: $s(150) = \$9.01$.

Next, let us consider the revenue side of operations and suppose that the firm can sell all it produces at $9 per unit. Inasmuch as an extra unit of sales increases total revenue by $9, *marginal revenue* is constant at $9. Marginal cost, however, increases as output increases, with each unit produced adding a little more to cost than the previous unit. We saw that the marginal cost of the 101st unit was $8.01, and that of the 151st unit was $9.01. If the firm makes and sells the 101st unit, its profit increases because marginal revenue exceeds marginal cost; that is, sale of the unit adds $9 to total revenue, but its production adds only $8.01 to total cost. On the other hand, the marginal cost of the 151st unit was $9.01, which exceeds marginal revenue of $9, so production and sale of this unit would reduce profit. Because the cost curve gets continually steeper as shown in Figure 9–1, marginal cost increases continually as output increases. Clearly, profit increases as output increases to 150 units, and decreases thereafter. Maximum profit occurs at $x = 150$ units. This illustrates the general economic principle that:

> *If marginal cost is increasing and marginal revenue is constant, maximum profit occurs when marginal cost equals marginal revenue.*

To obtain the optimal output directly, recall that $s(x) = C(x + 1) - C(x) = 0.02x + 6.01$ is the expression for the marginal cost of the $(x + 1)$st unit. Setting this equal to marginal revenue, 9, we have

$$0.02x + 6.01 = 9,$$

yielding $x = 149.5$, which we shall here arbitrarily round up to 150 units. The profit function, $P(x)$, is total revenue minus total cost, or

$$P(x) = R(x) - C(x).$$

The maximum profit is $P(150)$ which is

$$P(150) = R(150) - C(150) = 9(150) - [100 + 6(150) + 0.01(150^2)]$$
$$= \$125.$$

To summarize our example, recall that we started with

$$C(x) = 100 + 6x + 0.01x^2 \quad \text{and} \quad R(x) = 9x.$$

Marginal cost of the $(x + 1)$st unit was found as $C(x + 1) - C(x) = 0.02x + 6.01$. Setting this expression for marginal cost equal to marginal revenue of 9 led to the optimum output, $x = 150$. We then found maximum profit as $P(150) = R(150) - C(150) = \$125$.

〰〰

Exercise. Revenue is \$20.02 per unit and the cost of making x units is $50 + 0.02x^2$. *a*) What output leads to maximum profit? *b*) What is maximum profit? Answer: *a*) 500 units. *b*) \$4960.

〰〰

9.4 GENERAL SECANT SLOPE: THE DIFFERENCE QUOTIENT

In planning operations, it would be quite natural for an executive to be interested in the rate at which a firm's sales have been changing in recent years. For example, if from 1970 to 1975 sales increased from 100,000 to 120,000 units, the executive would note that the increased was $120,000 - 100,000 = 20,000$ in $1975 - 1970 = 5$ years, for an average increase of $20,000/5 = 4000$ units per year. We hasten to add that the executive will consider many other pieces of information in forming his judgment about expected future levels of sales, but the point is that average rates of change are important considerations in management decision making.

The average rate of change in a function per unit change in x when x increases by a to $x + a$ is the change in the function, $f(x + a) - f(x)$, divided by the change in x, which is $x + a - x = a$ or

$$\frac{f(x + a) - f(x)}{a}.$$

The last is called the *difference quotient.* As we saw in the previous section, if $f(x)$ is a cost function and $a = 1$, then

$$\frac{f(x + 1) - f(x)}{1}$$

is interpreted as the marginal cost of the $(x + 1)$st unit produced and, as shown in Figure 9–1, it is the slope of the secant line joining points on the curve obtained when x changes to $x + 1$. It follows that the difference quotient

$$\frac{f(x + a) - f(x)}{a}$$

is the slope of the secant line joining points on the curve obtained when x changes to $x + a$.

It is conventional in mathematics to use the symbol Δ (delta) to mean *the change in.* Thus, Δx means the change in x and is used in place of a in writing the difference quotient as

$$\frac{f(x + \Delta x) - f(x)}{\Delta x}.$$

If we are given $g(x) = x^2$, $x = 2$, and $\Delta x = 3$, the difference quotient is

$$\frac{g(x + \Delta x) - g(x)}{\Delta x} = \frac{g(2 + 3) - g(2)}{3} = \frac{5^2 - 2^2}{3} = 7.$$

Thus, 7 is the slope of the secant line joining points on the curve where $x = 2$ and $x = 2 + \Delta x = 5$. It is the average change in the function per unit change in x when x goes from 2 to 5.

Exercise.   Compute the difference quotient given $g(x) = x^2$, $x = 1$, and $\Delta x = 2$. Interpret the results. Answer: The quotient is 4. This is the slope of the secant line joining points where $x = 1$ and $x = 3$ and is the average change in the function per unit change in x as x goes from 1 to 3.

As another example, consider

$$f(x) = 2^x$$

for which the difference quotient is

$$\frac{2^{x+\Delta x} - 2^x}{\Delta x}.$$

Figure 9–2 shows the points P and Q, where P is $(x, 2^x)$ and Q is $[(x + \Delta x),$

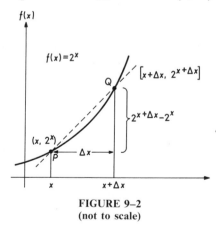

FIGURE 9–2
(not to scale)

$2^{x+\Delta x}]$. Recalling that the slope of a line is the difference of the vertical coordinates, $2^{x+\Delta x} - 2^x$, divided by the difference of the horizontal coordinates, Δx, we see again that the difference quotient is the slope of the secant line, PQ. If we choose $x = 1$ and $\Delta x = 0.5$, then (see Figure 9–3), the secant slope is

$$\frac{2^{1.5} - 2^1}{0.5} = \frac{2^{3/2} - 2}{0.5} = \frac{2\sqrt{2} - 2}{0.5} = \frac{2(1.4142) - 2}{0.5} - 1.6568.$$

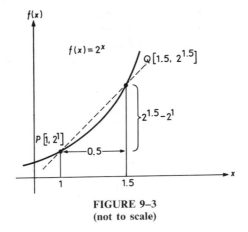

FIGURE 9–3
(not to scale)

9.5 NUMERICAL DIFFERENTIATION

A central problem of differential calculus is to find the slope of a *tangent* to a curve at a point. The problem is illustrated in Figure 9–4 where we note that we cannot by present means find the slope of the tangent at $P(2, 4)$

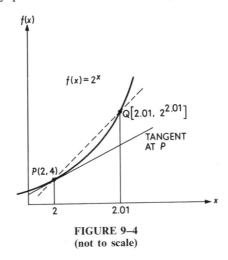

FIGURE 9–4
(not to scale)

because two points are required to compute the slope of a line and we know only the one point, P, on the tangent line. However, if we take a second point on the curve, near P, we can determine its coordinates because it is on the curve. The slope of the secant line will *approximate* the desired tangent slope, and the closer Q is to P, the better will be the approximation. In Figure 9–4, Δx is 0.01 and the secant slope is

$$\frac{2^{2.01} - 2^2}{0.01} = \frac{4.027 - 4}{0.01} = 2.7.[2]$$

---

[2] $2^{2.01} = $ antilog $[(2.01) \log 2] = $ antilog $(2.01)(0.3010) = $ antilog $(0.6050) = 4.027.$

The process of approximating the tangent slope by use of the difference quotient is called *numerical differentiation.* We see that the tangent slope at a point can be approximated by the difference quotient by using a small Δx, and the smaller the Δx (the closer it is to zero), the better will be the approximation. It will be important to recall the last sentence when we derive methods for finding the tangent slope exactly rather than approximately. Numerical differentiation can be done easily on an electronic computer, but it can be tedious, arithmetically, if done by hand. We shall restrict our practice examples to functions which do not require extensive calculations.

||||||||||||||||||||||||||||||||||||||||||||||||||||||||||||||||||||||||||||||||||||||||||||||||||||||||||||||||||||||||||||||||

Exercise.   By numerical differentiation, using $\Delta x = 0.1$, approximate the slope of the tangent to $f(x) = x^2$ at $x = 1$. Answer: 2.1.

||||||||||||||||||||||||||||||||||||||||||||||||||||||||||||||||||||||||||||||||||||||||||||||||||||||||||||||||||||||||||||||||

9.6 PROBLEM SET 1

1. If A is the area of a square of side x, write the formula for area in functional symbols.

2. If I is the amount of simple interest earned in t years on P dollars invested at rate r, then $I = Prt$. Write the functional form if we consider
   a) P and r as constant.   b) P and t as constant.

3. Given $f(x) = 3x^2 - 2$, $g(x) = 2^x$, and $h(x) = \log x$, evaluate the following:
   a) $f(0)$.   b) $g(0)$.   c) $h(20)$.   d) $h(1)$.   e) $f(-1)$.
   f) $g(-1)$.   g) $g(\frac{1}{2})$.

4. See Problem 3. Evaluate:
   a) $[f(2)][h(4)]$.   b) $g(3) - g(2)$.   c) $h(21) - h(20)$.
   d) $f(2.1) - f(2)$.   e) $h(3.1) - h(3)$.

5. If $g(x) = x^2y - y^2$, write the expression for $g(a)$.

6. If $f(y) = x^2y - y^2$, write the expression for $f(a)$.

7. Given the function $g(x)$, what represents
   a) The value of the function when $x = 4$?
   b) The value of the function when $x = b$?
   c) The change in the function when x changes to $x + 1$?

8. If $p(x) = 3x + 2k$, how much will $p(x)$ change when x increases by 2?

9. If $g(y) = 2y^2 - 3y$,
   a) How much will $g(y)$ change if y goes from 2 to 3?
   b) What is the expression for the change in $g(y)$ if y increases by 1 to $y + 1$?

10. If the cost of making x units of product is $C(x) = 100 + 3x + 0.01x^2$ and each unit made is sold at $11 per unit,
    a) What is the expression for the marginal cost of the $(x + 1)$st unit?
    b) What is the marginal cost of the 101st unit?
    c) What output leads to maximum profit?
    d) What is the maximum profit?

11. If the cost of making x units is $C(x) = 50 + x + 0.1x^2$ and each unit made is sold at $25 per unit,

a) What is the expression for the marginal cost of the $(x + 1)$st unit.
b) What is the marginal cost of the 11th unit?
c) What output leads to maximum profit?
d) What is the maximum profit?

12. What is the expression for the difference quotient (slope of the secant line) as x increases to $x + \Delta x$ on
   a) $h(x)$.   *b*) 3^x.   *c*) $\log x$.

13. Using $\Delta x = 0.1$, approximate by numerical differentiation the slope of the line tangent to the curve at the stated point. Note: The number in parentheses is the slope correct to the stated number of digits.
   a) $f(x) = 2x + 3$ at $x = 5$. $(m = 2)$.
   b) $g(x) = x^2 - 2x$ at $x = 1$. $(m = 0)$.
   c) $h(x) = \log x$ at $x = 2.5$. $(m = 0.1737)$.
   d) $f(x) = 2^x$ at $x = 1$. $(m = 1.3863)$. Note: $2^{1.1}$ is evaluated as antilog $[1.1(\log 2)]$ = antilog (0.3311).

14. Repeat Problem 13 *a*) and *b*) using $\Delta x = 0.01$.

9.7 CHARACTERISTICS OF COMMONLY USED FUNCTIONS

In Sections 9.8–9.13 we shall discuss some general characteristics of functions and particular characteristics of quadratic, cubic, exponential, and logarithmic functions. We shall learn how to sketch graphs of these functions quickly, without having to determine numerous points by laborious computation. Inasmuch as our interest is in general characteristics, we shall plot only certain significant points, and graphs will not be drawn to scale. Typical applications will be given for each function studied.

9.8 ASYMPTOTES

In the gasoline shortage of early 1974, drivers in Massachusetts saw the per gallon price almost double, but their demand for gasoline remained high. For other commodities, price increases may lead to sharp decline in demand. This price-demand aspect of a commodity is described in economics by a measure called the price elasticity of demand. We start this section by considering the special case of unitary price elasticity of demand, the case in which price and demand behave in inverse proportion so that the product (price per unit) (number of units demanded) is constant. For example, in

$$pq = 100,$$

at $q = 200$ units demand, p is \$0.50 per unit so that total sales volume is $pq = (0.50)(200) = \$100$. If demand changes to $(1/4)(200) = 50$ units, then price per unit is $(4)(0.50) = \$2.00$, and $pq = (2)(50) = \$100$ again. In functional form we have

$$p(q) = \frac{100}{q}.$$

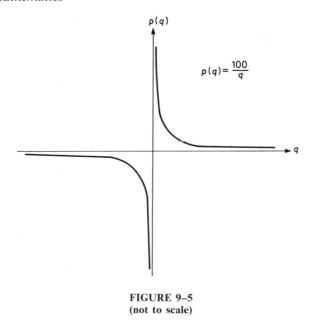

FIGURE 9–5
(not to scale)

The graph of $p(q)$ is shown in Figure 9–5 where, for practice purposes, we have included negative values of q even though they have no meaning in the applied sense.

To obtain the graph, we note first that q cannot be zero in $p(q) = 100/q$ because division by zero is *not defined*.[3] Next we see that if q moves to the right through values 10, 100, 1000, 10,000, and so on, increasing indefinitely, $p(q) = 100/q$ takes on the values 10, 1, 0.1, 0.01, and so on, continually approaching but never reaching zero. This behavior is described by saying that $p(q)$ approaches the horizontal axis *asymptotically* from above as we move indefinitely to the right, or by saying the horizontal axis is an asymptote of the curve. The connotation of asymptote (from the Greek word meaning *never touching*) is approaching, but never touching. To be complete here, we add that some curves intersect their asymptotes before settling down and approaching them. We shall not encounter such behavior in this text, but the reader may find it instructive to sketch a curve having this behavior.

Similarly, moving to the left through q values of 10, 1, 0.1, 0.01, and so on approaching zero, $p(q) = 100/q$ takes on the values 10, 100, 1000, 10,000, increasing without limit and the curve approaches the positive vertical axis as an asymptote as q approaches zero.

---

[3] *Not defined* means it is impossible; it cannot be done. If the reader has any question on this point, he should resolve it now because the point is crucial. For reference, see "zero" in the index.

Exercise. Describe the behavior of the branch of the curve in the third quadrant *a*) as *q* moves off to the left through values -1, -10, -100, -1000 and so on. *b*) As *q* approaches 0 from the left through values -1, -0.1, -0.01, -0.001, and so on. Answer: *a*) The curve approaches the horizontal axis asymptotically from below as *q* moves indefinitely to the left. *b*) The curve approaches the negative vertical axis asymptotically from the left as *q* approaches zero.

For shorthand, we shall use $\longrightarrow$ to mean approaches, and $\longrightarrow\infty$ (approaches infinity) to mean moving indefinitely to the right or upward, and $\longrightarrow -\infty$ to mean moving indefinitely to the left or downward. We would then say that $p(q) \longrightarrow -\infty$ as $q \longrightarrow 0$ from the left, and $p(q) \longrightarrow 0$ from below as $q \longrightarrow -\infty$.

Exercise. Use $\longrightarrow$ to describe the behavior of the first quadrant branch of the curve in Figure 9–5. Answer: $p(q) \longrightarrow 0$ from above as $q \longrightarrow \infty$; $p(q) \longrightarrow \infty$ as $q \longrightarrow 0$ from the right.

As we proceed, we shall see that curves with asymptotic behavior play an important role in applications.

9.9 CONTINUITY

In Figure 9–5 we may place a pencil on the branch of the curve in the first quadrant and draw as much as we want of that branch without lifting the pencil from the paper. However we cannot, without lifting the pencil, move to the branch in the second quadrant because of the break or *discontinuity* at $q = 0$. We shall be a little more explicit about continuity later in the chapter, but it will be sufficient for most of our purposes to say that a curve is continuous over an interval if we can draw the curve for that interval without lifting our pencil from the paper. As another example of a discontinuity, consider

$$f(x) = \frac{x^2 - 25}{x - 5}.$$

Factoring the difference of squares in the numerator yields

$$f(x) = \frac{(x - 5)(x + 5)}{x - 5}.$$

For any value of x except $x = 5$ (which would be division by zero), we have by cancellation

$$f(x) = x + 5, \quad x \neq 5.$$

Thus, the graph of the original function is the graph of the straight line $f(x) = x + 5$, with a hole or discontinuity at $x = 5$. The graph is shown in Figure 9–6 with an open circle indicating the discontinuity at $x = 5$.

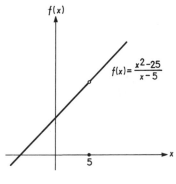

FIGURE 9–6

9.10 SMOOTH CURVES

Figure 9–7 shows a curve which satisfies our notion of continuity, but it has a sharp change of direction at $x = 3$. It appears reasonable to say that at point A and every other point except where $x = 3$, a *unique tangent* can be

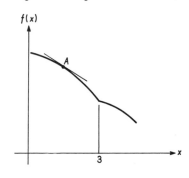

FIGURE 9–7

drawn to the curve. At $x = 3$, however, it appears that the left portion of the curve has a different tangent than the right portion, so there is not a unique tangent at this point. We shall say a curve is *smooth* over an interval if it has a unique tangent at each point in the interval. A curve cannot have a unique tangent at a point of discontinuity so if a curve has a discontinuity in an interval it cannot be smooth over that interval. However, as shown in

Figure 9–7, a curve which is continuous over an interval may not be smooth over the interval. In brief, smoothness guarantees continuity, but continuity does not guarantee smoothness.

9.11 TANGENTS

With modifications, our intuitive idea that the tangent to a curve at a point is a straight line just touching the curve at that point will be sufficient for present purposes. The modifications are that the tangent extended may touch, or cut, the curve at another point as shown by the tangents at *P* and *R* in Figure 9–8. To understand the nature of the second modification, see

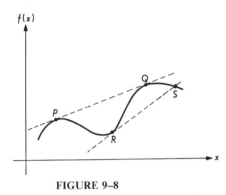

FIGURE 9–8

Figure 9–9. In 9–9A, the tangent at *P* is a horizontal line. If we now take the left half of the curve and rotate it using *P* as a pivot to give Figure 9–9B, the horizontal line is still tangent to the curve at *P* even though it now crosses the curve at *P*.

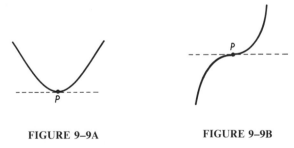

FIGURE 9–9A   **FIGURE 9–9B**

Exercise. Is the curve in Figure 9–9B smooth over an interval including point *P*? Explain. Answer: Yes, whether we think of *P* as being on the left or right portions of the graph, the tangent is the same horizontal line, so the tangent is unique at this point.

9.12 CONCAVITY

We note that the curve in Figure 9–9A is shaped like a cup which will hold water and we shall say that a curve, or any section of a curve, which has this shape is *concave upward.* The curve in Figure 9–9C would not hold water. It is *concave downward.* Now return to Figure 9–9B and note that to

FIGURE 9–9C   FIGURE 9–9D

the left of *P* the curve is concave downward and, to the right, concave upward. The point *P* where concavity changes from downward to upward is called an *inflection point* or a point of inflection. The tangent at an inflection point crosses the curve. Similarly, the concavity in Figure 9–9D changes from up to down at the inflection point *R* where the tangent crosses the curve at the point of tangency.

Inflection points have important implications in some applied areas. For example, suppose $C(x)$ in Figure 9–10 is the total cost of making x units of

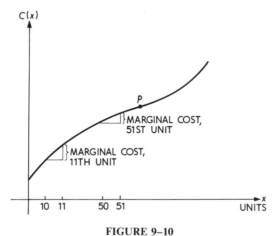

FIGURE 9–10

a product. Marginal cost, as we have discussed earlier, is change (vertical) in total cost when output, x, increases by one unit. Figure 9–10 shows that the marginal cost going from 10 to 11 units, that is the marginal cost of the 11th unit, is greater than the marginal cost of the 51st unit. In other words, marginal cost decreases with additional output if the cost curve is concave downward. Beyond *P* in Figure 9–10, the curve is concave upward and marginal cost is increasing. The inflection point, *P,* is the point where marginal cost changes from decreasing to increasing.

9.13 INCREASING AND DECREASING FUNCTIONS: STATIONARY POINTS

In Figure 9–11, the curve is falling (the function is decreasing) over the interval between *A* and *C*. Tangents to the curve in this interval, as at *B*, slant downward, so the tangent slopes are negative. We shall say a curve is

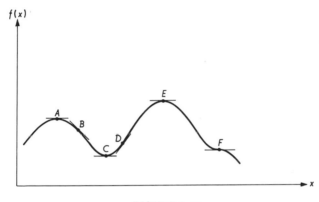

FIGURE 9–11

falling (function is decreasing) at a point if the tangent slope is negative at that point. Correspondingly, a curve is rising (function increasing) at a point if the tangent slope at that point is positive, as at point *D* in Figure 9–11. At point *A*, the tangent is horizontal, slope 0, and the curve is neither rising nor falling. We call a point where the tangent slope is zero a *stationary* point. In Figure 9–11, points *A, C, E,* and *F* are stationary points. Point *A* is called a *local maximum* because it is the highest point in its immediate vicinity, with the qualifier *local* serving to caution that there may be (and are in this case) higher points on the curve. Assuming the curve continues downward forever at both ends, point *E* is a *global maximum* because it is the highest point anywhere on the curve. Point *C*, on the other hand, is a *local minimum*. Next, note that point *F* is both a stationary point and an inflection point, so that we call it a *stationary inflection* point. Finally, and most importantly for further work, observe that at a local maximum such as *A* in Figure 9–11, the curve is stationary and concave downward, whereas at a local minimum, such as *C*, the curve is stationary and concave upward. Now, keeping point *F* in mind:

‖‖‖

Exercise.   *a*) Is it necessary that a local maximum or minimum be a stationary point? *b*) If we have a stationary point, is it necessarily a local maximum or minimum point? Answer: *a*) Yes. *b*) No. Read on.

‖‖‖

The tangent must be horizontal at a local maximum or minimum point, so the point must be stationary. However, F is a stationary point which is neither a maximum nor a minimum, so a stationary point is not necessarily a maximum or minimum. A major part of our applications of calculus has to do with finding maximum profit, minimum cost, and so on, so we will be interested first in finding stationary points, then in determining whether they are maximum or minimum points, or neither. We shall use the presence of downward or upward concavity, or inflection, in making this determination.

After pausing for a problem set, we shall return and apply the general characteristics we have been studying to some commonly used functions.

9.14 PROBLEM SET 2

1. What is the connotation of the word asymptote?
2. Letting x be 1, 2, 3, 4, and so on, state why the x-axis is an asymptote of $f(x) = 2^{-x}$.
3. Sketch the curve of a function which starts at a point in the second quadrant, falls smoothly, passing through the point (0, 5) as we move to the right, and approaches the x-axis as an asymptote from above.
4. Draw the horizontal line $f(x) = 10$. Then, starting from the origin, sketch a smooth curve of a function which approaches $f(x) = 10$ as an asymptote from below as $x \longrightarrow \infty$.
5. Start at a point in the first quadrant and sketch the curve of a function which falls smoothly as we move to the left, goes through (1, 0), and approaches the negative vertical axis as an asymptote as $x \longrightarrow 0$.
6. *a*) If $f(x) = \log x$, what is $f(1)$?
   b) What is $f(x)$ if $x = 10, 100, 1000$?
   c) What is $f(x)$ if $x = 0.1, 0.01, 0.001$?
   d) Make a sketch showing the general characteristics of the graph.
7. With respect to drawing a graph, what is the intuitive idea of
   a) Continuity?   *b*) Smooth curve?
8. The logarithmic function sketched in Problem 6 is defined only for the interval $x > 0$. Is the curve smooth and continuous over this interval? Explain.
9. Does the following function have any discontinuities? If so, where?

$$f(x) = \frac{x}{4 - x^2}.$$

10. In the first quadrant, sketch a section of a graph of a smooth curve which
    a) Rises to the right, is first concave up, then changes to concave down.
    b) Rises to the right, is first concave down, then changes to concave up.

11. a) In the first quadrant, sketch a section of a smooth curve which first declines
       to a minimum at *P*, then rises to a stationary inflection at *Q*, then rises to a
       maximum at *R*, then falls forever after.
    b) How many inflection points are there in the sketch for Part *a*)?

12. a) If a point is a local maximum, what two conditions must be satisfied?
    b) If a point is a local minimum, what two conditions must be satisfied?
    c) A smooth curve has a stationary point which is neither a maximum nor a
       minimum. How can this be?

13. $C(x)$ is the cost of making x units of product. It starts out concave upward, rises
    through an inflection where $x = 1000$, and continues to rise, concave downward.
    a) Sketch $C(x)$.
    b) Interpret $C(x)$ in terms of marginal cost.
    c) Economists would say that this is an unlikely cost function. Can you see why?

9.15 QUADRATIC FUNCTIONS

The general form of the quadratic functions we shall consider is

$$q(x) = sx^2 + fx + k,$$

where s, f, and k are constants and s is the coefficient of the *second* degree
term, f is the coefficient of the *first* degree term, and k is a constant term. A
quadratic function must have an x^2 term, so s cannot be zero. Examples of
quadratic functions are

$$g(x) = x^2,$$
$$h(x) = 3x^2 - 2x,$$
$$f(x) = 5 - 2x + 0.01x^2.$$

Quadratic functions are perhaps the simplest functions beyond the
straight line, and they are used frequently as illustrations in developing the
mathematical tools of the calculus. Such functions also arise quite naturally
in some applications. For example, suppose we conduct a penny-off sale on
a product, quoting the price as $9.99 for the purchase of one unit, $9.98 per
unit for the purchase of two units, $9.97 per unit for the purchase of three
units, and so on, so that the unit price for the purchase of x units would be
$\$(10 - 0.01x)$. The dollar amount of the sale to a customer who buys x units
would be $q(x)$, where

$$q(x) = (x \text{ units})(10 - 0.01x \text{ dollars per unit})$$
$$q(x) = x(10 - 0.01x)$$
$$q(x) = 10x - 0.01x^2$$

which is a quadratic we shall return to after we examine the characteristics of
graphs of quadratic functions.

QUADRATIC FUNCTIONS

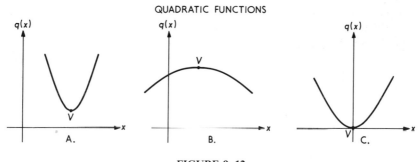

FIGURE 9–12

Quadratics of the form $q(x) = sx^2 + fx + k$ have graphs which are vertical parabolas such as those illustrated in Figure 9–12. The parabolas are smooth curves which are symmetrical about a vertical line, as illustrated in Figure 9–12C where the vertical line of symmetry is the vertical axis. A vertical parabola is either everywhere concave up or everywhere concave down. The high (low) point is called the *vertex, V,* and is either a global maximum or a global minimum.

As $x \longrightarrow \infty$ or $x \longrightarrow -\infty$, the term sx^2 ultimately overpowers the other terms in absolute size, so if the coefficient of x^2, s, is positive, the curve must open upward, and if s is negative, the curve opens downward. It is easy to show[4] that

$$x = -\frac{f}{2s} \text{ is the x-coordinate of the vertex of a vertical parabola.}$$

To sketch the graph of the parabola

$$q(x) \overset{e}{=} x^2$$

we note first that it opens upward because the x^2 term has the positive coefficient, $s = 1$. It has no first degree term, so $f = 0$. Hence, the vertex is at $x = -f/2s = 0$, and $q(0) = 0$. The vertex is at the origin, $(0, 0)$, and we have a parabola like that in Figure 9–12C. Again, the parabola

$$q(x) = 10 + 2x - 0.01x^2$$

has $s = -0.01$, so it opens downward. The coefficient of the first degree term is $f = 2$. Hence $x = -f/2s = -2/2(-0.01) = 100$ is the x-coordinate of the vertex, and $q(100) = 10 + 2(100) - 0.01(100)^2 = 110$, so the vertex point is $(100, 110)$ and we have a parabola like the one shown in Figure 9–12B.

---

[4]The vertex is on the vertical line of symmetry. Letting x be the coordinate of the vertex, $f(x)$ must be the same 1 unit on either side of x; that is at $x - 1$ and at $x + 1$. Setting $f(x - 1) = f(x + 1)$, where $f(x) = sx^2 + fx + k$ and simplifying, we find $x = -f/2s$.

Exercise. Describe the graph of the function $q(x) = 0.02x^2 - 2x + 75$. Answer: It is a parabola having its vertex at (50, 25) and opening upward as in Figure 9–12A.

Returning to our penny-off sale where $q(x)$ is the amount a customer pays for x units and

$$q(x) = 10x - 0.01x^2,$$

we find that this parabola opens downward, has its vertex at $x = -f/2s = -10/2(-0.01) = 500$ units, and $q(500) = 10(500) - 0.01(500)^2 = \$2500$. The vertex, (500, 2500), is a global maximum, so that the maximum dollar value of a sale is $2500 and occurs when 500 units are purchased.

Exercise. If $A(x) = 0.02x^2 - 2x + 75$ represents the average cost per unit made when x units are made, what interpretation would be given to the coordinates of the vertex, (50, 25)? Answer: The minimum average cost per unit is $25 and occurs when output is 50 units.

9.16 CUBIC FUNCTIONS

We shall consider cubics of the form

$$c(x) = tx^3 + sx^2 + fx + k,$$

where k is a constant term and t, s, and f represent, respectively, the coefficients of the *t*hird, *s*econd, and *f*irst degree terms. Cubic shapes are shown in Figure 9–13. The third degree term, tx^3, ultimately controls the function

CUBIC FUNCTIONS

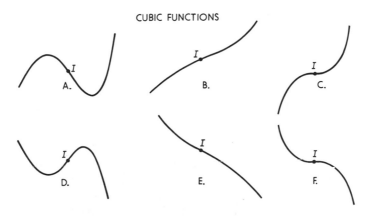

FIGURE 9–13

as $x \longrightarrow \infty$ and $x \longrightarrow -\infty$ so the curve will ultimately rise to the right (and fall to the left) if t is positive, as in Figure 9–13A, B, and C. If t is negative, the curve will ultimately fall to the right (and rise to the left) as in Figure 9–13D, E, and F. The cubic is a smooth curve which always has one, and only one, point of inflection, indicated by I in Figure 9–13. The point of inflection may be between a local maximum and a local minimum as in A and D, it may be an ordinary inflection as in B and E, or it may be a stationary inflection as in C and F. Observe that for curves ultimately rising to the right (t positive), the concavity goes from concave down to concave up, and the opposite concavity change applies for curves which ultimately fall to the right (t negative).

To sketch a cubic, we make use of formulas we will derive in a simple manner in the next chapter; namely,

$$x = -\frac{s}{3t} \text{ is the x-coordinate of the inflection point}$$

and

$$x = \frac{-s \pm \sqrt{s^2 - 3ft}}{3t} \text{ are the x-coordinates of stationary points, if any exist.}$$

In the last expression, if $s^2 - 3ft$ is positive, we take both roots and have the maximum-minimum shape of Figure 9–13A or D. If $s^2 - 3ft$ is negative, we have the ordinary inflection as in B or E. If $s^2 - 3ft$ is zero, we have the case of a stationary inflection as in C or F.

Consider the cubic

$$c(x) = 0.1x^3 - 9x^2 + 150x + 2600$$

where $t = 0.1$, $s = -9$, $f = 150$, and $k = 2600$. The inflection occurs at $x = -s/3t = -(-9)/3(0.1) = 30$. $c(30)$ is found to be 1700, so the inflection point is (30, 1700). Examining

$$x = \frac{-s \pm \sqrt{s^2 - 3ft}}{3t} = \frac{9 \pm \sqrt{81 - 3(150)(0.1)}}{3(0.1)} = \frac{9 \pm \sqrt{36}}{0.3}$$

$$= 50 \text{ and } 10,$$

we find we have the maximum-minimum case with stationary points at $x = 50$ and $x = 10$. Substituting, $c(50) = 100$ and $c(10) = 3300$. The larger function value, 3300, is the maximum, so we have a local maximum at (10, 3300), a local minimum at (50, 100), with the inflection point between them at (30, 1700). A sketch of the curve is shown in Figure 9–14.

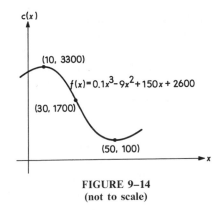

FIGURE 9–14
(not to scale)

Exercise. Where is the inflection point of the simple cubic $c(x) = x^3$. Which part of Figure 9–13 does its graph resemble? Answer: Here $t = 1$ and s and f are zero. The inflection is at $(0, 0)$. Also, $s^2 - 3ft = 0$, so we have a stationary inflection and the graph has the form of Figure 9–13C.

As another example, consider

$$c(x) = -2x^3 + 12x^2 - 30x + 40$$

where $t = -2, s = 12, f = -30$, and $k = 40$. Since t is negative, the curve ultimately falls to the right (and rises to the left). The inflection is at $x = -s/3t = -12/3(-2) = 2$, and $c(2) = 12$, or the point $(2, 12)$. We find that $s^2 - 3ft = 144 - 3(-30)(-2) = -36$, so the curve has only an ordinary inflection. Moreover, since the curve falls to the right, concavity must change from up to down at the inflection point, as shown in Figure 9–15.

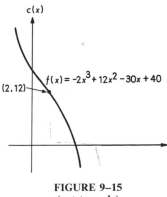

FIGURE 9–15
(not to scale)

We have noted earlier that a cost curve which starts concave down, inflects and becomes concave up, depicts a situation where marginal cost first decreases as output rises, then increases after output passes the inflection in the cost curve. To have this form, illustrated in Figure 9–13B, the curve must rise to the right so t must be positive, and it should have an ordinary[5] inflection, so $s^2 - 3ft$ must be negative.

||||||||||||||||||||||||||||||||||||||||||||||||||||||||||||||||||||||||||||||||||||||||||||||||||||||||||||||||||||||||||||||||||||||||||||||||

> **Exercise.**   If $C(x)$ is the cost of making x units of a product and $C(x) = 0.01x^3 - 0.3x^2 + 4x + 10$, at what output does marginal cost stop decreasing and start increasing? Answer: $x = 10$ units.

||||||||||||||||||||||||||||||||||||||||||||||||||||||||||||||||||||||||||||||||||||||||||||||||||||||||||||||||||||||||||||||||||||||||||||||||

9.17 EXPONENTIAL FUNCTIONS

The independent variable, x, in

$$f(x) = 2^x \quad \text{and} \quad g(x) = 3^{1-2x}$$

appears in the exponent and the functions are called *exponential* functions. Such functions have a variety of applications as we shall see after examining their characteristics by reference to Figure 9–16.

Each of the exponentials in Figure 9–16 has an asymptote. Also marked on each curve is a point for which the quantity in the exponent is zero. Thus, for $f(x) = 2^x$, the exponent is zero when $x = 0$, $f(x) = 2^0 = 1$, and the point $(0, 1)$ is marked on its graph, Figure 9–16A. We shall use this point for orientation when sketching the exponential and refer to it as the *zero exponent point*. The choice is expedient, of course, because any base (except zero) to the zero power is 1.

||||||||||||||||||||||||||||||||||||||||||||||||||||||||||||||||||||||||||||||||||||||||||||||||||||||||||||||||||||||||||||||||||||||||||||||||

> **Exercise.**   What is the zero exponent point for $f(x) = 3^{2-x}$? Answer: See Figure 9–16C.

||||||||||||||||||||||||||||||||||||||||||||||||||||||||||||||||||||||||||||||||||||||||||||||||||||||||||||||||||||||||||||||||||||||||||||||||

Each of the functions in Figure 9–16 is of the form

$$f(x) = a + bc^{dx+e}$$

where all but x are constants which may have any value, except b and d may not be zero, and c must be greater than 1. If the coefficient of x, which is d, is positive (as in Figure 9–16A), the curve approaches its asymptote as

---

[5]With an ordinary inflection, marginal cost decreases from, say $5 to $3, then increases. With a stationary inflection, the implication is that marginal cost decreases to zero (which is not reasonable) before increasing.

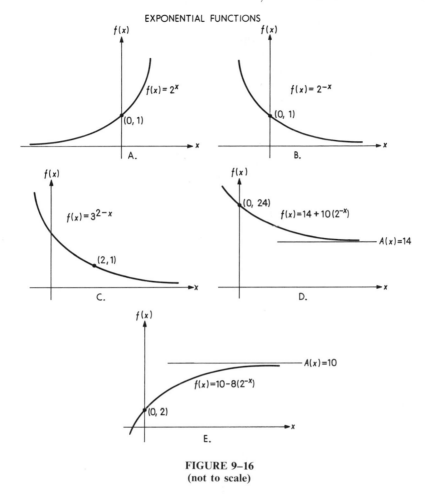

FIGURE 9–16
(not to scale)

$x \longrightarrow -\infty$. If x has a negative coefficient, as in Figure 9–16C, the curve approaches its asymptote as $x \longrightarrow \infty$. In either case, the term $bc^{dx+e} \longrightarrow 0$, so the asymptote is the remaining term, the constant a. Using $A(x)$ to represent the asymptote, we have

$$A(x) = a.$$

Thus, given

$$f(x) = 10 - 8(2^{-x}),$$

the asymptote is the horizontal line $A(x) = 10$. Since x has a negative coefficient, the asymptote is approached as $x \longrightarrow \infty$; that is, as we move to the right. The zero exponent point is at $-x = 0$ or $x = 0$, $f(0) = 10 - 8(2)^0 = 10 - 8(1) = 2$, so its coordinates are (0, 2). We sketch the curve by first drawing the horizontal asymptote at $A(x) = 10$. Then, placing our pencil at $(0, 2)$ which is below the asymptote, we draw a smooth curve approaching the asymptote from below as $x \longrightarrow \infty$, as shown in Figure 9–16E. Going from the zero exponent point toward the asymptote, the curve gets increasingly flatter. Going in the opposite direction from the zero exponent point, the curve gets increasingly steeper.

||||||||||||||||||||||||||||||||||||||||||||||||||||||||||||||||||||||||||||||||||||||||||||||||||||||||||||||||||||||||||||||||||||||||||||||||||||

Exercise. Describe the graph of $f(x) = 20 - 12(3^{5-2x})$. Answer: The asymptote is $A(x) = 20$. Starting at the zero exponent point, $(5/2, 8)$, the curve approaches the asymptote from below as $x \longrightarrow \infty$.

||||||||||||||||||||||||||||||||||||||||||||||||||||||||||||||||||||||||||||||||||||||||||||||||||||||||||||||||||||||||||||||||||||||||||||||||||||

In everyday language, exponential growth connotes explosive growth as in the statement that a country's population is increasing exponentially. In our study, exponential has a broader meaning which includes the various growth patterns exhibited in Figure 9–16. For arithmetic ease in our examples, we use integers for the base of exponentials, as in $f(x) = 2^x$. This function doubles when x increases by 1; that is, for $x = 1, 2, 3$, and so on, $f(x)$ is 2, 4, 8, and so on. However, many applied exponentials have more modest growth rates as, for example,

$$A(n) = (1.08)^n$$

which is the amount \$1 at 8 percent interest compounded annually will grow to in n years. We saw in Chapter 8 that at i percent, \$1 will double in about $69/i$ years, so at 8 percent it will take $A(n)$ about $69/8$ or 8.6 years to double.

In addition to functions which rise ever more rapidly, we have use for those which rise ever more slowly (Figure 9–16E) and those which decline ever more slowly (Figure 9–16D). Functions similar to E are used as models for *learning curves* and *response functions*. Response functions may also take the form of D. To interpret E as a learning curve, we let x be the number of weeks a worker has been on an assembly line, and $f(x)$ be the worker's output per hour. At the time of hiring, $x = 0$, output is 2 units per hour. The worker learns most rapidly at the beginning and at $x = 1$ week his output has risen to $10 - 8(2^{-1}) = 6$ units per hour. His performance improves ever more slowly as time goes on and approaches his potential of 10 units per hour, the asymptote.

To interpret Figure 9–16D as a response function, suppose that a company contracts to buy advertising time on a television series. Letting x be in, say, weeks, and $f(x)$ be the *rise* in level of sales per week, we see that the initial response is highest, then response tapers off toward a continuing level of 14 (say \$1400) per week. The rate of decline in response is highest at the beginning and drops off as time goes on.

|||||||||||||||||||||||||||||||||||||||||||||||||||||||||||||||||||||||||||||||||||||||||||||||||||||||||||||||||||||||||||||||||||||||||||||||||||||||||

Exercise. If the proportion of potential customers responding to an advertisement after x weeks is $p(x) = 0.4 - 0.4(2^{-x})$, what proportion will have responded by 1, 2, and 3 weeks? What level of response is being approached as time goes on? Answer: 0.2, 0.3, 0.35, approaching 0.4.

|||||||||||||||||||||||||||||||||||||||||||||||||||||||||||||||||||||||||||||||||||||||||||||||||||||||||||||||||||||||||||||||||||||||||||||||||||||||||

One exponential is so important in applications that it is referred to as *the* exponential function. It is

$$f(x) = e^x,$$

where e is about 2.71828. The special characteristic of the function is that the number e, which we encountered in Chapter 8 and will meet often as we proceed, is the base of the exponential. This function occurs so frequently that extensive tables of values of e^x and e^{-x} have been prepared for use in calculations. Table IX at the end of the book is an abbreviated table included for reference in our discussions. From this table we see that if $x = 0.46$, then $e^{0.46} = 1.5841$ and $e^{-0.46}$, the reciprocal of $e^{0.46}$, is 0.6313.

|||||||||||||||||||||||||||||||||||||||||||||||||||||||||||||||||||||||||||||||||||||||||||||||||||||||||||||||||||||||||||||||||||||||||||||||||||||||||

Exercise. If $f(x) = e^{0.05x}$, find $f(10)$ and $f(-10)$. Answer: $e^{0.5} = 1.6487$ and $e^{-0.5} = 0.6065$.

|||||||||||||||||||||||||||||||||||||||||||||||||||||||||||||||||||||||||||||||||||||||||||||||||||||||||||||||||||||||||||||||||||||||||||||||||||||||||

Like π, the constant e is irrational so that it cannot be expressed exactly by any finite number of digits, but both numbers can be written to as many digits as may be needed for the accuracy required in a problem. Readers may remember using $22/7$ to approximate π. Correspondingly, e is about $3/7$ less than π, or about $19/7$.[6] More accurately, π is 3.141593 and e is 2.718282.

We encountered e in Chapter 8 where we introduced the concept of compounding interest continuously rather than at discrete intervals of time, such as semi-annually or monthly. There we used the exponential

$$A = e^{iy}$$

to compute the amount \$1 compounded continuously at the quoted interest i would grow to in y years. Inasmuch as banks which compound continuously quote *nominal* interest rates (that is, rates that would apply if interest were compounded annually), it is helpful to determine the *effective* rate when the

---

[6] It may be useful to know that $355/113$ approximates π correctly to six decimal places and $1900/699$ approximates e correctly to three decimal places.

nominal rate is compounded continuously. This is easy to do. Thus, at $i = 0.08$ for $y = 1$ year, compounded continuously, a dollar will grow to $e^{0.08} = 1.0833$, so $1 grows to $1.0833 and the effective rate interst is 0.0833 or 8.33 percent. In general, a quoted nominal rate i, when compounding is continuous, is an effective rate of r, where

$$r = e^i - 1.$$

‖‖

Exercise.   What is the effective rate if interest is quoted at 10 percent compounded continuously? Answer: 10.52%.

‖‖

Continuous compounding has been applied in financial research for many years. In the tight money situation of 1973, investment funds were in high demand and some banks sought to attract deposits by offering the highest rate of interest permitted by law,[7] compounded continuously. As time goes on, it will be interesting to see whether continuous compounding is applied to investments other than demand deposits.

9.18 LOGARITHMIC FUNCTIONS

The reader may wish to review logarithms by verifying the entries in the following table.

x	$\log_{10} x$
0.0001	$-4.$
0.001	$-3.$
0.01	$-2.$
0.05	$0.6990 - 2 = -1.301$
0.1	$-1.$
0.4	$0.6021 - 1 = -0.3979$
0.5	$0.6990 - 1 = -0.301$
1.	0
2.	0.3010
10.	$1.$
100.	$2.$
200.	2.3010
1000.	$3.$

Starting in the table at $x = 1$, $\log x = 0$,[8] and moving up, we find that $\log x \longrightarrow -\infty$ as $x \longrightarrow 0$, so that the negative vertical axis is an asymptote.

---

[7] In addition, 1 year was replaced by $365/360 = 1.0139$ on the basis of the practice of using 360 days as a bank year, rather than 365 days. With this modification, 10% nominal becomes 10.67% effective as compared with 10.52% without the modification.

[8] It will be helpful to keep in mind that whatever the base of the system of logarithms, $\log 1$ is zero.

To the right of (1, 0), the curve rises slowly, but it always rises. The graph is shown in Figure 9–17. It has the exponential form, but the asymptote is vertical rather than horizontal. From our work with exponentials, it is clear that 10 to any power, positive or negative, is always positive and keeping in mind that the logarithm of a number is the *power* to which 10 must be raised to give the number, it follows that no power of 10 will give zero or a negative number, so the logarithms of zero and negative numbers do not exist. As Figure 9–17 shows, $f(x) = \log x$ is defined only for $x > 0$.

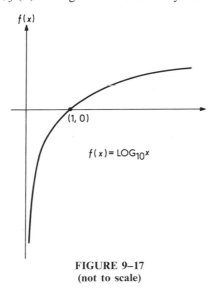

FIGURE 9–17
(not to scale)

The base 10 has computational advantages because it permits a single tabular entry (the mantissa) for a given set of digits no matter where the decimal point is. Thus 0.1959 (Table I) is the tabular entry for 157, 15.7, 0.0157, 15700, and so on, and we supply the characteristic (see Chapter 7) to take care of the decimal point. We have log $157 = 2.1959$, log $0.0157 = 0.1959 - 2$, and so on. Of course, any positive number except 1 can serve as the base of a system of logarithms. Because of the property of the characteristic, base 10 logarithms are fine for computation, but it probably will come as no surprise to the reader that the natural choice for mathematical analysis is the base e. This is the case, and logarithms to the base e are called *natural* logarithms. Hereafter, when no ambiguity can arise, we shall use ln N (where *ln* rhymes with *win*) to represent natural logarithms and log N to represent base 10 or *common* logarithms.

The principle of the characteristic does not apply for natural logarithms, so a useful table of natural logarithms would have separate entries for 15.7, 0.0157, 0.000157, and so on, and the entries would fill a very large volume. Table X at the end of the book contains entries suitable for most of our purposes. Thus, we find ln $0.4 = -0.9163$ and ln $12.8 = 2.5494$.

Exercise. Find ln 0.5 and ln 10. Answer: -0.6931 and 2.3026.

If, on occasion, we need a value of ln N which is not tabulated, we can obtain it by multiplying the common logarithm of N (Table I) by the conversion factor 2.3026. That is,

$$\ln N = 2.3026 \ (\log N).$$

To see where this factor comes from, recall that $p = \log_{10} N$ means $10^p = N$ and $q = \ln_e N$ means $e^q = N$, so

$$e^q = 10^p.$$

Next recall the rule of logarithms which says that $\log (a^b) = b(\log a)$, for any system of logarithms. Taking the natural logarithm of both sides of our equation, we have

$$\ln(e^q) = \ln(10^p)$$

so

$$q(\ln e) = p(\ln 10)$$

and

$$q = p(\ln 10) = p(2.3026)$$

because ln e means $\ln_e e$ and in any system of logarithms, the logarithm of the base of the system is 1. Now q is ln N and p is log N, so the last statement says

$$\ln N = 2.3026 \ (\log N).$$

To compute ln 157, we find log 157 $= 2.1959$, and

$$\ln 157 = 2.3026(2.1959) = 5.0563,$$

whereas

$$\ln 0.0157 = 2.3026(0.1959 - 2) = 2.3026(-1.8041) = -4.1541.$$

Exercise. Find ln 520 and ln 0.00520. Answer: 6.2539, -5.2591.

The rules of logarithms are the same whatever the base of the system. These are:

$$\ln (ab) = \ln a + \ln b$$

$$\ln\left(\frac{a}{b}\right) = \ln a - \ln b$$

$$\ln a^b = b \ (\ln a).$$

To practice with the rules, let us compute ln 16 as

$$\ln 16 = \ln(2)(8) = \ln 2 + \ln 8 = 0.6931 + 2.0794 = 2.7725$$

which compares with ln 16 = 2.7726 in Table X. Alternatively,

$$\ln 16 = \ln (4)^2 = 2 (\ln 4) = 2(1.3863) = 2.7726 \text{ or}$$

$$\ln 16 = \ln \left(\frac{8}{0.5}\right) = \ln 8 - \ln 0.5 = 2.0794 - (-0.6931) = 2.7725.$$

Exercise. *a)* Compute ln 11 using ln 22 and ln 2 and compare with the Table X entry for ln 11. *b)* Compute ln 100 from ln 10. Answer: *a)* 2.3979 which is the tabular entry for ln 11. *b)* ln 100 = ln 10^2 = 2(ln 10) = 2(2.3026) = 4.6052.

Rules of logarithms play a prominent role in applications involving exponential functions. We have learned that at interest rate i compounded continuously, $1 will grow to e^{iy} in y years. If we wish to know how long it will take for money to double at 8 percent compounded continuously, ($1)e^{0.08y}$ must grow to $2 so that

$$e^{0.08y} = 2.$$

We solve by taking the natural logarithm of both sides

$$\ln (e^{0.08y}) = \ln 2$$

$$0.08y \ (\ln e) = 0.6931$$

$$0.08y = 0.6931$$

$$y = \frac{0.6931}{0.08} = \frac{69.31}{8} = 8.66 \text{ years.}$$

The last, of course, is our rule that at i percent, money will double in about $69/i$ years.

Exercise. If a country has continuous inflation at the rate of 20 percent a year, how long will it take for the price level to triple? Answer: 5.5 years.

A price index (such as the *Consumers' Price Index* in the United States) measures the current average price level relative to a base year where the index is set at 100% = 1.00. If, a year later, the index is 125 percent = 1.25, the average price level has risen by 25 percent. The reciprocal of the index is called the *purchasing power of the dollar*. At an index level of 125 percent = 1.25, the purchasing power of the dollar is $\frac{1}{1.25}$ = 0.80 or $0.80, so purchasing power changed from 1.00 to 0.80, a change of −0.20 or −20 percent.

Exercise. If prices decrease by 25 percent, what happens to the purchasing power of the dollar? Answer: The index would change from 1(100%) to 0.75(75%), and the purchasing power of the dollar would be $\frac{1}{0.75} = \frac{4}{3}$, an *increase* of $\frac{1}{3}$ or $33\frac{1}{3}$ percent.

If inflation is at the continuous rate of 12 percent per year and we wish to determine how long it will take for the purchasing power of the dollar to drop to 0.40, we note that this requires the price index to be $\frac{1}{0.40}$. That is, we seek the number of years, y, such that

$$e^{0.12y} = \frac{1}{0.40} \quad \text{or} \quad e^{-0.12y} = 0.40.$$

Using the latter expression, we find

$$\ln (e^{-0.12y}) = \ln 0.40$$

$$-0.12y \ln e = \ln 0.40$$

$$-0.12y = \ln 0.40$$

$$y = \frac{\ln 0.40}{-0.12} = \frac{-0.9163}{-0.12} = 7.64 \text{ years.}$$

As another example, if we are told that the amount of oil that can be pumped from an oil field decreases by 20 percent per year and asked how long it will take for output to fall to 30 percent of its present level, we have,

$$e^{-0.2y} = 0.3.$$

Exercise. Solve the last equation for y. Answer: About six years.

9.19 PROBLEM SET 3

By means of sketches, show the pertinent characteristics of each of the following:

1. $q(x) = x^2 + 10$.
2. $q(x) = 4x - x^2 + 5$.
3. $q(x) = x^2 - 6x + 25$.
4. $q(x) = 25 + x^2$.

In words, describe the pertinent characteristics of each of the following:

5. $q(x) = 3x^2 - 12x + 15$.
6. $q(x) = 16x - 2x^2 - 12$.

If $A(x)$ is average cost per unit when x units are produced, what output yields minimum average cost per unit, and what is this minimum?

7. $A(x) = 50 - 0.8x + 0.01x^2$.
8. $A(x) = 175 - 3x + 0.015x^2$.

By means of sketches, show the pertinent characteristics of each of the following:

9. $c(x) = x^3 - 3x^2 + 5$.
10. $c(x) = x^3 - 9x^2 + 24x$.

11. $c(x) = -0.1x^3 + 0.6x^2 - 1.5x + 5$. 12. $c(x) = x^3 - 3x^2 + 3x$.

In each of the following, $C(x)$ is the total cost of making x units of product. Find the output, x, where marginal cost stops decreasing and starts increasing.

13. $C(x) = 0.02x^3 - 3x^2 + 156x + 100$.

14. $C(x) = 0.15x^3 - 9x^2 + 200x + 500$.

By means of sketches, show the pertinent characteristics of each of the following:

15. $f(x) = 2^{3-x}$.

16. $f(x) = 10 - 5(3^x)$.

17. $f(x) = 20 - 10(2^{5-x})$.

18. $f(x) = 10 + 20(2^{4-2x})$.

19. Compute the following:

   a) 2^3.   b) 2^{-3}.   c) e^2.   d) e^{-2}.   e) $e^{-0.01}$.   f) $e^{1.5}$.

20. If the proportion of potential customers responding to an advertisement after t weeks is $r(t) = 0.7 - 0.7e^{-0.1t}$, what percent of potential customers will have responded after a) 1 week? b) 10 weeks? c) What is the level of response approaching as time goes on?

21. The monthly demand for a new product t months ($t \geq 1$) after placing it on the market is $D(t) = 1000 + 500e^{-0.05(t-1)}$. What is the demand a) After 1 month? b) After 21 months? c) Toward what level is demand falling as time goes on?

What is the effective interest rate if interest is

22. 5 percent compounded continuously?

23. 10 percent compounded continuously?

24. π is about $22/7$. About how much is e?

25. Find the following:

   a) $\log 1$.   b) $\ln 1$.   c) $\log 0.5$.   d) $\ln 0.5$.
   e) $\log 24$.   f) $\ln 24$.   g) $\ln e$.   h) $\log 10$.   i) $\ln 10$.

26. a) What multiplier of $\log N$ gives $\ln N$?
   b) What multiplier of $\ln N$ gives $\log N$?
   c) Compute $\ln 37.5$.   d) Compute $\ln 0.375$.

27. How can $\ln 36$ be computed from
   a) $\ln 12$ and $\ln 3$?   b) $\ln 6$?
   c) How can $\ln 0.05$ be computed from $\ln 0.5$ and $\ln 10$? Compute $\ln 0.05$ by this method.

28. a) Write $\ln x^3$ without using an exponent.
   b) Write $\ln x - \ln y$ using the symbol $\ln$ just once.
   c) Write the equivalent of $(1/a)\ln b$ with $(1/a)$ in another position.
   d) Write $\ln a^2 - \ln a$ using the symbol $\ln$ just once.
   e) Write $\ln (1/x)$ without using a fraction, and using $\ln$ just once.

29. Solve $e^{0.2x} = 6$ for x.

30. Solve $e^{-2x} = 0.8$ for x.

31. How long will it take for money to increase by 50 percent at interest of 10 percent compounded continuously?

32. If the index of consumers' prices rises 10 percent a year compounded continuously,
   a) How long will it take for the index to rise by 75 percent?
   b) What will be the purchasing power of the dollar when the index has risen by 75 percent?

33. If the amount of oil pumped from an oil field declines continuously at a rate of 20 percent per year, when will the amount pumped fall to 1/5 of the present amount?

34. If consumer prices rise continuously at the rate of 12 percent per year, how long will it take for the purchasing power of the dollar to decline to 50 cents?

9.20 SEQUENCES AND LIMITS

One dollar at 8 percent compounded annually grows to $(1.08) at the end of 1 year, $(1.08)^2$ at the end of 2 years, and so on to $(1.08)^n$ at the end of n years. If we write

$$\begin{array}{cccccc}
\text{Term Number} & 1 & 2 & 3 & \cdots & n \\
\text{Term} & (1.08), & (1.08)^2, & (1.08)^3, & \cdots, & (1.08)^n
\end{array} \tag{1}$$

the set of numbers (1) is an ordered set because its members can be designated by the term numbers 1, 2, 3, 4, and so on. An ordered set of numbers is called a *sequence*. The sequence (1) is an example of a *geometric* sequence which means that the ratio of each term to the immediately preceding term is constant, the constant being 1.08. In general, a geometric sequence is defined by stating its first term and the common ratio. Thus, if a is the first term and r is the common ratio, the sequence is

$$a, \; ar, \; ar^2, \; ar^3, \; \cdots, \; ar^{n-1}$$

where ar^{n-1} is called the *general* term, so that, for example, the $n = $ 10th term is ar^9.

‖‖‖

Exercise. If $a = 1$, r is 2, what is the 5th term? Answer: 16.

‖‖‖

Examples of geometric sequences are

$$1, \; 2, \; 4, \; 8, \; 16, \; \cdots \text{ whose general term is } 2^{n-1} \tag{2}$$

and

$$\frac{1}{2}, \; \frac{1}{4}, \; \frac{1}{8}, \; \frac{1}{16}, \; \cdots \text{ whose general term is } \left(\frac{1}{2}\right)^n \tag{3}.$$

‖‖‖

Exercise. What is the next term in sequence (2); sequence (3)?
Answer: 32; $\frac{1}{32}$.

‖‖‖

We will be interested in the behavior of the terms of a sequence as n gets larger and larger; that is, $n \longrightarrow \infty$. In the case of (3), it is clear that the terms get increasingly small and approach 0 as a limiting value as $n \longrightarrow \infty$. That is, as $n \longrightarrow \infty$, $(1/2)^n = 2^{-n} \longrightarrow 0$ as a limit. Of course, 2^{-n} will never equal zero no matter how large n is, but the limiting value is zero and we say that "the limit of 2^{-n} as $n \longrightarrow \infty$ is zero." This is expressed symbolically as

$$\lim_{n\to\infty} 2^{-n} = 0.$$

Sequence (3) is said to be *convergent,* meaning it has a limiting value as $n \longrightarrow \infty$. Sequence (2), on the other hand, is *divergent* because as $n \longrightarrow \infty$, 2^{n-1} gets increasingly large and does not approach a limit. Thus,

$$\lim_{n\to\infty} 2^{n-1} \quad \text{does not exist.}$$

In general, a geometric sequence is convergent to the limiting value zero if the common ratio between successive terms, r, is between -1 and $+1$. We shall introduce some applications of sums of terms of geometric sequences in the next section.

We now return to the familiar constant e. We introduced this constant in Chapter 8 by noting that \$1 at 100% interest would become $(1 + 1)^1 = \$2$ at the end of a year. Compounded twice a year, the year-end amount would be $(1 + 1/2)^2 = \$2.25$, and so on. The numbers generate the following sequence:

Term Number	1	2	3		n
Term	$(1 + 1)^1,$	$(1 + 1/2)^2,$	$(1 + 1/3)^3, \cdots, (1 + 1/n)^n$		

or Term $\left(\dfrac{2}{1}\right)^1,$ $\left(\dfrac{3}{2}\right)^2,$ $\left(\dfrac{4}{3}\right)^3,$ $\cdots, \left(\dfrac{n+1}{n}\right)^n.$

The limit of this sequence as $n \longrightarrow \infty$ is the amount \$1 would grow to in a year at 100 percent interest compounded continuously. The following table shows some of the terms of this sequence rounded to seven decimal places.

n (Term Number)	Expression	Value of the Term
1	$(2)^1$	2.000 0000
2	$(3/2)^2$	2.250 0000
3	$(4/3)^3$	2.370 3704$^-$
4	$(5/4)^4$	2.441 4062$^+$
5	$(6/5)^5$	2.488 8320
.	.	.
100	$(101/100)^{100}$	2.704 8138$^+$
.	.	.
1000	$(1001/1000)^{1000}$	2.716 9238$^+$
10,000	$(10,001/10,000)^{10000}$	2.718 1459$^+$
.	.	.
100,000	$(100,001/100,000)^{100000}$	2.718 2546$^+$
.	.	.
1,000,000	$(1,000,001/1,000,000)^{1000000}$	2.718 2818$^+$
$\downarrow$		$\downarrow$
∞		?

The behavior shown in the table of the values of the terms in the sequence as $n \longrightarrow \infty$ is perplexing. On the one hand, as n goes from 10,000 to 100,000 to 1,000,000, the first four digits remain at 2.718. Indeed all eight digits in the final table entry would remain fixed no matter how large an n we used. On the other hand, no matter how much we increase n, new digits appear further out in the sequence and we never come to a stopping point. For example, as we increase n further, the digits 2.71828 18284 59045 become fixed, but there are more to come *ad infinitum*. This behavior gives rise to the mathematical concept that a sequence may be said to have a limit, even though the numerical value of that limit cannot be written exactly with any finite number of digits. The behavior we have observed occurs with all irrational numbers, such as π, $\sqrt{2}$, and $\sqrt{3}$, so the concept is vital because such numbers are just as important as integers and rational numbers in what is called the *real* number system. Just as $\sqrt{2}$ is not 1.41 nor 1.414 but is the limit of an unending sequence as $n \longrightarrow \infty$, so e is not 2.7, nor 2.718, but is

$$e = \lim_{n \to \infty} \left(1 + \frac{1}{n} \right)^n = \lim_{n \to \infty} \left(\frac{n + 1}{n} \right)^n.$$

When we encounter this limit expression again, as we shall, we simply replace it by e.

As another example of a sequence which has a familiar irrational number as a limit, consider

$$\frac{3}{2}, \ \frac{17}{12}, \ \frac{577}{408}, \ \ldots$$

This sequence has been generated by an *iterative* process, meaning by a rule which tells how to get the next term in the sequence from the present term. The rule here is to add $\frac{1}{2}$ the present term to the reciprocal of the term, given the starting number $\frac{3}{2}$. Using $\frac{3}{2}$, we compute

$$\left(\frac{1}{2} \right)\left(\frac{3}{2} \right) + \left(\frac{2}{3} \right) = \frac{3}{4} + \frac{2}{3} = \frac{17}{12}$$

which is the next term. Then using 17/12 we find

$$\frac{1}{2}\left(\frac{17}{12} \right) + \left(\frac{12}{17} \right) = \frac{17}{24} + \frac{12}{17} = \frac{577}{408}.$$

Exercise. Write the beginning expression for the next term.

Answer: $\dfrac{577}{2(408)} + \dfrac{408}{577}$. For reference, this is $\dfrac{665,857}{470,832}$.

Carrying out the divisions, the term values are

Term Number (n)	Expression	Value
1	3/2	1.5000 0000
2	17/12	1.4166 6666
3	577/408	1.4142 1568+
4	665,857/470,832	1.4142 1356+

The limit of this sequence of values is the irrational number, $\sqrt{2}$, and further calculation (larger n) shows that all nine digits in the last entry of the table remain fixed as n increases. (See the next Problem Set for practice exercises in computing square roots by the iterative procedure).

We have seen sequences which have 0 as a limit, an irrational number as a limit, and one which does not have a limit. For practice, consider

$$3, \quad \frac{5}{2}, \quad \frac{7}{3}, \quad \frac{9}{4}, \quad \ldots, \quad \frac{2n+1}{n}.$$

It is perhaps obvious that the limit as $n \longrightarrow \infty$ is 2, either by examining the successive terms of the sequence, or by noting that for large n, $2n + 1$ is practically the same as $2n$, so $(2n + 1)/n$ is practically the same as $2n/n = 2$. Or, we may write

$$\lim_{n \to \infty} \frac{2n+1}{n} = \lim_{n \to \infty} \left(\frac{2n}{n} + \frac{1}{n} \right) = \lim_{n \to \infty} \left(2 + \frac{1}{n} \right) = 2$$

where the final step follows from the fact that as $n \longrightarrow \infty$, 2 does not change, but $1/n$ approaches 0.

Exercise. Write the first five terms and the 100th term of the sequence whose general term is $(n + 2)/(n + 1)$. What is the limit of this sequence as $n \longrightarrow \infty$? Answer: $\frac{3}{2}, \frac{4}{3}, \frac{5}{4}, \frac{6}{5}, \frac{7}{6}, \frac{102}{101}$, approaching 1 as a limit.

9.21 SERIES AND LIMITS

A series is the *sum* of an ordered set of terms. The sum of the first n terms is written as S_n. For example,

$$S_n = 1 + \frac{1}{2} + \frac{1}{4} + \frac{1}{8} + \cdots + \frac{1}{2^{n-1}}$$

is a geometric series. $S_1 = 1$, S_2 is the sum of the first two terms which is $\frac{3}{2}$, $S_3 = \frac{7}{4}$, and so on, so the sums form the sequence $1, \frac{3}{2}, \frac{7}{4}, \frac{15}{8}$, and we might rightly guess that the limit of this sequence is 2. We shall verify this guess shortly.

Geometric series arise in finance and many other applied fields, and it is useful to have a formula for determining the sum of n terms, S_n. If we start with

$$S_n = 1 + r + r^2 + r^3 + \cdots + r^{n-1},$$

then from this subtract $r(S_n)$ which is

$$rS_n = \qquad r + r^2 + r^3 + \cdots + r^{n-1} + r^n$$

all but the first and last terms on the right disappear, leaving the difference

$$S_n - rS_n = 1 - r^n.$$

Factoring on the left yields

$$S_n(1 - r) = 1 - r^n \quad \text{or} \quad S_n = \frac{1 - r^n}{1 - r}.$$

If the first term of the series had been a rather than 1, the formula would be

$$S_n = a\left(\frac{1 - r^n}{1 - r}\right).$$

We applied formulas for the sums of finite geometric series in Chapter 8 when we worked with annuities, so we shall concentrate our attention here on applications involving *infinite* geometric series. First we note that if

$$S_n = 1 + 2 + 2^2 + 2^3 + \cdots + 2^{n-1},$$

the limit of S_n as $n \longrightarrow \infty$ does not exist because the terms become larger and larger in value. This is the case whenever the common ratio, here 2, between successive terms exceeds 1 in absolute value. S_n does approach a limit as $n \longrightarrow \infty$ if r is between -1 and $+1$. To see what this limit is, consider

$$\lim_{n \to \infty} \frac{1 - r^n}{1 - r}.$$

Now if r is any number, such as $\frac{1}{2}$, which is between -1 and 1, r^n becomes smaller, approaching zero as a limit, as $n \longrightarrow \infty$, and the r^n term is the only term affected as $n \longrightarrow \infty$. Hence,

$$\lim_{n \to \infty} S_n = \frac{1}{1 - r},$$

or, if the first term is a, rather than 1,

$$\lim_{n \to \infty} S_n = \frac{a}{1 - r}.$$

As a first application, let us suppose we are building a house. Each dollar we spend becomes income to a recipient. Let us suppose recipients spend $\frac{3}{4}$ of each dollar on consumption and invest $\frac{1}{4}$ in, say, a savings account, a supposition which is described in economics by saying the *marginal propensity to consume* is $\frac{3}{4}$ and the *marginal propensity to save* is $\frac{1}{4}$. The recipient of the first $1 spends $¾, which becomes income to another recipient who spends $(¾)(¾) = $⁹⁄₁₆, and so on, generating an income stream which is the geometric series

$$S_n = 1 + \frac{3}{4} + \frac{9}{16} + \frac{27}{64} + \cdots$$

with a common ratio $r = ¾$. The limit of S_n as $n \to \infty$ is

$$\lim_{n \to \infty} S_n = \frac{1}{1 - \frac{3}{4}} = \frac{1}{\frac{1}{4}} = 4.$$

The limit, $4, is called the *multiplier*. It means that an expenditure of $1 generates income of $4 if the marginal propensity to consume (MPC) is ¾. In general, we have

$$\text{Multiplier} = \frac{1}{1 - \text{MPC}}.$$

Exercise. If people spend 90 percent of their income on consumption, how much income is generated by a $100 expenditure? Answer: MPC = 0.9 and the multiplier is 10, so $1000 is generated.

In Chapter 8 we showed that if we seek to invest a sum of money now, A_n, at interest i per period, so that the account will provide an income of R per period for n periods (first payment at the end of the first period), then

$$A_n = R\left[\frac{1 - (1 + i)^{-n}}{i}\right].$$

Now suppose we wish to provide R per period *forever*. We should invest now the amount

$$A_\infty = \lim_{n \to \infty} A_n = \lim_{n \to \infty} R\left[\frac{1 - (1 + i)^{-n}}{i}\right].$$

The only part of the expression affected as n changes and $\to \infty$ is $(1 + i)^{-n}$, which is $1/(1 + i)^n$. Suppose $i = 0.08$ or any other usual interest rate. Then $(1 + i)$ is greater than 1 and $(1 + i)^n$ gets ever larger as $n \to \infty$, so $1/(1 + i)^n$ gets ever smaller and approaches zero as a limit. We see that $(1 + i)^{-n}$ approaches 0 as a limit and therefore

$$A_\infty = \lim_{n \to \infty} R\left[\frac{1 - (1 + i)^{-n}}{i}\right] = R\left(\frac{1}{i}\right) = \frac{R}{i}.$$

Thus, $A_\infty = R/i$ will provide R per year forever with the first R available one year after A_∞ has been invested at i percent. If the first R is to be available immediately, we would have to add R to R/i.

As an example, if an alumnus of a college wishes to establish an endow-ment which will provide a $1000 annual scholarship forever and funds can be invested at 5 percent for this purpose, the amount of the endowment would be

$$\frac{1000}{0.05} = \$20,000$$

if the first scholarship is granted a year after the investment is made or $21,000 if the first scholarship is to be granted immediately.

An annuity of the type under discussion, one which continues forever, is called a *perpetuity*. We have found the formula for a perpetuity by taking a limit as $n \longrightarrow \infty$. A simpler way of deriving the formula is to note that the amount A_∞ at interest for one year must provide R interest, and after R is withdrawn the amount A_∞ must remain to accumulate another R during the next year. Thus, if we start with A_∞ it will grow in a year to $A_\infty(1 + i)$. Withdrawing R leaves $A_\infty (1 + i) - R$, and the amount left must be A_∞. Hence,

$$A_\infty(1 + i) - R = A_\infty$$

$$A_\infty + A_\infty i - R = A_\infty$$

$$A_\infty i = R$$

$$A_\infty = \frac{R}{i},$$

as before.

If a perpetuity is to provide R every m years rather than every year, A_∞ must be such that at the end of m years, when it has grown to $A_\infty(1 + i)^m$, R can be withdrawn and A_∞ remain. In this case,

$$A_\infty(1 + i)^m - R = A_\infty$$

$$A_\infty(1 + i)^m - A_\infty = R$$

$$A_\infty[(1 + i)^m - 1] = R$$

$$A_\infty = \frac{R}{(1 + i)^m - 1}.$$

The *capitalized cost* of an asset such as a machine is the first cost, R, plus an amount which will provide R for replacement every m years forever. This is R for the first cost, plus A_∞ for the perpetuity. Adding R to the foregoing expression for A_∞, we have the capitalized cost

$$C_\infty = R + A_\infty = R + \frac{R}{(1 + i)^m - 1} = R\left[1 + \frac{1}{(1 + i)^m - 1}\right]$$

$$= R\left[\frac{1 + (1 + i)^m - 1}{(1 + i)^m - 1}\right] = R\left[\frac{(1 + i)^m}{(1 + i)^m - 1}\right].$$

Another form of this result can be obtained by multiplying the numerator and denominator of the last expression by $(1 + i)^{-m}$. This is

$$C_\infty = \frac{R}{1 - (1 + i)^{-m}},$$

where $(1 + i)^{-m}$ can be found from Table III at the end of the book. As an example, the capitalized cost at 8 percent of a machine purchased for $R = \$5000$ and having a life of $m = 5$ years is

$$C_\infty = \frac{5000}{1 - (1.08)^{-5}} = \frac{5000}{1 - 0.680583} = \frac{5000}{0.319417} = \$15,654.$$

Thus, $15,654 will pay the original cost of $5000 and provide replacement amounts of $5000 every five years, forever.

Exercise. Suppose a machine with a life of 10 years can be purchased at $10,000 to perform the same function as the machine in the foregoing example. Find the capitalized cost at 8 percent of this machine. Answer: $18,629.

Comparison of the last example and exercise leads to the conclusion that the $5000 machine with a five-year life is less costly than the $10,000 machine with a ten-year life. Such comparisons are useful in capital planning decision making. We should note in passing that an asset such as a machine may have a scrap value at the end of its useful life. If such were the case, capitalized cost would be the purchase cost, plus A_∞ computed using the purchase cost minus the scrap value as R.

Returning to

$$e = \lim_{n \to \infty} \left(1 + \frac{1}{n}\right)^n,$$

the binomial $(1 + 1/n)^n$ can be expanded by the binomial theorem into an infinite series and by advanced analysis it can be shown that the limit of this series is the *series expansion for e*

$$e = 1 + 1 + \frac{1}{2!} + \frac{1}{3!} + \frac{1}{4!} + \frac{1}{5!} + \cdots$$

where ! is the factorial symbol and $4! = (4)(3)(2)(1)$, $3! = (3)(2)(1)$, and similarly for other factorials. This series converges rapidly to accurate approximations of e. For example, the sum of the first 10 terms is 2.71828, which is e correct to five decimal places.

Exercise. Compute the sum of the first six terms in the series expansion for e. Answer: $1 + 1 + \frac{1}{2} + \frac{1}{6} + \frac{1}{24} + \frac{1}{120} = \frac{326}{120} = 2.71666$ or, rounded, 2.717 as contrasted to 2.718 which is the correct value for e rounded to three decimal places.

The series expansion for e provides the value for e^1. If the exponent is x, so that we have the function e^x, it can be shown that the series expansion is

$$e^x = 1 + x + \frac{x^2}{2!} + \frac{x^3}{3!} + \frac{x^4}{4!} + \cdots .$$

This series converges (that is, the sequence of successive sums will have a limit) for any value of x, but the convergence is very slow if x is large. Convergence is rapid if x is near zero. For example, if we approximate $e^{0.2}$ as

$$e^{0.2} = 1 + 0.2 + \frac{0.04}{2} + \frac{0.008}{6} + \frac{0.0016}{24}$$

$$= 1 + 0.2 + 0.02 + 0.001333 + 0.00006667 = 1.2214,$$

the result is the same as the entry for $e^{0.2}$ in Table IX at the end of the book. Similarly, values of e^{-x} can be computed from the series

$$e^{-x} = 1 - x + \frac{x^2}{2!} - \frac{x^3}{3!} + \frac{x^4}{4!} + \cdots .$$

||||||||||||||||||||||||||||||||||||||||||||||||||||||||||||||||||||||||||||||||||||||||||||||||||||||||||||||||||||||||||||||||||||||||||||||||||||

Exercise. *a*) From Table IX, what is $e^{-0.1}$? *b*) Approximate $e^{-0.1}$ using the first three terms of the series for e^{-x}. Answer: *a*) 0.9048. *b*) 0.905.

||||||||||||||||||||||||||||||||||||||||||||||||||||||||||||||||||||||||||||||||||||||||||||||||||||||||||||||||||||||||||||||||||||||||||||||||||||

9.22 LIMITS OF FUNCTIONS

Suppose the producer of a liquid product charges a customer $100 for transportation plus $10 per gallon for orders up to and including 50 gallons. For orders exceeding 50 gallons, an overall 10 percent discount is granted. The charge is thus $100 + 10x$ for $x \le 50$, and a discount of 10 percent leaves $90 + 9x$ for $x > 50$. Letting $C(x)$ be the charge for x gallons, we have

$$C(x) = 100 + 10x; \ x \le 50$$
$$C(x) = 90 + 9x; \ x > 50.$$

$C(x)$ is shown in Figure 9–18. Using terminology introduced earlier in the chapter, we say that $C(x)$ has a discontinuity at $x = 50$. We wish now to describe continuity more precisely than we did in earlier discussions. To do this, we shall say a function is continuous at a point $x = a$ if for *any* sequence of values of x approaching a, the corresponding sequences of values of $f(x)$ have a limit, and that limit is $f(a)$. Clearly, the function $C(x)$ does not satisfy this criterion at $x = a = 50$, because for any sequence of x values approaching 50 on the left part of the graph, where $C(x) = 100 + 10x$, the corresponding sequence of values of $C(x)$ approach $100 + 10(50) = 600$ as a limit, whereas if we move in toward 50 on the right part of the graph, where $C(x) = 90 + 9x$, the values of $C(x)$ approach $90 + 9(50) = 540$ as a limit. The limits of the two sequences of values for $C(x)$ as

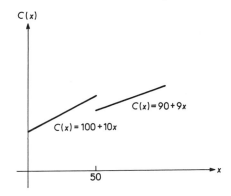

FIGURE 9–18 .

$x \longrightarrow 50$ are different, so $C(x)$ is discontinuous at $x = 50$. It is intuitively obvious that $x = 50$ is the only discontinuity, so we shall simply state that $C(x)$ is continuous for all x except $x = 50$.

We shall say a function $f(x)$ is continuous *at* $x = a$ if for *any* sequence of values of x approaching a

$$\lim_{x \to a} f(x) = f(a);$$

that is, the limit exists and is $f(a)$. We have observed intuitively, and mathematicians can prove rigorously, that functions such as $f(x) = 2x$, $g(x) = x^3 + 3x^2 - 5$, $h(x) = e^{0.05x}$ are continuous at all points. The practical importance, to us, of the definition of continuity is that if we know a function is continuous at a point, we can find the limit of the function by *substitution;* that is, assured of continuity, $\lim f(x)$ as $x \longrightarrow a$ is $f(a)$. For example, if $f(x) = x^2 - 3x + 2$, the

$$\lim_{x \to 1} f(x) = f(1) = 1^2 - 3(1) + 2 = 0.$$

Exercise. Find $\lim_{x \to 0}(2x + e^x + 5)$. Answer: $0 + e^0 + 5 = 1 + 5 = 6$.

In chapters to come, we shall rarely be concerned with continuity because our applications will involve functions which are continuous at all points of interest to us. However, the fundamental definition of the differential calculus involves an expression whose limit cannot be determined by substitution, because substitution leads to a zero denominator, or because we cannot substitute ∞, which is not a number, in evaluating a limit as $x \longrightarrow \infty$. For example if $f(x) = 1/x$, then

$$\lim_{x \to \infty} f(x) = \lim_{x \to \infty}\left(\frac{1}{x}\right) = 0.$$

We cannot substitute ∞ here, but we can (actually or mentally) construct sequences of values of x becoming ever larger and convince ourselves that $1/x$ approaches 0 as a limit as $x \to \infty$. On the other hand,

$$\lim_{x \to 0} f(x) = \lim_{x \to 0} \frac{1}{x} \quad \text{does not exist}$$

because for any sequence of x values approaching 0, $1/x$ gets ever larger in absolute value and has no limiting value.

Next, consider $g(x) = x^2/x$ which has a discontinuity at $x = 0$. Here it is important to understand the precise meaning of the symbol $x \to a$. It is that x approaches a through any sequence of values, *not including a itself*. Now, if we exclude 0, the function x^2/x is simply x. Hence, since the limit symbol excludes 0, we may write

$$\lim_{x \to 0} \left(\frac{x^2}{x}\right) = \lim_{x \to 0} (x) = 0.$$

In effect, we have changed a function whose limit cannot be determined by substitution to one whose limit can be found by substitution. As another example, noting that $x^3 - 8 = (x - 2)(x^2 + 2x + 4)$, we can see that

$$\lim_{x \to 2} \frac{x^3 - 8}{x - 2} = \lim_{x \to 2} \frac{(x - 2)(x^2 + 2x + 4)}{x - 2} = \lim_{x \to 2} (x^2 + 2x + 4) = 12.$$

Exercise.   *a*) Factor $x^2 - 9$.   *b*) Find $\lim_{x \to 3} \frac{x^2 - 9}{x - 3}$.

Answer: *a*) $(x - 3)(x + 3)$.   *b*) 6.

There are some important limits which cannot be determined by the elementary methods used thus far. We can, by computation of a sequence of values for the function at hand in such a case, make a good guess as to whether a limit exists and, if so, what the value or approximate value of the limit is. However, mathematical techniques beyond the scope of this text are required to prove whether or not our guesses are correct. A case in point is $f(x) = (1 + 1/x)^x$, for which it can be proved that

$$\lim_{x \to \infty} \left(1 + \frac{1}{x}\right)^x = e.$$

Earlier in the chapter, our tabulation of values of this function suggested that more and more decimal digits stabilized as x increased, but it would require advanced methods to prove that this is true, and that nothing unexpected happens in the far decimal places. Let us guess at another limit which will be of importance to us shortly. It is

$$\lim_{x \to 0} \left(\frac{e^x - 1}{x}\right).$$

We find from Table IX that for $x = 1$, 0.1, and 0.01, e^x is 2.7183, 1.1052, and 1.0101. The following table shows values of $(e^x - 1)/x$ for these values of x, and some results of further computations have been included.

x	*Expression*	*Value*
1.00000	$(e^1 - 1)/1$	1.7183
0.100000	$(e^{0.1} - 1)/0.1$	1.052
0.010000	$(e^{0.01} - 1)/0.01$	1.01
0.001		1.0005
0.0005		1.00025
0.00005		1.00002
↓		↓
0		?

Exercise.   Guess what limit the right hand column in the fore-going table is approaching. Answer: Read on.

It can be proved rigorously that the limit at hand is 1, as suggested by the tabulated values. We shall refer to this fact in the next section.

9.23 DERIVATIVES OF THE FUNCTIONS x^n, e^x, and ln x

We are now at the threshold of the differential calculus. In this closing section of the chapter we state and interpret the fundamental definition of the differential calculus, the definition of the derivative of a function, and proceed to derive rules for finding the derivatives of the functions x^n, e^x, and ln x.

Earlier in the chapter, we introduced the difference quotient which was the change in the function divided by the change in x when x changed by an amount Δx. The ratio was shown to be the slope of the secant line joining two points on the curve whose x coordinates differed by Δx. The ratio was

$$\text{Secant slope} = \frac{\text{Change in } f(x)}{\text{Change in } x} = \frac{f(x + \Delta x) - f(x)}{\Delta x}.$$

The basic definition of the differential calculus states that the *derivative of a function of x with respect to x* is another function, called "*f* prime of *x*" and abbreviated $f'(x)$, where

$$f'(x) = \lim_{\Delta x \to 0} \frac{f(x + \Delta x) - f(x)}{\Delta x}.$$

To obtain a geometric interpretation of this definition, see Figure 9–19 which shows a secant line, PQ, for one value of Δx. Now think of keeping P fixed and taking smaller values for Δx. This would bring Q down the

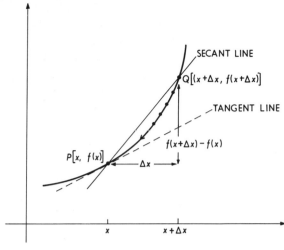

FIGURE 9–19

curve through a sequence of points illustrated by dots in Figure 9–19. If we were to draw new secant lines as Q moves toward P (that is, $\Delta x \longrightarrow 0$), they would rotate about the fixed point P as a pivot and become ever closer to the tangent line at P. It follows that as Δx approaches 0 as a limit, the secant line approaches the tangent line as a limit, and the slope of the secant line approaches the slope of the tangent line as a limit. Inasmuch as the difference quotient is the slope of the secant line, it follows that as $\Delta x \longrightarrow 0$, the difference quotient approaches the tangent line slope as a limit. Consequently, we define

$$f'(x) = \lim_{\Delta x \to 0} \frac{f(x + \Delta x) - f(x)}{\Delta x}$$

to be the slope of the line tangent to the curve at a point whose horizontal coordinate is x. Thus, if we think of $f(x)$ as a formula for finding points on a curve, $f'(x)$ is the formula for finding the slopes of lines tangent to the curve. For example, if $f(x) = x^2$, then as we shall soon prove, $f'(x) = 2x$. When $x = 0.6$, the corresponding point on the curves is $(0.6, 0.36)$ and the slope of the tangent line at this point is $f'(0.6) = 2(0.6) = 1.2$.

Our first problem is to find $f'(x)$ for the *power* function, $f(x) = x^n$. We have

$$f'(x) = \lim_{\Delta x \to 0} \frac{f(x + \Delta x) - f(x)}{\Delta x} = \lim_{\Delta x \to 0} \frac{(x + \Delta x)^n - x^n}{\Delta x}.$$

This limit cannot be found by substituting $\Delta x = 0$ because the result would be $0/0$, but we can change the expression to one whose limit can be found by setting $\Delta x = 0$. To do this, we shall expand $(x + \Delta x)^n$ and change its form. Let us practice by expanding $(x + \Delta x)^4$. We have,

$$(x + \Delta x)^4 = (x + \Delta x)^2 (x + \Delta x)^2$$
$$= (x^2 + 2x\overline{\Delta x} + \overline{\Delta x^2})(x^2 + 2x\overline{\Delta x} + \overline{\Delta x^2})$$

and patiently continuing the multiplication we obtain

$$x^4 + 4x^3\overline{\Delta x} + 6x^2\overline{\Delta x^2} + 4x\overline{\Delta x^3} + \overline{\Delta x^4},$$

which we can change to

$$x^4 + \Delta x[4x^3 + (6x^2\overline{\Delta x} + 4x\overline{\Delta x^2} + \overline{\Delta x^3})].$$

To serve our purpose, we shall express the last as

$$x^4 + \Delta x[4x^3 + (\text{terms having } \Delta x \text{ as a factor})].$$

Exercise. Expand $(x + \Delta x)^2$ in the form of the last statement.
Answer: $x^2 + \Delta x[2x + (\text{terms having } \Delta x \text{ as a factor})].$

Continuing, we would find by carrying out the foregoing procedure that

$$(x + \Delta x)^3 = x^3 + \Delta x[3x^2 + (\text{terms having } \Delta x \text{ as a factor})].$$
$$(x + \Delta x)^6 = x^6 + \Delta x[6x^5 + (\text{terms having } \Delta x \text{ as a factor})].$$

Note in each case the first term in the bracket. For $(x + \Delta x)^4$ it is $4x^3$; for $(x + \Delta x)^2$ it is $2x^1$; and for $(x + \Delta x)^6$ it is $6x^5$.

Exercise. What would be the first term in the bracket for
a) $(x + \Delta x)^7$? b) $(x + \Delta x)^{10}$? c) $(x + \Delta x)^n$? Answer: a) $7x^6$.
b) $10x^9$. c) nx^{n-1}.

Exercise answer c) is the form we need. We see that

$$f'(x) = \lim_{\Delta x \to 0} \frac{(x + \Delta x)^n - x^n}{\Delta x}$$

$$= \lim_{\Delta x \to 0} \frac{x^n + \Delta x[nx^{n-1} + (\text{terms having } \Delta x \text{ as a factor})] - x^n}{\Delta x}$$

$$= \lim_{\Delta x \to 0} \frac{\Delta x(nx^{n-1} + \text{terms with } \Delta x \text{ as a factor})}{\Delta x}$$

$$- \lim_{\Delta x \to 0} [nx^{n-1} + (\text{terms having } \Delta x \text{ as a factor})] = nx^{n-1}$$

because all terms having Δx as a factor approach zero as $\Delta x \to 0$. We have found our first rule for the differential calculus. It is:

Power Rule: If $f(x) = x^n$, then $f'(x) = nx^{n-1}$.

For example, if $f(x) = x^4$, then $f'(x) = 4x^3$. If $g(x) = x^8$, then $g'(x) = 8x^7$. The rule can be shown to hold for any constant exponent. Thus, if

$$f(x) = x^{1/2}, \text{ then } f'(x) = \frac{1}{2}x^{1/2-1} = \frac{1}{2}x^{-1/2}.$$

Exercise.   Find $h'(x)$ if $h(x)$ is *a*) x^5. *b*) x^2. *c*) $x^{3/2}$. Answer: *a*) $5x^4$. *b*) $2x$. *c*) $(3/2)x^{1/2}$.

This section is concerned with the derivation of rules, and we leave the application of these rules for the next chapter. Let us now turn to the exponential function, $f(x) = e^x$. Its derivative is

$$f'(x) = \lim_{\Delta x \to 0} \frac{f(x + \Delta x) - f(x)}{\Delta x} = \lim_{\Delta x \to 0} \frac{e^{x+\Delta x} - e^x}{\Delta x},$$

where again substitution of $\Delta x = 0$ leads to $0/0$. Moreover, there is no way to change the expression so that the limit can be found in the simple manner applied in the case of the power function. Consequently, we shall use a sequence of Δx values approaching zero to try to learn what the limit is. Before doing so, we note that $e^{x+\Delta x} = e^x(e^{\Delta x})$ by a law of exponents, so

$$\lim_{\Delta x \to 0} \frac{e^{x+\Delta x} - e^x}{\Delta x} = \lim_{\Delta x \to 0} \frac{e^x(e^{\Delta x}) - e^x}{\Delta x} = \lim_{\Delta x \to 0} \frac{e^x(e^{\Delta x} - 1)}{\Delta x}.$$

In considering this limit, only Δx changes, so the factor e^x remains as e^x in the limit and can be placed outside, so

$$f'(x) = e^x \left[\lim_{\Delta x \to 0} \frac{e^{\Delta x} - 1}{\Delta x} \right] = e^x$$

because, as discussed at the end of Section 9–22,

$$\lim_{\Delta x \to 0} \frac{e^{\Delta x} - 1}{\Delta x} = 1.$$

We have our second rule which says that the exponential function is its own derivative. Thus,

Exponential Rule: If $f(x) = e^x$, then $f'(x) = e^x$.

Finally, we find the rule for the derivative of $f(x) = \ln x$. Thus,

$$f'(x) = \lim_{\Delta x \to 0} \frac{\ln (x + \Delta x) - \ln x}{\Delta x}.$$

Experimentation has shown how to change this to a form in which the limit to be evaluated is $\lim(1 + 1/n)^n$ as $n \longrightarrow \infty$, which is e. Here is how it is done. First, we know by a rule of logarithms that $\ln [(x + \Delta x)/x]$ is the logarithm of the numerator minus the logarithm of the denominator, so in the expression for $f'(x)$, $\ln(x + \Delta x) - \ln x$ can be changed to $\ln[(x + \Delta x)/x]$, which is the same as $\ln(1 + \Delta x/x)$. Hence,

$$f'(x) = \lim_{\Delta x \to 0} \frac{\ln\left(1 + \frac{\Delta x}{x}\right)}{\Delta x} = \lim_{\Delta x \to 0} \left(\frac{1}{\Delta x}\right)\ln\left(1 + \frac{\Delta x}{x}\right).$$

To progress toward the form we seek, we next multiply the expression by $(1/x)x$, which is permissible because it is multiplication by 1. We place $(1/x)$ outside and x inside the limit symbol, and this is permissible because x is not affected when $\Delta x \to 0$. We have

$$f'(x) = \left(\frac{1}{x}\right)\lim_{\Delta x \to 0}\left(\frac{x}{\Delta x}\right)\ln\left(1 + \frac{x}{\Delta x}\right).$$

We are now going to substitute n for $x/\Delta x$. In so doing, we must note that as $\Delta x \to 0$, $x/\Delta x$, which is n, $\to \infty$. Also, if $x/\Delta x = n$, then $\Delta x/x = 1/n$. Hence,

$$f'(x) = \left(\frac{1}{x}\right)\lim_{n \to \infty} n\left[\ln\left(1 + \frac{1}{n}\right)\right] = \left(\frac{1}{x}\right)\lim_{n \to \infty} \ln\left(1 + \frac{1}{n}\right)^n,$$

where, at the end, we used a rule of logarithms to write n as an exponent. The limit part of the last expression is the natural logarithm of e, which is 1. Hence,

Logarithmic Rule: If $f(x) = \ln x$, then $f'(x) = \frac{1}{x}.$

The rules we have developed will be applied over and over again in the chapters to come. The purpose of this section was to build an understanding of the source of the rules—the difficult limit process—and get this behind us so that as we move on we can concentrate on applications. It is not difficult to learn how to *do* calculus, for all that is necessary is the manipulation of expressions according to some relatively simple rules. However, *doing* and *understanding* are quite different matters, and many students feel uncomfortable doing what they do not understand. Those who wish to build an understanding of this underlying theory of differential calculus should study the material in this section until they feel they can reproduce the development of the rules for finding the derivatives of x^n, e^x, and $\ln x$.

9.24 PROBLEM SET 4

In each of the following, write the first five terms of the sequence whose general term is given; then, using as many terms as needed, find the limit of the sequence as $n \to \infty$, if one exists.

1. $\frac{n^2 + 1}{n^2}.$

2. $\frac{n^2 - 10}{n}.$

3. $\left(1 + \frac{1}{n}\right)^n.$

4. $3^n.$

5. $2^{-n}.$

6. $\frac{n^2}{2^n}.$

The square root of n can be found by guessing an approximate value to use as the first term of a sequence of approximations. To get the next term, add $\frac{1}{2}$ the present term to $n/2$ times the reciprocal of the present term. For example, if $n = 28$, $\sqrt{n}$ is about 5, so the first term is 5. The second term is $(\frac{1}{2})5 + (\frac{28}{2})(\frac{1}{5}) = \frac{53}{10}$. The third term is $\frac{53}{20} + (\frac{28}{2})(\frac{10}{53}) = \frac{5609}{1060} =$ about 5.2915094 as compared to the more accurate $\sqrt{28} = 5.2915026$. Approximate the following square roots by means of the third term of a sequence and compare the result with the given value.

7. $\sqrt{10} = 3.1622777$.   8. $\sqrt{6} = 2.4494897$.

By use of the formula for S_n, find the following sums:

9. $1 + \frac{2}{3} + \frac{4}{9} + \cdots$ to $n = 5$ terms.

10. $1 + 2 + 4 + 8 + \cdots$ to $n = 8$ terms.

11. $\lim_{n \to \infty} \left(1 + \frac{1}{2} + \frac{1}{4} + \frac{1}{8} + \cdots\right)$.

12. $\lim_{n \to \infty} \left(1 + \frac{2}{3} + \frac{4}{9} + \cdots\right)$.

MPC is the marginal propensity to consume. How much income will be generated by a $1000 expenditure if

13. MPC $= 0.92$?   14. MPC $= 0.8$?

15. If we deposit $\$A$ now at i percent per year, $\$R$ can be withdrawn each year for n years, starting at the end of the year of deposit. The relevant formula (See Chapter 8) is

$$R_n = A\left[\frac{i}{1 - (1 + i)^{-n}}\right].$$

a) What is $\lim_{n \to \infty} R_n$?

b) What yearly return, forever, would be provided by a deposit of $A = \$1000$ at 8 percent compounded annually?

16. Determine $\lim_{n \to \infty} A_n$ if

$$A_n = R\left[\frac{1 - (1 + i)^{-n}}{i} + 1\right].$$

17. An endowment invested at $i = 8$ percent interest per year is to provide an income of $6000 per year, forever. What should be the amount of the endowment if:
a) The first $6000 is to be available at the end of the first year?
b) The first $6000 is to be available at once?

18. Machine A costs $4000 and has a life of 20 years. Machine B, which serves the same function as A, costs $2500 and lasts 10 years. Neither machine has any scrap value. At 8 percent,
a) What is the capitalized cost of machine A?
b) Is machine B a better buy? Explain.

19. a) What is the series expansion for e?
b) Approximate e by using the first four terms of the series.

20. a) What is the series expansion for e^{-x}?
b) Approximate $e^{-0.2}$ by using the first three terms of the series.

21. If $C(x)$ is the cost of x units and $C(x) = 10 + 3x$ for $x \leq 50$, but $C(x) = 10 + 2x$ for $x > 50$, is $C(x)$ continuous at $x = 50$? Explain.

22. If $C(x)$ is the cost of x units and $C(x) = 10 + 3x$ when $x \leq 50$, but $C(x) = 60 + 2x$ when $x > 50$, is $C(x)$ continuous at $x = 50$? Explain.

23. In considering $\lim_{x \to a} f(x)$, why is it helpful to know $f(x)$ is continuous at $x = a$?

The functions x, x^2, e^x, positive $x^{1/2}$, $x^{1/3}$, and $\ln x$ are continuous for all positive values of x. Find:

24. $\lim_{x \to 2} (e^x - 2)$.   25. $\lim_{x \to 2} (2x - 3 \ln x)$.   26. $\lim_{x \to 2} (2x - x^2)$.

27. $\lim_{x \to 64} (2x^{1/3} + 3x^{1/2})$.

Find the following limits if they exist.

28. $\lim_{x \to 0} \dfrac{x^3}{x^2}$.   29. $\lim_{x \to 0} \dfrac{x^2}{x^3}$.   30. $\lim_{x \to \infty} \dfrac{1}{x}$.

31. $\lim_{x \to 2} \dfrac{x^2 - 4}{x - 2}$.   32. $\lim_{x \to a} \dfrac{x^2 - a^2}{x - a}$.

Write the definition for the following derivatives in limit symbols.

33. $f(x) = x^n$.   34. $g(x) = e^x$.   35. $h(x) = \ln x$.

State the following rules of the differential calculus:

36. The Power Rule.   37. The Exponential Rule.

38. The Logarithmic Rule.

9.25 REVIEW PROBLEMS

1. Temperature can be expressed in degrees Fahrenheit, F, or degrees Centigrade, C. If $C(F) = \left(\dfrac{5}{9}\right)(F - 32)$, what would be $F(C)$?

2. At interest rate i compounded n times a year for y years, a deposit of $\$P$ will grow to $\$A$, where

$$A = P\left(1 + \frac{i}{n}\right)^{ny}.$$

Write the functional form of the statement if
a) P, n, and y are considered as constants.
b) p, i, and n are considered as constants.

3. Given $p(x) = 2x + 3$, $r(x) = 3^x$, and $s(x) = \log x$. Evaluate
a) $p(0)$.   b) $r(0)$.   c) $s(10)$.   d) $p(5)$.
e) $r(-1)$.   f) $s(1)$.

4. See Problem 3. Evaluate
a) $p(3) - p(2)$.   b) $[r(2)][p(5)]$.   c) $r(2) - r(1)$.
d) $r(-1) - r(-2)$.   e) $s(6.2) - s(6.1)$.

5. Evaluate $p(3)$ if $p(r) = xr^2 - 9x$.

6. If $g(h) = 3h - 2k^2 + 5$, how much will $g(h)$ change if h increases by 1?

7. If the cost of making x units of product is $C(x) = 200 + 2x + 0.005x^2$ and each unit made is sold at $\$10$ per unit,
a) What is the expression for the marginal cost of the $(x + 1)$st unit?
b) What is the marginal cost of the 1000th unit?
c) What output leads to maximum profit?
d) What is the maximum profit?

8. What is the expression for the difference quotient (slope of the secant line) as x increases to $x + \Delta x$ in
   a) $p(x)$.   b) 2^{-x}.   c) $2^x(\log x)$.

9. Using $\Delta x = 0.1$, approximate the slope of the tangent to the curve at the stated point by numerical differentiation. Note: The number m in parentheses at the right is the slope correct to the stated number of digits.
   a) $f(x) = x^2$ at $x = 2$.   ($m = 4$).
   b) $g(x) = 2x^2 - 3x$ at $x = 0$.   ($m = -3$).
   c) $h(x) = \log x$ at $x = 2$.   ($m = 0.2171$).
   d) $p(x) = 2^{-x}$ at $x = 1$.   ($m = -0.3466$).

10. Approximate the slope of the tangent to $f(x) = x^2$ at $x = 1$, using $\Delta x = 0.01$.

11. Does $f(x) = 3^x$ have an asymptote? Explain. Hint: Try x values of 1, 2, 3, and so on, then $-1, -2, -3$, and so on.

12. Draw a pair of coordinate axes and put in the line $x = 10$. Place your pencil at a point to the right of $x = 10$ in the first quadrant and sketch a smooth curve which rises toward $x = 10$ as an asymptote and falls toward the x-axis as an asymptote.

13. a) Sketch a curve which is continuous but not smooth over an interval.
    b) Sketch a curve which has a discontinuity.

14. A smooth curve has a local maximum at (2, 4), a local minimum at (5, 3), and a stationary inflection at (9, 6). It has no inflections to the left of (2, 4) nor to the right of (9, 6). Sketch the curve.

15. What do we mean when we say a curve is rising or falling at a point?

By means of sketches, show the pertinent characteristics of

16. $q(x) = 6x - 0.3x^2$.   17. $q(x) = 0.5x^2 - 4x + 15$.

18. In a nickel-off sale, a customer who buys x units of an item pays $10 - 0.05x$ dollars per unit.
    a) What is the dollar cost, $D(x)$, to a customer who buys x units?
    b) What is the maximum amount a customer can spend on one purchase of x units?

By means of sketches, show the pertinent characteristics of

19. $C(x) = x^3 - 6x^2 + 34$.   20. $C(x) = x^3 + x + 4$.

21. If the cost of making x units of product is $C(x) = 0.2x^3 - 6x^2 + 100x + 50$, at what output level does marginal cost start increasing?

22. Find the value of
    a) 2^{-1}.   b) 3^{-2}.   c) $(0.5)^{-2}$.   d) $e^{-1.5}$.   e) $e^{0.08}$.   f) e^0.

By means of sketches, show the pertinent features of

23. $f(x) = 50 - 40e^{-0.5x}$.   24. $f(x) = 50 + 50(2^{5-x})$.

25. If workers' output per day changes due to learning, and output per day at t days is
    $$g(t) = 15 - 10e^{-0.1t},$$
    a) What is the beginning output rate at $t = 0$?
    b) What is the rate at $t = 10$ days?
    c) What potential output rate is being approached as time goes on?

26. What is the effective rate if interest is quoted at
    a) 9 percent compounded continuously?

b) 15 percent compounded continuously?

27. Find the following:
    a) log 15.
    b) ln 15.
    c) log 0.3.
    d) ln 0.3.
    e) ln 126.
    f) ln 0.25.

28. How can ln 64 be computed from
    a) ln 8?
    b) ln 4?
    c) ln 16 and ln 4?
    d) How can ln 0.25 be computed from ln 2.5 and ln 10? Do the computation.

29. a) Write ln x^2 − ln y^2 without exponents and using *ln* just once.
    b) Write ln (e/x) without fractions and using *ln* just once.
    c) Write ln 2 + ln a using *ln* just once.

30. Solve $e^{-0.1x} = 0.8$ for x.

31. Solve $e^{2x} = 20$ for x.

32. a) If a price index, now 100 percent or 1.00, rises continuously at a rate of 9 percent per year, what will the index be 10 years hence?
    b) What will be the purchasing power of the dollar 10 years hence?

33. If the Consumers' Price Index rises continuously at a rate of 11 percent per year, how long will it take for the value of the present dollar to fall to 20 cents?

Write the first five terms of the sequence whose general term is given and then state the limit of the sequence as $n \longrightarrow \infty$, if it exists.

34. $\dfrac{n + 10}{n^2}$.

35. $\dfrac{2n + 1}{n}$.

36. $\left(\dfrac{n + 1}{n}\right)^n$.

37. $\left(\dfrac{1}{3}\right)^n$.

38. $\left(\dfrac{1}{3}\right)^{-n}$.

By the method described prior to Problem 7 of Set 4, find the following square roots and compare results with the given root.

39. $\sqrt{34} = 5.83095$.

40. $\sqrt{0.05} = 0.22361$.

41. By the formula for S_n, find the following sum:
    $$1 + \frac{1}{4} + \frac{1}{16} + \cdots \text{ to } n = 4 \text{ terms.}$$

42. If $S_n = (1 + 0.8 + 0.64 + 0.512 + \cdots)$, find $\lim\limits_{n \to \infty} S_n$.

43. What income will be generated by a $1 expenditure if the marginal propensity to consume is 3/4?

44. An endowment invested at 6 percent per year is to provide a scholarship of $1800 each year forever, the first grant being made at the time of deposit of the endowment. How much should be deposited?

45. Machine A costs $5000 and lasts 8 years, whereas B, which serves the same function, costs $3450 and lasts 5 years. At 8 percent,
    a) What is the capitalized cost of machine A?
    b) Is B a better buy? Explain.

Find the following limits if they exist.

46. $\lim\limits_{x \to 2} \dfrac{x + 2}{x - 2}$.

47. $\lim\limits_{x \to 0} \dfrac{x^2 + 2x}{x}$.

48. $\lim\limits_{a \to b} \dfrac{a^2 - b^2}{a - b}$.

49. $\lim\limits_{x \to 0} \dfrac{(x + \Delta x)^7 - x^7}{\Delta x}$.

50. Given any function, $p(x)$, what is the definition of its derivative, $p'(x)$?

10

Differential Calculus

10.1 INTRODUCTION

MANY APPLIED PROBLEMS in differential calculus are concerned with optimization; that is, finding maximum or minimum values of a function. To set the stage for our introduction to optimization, we consider the problem faced by a Massachusetts manufacturer who ships a dry chemical via United Parcel Service (UPS), a company which will deliver packages whose length plus girth[1] does not exceed 108 inches.

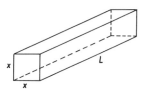

FIGURE 10–1

The shipper uses containers which are rectangular boxes with square cross section as shown in Figure 10–1. The volume of a box is

$$V = (\text{length})(\text{width})(\text{height}) = Lwh.$$

For the box in the Figure, this is

$$V = x \cdot x \cdot L = x^2 L.$$

---

[1]The length plus girth of this book, when closed, is found by adding the vertical length of the cover to the distance all the way around the book.

The length plus girth is $L + 4x$ and because the shipper wants to maximize the box volume, he will use all 108 inches. Thus,

$$L + 4x = 108 \quad \text{or} \quad L = 108 - 4x.$$

Substituting the last expression for L into the volume expression, we have

$$V = x^2L = x^2(108 - 4x) = 108x^2 - 4x^3$$

or

$$V(x) = 108x^2 - 4x^3.$$

Thus, the volume is a function of x, the side of the square cross section. If $x = 10$ and $x = 20$

$$V(10) = 108(100) - 4(1000) = 6{,}800 \text{ cubic inches}$$

and

$$V(20) = 108(400) - 4(8000) = 11{,}200 \text{ cubic inches},$$

so it is clear that the volume changes as x changes. The shipper wants to know the value of x which yields the maximum volume.

The shape of the graph of $V(x)$ is shown in Figure 10–2. We observe that the maximum $V(x)$ is at the peak of the curve. The line tangent to $V(x)$

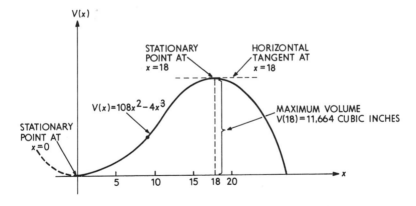

FIGURE 10–2

at the peak is horizontal (slope 0), and the peak is called a *stationary* point. At the end of Chapter 9, we found that the derivative, $V'(x)$, gives the slope of the curve $V(x)$ at the point whose horizontal coordinate is x. We want the value of x where the tangent slope, $V'(x)$, is zero. Recalling from Chapter 9 that the power rule says

the derivative of x^n is nx^{n-1}

we see that the derivatives of x^2 and x^3 are, respectively, $2x$ and $3x^2$. [2] Hence,

$$V'(x) = 108(2x) - 4(3x^2) = 216x - 12x^2.$$

This derivative will be zero when

$$216x - 12x^2 = 0$$

or, factoring,

$$12x(18 - x) = 0$$

from which

$$x = 0 \quad \text{and} \quad x = 18. \text{ [3]}$$

Referring to Figure 10–2, we note that the stationary point at $x = 0$ is a local *minimum,* and the local *maximum* is at $x = 18$. Recalling that

$$L = 108 - 4x$$

we have

$$L = 108 - 4(18) = 36 \text{ inches,}$$

so the optimum box dimensions, $(x)(x)(L)$, are 18 by 18 by 36 inches and the maximum volume is

$$V = 18(18)(36) = 11{,}664 \text{ cubic inches.[4]}$$

Our study of calculus will be directed toward the solution of applied problems. Before continuing this study, it will be helpful to practice the calculus rules developed at the end of Chapter 9.

10.2 SIMPLE RULES AND OPERATIONS

Power Rule: If $f(x) = x^n$, then $f'(x) = nx^{n-1}$.
Exponential Rule: If $f(x) = e^x$, then $f'(x) = e^x$.
Logarithm Rule: If $f(x) = \ln x$, then $f'(x) = 1/x$.

In each case, the geometric interpretation is that $f(x)$ is the formula for

---

[2] We shall offer extensive practice with the derivative rules shortly.
[3] See Index for reference to the solution of quadratic equations.
[4] To be precise, this is the maximum volume of the chemical plus the material from which the box is made.

finding points on the curve, and $f'(x)$ is the formula for finding the slope of the line tangent to the curve at these points. For example, if

$$f(x) = \ln x \text{ so that } f'(x) = 1/x,$$

then when $x = 2$ we find $f(2) = \ln 2 = 0.6931$ from Table X, so the point is (2, 0.6931). The slope of the tangent line at this point is $f'(2) = 1/2 = 0.5$. See Figure 10–3.

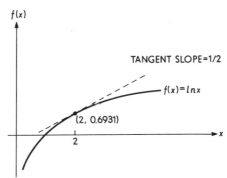

FIGURE 10–3

|||||||||||||||||||||||||||||||||||||||||||||||||||||||||||||||||||||||||||||||||||||||||||||||||||||||||||||||||||||||||||||||||

Exercise. If $f(x) = e^x$, a) What are the coordinates of the point on the curve where $x = 0$, and b) What is the slope of the tangent line at this point? Answer: a) $f(0) = e^0 = 1$, so the point is (0, 1). b) $f'(x) = e^x$, so $f'(0) = e^0 = 1$ and the tangent slope is 1.

|||||||||||||||||||||||||||||||||||||||||||||||||||||||||||||||||||||||||||||||||||||||||||||||||||||||||||||||||||||||||||||||||

Convention. From now on we shall usually speak of the *slope of a curve at a point* rather than use the longer correct statement, *slope of the line tangent to the curve at a point.*

The power rule, which states that the derivative of x^n is nx^{n-1}, applies for any constant exponent. For example, if

$$f(x) = x^{1/3}$$

then $n = 1/3$ and

$$f'(x) = \frac{1}{3}x^{1/3-1} = \frac{1}{3}x^{-2/3} = \frac{1}{3x^{2/3}} \cdot {}^{5}$$

The latter form, with the positive exponent, is needed if we wish to evaluate $f'(x)$. Thus, at $x = 8$, we have

$$f'(8) = \frac{1}{3}(8)^{-2/3} = \frac{1}{3(8)^{2/3}} = \frac{1}{3(2)^2} = \frac{1}{12} \cdot$$

---

[5]It may be helpful here to review the material on exponents in Appendix 2.

Exercise. What is the slope of $f(x) = x^{1/2}$ at the point where x is 9? Answer: $f'(9) = 1/6$.

The derivative of cf(x) = cf'(x), if c is a constant. To illustrate the meaning of the italicized statement, recall that $y = x$, or $f(x) = x$, is a straight line of slope 1. Again, $g(x) = 2x$ is a straight line of slope 2. Thus, the slope (or derivative) of $2x$ is twice the slope (or derivative) of x. We surmise correctly that the derivative of a constant times a function is the constant times the derivative of the function. Hence,

$$\text{if } f(x) = 3x^2, \text{ then } f'(x) = 3(2x) = 6x.$$

Again,

$$\text{if } f(x) = 5e^x, \text{ then } f'(x) = 5e^x.$$

Exercise. Find $f'(4)$ if $f(x) = 2 \ln x$. Answer: $1/2$.

Next, observe that

$$\text{if } f(x) = x = x^1, \text{ then } f'(x) = 1x^0 = 1$$
$$\text{and if } g(x) = 3x, \text{ then } g'(x) = 3x^0 = 3.$$

The derivative of cx is c, if c is a constant.

Exercise. What is $f'(x)$ if $f(x)$ is a) $5x$, b) $2x$, c) $-10x$, d) $x/3$? Answer: a) 5.   b) 2.   c) -10.   d) $1/3$.

The derivative of a constant is zero. The line $y = 10$, or $f(x) = 10$, is horizontal, with slope 0, so that $f'(x) = 0$. Similarly, $f(x) = -6, f(x) = \pi$, or $f(x) = c$ where c is any constant, are horizontal lines whose slope (derivative) is zero. We summarize this finding by saying the derivative of a constant is zero.

Exercise. In $f(x) = a$, how do we know a is a constant? What is $f'(x)$? Answer: The x in $f(x)$ specifies the x is the independent variable, so a is treated as a constant and $f'(x) = 0$.

The derivative of a sum is the sum of the derivatives. That is, the derivative may be taken term by term. To illustrate the point, note that $f(x) = 3x^2 + 2x^2$ is $f(x) = 5x^2$ and its derivative is $10x$. If we take the sum of derivatives and write $f'(x) = 6x + 4x$ we again obtain $10x$, and we conjecture correctly that the derivative may be taken term by term. Thus,

$$\text{if } f(x) = 3x^2 - 2x + 5e^x - 2 \ln x - 6,$$

then

$$f'(x) = 6x - 2 + 5e^x - \frac{2}{x}.$$

Exercise. Find $f'(x)$ if $f(x) = 2x^{3/2} - 4x^2 + 5 \ln x + 12x + 6$.
Answer: $f'(x) = 3x^{1/2} - 8x + (5/x) + 12$.

We shall defer until later the problems of finding derivatives of products and quotients because they cannot be taken in a simple manner. To see why, consider $f(x) = (x^2)(x^3)$. The conjecture that $f'(x)$ is the derivative of x^2 times the derivative of x^3 is easily seen to be false.

Exercise. a) Since $f(x) = (x^2)(x^3) = x^5$, what is $f'(x)$? b) What incorrect result occurs if we multiply the derivatives of x^2 and x^3?
Answer: a) $f'(x) = 5x^4$. b) $(2x)(3x^2) = 6x^3$.

Attention should be called to the fact that application of the power rule requires that the function be in the form x^n. We shall often encounter expressions such as $1/x$, $1/x^2$, or $1/x^{1/2}$. Before the power rule is applied, these must be changed to x^{-1}, x^{-2}, $x^{-1/2}$, respectively. For example, we change

$$f(x) = \frac{3}{x^2} - \frac{2}{x^{1/2}} \quad \text{to} \quad f(x) = 3x^{-2} - 2x^{-1/2}$$

so

$$f'(x) = -6x^{-3} + x^{-3/2} = -\frac{6}{x^3} + \frac{1}{x^{3/2}}.$$

Again, if

$$f(x) = 3x(2x - 4)$$

we must expand the expression to get the proper power forms

$$f(x) = 6x^2 - 12x.$$

Then

$$f'(x) = 12x - 12.$$

Finally, we note that we have used the *natural* (base e) logarithmic function, $f(x) = \ln x$, and the natural exponential function, $f(x) = e^x$. Base 10 logarithms, log x, are seldom encountered in calculus. If the log x function does arise, we recall from Chapter 9 that

$$\log x = \frac{1}{\ln 10} \ln x = \frac{1}{2.3026} \ln x = 0.4343 \ln x.$$

It follows that

$$\text{if } f(x) = \log x, \ f'(x) = \frac{0.4343}{x}.$$

It will be shown near the end of Section 10.11 that if we have an exponential with a base other than e, say $f(x) = a^x$, then the derivative is a^x times a conversion factor which is $\ln a$. That is

$$\text{if } f(x) = a^x, \text{ then } f'(x) = a^x \ln a.$$

For example,

$$\text{if } f(x) = 5^x, \ f'(x) = 5^x \ln 5 = 5^x(1.6094).$$

Exercise. Find $f'(x)$ and $f'(3)$ if $f(x) = 2^x$. Answer: $f'(x) = 2^x \ln 2$. $f'(3) = 2^3(0.6931) = 5.5448$.

Summary

Note: These formulas may also be found in Table XII at the end of the book.

1. *a)* $f(x)$ is the formula for finding points on a curve.
   b) $f'(x)$ is the formula for finding the slope of (a line tangent to) the curve at a point.
2. Power Rule: If $f(x) = x^n$, $f'(x) = nx^{n-1}$.
   Example: If $f(x) = x^3$, $f'(x) = 3x^2$.
3. Exponential Rule:
   a) If $f(x) = e^x$, then $f'(x) = e^x$.
   b) If $f(x) = a^x$, then $f'(x) = a^x \ln a$.
   Example: If $f(x) = 3^x$, $f'(x) = 3^x \ln 3 = 3^x(1.0986)$.
4. Logarithmic Rule: If $f(x) = \ln x$, then $f'(x) = 1/x$.
   Example: If $f(x) = \ln x$, then $f'(10) = \frac{1}{10} = 0.1$.
5. The derivative of a constant is zero.
   Examples: If $f(x) = 10$, $f'(x) = 0$; if $f(x) = \pi$, $f'(x) = 0$.
6. The derivative of a constant times x is the constant.
   Examples: If $f(x) = cx$, $f'(x) = c$.
   If $f(x) = x$, $f'(x) = 1$.
   If $f(x) = 3x$, $f'(x) = 3$.
7. The derivative of a constant times a function is the constant times the derivative of the function; that is, the derivative of $cf(x)$ is $cf'(x)$.
   Examples: If $f(x) = 3 \ln x$, $f'(x) = 3(1/x) = 3/x$.
   If $f(x) - 2(4)^x$, $f'(x) = 2(4^x)\ln 4$.

8. The derivative may be taken term by term.
   Example: If $f(x) = 3x^{4/3} - 2x^3 + 5x - 7 \ln x + 10$,
   then $f'(x) = 4x^{1/3} - 6x^2 + 5 - (7/x)$.

10.3 PROBLEM SET 1

Find $f'(x)$ if

1. $f(x) = x^3$.
2. $f(x) = 3 \ln x$.
3. $f(x) = 3^x$.
4. $f(x) = 2e^x$.
5. $f(x) = 3x^2 + 2x - 6$.
6. $f(x) = 5x^3 - 4x^2 + 3x + 20$.
7. $f(x) = 4x + 3$.
8. $f(x) = 4$.
9. $f(x) = x$.
10. $f(x) = x - \dfrac{1}{x}$.
11. $f(x) = \dfrac{1}{x} - \dfrac{1}{x^2}$.
12. $f(x) = x^{5/3}$.
13. $f(x) = 3x^{1/3}$.
14. $f(x) = x^{1/2}$.
15. $f(x) = \dfrac{1}{x^{1/2}}$.
16. $f(x) = 2x^{-1/2} + 3x^{-1/3} + (2)^{1/5}$.
17. $f(x) = (0.9)^x$.
18. $f(x) = 3x^4 + 2x^{1/2} + K$.
19. $f(x) = 3(2)^x$.
20. $f(x) = e^x + 5 \ln x$.

Write the coordinates of the point on the curve for the stated value of x and find the slope of the curve at this point.

21. $f(x) = 3x^2$ at $x = \frac{1}{2}$
22. $f(x) = 2x^{3/2} - 3x + 5$ at $x = 4$.
23. $f(x) = e^x + x^2 - 3x + 5$ at $x = 0$.
24. $f(x) = x^{1/2}$ at $x = \frac{1}{4}$.
25. $f(x) = 5x - 2 \ln x$ at $x = 1$.
26. $f(x) = 2x - \dfrac{4}{x}$ at $x = 2$.
27. $f(x) = (0.9)^x$ at $x = 1$.
28. $f(x) = e^x - 3(2)^x$ at $x = 0$.
29. $f(x) = x - \ln x$ at $x = 2$.
30. $f(x) = 3e^x$ at $x = 0.05$.

10.4 CONCAVITY: SECOND DERIVATIVE TEST FOR MAXIMA AND MINIMA

Returning to the introductory example of Section 10.1, recall that the problem was to maximize $V(x)$ where

$$V(x) = 108x^2 - 4x^3.$$

Geometrically, the maximum volume occurs at a peak in the curve, and a peak is a stationary point; that is, a point where the tangent is horizontal, slope 0. Inasmuch as the tangent slope formula is $V'(x)$, the problem is to find where $V'(x) = 0$. We proceed to find

$$V'(x) = 216x - 12x^2$$

and set it equal to zero,

$$216x - 12x^2 = 0$$

so that

$$12x(18 - x) = 0$$

and therefore $x = 0$ and $x = 18$ are the x-coordinates of stationary points. We saw from the graph, Figure 10–2, that the stationary point at $x = 0$ was a local minimum, and the one at $x = 18$ was a local maximum. The question we now address is that of determining whether a stationary point is a maximum or a minimum (or neither) *without* having a graph of the function available. We start by noting that at points on $V(x)$ where $V'(x)$ is positive, tangents rise to the right and have positive slope numbers. We shall say that a function is *increasing* (curve is rising) at points where its derivative is *positive*. Similarly a function is *decreasing* (curve is falling), at points where its derivative is *negative* and, of course, a function is *stationary* at points where its derivative is *zero*. For the function $V(x)$,

$$V'(x) = 216x - 12x^2,$$

at $x = 10$

$$V'(10) = 2160 - 1200 = +960$$

so the curve is rising at $x = 10$ as may be verified from Figure $10-2$.

||||||||||||||||||||||||||||||||||||||||||||||||||||||||||||||||||||||||||||||||||||||||||||||||||||||||||||||||||||||||||||||||||||||||||||

Exercise.   Determine by the derivative whether $V(x)$ is increasing or decreasing at $x = 20$. Answer: $V'(20) = -480$, so $V(x)$ is decreasing.

||||||||||||||||||||||||||||||||||||||||||||||||||||||||||||||||||||||||||||||||||||||||||||||||||||||||||||||||||||||||||||||||||||||||||||

We repeat that a function is increasing at a point where the derivative is positive. *Let us apply this statement to the slope function*

$$V'(x) = 216x - 12x^2.$$

Its derivative, which we write with a double prime as

$$V''(x) = 216 - 24x,$$

is called the *second* derivative of $V(x)$. Now if the derivative of the slope function is *positive*, it means that the *slope is increasing* which indicates that the curve is *concave upward,* having a shape like a cup which holds water, as in Figure 10–4A. Again, in Figure 10–4B, (remembering that a slope

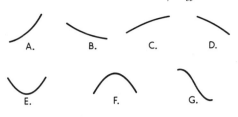

FIGURE 10–4

sequence such as $-3, -2, -1$ is *increasing*) the curve is concave upward so the second derivative is positive. Parts C and D of Figure 10–4 show *downward* concavity, slopes are decreasing, and the second derivative is *negative*. When viewing the curve D, remember that a slope sequence such as -1, $-2, -3, -4$ is a decreasing sequence.

Returning to $V(x) = 108x^2 - 4x^3$, we find

$$V'(x) = 216x - 12x^2$$
$$V''(x) = 216 - 24x.$$

We have

$$V''(15) = 216 - 24(15) = -144$$

so $V(x)$ is concave downward at $x = 15$, as may be verified by reference to Figure 10–2.

Exercise. Find $V''(5)$ and interpret the result. Answer: $V''(5) = +96$. The curve is concave upward at $x = 5$.

Observe that $V''(9) = 216 - 24(9) = 0$. We cannot describe the curve as either concave upward or concave downward because $x = 9$ is the point in Figure 10–2 where concavity changes from upward to downward. It is an *inflection* point. Again, concavity changes from downward to upward at the inflection in Figure 10–4G. We shall return to this matter shortly.

Our main interest in concavity is to observe that at a *local maximum*, as in Figure 10–4F, a curve is concave downward so the *second derivative is negative*. At a local *minimum*, as in Figure 10–4E, the curve is concave upward, so the *second derivative is positive*. Thus, for our function $V(x)$, for which

$$V''(x) = 216 - 24x,$$

the stationary point at $x = 18$ yields

$$V''(18) = 216 - 24(18) = -216$$

so there is a local maximum at $x = 18$, as verified by Figure 10–2.

||||||||||||||||||||||||||||||||||||||||||||||||||||||||||||||||||||||||||||||||||||||||||||||||||||||||||||||||||||||||||||||||

Exercise. Verify that $V(x)$ has a local minimum at the stationary point where $x = 0$. Answer: $V''(0) = +216$ and the positive second derivative shows the stationary point is a minimum.

||||||||||||||||||||||||||||||||||||||||||||||||||||||||||||||||||||||||||||||||||||||||||||||||||||||||||||||||||||||||||||||||

Second Derivative Test for Maxima and Minima. What we have learned thus far is that we can locate stationary points on $f(x)$ by setting the first derivative, $f'(x)$, equal to zero and solving for x. If we then substitute the resultant x into the second derivative, $f''(x)$, and the second derivative is positive (negative), the stationary point is a local minimum (maximum). To apply this, suppose that the total cost, $C(q)$, of making q units of a product, is

$$C(q) = 9 + 2q + 0.01q^2.$$

If we make 10 units,

$$C(10) = 9 + 2(10) + 0.01(100) = \$30.$$

The *average cost per unit* then would be total cost divided by number of units which, for 10 units, would be $C(10)/10 = 30/10$ or \$3 per unit. In general the average cost per unit, $A(q)$, is

$$A(q) = \frac{C(q)}{q} = \frac{9 + 2q + 0.01q^2}{q} = \frac{9}{q} + 2 + 0.01q.$$

To examine $A(q)$ for possible maxima or minima, we find

$$A'(q) = -\frac{9}{q^2} + 0.01.$$

Setting $A'(q) = 0$, we have

$$-\frac{9}{q^2} + 0.01 = 0 \quad \text{or} \quad q^2 = 900$$

from which it follows that there are stationary points where $q = 30$ and $q = -30$. Taking the second derivative, we have

$$A''(q) = \frac{18}{q^3}.$$

At $q = 30$, $A''(30)$ is *positive*, so we have a local *minimum*. There is a local maximum at $q = -30$, but this is not of interest in the applied setting of this example. The minimum average cost per unit occurs at $q = 30$ and is

$$A(30) = \frac{9}{30} + 2 + 0.01(30) = 0.3 + 2 + 0.3 = 2.6$$

or \$2.60 per unit.

||||||||||||||||||||||||||||||||||||||||||||||||||||||||||||||||||||||||||||||||||||||||||||||||||||||||||||||||||||||||||||||||||

Exercise.   Find the stationary point on $f(x) = 32x - x^4$ and what type of point it is. Answer: (2, 48) is a local maximum.

||||||||||||||||||||||||||||||||||||||||||||||||||||||||||||||||||||||||||||||||||||||||||||||||||||||||||||||||||||||||||||||||||

Turning now to the case where the second derivative is zero at a stationary point, that is, both $f'(x)$ and $f''(x) = 0$, consider Figure 10–5. It is easily demonstrated that for each of the four functions, both the first and second derivatives are zero when $x = 0$.

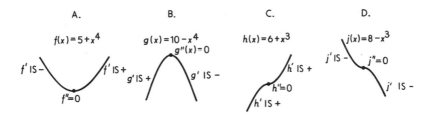

FIGURE 10–5

||||||||||||||||||||||||||||||||||||||||||||||||||||||||||||||||||||||||||||||||||||||||||||||||||||||||||||||||||||||||||||||||||

Exercise.   Show that if $h(x) = 6 + x^3$ as in Figure 10–5C, then $h'(x)$ and $h''(x)$ are zero at $x = 0$. Answer: $h'(x) = 3x^2$ and $h''(x) = 6x$ are zero when $x = 0$.

||||||||||||||||||||||||||||||||||||||||||||||||||||||||||||||||||||||||||||||||||||||||||||||||||||||||||||||||||||||||||||||||||

We see from Figure 10–5 that if $f''(x) = 0$ *at a stationary point*, the point could be a local maximum, minimum, or a stationary inflection, so the second derivative test fails and we shall have to use another method to determine the type of stationary point at hand. One such method is to evaluate the first derivative, $f'(x)$, a little[6] to the left and right of the candidate point. If the resulting signs of $f'(x)$ go from $-$ to $+$, the curve falls, then rises, so the candidate marks a local minimum, as in Figure 10–5A. If the signs go from $+$ to $-$, the curve rises, then falls, so we have a local maximum as in B. If the signs of $f'(x)$ go from $+$ to $+$ as in C, or from $-$ to $-$ as in D, the candidate marks a *stationary inflection* point. For example, if we examine

$$f(x) = x^3 - 6x^2 + 12x$$

we find

$$f'(x) = 3x^2 - 12x + 12.$$

---

[6]A *little* means that the points chosen are closer to the candidate point than other candidate points.

Setting $f'(x) = 0$,

$$3x^2 - 12x + 12 = 0 \quad \text{or} \quad x^2 - 4x + 4 = 0$$

which factors into

$$(x - 2)(x - 2) = 0$$

so we have a stationary point at $x = 2$. The second derivative is

$$f''(x) = 6x - 12 \quad \text{and} \quad f''(2) = 0$$

so the test fails. Taking $x = 1$ and $x = 3$ as points to the left and right of the candidate $x = 2$, we evaluate the first derivative, $f'(x) = 3x^2 - 12x + 12$ as

$$f'(1) = 3 \quad \text{and} \quad f'(3) = 3.$$

The sign of $f'(x)$ did not change, so we have a *stationary inflection* at $x = 2$, $f(x) = 8$.

Exercise. Find the stationary point on $g(x) = 10 - x^4$ and determine its type. Answer: $g'(x) = 0$ at $x = 0$, and $g''(0) = 0$ so the test fails. Evaluating $g'(x) = -4x^3$ to the left ($x = -1$) and right ($x = 1$) of $x = 0$, we find $g'(-1) = 4$ and $g'(1) = -4$, so the curve rose, then fell, and there is a maximum at $x = 0$, $g(x) = 10$. See Figure 10–5B.

At times we are interested in finding inflection points, stationary (zero slope) or otherwise (nonzero slope). To this end, we note that concavity changes from upward to downward, or vice versa, at an inflection point, so $f''(x)$ changes from $-$ to $+$ or $+$ to $-$ which means $f''(x)$ must be zero at an inflection.[7] Thus, in seeking inflection points, we set the second derivative, $f''(x)$, equal to zero, solve, and then test the result using points a little to the left and right in $f''(x)$. If $f''(x)$ changes sign, we have an inflection point. For example, let us examine

$$f(x) = -2x^3 + 12x^2 - 18x + 40$$

for stationary points, and then for inflection points. We have

$$f'(x) = -6x^2 + 24x - 18$$

and $f'(x) = 0$ when

$$-6x^2 + 24x - 18 = 0 \quad \text{or} \quad x^2 - 4x + 3 = 0.$$

Factoring,

$$(x - 1)(x - 3) = 0,$$

---

[7]The converse is not necessarily true. That is, as shown in Figures 10–5A and B, if the second derivative is zero, the point is not necessarily an inflection.

so $x = 1$ and $x = 3$ mark stationary points. To test these, we find

$$f''(x) = -12x + 24; \quad f''(1) = 12 \quad \text{and} \quad f''(3) = -12,$$

so $x = 1$ marks a local minimum and $x = 3$ a local maximum. Next, searching for inflection, we set $f''(x) = 0$. Thus,

$$-12x + 24 = 0.$$

Solving,

$$x = 2.$$

Using $x = 1.5$ and $x = 2.5$ in $f''(x) = -12x + 24$, we find

$$f''(1.5) = 6 \quad \text{and} \quad f''(2.5) = -6.$$

The sign changed, so $x = 2$ marks an inflection. In summary, $f(x)$ has an inflection at (2, 36), a local minimum at (1, 32), and a local maximum at (3, 40).

Exercise. Find the local maximum, minimum, and inflection points for $f(x) = x^3 - 6x^2 + 17$. Answer: (0, 17); (4, -15); (2, 1).

Summary

1. We learn the following about the behavior of a curve at a point if we substitute the coordinates of the point into the first and second derivatives and observe the signs of these derivatives

a) Change in the function or direction of the curve.

Sign of First Derivative	Change in the Function	Direction of the Curve
+	Increasing	Rising
0	Stationary	Stationary
−	Decreasing	Falling

b) Concavity

Sign of the Second Derivative	Concavity
−	Downward
+	Upward

The second derivative is zero at an inflection point.

2. Stationary Points: Local Maxima and Minima
   Stationary points are found by setting $f'(x)$ equal to zero and solving. These points are candidates for maxima–minima.
3. The Second Derivative Test for Maxima and Minima
   a) Substitute stationary points into $f''(x)$. If $f''(x)$ is negative (positive), the candidate yields a local maximum (minimum).
   b) If $f''(x) = 0$ at a stationary point, the test fails and we evaluate the *first derivative*, $f'(x)$, a little to the left and right of the stationary point. If the results are $+$ to $-$ ($-$ to $+$), the point marks a maximum (minimum). If the sign does not change the point is a stationary inflection.
4. Inflection Points
   Find points where the second derivative, $f''(x)$, equals zero. Evaluate the *second derivative* to the left and right of a point where $f''(x) = 0$. The point is an inflection if $f''(x)$ changes sign.

10.5 PROBLEM SET 2

1. If the average cost per unit when x units are made is

$$A(x) = \frac{10000}{x} + 4x + 100,$$

   a) Find the x which will minimize average cost, and prove it is a minimum by the second derivative test.
   b) Compute the minimum average cost per unit.
2. If we use 100 feet of fence to enclose a rectangular plot of land, we need $2x$ feet for one pair of sides (or x feet for one of these sides), leaving $100 - 2x$ feet for the other pair of sides (or $50 - x$ for one of these sides). The plot then is x feet by $50 - x$ feet or

$$A(x) = x(50 - x) = 50x - x^2.$$

   a) Find the value of x which maximizes the area, $A(x)$, and prove it is a maximum by the second derivative test.
   b) What are the dimensions of the maximum area plot?
   c) What is the maximum area that can be enclosed?
3. In many sampling surveys, published results state the number of people, n, who were questioned and p, the proportion who answered yes. The remaining proportion, $1 - p$, did not answer yes. Inherently, samples from larger groups contain sampling error which statisticians measure by the function

$$V(p) = \frac{p}{n} - \frac{p^2}{n}.$$

   a) Treating n as constant, find the value of p which maximizes $V(p)$ and prove it is a maximum.
   b) Taking $n = 10$, find $V(0.5)$ and $V(0.1)$.

4. The cost per unit assembled, $c(x)$, when an assembly line operates at x units per hour is estimated as

$$c(x) = \frac{100}{x} + \ln x - 4.$$

a) Find the rate x which minimizes cost per unit assembled.
b) Prove a) does yield a minimum.
c) What is the minimum cost per unit? (Note: Recall that $\ln N = 2.3026 \log N$.)

For the following, find coordinates of local maxima or minima. In each case, state the type of point. (Use the testing procedure given in the summary at the beginning of this Problem Set.)

5. $f(x) = x^3 - 27.$
6. $f(x) = 5x - x^2.$
7. $f(x) = x^3 - 4x^2.$
8. $f(x) = x^4 - 32x + 48.$
9. $f(x) = x^3 - 12x^2 + 12.$
10. $f(x) = x^3 - 3x^2 + 2.$
11. $f(x) = x^2 - 32 \ln x$, for $x > 0.$
12. $f(x) = x^2 + \dfrac{128}{x}.$
13. $f(x) = x - e^x.$
14. $f(x) = x - \ln x.$
15. $f(x) = e^3 x - e^x$, where $e^3 = (2.71828 \cdot \cdot)^3.$
16. $f(x) = 2^x - 5.$

Find the coordinates of maxima, minima, and inflection points, if any exist.

17. $f(x) = x^3 - 12x^2 + 12.$
18. $f(x) = 2x^4 - x.$
19. If we have $f(x) = (1/3)x^3 + x^2 - 3x$ so that

$$f'(x) = x^2 + 2x - 3 = (x + 3)(x - 1)$$

and

$$f''(x) = 2x + 2,$$

make a table with x values $-4, -3, -2, -1, 0, 1, 2, 3$ in the first column, the sign (or 0) of f' in the second column, the sign (or 0) of f'' in the third column. In the fourth column, indicate whether the curve is rising, falling, or stationary at the x value. In the fifth column, indicate whether concavity is upward, downward, or neither (inflection) at the point in question.

20. At any point, a polynomial is in one of the eight conditions shown in the tabular boxhead below.

		Rising			Falling		
Maximum	Minimum	Concave Up	Concave Down	Inflection	Concave Up	Concave Down	Inflection

In which of these conditions is the curve

$$y = \frac{x^4}{4} + \frac{2x^3}{3} - \frac{15x^2}{2}$$

at the following values of x: $-6, -5, -4, -3, -2, -1, 0, 1, 5/3, 2, 3, 4, 5?$

21. In the cubic function

$$c(x) = tx^3 + sx^2 + fx + k,$$

k is a constant and t, s, and f stand for the constant coefficients of the *t*hird, *s*econd, and *f*irst power terms, and t must not be zero.

a) All cubics have exactly one inflection point. Using $c''(x)$, find the expression for the x-coordinate of the inflection point.

b) Using the result of a) find the x-coordinate of the inflection point of

$$c(x) = x^3 - 9x^2 + 4x + 10$$

without taking derivatives.

10.6 MAXIMA AND MINIMA: APPLICATIONS

In this section we present a series of examples which involve the application of calculus rules and procedures which have been developed in earlier sections.

Example 1. A rectangular warehouse with a flat roof is to have a floor area of 9600 square feet. The interior is to be divided into storeroom and office space by an interior wall parallel to one pair of the sides of the building. The roof and floor areas will be 9600 square feet for any building, but the total wall length will vary for different dimensions. For example, a 96 by 100 foot building could have a 96-foot interior wall, two 96-foot and two 100-foot exterior walls for a total length of $3(96) + 2(100) = 488$ feet.

||||||||||||||||||||||||||||||||||||||||||||||||||||||||||||||||||||||||||||||||||||||||||||||||||||||||||||||||||||||||||||||||||||||||||||||||||||

Exercise. If the building dimensions are $40 \times 240 = 9600$ square feet, what will be the total length of wall if the interior wall is 40 feet long? Answer: $3(40) + 2(240) = 600$ feet.

||||||||||||||||||||||||||||||||||||||||||||||||||||||||||||||||||||||||||||||||||||||||||||||||||||||||||||||||||||||||||||||||||||||||||||||||||||

The problem is to find the dimensions which minimize the total amount

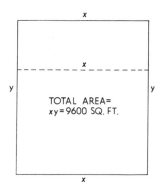

FIGURE 10–6

of wall. Letting x and y be the dimensions as in Figure 10–6, we see that the total amount of wall, w, is

$$w = 3x + 2y,$$

subject to the constraint that

$$xy = 9600 \quad \text{or} \quad y = \frac{9600}{x}.$$

As first stated, w depends upon *two* variables, x and y, and our calculus to this point deals with only *one* independent variable. However, since $xy = 9600$, we can replace y by $9600/x$ and write w as a function of the single variable, x. Thus,

$$w(x) = 3x + 2\frac{(9600)}{x}.$$

We next find

$$w'(x) = 3 - \frac{2(9600)}{x^2}$$

and determine stationary points by setting $w'(x) = 0$, thus:

$$3 - \frac{2(9600)}{x^2} = 0 \quad \text{so} \quad x^2 = \frac{2(9600)}{3} = 6400.$$

$$x = \pm 80.$$

The value $x = -80$ is meaningless in this problem. To determine the type of stationary point at $x = 80$, we find the second derivative,

$$w''(x) = \frac{4(9600)}{x^3}$$

and note that $w''(80)$ is positive, so we have a minimum. Recalling that $y = 9600/x = 9600/80 = 120$, we find that the building dimensions should be 80×120, with the interior wall being 80 feet long. The minimum total wall length is

$$w = 3x + 2y = 3(80) + 2(120) = 480 \text{ feet.}$$

Exercise. Suppose the floor area is to be 10,000 square feet, with no interior wall. *a*) What is the function $w(x)$? *b*) What dimensions minimize $w(x)$? Answer: *a*) $w(x) = 2x + 20,000/x$. *b*) A square building, 100 by 100 feet.

Example 2. A rectangular plot of land is to be enclosed by a fence. Fence for the horizontal sides costs $10 per running foot, while that for the vertical sides costs $5 per running foot. What is the maximum area which can be enclosed if $1500 is available for purchasing the fence?

We shall carry through the solution of this problem by listing questions and answers in a step-by-step sequence in order to provide a reference framework for solving optimization problems. We shall use x and y, respectively, for the horizontal and vertical dimensions, and A for the area.

1) What quantity is to be maximized? Answer: The area, A.
2) What is the formula for this quantity? Answer: $A = xy$.
3) How can this quantity be made a function of a single variable? Answer: By expressing y in terms of x, or vice versa and substituting into (2).
4) What information is available to accomplish (3)? Answer: We know the total amount to be spent is $1500 and this must equal the cost of the two horizontal sides, which will be 2($10x$) plus the cost of the vertical sides which will be 2($5y$); hence,

$$2(10x) + 2(5y) = 1500 \quad \text{or} \quad 20x + 10y = 1500.$$

Hence,

$$y = \frac{1500 - 20x}{10} = 150 - 2x.$$

5) Express the quantity to be optimized as a function of a single variable. Answer: $A(x) = x(150 - 2x) = 150x - 2x^2$.
6) Find the first derivative, set it equal to zero, and solve. Answer: $A'(x) = 150 - 4x$ is zero when $x = 150/4 = 37.5$.
7) Test the result in (6) by the second derivative. Answer: $A''(x) = -4$ is always negative, so we have a maximum.
8) Evaluate the remaining variable, and find the optimum. Answer: $y = 150 - 2(37.5) = 75$. The dimensions of the maximum area enclosure are 37.5 by 75 feet, and the maximum area is 2,812.5 square feet.

Example 3. *Parameterizing a Model.* When we write $y = mx + b$ to represent the equation of a straight line, the arbitrary letters m and b represent constants (slope and y-intercept) which distinguish one line from another. Such *arbitrary constants* (that is, letters standing for constants) are called *parameters.* Now return to Example 2 and let $h (instead of $10) be the cost per horizontal foot, and $v (instead of $5) be the cost per vertical foot.

Similarly, let $C (instead of $1500) represent the total cost. We would then have

$$C = 2x(\$h) + 2y(\$v)$$

from which

$$y = \frac{C - 2xh}{2v}$$

and

$$A(x) = x\left(\frac{C - 2xh}{2v}\right).$$

The model for $A(x)$ has now been parameterized, and we seek the optimal solution in terms of the parameters C, h, and v. We have

$$A(x) = \frac{xC}{2v} - \frac{x^2h}{v}.$$

Remembering that the parameters represent constants, we find

$$A'(x) = \frac{C}{2v} - \frac{2xh}{v}$$

and $A'(x) = 0$ when

$$\frac{C}{2v} - \frac{2xh}{v} = 0.$$

Multiplying by $2v$, we have

$$C - 4xh = 0 \quad \text{or} \quad x = \frac{C}{4h}.$$

We obtain

$$y = \frac{C - 2xh}{2v} = \frac{C - 2\left(\frac{C}{4h}\right)h}{2v} = \frac{4Ch - 2Ch}{2v(4h)}$$

$$y = \frac{2Ch}{2v(4h)} = \frac{C}{4v}.$$

The maximum area, xy, is

$$A = \left(\frac{C}{4h}\right)\left(\frac{C}{4v}\right) = \frac{C^2}{16vh}.$$

The obvious advantage of using parameters instead of specific numbers in solving a problem of a given type is that the solution of the parameterized model is a *general* solution which covers all specific cases. In the present problem type, we see that if $C = \$1500$ is to be spent with $h = \$10$ and $v = \$5$, then optimal horizontal length is $x = C/4h = 1500/4(10) = 37.5$ feet and $y = C/4v = 1500/4(5) = 75$ feet.

||||||||||||||||||||||||||||||||||||||||||||||||||||||||||||||||||||||||||||||||||||||||||||||||||||||

Exercise. If, in Example 2, the horizontal and vertical cost per foot are \$2 and \$3, and \$2400 is to be spent, what dimensions will maximize the area? Answer: 300 by 200 feet.

||||||||||||||||||||||||||||||||||||||||||||||||||||||||||||||||||||||||||||||||||||||||||||||||||||||

Example 4. A rectangular manufacturing plant with a floor area of 16,875 square feet is to be built on a straight road. The front of the plant must be set back 60 feet from the road, and buffer strips (grass and trees) 30 feet on each side and 20 feet at the back must be provided, as shown in

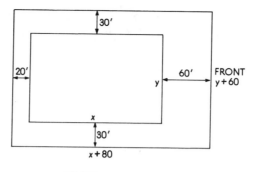

FIGURE 10–7

Figure 10–7. Because of the high cost of land, builders seek to minimize the total area, plant plus buffers. This area is

$$A = (x + 80)(y + 60).$$

Since the *plant* area is 16,875, we have

$$xy = 16875 \quad \text{so} \quad y = \frac{16875}{x}.$$

Hence,

$$A(x) = (x + 80)\left(\frac{16875}{x} + 60\right)$$

$$= 16875 + 60x + \frac{80(16875)}{x} + 4800.$$

We find

$$A'(x) = 60 - \frac{80(16875)}{x^2},$$

and $A'(x)$ is zero when

$$60 - \frac{80(16875)}{x^2} = 0$$

$$x^2 = \frac{80(16875)}{60} = 22500,$$

$$x = 150 \text{ feet.}$$

The second derivative test shows this dimension will minimize $A(x)$. It follows that

$$y = \frac{16875}{x} = \frac{16875}{150} = 112.5 \text{ feet.}$$

The plant dimensions should be 112.5′ at the front by 150 feet deep. The total land dimensions will then be

$$x + 80 = 230 \text{ feet by } y + 60 = 172.5 \text{ feet.}$$

The minimal total land area is $(230)(172.5) = 39,675$ square feet.

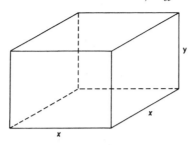

FIGURE 10-8

Example 5. Figure 10–8 shows a box with square top and bottom and rectangular sides. The total surface area of this box is $2x^2$ for top and bottom, plus $4xy$ for the four sides. Thus,

$$A = 2x^2 + 4xy.$$

The volume of the box is the area of the base times the altitude. Thus,

$$V = x^2y.$$

Suppose that we require a box of volume 2592 cubic inches, and we seek to minimize the cost of materials for the sides, top, and bottom. Side material costs 6 cents per square inch, and top and bottom material 9 cents per square inch. Total cost, C, will be

$C =$ (top and bottom area) at 9 cents $+$ (four side areas) at 6 cents
$C = (2x^2)(9) + (4xy)(6)$

From

$$V = x^2y$$

or

$$2592 = x^2y$$

we have

$$y = \frac{2592}{x^2}.$$

Substituting for y in the cost equation yields

$$C(x) = 18x^2 + \frac{24(2592)}{x}.$$

The derivative is

$$C'(x) = 36x - \frac{(2592)(24)}{x^2},$$

and $C'(x) = 0$ when

$$36x = \frac{(2592)(24)}{x^2}.$$

Multiplying by x^2 we find

$$36x^3 = (2592)(24)$$

or

$$x^3 = 1728.$$

Hence,

$$x = (1728)^{1/3} = 12 \text{ inches.}$$

It follows that

$$y = \frac{2592}{x^2} = \frac{2592}{144} = 18 \text{ inches.}$$

Exercise. For the foregoing: *a)* Find the cost of a 12 by 12 by 18 inch box. *b)* Prove this cost is a minimum by the second derivative test. Answer: *a)* 7776 cents, or \$77.76. *b)* $C''(x) = 36 + (48)(2592)/x^3$ is positive when $x = 12$.

Example 6. According to United Parcel Service requirements stated at the opening of the Chapter, the length plus girth of a package must not exceed 108 inches. Suppose packages are to be cylinders, as shown in Figure 10–9, and we seek the dimensions (length L and radius r) which will yield

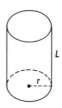

FIGURE 10–9

maximum volume. The volume of a cylinder is the circular base area, πr^2, times the height L, so

$$V = \pi r^2 L.$$

The girth of the cylinder is the perimeter of the circular cross-section, $2\pi r$, so length plus girth is $L + 2\pi r$ and

$$L + 2\pi r = 108 \quad \text{or} \quad L = 108 - 2\pi r.$$

Hence, substituting

$$V(r) = \pi r^2 (108 - 2\pi r) = 108\pi r^2 - 2\pi^2 r^3.$$

The derivative is

$$V'(r) = 216\pi r - 6\pi^2 r^2.$$

Setting $V'(r) = 0$,

$$6\pi r(36 - \pi r) = 0$$

from which

$$r = 0 \quad \text{and} \quad r = \frac{36}{\pi}.$$

The second derivative is

$$V''(r) = 216\pi - 12\pi^2 r$$

and

$$V''\left(\frac{36}{\pi}\right) = 216\pi - 12\pi^2\left(\frac{36}{\pi}\right) = -216\pi,$$

so $r = 36/\pi$ yields a maximum. Next,

$$L = 108 - 2\pi\left(\frac{36}{\pi}\right) = 36$$

so the optimal dimensions for maximum volume are a radius of $36/\pi$ (about 11.5 inches) and a height of 36 inches. The maximum volume will be

$$V = \pi\left(\frac{36}{\pi}\right)^2 (36) = \frac{(36)^3}{\pi} = \frac{46656}{\pi} = 14{,}851 \text{ cubic inches.}$$

For comparison, recall that the maximum volume for rectangular boxes was found at the beginning of the Chapter to be 11,664 cubic inches. If the goal is to maximize volume, cylinders are preferred over rectangular boxes.

***

Exercise. In the last example, suppose the limitation on length plus girth is the parameter, M inches. Express the dimensions r and L for maximum volume in terms of M. Answer: $r = M/3\pi$ and $L = M/3$.

***

Example 7. In normal operations, a plant employs 100 men working 8 hours a day for a total of $100(8) = 800$ man-hours of work per day. In normal operations, productivity averages 30 units per man-hour worked. Thus, in a normal day, production is

$$P = \text{(man-hours worked)(output per man-hour)}$$
$$= (800)(30) = 24{,}000 \text{ units.}$$

If the work level (man-hours) is raised, management estimates that average output per man-hour falls off at the rate of 2.5 units for each extra 100 man-hours, or by 0.025 units for each man-hour in excess of 800. For example, if 840 man-hours are worked, the excess would be $840 - 800 = 40$ and average productivity would be $30 - 40(0.025) = 29$ units per man-hour. Total output would then be

$$P = 840(29) = 24{,}360 \text{ units.}$$

The problem is to find the work level, x man-hours, which maximizes output. Reviewing the illustrative calculations, we find that for $x = 840$ man-hours,

$$P(840) = 840[30 \quad (840 \quad 800)(0.025)].$$

In general,

$$\begin{aligned}
P(x) &= x[30 - (x - 800)(0.025)] \\
&= 30x - 0.025x^2 + 20x \\
&= 50x - 0.025x^2.
\end{aligned}$$

We find

$$P'(x) = 50 - 0.05x$$

and

$$P'(x) = 0 \quad \text{when} \quad x = 1000 \text{ man-hours.}$$

Also

$$P''(x) = -0.05$$

so we have a maximum which is

$$P(1000) = 50(1000) - 0.025(1000)^2 = 25000 \text{ units.}$$

10.7 EVALUATION OF PARAMETERS

Students often ask where the values for parameters used in mathematical models are obtained. Even a partial answer to this question would require more pages than are appropriate in this text, so we shall confine ourselves to a few comments on matters suggested by the last example. Here it was assumed that a single product was involved, so reasonably accurate data for man-hours worked and output would be available in regularly kept payroll and production records. These and other regularly kept records may be direct sources for parameter values. However, we often have to adapt available data to our purposes, or arrange for the collection of data which is not part of the regular information system. Thus, in the last example, it is not likely that the parameter 0.025, representing productivity decline in response to increasing work level, would be available in the information system. We could, however, secure available data on man-hours and output for periods of different work (man-hour) levels and try to learn what happens to output per man-hour as the work level changes. A standard method for doing this is to graph the data as illustrated in Figure 10–10.

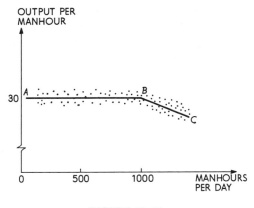

FIGURE 10–10

As in the case of most applied problems, the plotted points are scattered and we try to represent the central trend of the scatter using, in this case, two straight line segments.[8] The trend of segment AB on Figure 10–10 is horizontal at 30 units per man-hour, and the slope of the trend segment BC would be the estimate of the productivity decline which was −0.025 in the last example.

The thumbnail sketch just given has in it a number of elements on parameter evaluation, but it oversimplifies what can be a difficult problem. The parameter values we are able to obtain range from very rough to highly accurate approximations and, at times, properly could be described as guesses. It is axiomatic that good information is needed for good decision making, and one of the challenges that face managers of the future as computer technology moves forward is to provide the data base from which reliable estimates of key parameters can be obtained efficiently.

10.8 AN INVENTORY MODEL

If a manufacturer has to make 98,000 units of an item during a year, he has his choice of a number of different production schedules. He could make the entire lot early in the year in one large production run. By so doing, he might hold production costs down because of the economies of mass production; however, he would have to carry much of the production run in inventory, and the cost of carrying the inventory might overbalance the savings arising from mass-production economies.

To make the problem more specific, let us suppose that it costs $500 to make the factory ready for a production run of the item, then $5 for each

---

[8]Fitting lines or curves to scattered data and measuring the error in such fits is an important technique which may be found in textbooks on numerical methods and statistics. A method for fitting a straight line is given in Chapter 6, Section 28 of this book.

unit made after the factory has been readied, and that it costs 50 cents per year for each unit carried in inventory. Let L, the lot size, be the number of units made in each production run. We shall assume that after a lot has been made, the L units are placed in inventory and then used up (withdrawn from inventory) at a uniform rate such that inventory is zero when the next lot appears. This last assumption permits us to use the average, $L/2$, to formulate inventory cost.

We shall assign parameters as follows:

Number of units to be made in a year: $N = 98,000.$
Variable cost per unit made[9]: $v = \$5.$
Fixed setup cost per lot or batch: $F = \$500.$
Annual inventory cost per unit in average
    inventory: $i = \$0.50.$
Number of units made per lot: L (unknown).
Average inventory: $L/2.$

We may then state

Number of lots to be made in a year

$$= \frac{\text{Number of units per year}}{\text{Number of units per lot}} = \frac{N}{L} = \frac{98,000}{L}.$$

Total setup cost per year

$$= (\text{Lots per year})(\text{Setup cost per lot}) = \frac{N}{L}(F) = \frac{98,000(500)}{L}.$$

Inventory cost per year

$$= (\text{Units in average inventory})(\text{Inventory cost/unit})$$

$$= \frac{iL}{2} = \frac{0.5L}{2}.$$

Total variable cost per year
$$= (\text{Units per year})(\text{Variable cost per unit}) = N(v) = 98,000(5).$$
Finally, the total cost of production and inventory for a year will be

$$C(L) = \text{Setup cost} + \text{variable cost} + \text{inventory cost}$$

$$C(L) = \frac{98,000(500)}{L} + 98,000(5) + \frac{0.5L}{2}$$

or, parameterized,

$$C(L) = \frac{NF}{L} + vN + \frac{iL}{2}.$$

---

[9]We assume here that v is constant at various production levels.

Seeking to minimize $C(L)$, we find

$$C'(L) = -\frac{NF}{L^2} + 0 + \frac{i}{2}$$

and $C'(L) = 0$ when

$$-\frac{NF}{L^2} + \frac{i}{2} = 0 \quad \text{so} \quad L = \sqrt{\frac{2NF}{i}}.$$

Inasmuch as

$$C''(L) = \frac{2NF}{L^3}$$

is positive for any applied problem (that is, N, F, and L are positive), we have a minimum. For the illustrative parameter values,

$$L = \sqrt{\frac{2(98,000)(500)}{0.5}} = \sqrt{196,000,000} = 14,000 \text{ units/lot.}$$

It follows that the firm would make $98,000/14,000 = 7$ lots each year or a lot every $365/7$ days, that is, a lot every 52 days.

||||||||||||||||||||||||||||||||||||||||||||||||||||||||||||||||||||||||||||||||||||||||||||||||||||||||||||||||||||||||||||||||||

Exercise.   The annual requirement for another product made by the foregoing firm is 7,200 units. The setup cost per batch is $100 and the inventory cost is $1 per unit in average inventory. a) What batch size will minimize total annual cost? b) How often should a lot be made? Answer: a) 1,200 per lot. b) Six lots will be made in a year, or a lot every two months.

||||||||||||||||||||||||||||||||||||||||||||||||||||||||||||||||||||||||||||||||||||||||||||||||||||||||||||||||||||||||||||||||||

In the model

$$C(L) = \frac{NF}{L} + vN + \frac{iL}{2},$$

note that vN is constant, so its derivative is zero. Consequently, the parameter v does not appear in the optimal solution. Notice also that when L is small, the setup cost term NF/L is large but the inventory cost $iL/2$ is small, and the reverse holds when L is large. The optimal balance of the cost terms occurs when

$$L = \sqrt{\frac{2NF}{i}}.$$

Finally, observe that the optimal L is not proportional to N. That is, if the annual requirement was reduced by 19 percent to 81 percent of its old value, so that the new requirement is $0.81N$, then

$$L = \sqrt{\frac{2(0.81N)F}{i}} = 0.9\sqrt{\frac{2NF}{i}}$$

so the optimal lot size is now 0.9 or 90 percent of (or 10 percent below) its old value.

Exercise. If the annual requirement is cut to $\frac{1}{4}$ its old value, how would this affect the optimal lot size? Answer: The optimal lot size would now be $\frac{1}{2}$ the old value.

10.9 PROBLEM SET 3

1. A rectangular warehouse is to have 3,300 square feet of floor area and is to be divided into two rectangular rooms by an interior wall. Cost per running foot is $125 for exterior walls and $80 for the interior wall.
   a) What dimensions will minimize total wall cost?
   b) What is the minimum cost?

2. (Similar to Problem 1, but the area is to be found.) If $49,500 has been allocated for walls,
   a) What are the dimensions of the largest warehouse that can be built?
   b) What is the floor area of this warehouse?

3. (Problem 1 in parameterized form.) The floor area is to be A square feet and the cost per running foot is e for exterior walls and i for the interior wall. Let x and y be the warehouse dimensions, with x being the length of the interior wall.
   a) Write the expression for total wall cost, $C(x)$.
   b) Find the expression for x which minimizes $C(x)$.

4. (Problem 2 in parameterized form.) If D are allocated for wall construction then, using the parameters in Problem 3,
   a) Write the expression for the enclosed area, $A(x)$.
   b) Find the expression for x which will minimize $A(x)$.

5. Both interior and exterior walls of a 13,500-square-foot rectangular warehouse cost $100 per running foot. The warehouse is to be divided into eight rooms by three interior walls running in the x direction and one running in the y direction.
   a) What dimension will lead to minimal total wall cost?
   b) What is this minimal cost?

6. A rectangular area of 1050 square feet is to be enclosed by a fence, then divided down the middle by another piece of fence. The fence down the middle costs $0.50 per running foot, and the other fence costs $1.50 per running foot. Find the minimum cost for the required fence.

7. Fence is required on three sides of a rectangular plot. Fence for the two ends costs $1.25 per running foot; fence for the third side costs $2 per running foot. Find the maximum area that can be enclosed with $100 worth of fence.

8. A rectangular cardboard poster is to contain 96 square inches of printed material and have a 2-inch border top and bottom, and a 3-inch border at the sides. Find the dimensions and the area of the minimum sized poster which fulfills these specifications. (Note: Let x and y be the dimensions of the 96-square inch area.)

9. A rectangular-shaped manufacturing plant with a floor area of 600,000 square feet is to be built in a location where zoning regulations require buffer strips 50 feet wide front and back, and 30 feet wide at either end. (A buffer strip is a

grass and tree belt which must not be built upon.) What plot dimensions will lead to minimum total area for plant and buffer strips? What is this minimum total area? (Note: Let x and y be the dimensions of the 600,000 square foot area.) If the plant dimensions were made 1500 by 400 feet rather than the dimensions leading to minimum area, by how much would the total plot area exceed the minimum area?

10. In the United Parcel Service example at the beginning of the Chapter, the length plus girth of a package was restricted to 108 inches. Suppose a shipper uses rectangular boxes with square ends made of a perforated material to provide ventilation to the box contents. To secure maximum ventilation, the shipper wants the total surface area to be as large as possible. Use x as the side of the square base and L as the length.
    a) Write the expression for the area, A, in terms of L and x.
    b) Express $A(x)$ as a function of x alone.
    c) What value of x maximizes $A(x)$? Prove this is a maximum.
    d) What are the dimensions of the maximum area box?

11. (See Problem 10.) Suppose the maximum length plus girth is M. Express the optimal x and L in terms of M.

12. A box with a square top and bottom is to be made to contain a volume of 64 cubic inches. What should be the dimensions of the box if its surface area is to be a minimum? What is this minimum surface area?

13. A box with a square bottom and no top is to be made to contain a volume of 500 cubic inches. What should be the dimensions of the box if its surface area is to be a minimum? What is this minimum surface area?

14. A box with a square top and bottom is to be made to contain 250 cubic inches. Material for top and bottom costs $2 per square inch and material for the sides costs $1 per square inch. What should be the dimensions of the box if its cost is to be a minimum? What is the minimum cost?

15. A box with a square bottom and no top is to be made to contain 100 cubic inches. Bottom material costs five cents per square inch and side materials costs two cents per square inch. Find the cost of the least expensive box that can be made.

16. A box with a square bottom and no top is to be made from a 6 by 6 inch piece of material by cutting equal-sized squares from the corners, then turning up the sides. What should the dimensions of the squares be if the box is to have maximum volume? (Note: The quadratic expression encountered in the solution is factorable.)

17. A box with a rectangular bottom and no top is to be made from a rectangular piece of material with dimensions 16 by 30 inches by cutting equal-sized squares from the corners, then turning up the sides. What should the dimensions of the squares be if the box is to have maximum volume? (Note: The quadratic expression encountered in the solution is factorable.)

18. A cylindrical storage tank is to contain $V = 16,000\pi$ cubic feet (about 400,000 gallons). The cost of the tank is proportional to its area, so the minimal cost tank will be the one with minimum area. The volume of a cylinder of radius r and height h is $V = \pi r^2 h$. Its surface area is the area of top and bottom, $2\pi r^2$, plus the side area, $2\pi rh$. Find the dimensions, r and h, of minimal area tank. (Note: We have used $V = 16,000\pi$ so that π will cancel out.)

19. Suppose the tank in Problem 18 is to be built into the ground to catch runoff water, so it needs no top. Suppose, further, that the cost of the base of the tank is $10 per square foot and the sides $8.64 per square foot. What dimensions will lead to the minimal cost tank?

20. A consulting firm conducts training sessions for employees of various companies. The charge to a company sending employees to a session is $50 per employee, less $0.50 for each employee in excess of 10. That is, for example, if 12 employees are sent, the charge per employee would be $49.00 and the total prorated charge to the company would be $12(49.00) = \$588.00$. The consulting firm further has a fixed total charge for groups of x or more, where x is the number which maximizes the prorated group charge. What should x be, and what is the maximum total group charge to a company?

21. A household appliance service organization has a parts stockroom and a garage at its central office location. Its trucks and drivers service customers in a roughly circular area of radius r miles around the central office. Customer density is approximately uniform in the area and each month the number of service calls per square mile is $80/\pi$ (about 25). Therefore the number of calls in a month is found by multiplying the number of customers per square mile by the number of square miles in the service area.
    a) What is the expression for the number of calls in a month?
    b) The company figures travel cost at $2 per mile and computes mileage per call at $r/2$ miles out from the garage plus $r/2$ miles back in, for a total of r miles per call, on the average. The travel charge per call, excluding parts and labor, is fixed at $24. What is the expression for the net travel income per call? ("Net" means after deducting mileage cost.)
    c) What is the expression for the total net monthly travel income?
    d) What service area radius will maximize total net monthly travel income?
    e) What is the maximum net monthly travel income?

22. (This is a variation on the inventory model discussed in an earlier section.) A retail firm orders a product from a supplier Q units at a time. During a year, the firm will order $N = 2400$ units. The cost per unit ordered is $u = \$4$, and the cost of preparing and handling is figured at $c = \$12$ per order. The annual cost of carrying the average inventory of $Q/2$ units is $p = 0.25$ times (or 25 percent of) the purchase cost of $Q/2$ units. What are the parameterized expressions for:
    a) The number of orders placed in a year?
    b) Total ordering cost per year?
    c) The cost of Q units?
    d) Inventory cost per year?
    e) $S(Q)$, the sum of ordering and inventory cost per year?
    f) The order quantity, Q, which minimizes $S(Q)$?
    g) What is the optimal order quantity for the parameter values given in the problem statement?

23. A manufacturing process generates $W = 4096$ cubic feet of waste per year. The waste is accumulated in a cubical container of side x feet (area $= 6x^2$, volume $= x^3$) which lasts one year. The container costs $K = \$2$ per square foot. When full, the container is emptied and the interior surface (also assumed to be $6x^2$

square feet) is decontaminated at a cost of $d = \$0.50$ per square foot. What are the parameterized expressions for:

a) The yearly container cost?
b) The number of decontaminations per year?
c) The yearly decontamination cost?
d) The yearly sum, $C(x)$, of container plus decontamination cost?
e) The container dimension, x, which minimizes $C(x)$?

Finally,

f) What is the optimum container dimension for the parameter values given in the problem statement?

10.10 d/dx NOTATION

The symbol d/dx, attributable to the mathematician Leibniz, means *the derivative with respect to x* (not d divided by d times $x!!$). Thus,

$$\frac{d}{dx}f(x)$$

means the derivative of $f(x)$ with respect to x, so

$$\frac{d}{dx}(3x^2 - 3x) = 6x - 3.$$

Exercise. Find d/dy $(e^y - \ln y)$. Answer: $e^y - 1/y$.

Although having the same meaning as $f'(x)$, the symbol d/dx is more descriptive because d suggests derivative and the fractional appearance of d/dx connotes the ratio (slope) character of the derivative. Further, $f'(x)$ is a bit awkward in that when we wish to state, say, the power rule, we need the compound statement·

$$\text{if } f(x) = x^n, \quad \text{then} \quad f'(x) = nx^{n-1},$$

but use of d/dx leads to the simple statement

$$\frac{d}{dx}x^n = nx^{n-1}.$$

We shall also find it useful to write expressions such as the last in the slightly different form

$$\frac{dx^n}{dx} = nx^{n-1}.$$

Exercise. State the rule for the derivative of $\ln x$ using d/dx. Answer: $(d/dx)\ln x = 1/x$.

10.11 THE CHAIN RULE

The gross profit of a retail store clearly is a function of sales volume, s, and we may use $g(s)$ to represent this function. Also, net profit (after expenses and so on), n, is a function of gross profit and we may use $n(g)$ to represent this function. Here we have the n function of g, where g is a function of s, and we may express this as $n[g(s)]$, which we read as the n function of the g function of s. This function of a function is an example of a *composite* function. The chain rule tells us how to express

$$\frac{d}{ds}n[g(s)]$$

in terms of the derivatives of the separate functions,

$$\frac{dn(g)}{dg} \quad \text{and} \quad \frac{dg(s)}{ds}.$$

To see the nature of the chain rule, let us suppose that $g(s) = 0.4s$ which says that gross profit is 40 percent of sales, and that $n(g) = 0.2g$ which says that net profit is 20 percent of gross profit. Thus, for sales of $1, gross profit is $0.40 and net profit is $0.2(0.4) = \$0.08$. Observe that

$$n(g) = 0.2g, \quad \text{so that} \quad \frac{dn(g)}{dg} = 0.2$$

and

$$g(s) = 0.4s, \quad \text{so that} \quad \frac{dg(s)}{ds} = 0.4$$

and that

$$\frac{dn(g)}{dg} \cdot \frac{dg(s)}{ds} = 0.2(0.4) = 0.08. \tag{1}$$

Observe next that we could have written

$$n[g(s)] = 0.2g(s) = 0.2(0.4)s = 0.08s$$

so that

$$\frac{dn[g(s)]}{ds} = 0.08. \tag{2}$$

Comparing (1) and (2) we find that in this example

$$\frac{dn[g(s)]}{ds} = \frac{dn(g)}{dg} \cdot \frac{dg(s)}{ds}. \tag{3}$$

Note: Very often, this last type of expression is abbreviated as

$$\frac{dn(g)}{ds} = \frac{dn(g)}{dg} \cdot \frac{dg}{ds}, \quad \text{or} \quad \frac{dn}{dg} \cdot \frac{dg}{ds},$$

it being assumed that if the derivative of $n(g)$ is to be taken with respect to s, (not g), then g *must be a function of* s.

The statement (3) is called the *chain rule*. We shall not prove it here, but the rule holds in general. If we see the expression

$$\frac{dh(w)}{dv},$$

the fact that the derivative is with respect to v and v does not appear in the symbol $h(w)$ means that w is assumed to be a function of v, and

$$\frac{dh(w)}{dv} = \frac{dh(w)}{dw} \cdot \frac{dw}{dv}.$$

Again, in

$$\frac{df(z)}{dx}$$

we assume z is a function of x. It is natural to take the derivative of $f(z)$ with respect to z. The chain rule says to do this and multiply the result by the derivative of z with respect to x. Thus we have the

Chain Rule:

$$\frac{df(z)}{dx} = \frac{df(z)}{dz} \cdot \frac{dz}{dx}.$$

Exercise. Complete $dq(p)/ds =$. Answer: $[dq(p)/dp]$ $[dp/ds]$.

The chain rule is an important tool. In the first place, it allows us to generalize our simple rules for power, exponential, and logarithmic functions. For example, if instead of the simple function $\ln x$ we have the logarithmic function of a function of x, $\ln f(x)$, we may write in abbreviated form

$$\frac{d \ln f}{dx} = \frac{d \ln f}{df} \cdot \frac{df}{dx}$$

$$= \frac{1}{f} \cdot \frac{df}{dx}$$

or, more completely,

$$\frac{d}{dx} \ln f(x) = \frac{1}{f(x)} \frac{df(x)}{dx}.$$

Thus, the derivative of the logarithm of a *function* of x is the reciprocal of the function, multiplied by the derivative of the function. The rule applies, of course, for the simple function $\ln x$, which we may view as being $\ln f(x)$ with $f(x) = x$. Thus

$$\frac{d}{dx} \ln x = \frac{1}{x} \frac{d(x)}{dx} = \frac{1}{x}(1) = \frac{1}{x}.$$

Again,

$$\frac{d}{dx}\ln(x^2 + 3x) = \frac{1}{x^2 + 3x}\frac{d}{dx}(x^2 + 3x) = \frac{2x + 3}{x^2 + 3x}.$$

Exercise.   Find $d/dx[\ln(3x + 2)]$. Answer: $3/(3x + 2)$.

Turning next to the power function of a function of x, that is $[f(x)]^n$, rather than simply x^n, we have, in abbreviated form,

$$\frac{d}{dx}f^n = \frac{df^n}{df}\cdot\frac{df}{dx} = nf^{n-1}\frac{df}{dx}$$

or, more completely

$$\frac{d}{dx}[f(x)]^n = n[f(x)]^{n-1}\frac{df(x)}{dx}.$$

For example,

$$\frac{d}{dx}(x^3 + 2x^2 + 5)^{100} = 100(x^3 + 2x^2 + 5)^{99}(3x^2 + 4x).$$

Exercise.   Find $(d/dx)(x^2 + 5x)^3$. Answer: $3(x^2 + 5x)^2(2x + 5)$.

Finally, for the exponential function of a function of x, that is $e^{f(x)}$ rather than simply e^x, we have in abbreviated form

$$\frac{d}{dx}e^f = \frac{de^f}{df}\cdot\frac{df}{dx} = e^f\cdot\frac{df}{dx}$$

or, more completely,

$$\frac{d}{dx}e^{f(x)} = e^{f(x)}\frac{df(x)}{dx}.$$

As an example, we find

$$\frac{d}{dx}e^{x^2} = e^{x^2}\frac{d}{dx}(x^2) = e^{x^2}(2x).$$

Exercise.   Find $d/dx(e^{0.2x+1})$. Answer: $(0.2)e^{0.2x+1}$.

The chain rule arises often in applications. We shall present two examples here. First, consider the function

$$1 - e^{-0.2t}.$$

We find that the function has the value 0 when $t = 0$. At $t = 5$, the value of the function is

$$1 - e^{-1} = 1 - 0.3679 = 0.6321.$$

Exercise.   What is $\lim\limits_{t \to \infty} (1 - e^{-0.2t})$? Answer: The limit is 1 because $e^{-0.2t} = 1/e^{0.2t}$ has 0 as a limit as $t \to \infty$.

The exercise shows that the function increases toward a limit of 1 as t increases. If we interpret the values of the function as response percentages, we may say that we have 63.21 percent response at time $t = 5$, and that the response approaches, but never reaches, 100 percent as time goes on. Now suppose that the number of people, N, who have responded to an advertising campaign at time t days is given by

$$N = 1,000,000(1 - e^{-0.2t}).$$

Here, the total potential audience is 1,000,000. The number responding at time $t = 0$ is zero, and the number responding approaches 1,000,000 as a limit as t increases.

Next, suppose that revenue averages \$5 per response, and that campaign costs are \$5000 fixed, plus \$10,000 per day. Profit at time t days will be

$$P = \text{Revenue} - \text{Cost},$$

or

$$P = 5(1,000,000)(1 - e^{-0.2t}) - (5000 + 10,000t).$$

We wish to maximize P, and to that end we find

$$\frac{dP}{dt} = (5)(1,000,000)(0.2e^{-0.2t}) - 10,000.$$

Setting dP/dt equal to zero yields

$$1,000,000e^{-0.2t} = 10,000$$

or

$$e^{-0.2t} = 0.01.$$

Taking natural logarithms of both sides, we have

$$-0.2t = \ln(0.01) = 2.3026(\log 0.01)$$

so

$$-0.2t = 2.3026(-2)$$
$$t = 23.026,$$

or about 23 days.

Exercise.   Prove that $t = 23.026$ yields a maximum by the second derivative test. Answer: $d^2P/dt^2 = -200,000e^{-0.2t}$. Inasmuch as e to any power is positive, the second derivative just shown is always negative.

As a second example involving the application of the chain rule, consider Figure 10–11. A road is to be constructed joining cities A and D. If construction costs were uniform, the least expensive procedure would be to make a

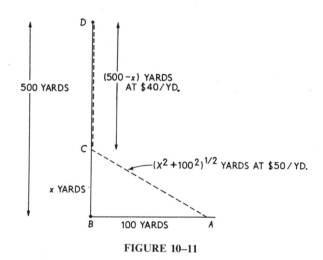

FIGURE 10–11

straight-line connection between A and D. However, construction costs are greater on the slant than on the vertical, so the minimal cost road will not be on the straight slant line $\overline{AD}$. Our problem is to find the point C which will provide the optimal balance between slant and vertical sections and lead to minimal cost.

The vertical distance $\overline{BD}$ is given as 500 yards. If we let $\overline{BC} = x$, then $\overline{CD}$ will be $500 - x$. The distance $\overline{BA}$ is given as 100 yards. Applying the right triangle relation, the hypotenuse, $\overline{AC}$, is the square root of the sum of the squares of the sides, so $\overline{AC}$ is $(x^2 + 100^2)^{1/2}$. The total cost, C, is $\overline{CD}$ yards at \$40 per yard, plus $\overline{AC}$ yards at \$50 per yard. Hence,

$$C = 40(500 - x) + 50(x^2 + 100^2)^{1/2}.$$

The condition for minimal C is that dC/dx be zero. Applying the chain rule to the second term on the right, we find

$$\frac{dC}{dx} = -40 + 50\left(\frac{1}{2}\right)(x^2 + 100^2)^{-1/2}(2x).$$

Equating dC/dx to zero yields

$$-40 + 50x(x^2 + 100^2)^{-1/2} = 0,$$

from which

$$50x(x^2 + 100^2)^{-1/2} = 40,$$

and

$$50x = 40(x^2 + 100^2)^{1/2}.$$

Squaring both sides,[10] we have

$$2500x^2 = 1600(x^2 + 100^2),$$
$$900x^2 = 1600(100^2).$$

Taking the square root of both sides, we find

$$30x = 40(100)$$

$$x = \frac{400}{3}.$$

Substitution of this value of x shows that it does satisfy the original equation which was to be solved, $50x = 40(x^2 + 100^2)^{1/2}$. The total cost when $x = 400/3$ yards may be computed to be $23,000.

To verify that $23,000 is the minimum, we return to

$$\frac{dC}{dx} = -40 + 50x(x^2 + 100^2)^{-1/2}$$

and proceed to obtain the second derivative. We note that the second term on the right is not in proper power form, but that we can obtain the proper form by placing the x inside the parentheses with the exponent -2. Thus,

$$\frac{dC}{dx} = -40 + 50[x^{-2}(x^2 + 100^2)]^{-1/2}$$

$$= -40 + 50(1 + 100^2x^{-2})^{-1/2}.$$

We may now apply the chain rule to find the second derivative, denoted by

$$\frac{d^2C}{dx^2}.$$

Thus,

$$\frac{d^2C}{dx^2} = 0 + 50\left(-\frac{1}{2}\right)(1 + 100^2x^{-2})^{-3/2}(100^2)(-2x^{-3})$$

$$= 50(100^2)(x^{-3})(1 + 100^2x^{-2})^{-3/2}.$$

---

[10]Squaring can introduce incorrect results, so we should be careful to check our final answer.

All factors in the last expression are positive when x is positive. Hence, the second derivative is positive at $x = 400/3$, and $C(400/3) = \$23{,}000$ is the minimum.

Higher Derivatives in Leibniz Notation. In the foregoing, the second derivative of C with respect to x, corresponding to $C''(x)$, was written as

$$\frac{d^2 C}{dx^2}.$$

Note that the 2 means *second.* The expression is read as "d second C, d x second." Higher derivatives, which we do not use in this book, are written in the same manner. Thus, the third derivative would be

$$\frac{d^3 C}{dx^3}.$$

Exercise. Translate the last expression. What would be the corresponding expression in prime symbols? Answer: "d third C, d x third," or $C'''(x)$.

We pause here to confirm the statement in Section 10.2 that

$$\frac{d}{dx} a^x = a^x \ln a.$$

To show this, we start with

$$f(x) = a^x$$

and take the natural logarithm of both sides, yielding

$$\ln f(x) = \ln a^x$$
$$\ln f(x) = x \ln a.$$

Next

$$\frac{d}{dx} \ln f(x) = \frac{d}{dx} x \ln a.$$

Applying the chain rule on the left, we have

$$\frac{1}{f(x)} \frac{df(x)}{dx} = \ln a$$

or

$$\frac{df(x)}{dx} = f(x) \ln a.$$

Replacing $f(x)$ by a^x, we have

$$\frac{da^x}{dx} = a^x \ln a,$$

which was to be shown.

In later applications, we shall make use of the general form of the chain rule,

$$\frac{dg(f)}{dx} = \frac{dg(f)}{df} \cdot \frac{df}{dx}$$

as well as its specific application. For example, we shall want to write expressions such as

$$\frac{de^y}{dx} = \frac{de^y}{dy} \cdot \frac{dy}{dx} - e^y \cdot \frac{dy}{dx}.$$

As another example, we apply the chain rule twice to obtain

$$\frac{de^{2y}}{dx} = \frac{de^{2y}}{dy} \cdot \frac{dy}{dx} = e^{2y}\frac{d(2y)}{dy} \cdot \frac{dy}{dx} = 2e^{2y}\frac{dy}{dx}.$$

||||||||||||||||||||||||||||||||||||||||||||||||||||||||||||||||||||||||||||||||||||||||||||||||||||||||||||||||||||||||

Exercise. Apply the chain rule twice to complete

$$\frac{de^{z^2}}{dx} =$$

Answer: $2ze^{z^2}\,(dz/dx)$.

||||||||||||||||||||||||||||||||||||||||||||||||||||||||||||||||||||||||||||||||||||||||||||||||||||||||||||||||||||||||

Summary of Rules — Composite Functions

(See also Table XII at the end of the book.)

1. General Chain Rule:

$$\frac{dg(f)}{dx} = \frac{dg(f)}{df} \cdot \frac{df(x)}{dx}.$$

Examples: $\dfrac{d}{dx}\ln y = \dfrac{d\ln y}{dy} \cdot \dfrac{dy}{dx} = \dfrac{1}{y}\dfrac{dy}{dx}.$

$$\frac{d}{dx}\ln (3y+2) = \frac{d}{dy}\ln (3y+2) \cdot \frac{dy}{dx} = \left(\frac{3}{3y+2}\right)\frac{dy}{dx}.$$

2. Power Rule:

$$\frac{d[f(x)]^n}{dx} = n[f(x)]^{n-1} \cdot \frac{df(x)}{dx}.$$

Example: $\dfrac{d}{dx}(x^2 + 3x)^{10} = 10(x^2 + 3x)^9(2x + 3).$

3. Exponential Rule:

$$\frac{de^{f(x)}}{dx} = e^{f(x)} \cdot \frac{df(x)}{dx} \quad \text{and} \quad \frac{d}{dx}a^{f(x)} = a^{f(x)} \ln a \cdot \frac{df(x)}{dx}.$$

Examples: $\dfrac{d}{dx}e^{x^2} = e^{x^2}(2x).$

$$\frac{d}{dx}(5)^{0.2x} = 5^{0.2x}(\ln 5)(0.2) = 0.32188(5^{0.2x}).$$

4. Logarithmic Rule:

$$\frac{d}{dx}\ln f(x) = \frac{1}{f(x)}\frac{df(x)}{dx}.$$

Example: $\dfrac{d}{dx}\ln(x^2 - 2x) = \dfrac{1}{x^2 - 2x}(2x - 2) = \dfrac{2x - 2}{x^2 - 2x}.$

10.12 PROBLEM SET 4

Find the derivative (d/dx) for each of the following.

1. $(3x - 2)^{10}$.
2. $3(2x^3 - 5x)^4$.
3. e^{3x+2}.
4. e^{-x}.
5. $3e^{2x^3}$.
6. $2e^{-0.05x}$.
7. 2^{3-x}.
8. $3^{(x^2)/2}$.
9. $4\ln(3x + 5)$.
10. $\ln(2x^2 + 3x)$.
11. $f(x) = (3x)^{1/2}$.
12. $f(x) = x + (2x - 3)^{-1}$.
13. $f(x) = (3x - 2)^{1/2}$.
14. $f(x) = (3 - x)^3 + (5 - 2x)^2$.
15. $f(x) = e^{(-x^2)/2} + 3e^x$.
16. $f(x) = \ln x$.

17. *a*) Complete the chain rule: $\dfrac{dH(w)}{dx} = \qquad .$

   b) In (*a*), what is assumed about *w*?

18. *a*) Complete the chain rule: $\dfrac{dP(g)}{dz} = \qquad .$

   b) In (*a*), what is assumed about *g*?

19. Given $(x^{2/3})^4 = x^{8/3}$

   a) Find the derivative with respect to *x* by applying the chain rule to the left-hand form of the expression.

   b) Find the derivative with respect to *x* of the right-hand form by the simple power rule and show the results in (*a*) and (*b*) are equal.

20. *a*) Expand $(3 - 2x)^2$ and find the derivative by the simple power rule.

   b) Find the derivative of $(3 - 2x)^2$ by the chain rule and show the results in (*a*) and (*b*) are equal.

21. The total potential audience for an advertising campaign is 10,000. Average revenue per response is $3, and the cost of the campaign is $300 per day, plus a fixed cost of $500. Find the number of days, *t*, the campaign should be continued to maximize profit if the response function is $1 - e^{-0.25t}$.

22. The circular roadway in Figure A has the equation $x^2 + y^2 = 10,000$, where *x* and *y* are in feet. An exit is planned at $P(60, 80)$. The exit path is to be tangent

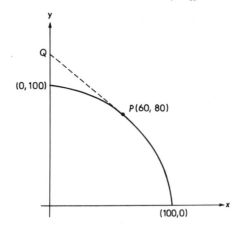

FIGURE A

to the circular path at P. Find where Q will be. (Find the tangent slope at P and reason from this number.)

23. See Figure B. A plant is to be constructed on a road (the x-axis) at a point which minimizes its distances from P and Q. That is, in

$$s = \overline{RP} + \overline{RQ},$$

the distance s is to be minimized.

a) What are the expressions for $\overline{RP}$ and $\overline{RQ}$?
b) Write the equation for s.
c) Find ds/dx.
d) Find value of x which minimizes s.
e) What is the value of the minimal s?

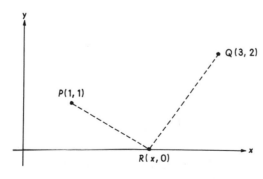

FIGURE B

24. Complete the following:

a) $\dfrac{d}{dx} e^v =$.

b) $\dfrac{d}{dx} e^{-3v} =$.

c) $\dfrac{d}{dI} e^{-v/10} =$.

d) $\dfrac{d}{dz} \ln x =$.

e) $\dfrac{d}{dh} \ln(2x + 3) =$.

f) $\dfrac{d}{dw} (2y + 3)^2 =$.

10.13 PRODUCT AND QUOTIENT RULES

The expression $(x^2)(x^3)$ equals x^5 and its derivative is $5x^4$. We can obtain the same result by

$$(x^2)\frac{d}{dx}(x^3) + (x^3)\frac{d}{dx}(x^2) = x^2(3x^2) + x^3(2x) = 5x^4.$$

Thus, if in the product $(x^2)(x^3)$ we think of x^2 as the first function and x^3 as the second function, the derivative of the product of the two functions is *the first function times the derivative of the second plus the second function times the derivative of the first*. This rule holds in general. Thus,
   Product Rule:

$$\frac{d}{dx}f(x)g(x) = f(x)g'(x) + g(x)f'(x).$$

For example,

$$\frac{d}{dx}x^2 e^x = x^2 \frac{d}{dx}e^x + e^x \frac{d}{dx}x^2$$

$$= x^2 e^x + e^x(2x) = xe^x(x+2).$$

||||||||||||||||||||||||||||||||||||||||||||||||||||||||||||||||||||||||||||||||||||||||||||||||||||||||||||||||||||||||||||||||||

Exercise.   Find $(d/dx)\, e^x \ln x$. Answer: $e^x[(1/x) + \ln x]$.

||||||||||||||||||||||||||||||||||||||||||||||||||||||||||||||||||||||||||||||||||||||||||||||||||||||||||||||||||||||||||||||||||

It may aid the reader to get a feeling for the source of the rule by applying the general definition of the derivative (Chapter 9) to the simple product

$$p(x) = (2x)(3x),$$

whose derivative is $12x$.
   According to the definition,

$$p'(x) = \lim_{\Delta x \to 0} \frac{p(x + \Delta x) - p(x)}{\Delta x}$$

$$= \lim_{\Delta x \to 0} \frac{[2(x + \Delta x)][3(x + \Delta x)] - (2x)(3x)}{\Delta x}$$

$$= \lim_{\Delta x \to 0} \frac{(2x + 2\Delta x)(3x + 3\Delta x) - (2x)(3x)}{\Delta x}$$

$$= \lim_{\Delta x \to 0} \frac{(2x)(3x) + 2x(3\Delta x) + 2\Delta x(3x) + 2\Delta x(3\Delta x) - (2x)(3x)}{\Delta x}$$

$$= \lim_{\Delta x \to 0} \frac{(2x)(3\Delta x) + 3x(2)\Delta x + 6(\Delta x)^2}{\Delta x}$$

$$= \lim_{\Delta x \to 0} [2x(3) + 3x(2) + 6\Delta x] = 2x(3) + 3x(2).$$

Recalling that $p(x) = (2x)(3x)$, we see that the limit just obtained is given by the product rule. Thus,

$$p'(x) = 2x\frac{d}{dx}(3x) + 3x\frac{d}{dx}(2x) = 2x(3) + 3x(2) = 12x.$$

For practice, we find

$$\frac{d}{dx}x^4e^x = x^4\frac{d}{dx}e^x + e^x\frac{d}{dx}(x^4) = x^4e^x + e^x(4x^3)$$

$$= x^3e^x(x + 4).$$

Again,

$$\frac{d}{dx}x(x^2 + 3)^{3/2} = x\frac{d}{dx}(x^2 + 3)^{3/2} + (x^2 + 3)^{3/2}\frac{d}{dx}(x)$$

$$= x\left(\frac{3}{2}\right)(x^2 + 3)^{1/2}(2x)* + (x^2 + 3)^{3/2}$$

$$= (x^2 + 3)^{1/2}(3x^2 + x^2 + 3) = (x^2 + 3)^{1/2}(4x^2 + 3),$$

where * calls attention to the use of the chain rule.

||||||||||||||||||||||||||||||||||||||||||||||||||||||||||||||||||||||||||||||||||||||||||||||||||||||||||||||||||||||||

Exercise. Find $(d/dx)(x^2 \ln x)$. Answer: $x(1 + 2 \ln x)$.

||||||||||||||||||||||||||||||||||||||||||||||||||||||||||||||||||||||||||||||||||||||||||||||||||||||||||||||||||||||||

The *quotient rule* states that the derivative of the quotient of two functions is the denominator function times the derivative of the numerator, minus the numerator function times the derivative of the denominator, all over the square of the denominator. Thus,

Quotient Rule:

$$\frac{d}{dx}\left[\frac{f(x)}{g(x)}\right] = \frac{g(x)f'(x) - f(x)g'(x)}{[g(x)]^2}.$$

For example,

$$\frac{d}{dx}\left[\frac{x^5}{x^2}\right],$$

which we know is the derivative of x^3, or $3x^2$, can be found by the quotient rule as:

$$\frac{x^2\frac{d}{dx}(x^5) - x^5\frac{d}{dx}(x^2)}{(x^2)^2} = \frac{x^2(5x^4) - x^5(2x)}{x^4} = \frac{3x^6}{x^4} = 3x^2.$$

Again,

$$\frac{d}{dx}\left[\frac{e^{x^2}}{x}\right] = \frac{x\frac{d}{dx}e^{x^2} - e^{x^2}\frac{d}{dx}(x)}{x^2} = \frac{xe^{x^2}(2x)* - e^{x^2}}{x^2} = \frac{e^{x^2}(2x^2 - 1)}{x^2},$$

where* calls attention to the use of the chain rule.

As another example,

$$\frac{d}{dx}\frac{\ln(3x+2)}{x} = \frac{x\frac{d}{dx}[\ln(3x+2)] - \ln(3x+2)\frac{d}{dx}(x)}{x^2}$$

$$= \frac{x\left(\frac{1}{3x+2}\right)(3) - \ln(3x+2)}{x^2}$$

$$= \frac{3x - (3x+2)\ln(3x+2)}{x^2(3x+2)}.$$

‖‖

Exercise. Find $(d/dx)(\ln x/e^x)$.

Answer: $\left(\frac{1}{x} - \ln x\right)/e^x = (1 - x \ln x)/xe^x$.

‖‖

The reader may wish to develop a feel for the source of the quotient rule by working problem 25 in the next problem set.

We shall pause here for practice with the new rules and the chain rule. Applications of these rules will follow shortly.

Summary

(See also Table XII at the end of the book.)

Product Rule:

$$\frac{d}{dx}[f(x)g(x)] = f(x)g'(x) + g(x)f'(x).$$

Quotient rule:

$$\frac{d}{dx}\left[\frac{f(x)}{g(x)}\right] = \frac{g(x)f'(x) - f(x)g'(x)}{[g(x)]^2}.$$

10.14 PROBLEM SET 5

Find the derivative with respect to x by the product rule.

1. $(2x + 1)(3x + 1)$.
2. $x^{1/2}(3x^2 + 2x)$.
3. $(3x + 1)^{4/3}(3 - 2x)^{3/2}$.
4. $x^2(3x^2 + 2x)$.
5. $x^{-2}(x + 1)^6$.
6. $x^{1/3}x^{2/3}$.

Find the derivative with respect to x by the quotient rule.

7. $\frac{x}{x}$.
8. $\frac{x}{1 - x}$.
9. $\frac{x - 2}{x + 1}$.
10. $\frac{x^2}{2x + 3}$.
11. $\frac{(2x + 5)^{1/2}}{x + 3}$.
12. $\frac{x}{(2x - 3)^{1/2}}$.

Find the derivative with respect to *x:*

13. $x^2 e^{2x}$.

14. $\dfrac{x^3}{e^x}$.

15. $x \ln x - x$.

16. $e^{-x^2} + 2 \ln(x - x^2)$.

17. 2^x.

18. $\dfrac{e^{2x}}{x}$.

19. $3e^{-x^{2/2}}$.

20. $\dfrac{e^x}{\ln x}$.

21. $5^x \ln 3x$.

22. $[\ln(x^2 - 5)]^3$.

23. Find the slope of $y = e^{-x^2/2}$ at the point where
   a) $x = 0$.   *b*) $x = 1$.
24. Find the slope of $e^{-x} \ln x$ at the point where
   a) $x - 1$.   *b*) $x = 2$.
25. Consider the quotient of the functions x^2 and $3x$

$$Q(x) = \frac{x^2}{3x}$$

which is $x/3$ and so has a derivative of $\frac{1}{3}$. Apply the limit definition for the derivative

$$Q'(x) = \lim_{\Delta x \to 0} \frac{Q(x + \Delta x) - Q(x)}{\Delta x},$$

simplify the result and show it can be written in the form prescribed by the quotient rule.

26. The probability that an event occurs on any trial is p and the probability that it does not occur is $1 - p$. The probability P that the event occurs *for the first time* on the fifth trial depends on p and is

$$P(p) = p(1 - p)^4.$$

What probability, p, maximizes the probability that the event will occur for the first time on the fifth trial? (Note: One would guess intuitively that this would be $p = \frac{1}{5}$. Prove by derivatives that this is the case.)

10.15 MARGINAL COST AND MARGINAL REVENUE

If we write the total daily cost, $C(m)$, of renting a car which costs $15 per day plus 20 cents per mile, m, driven as

$$C(m) = 15 + 0.2m,$$

then

$$C'(m) = 0.2.$$

We may state that the total cost $C(m)$ increases at the *rate* of $0.20 for each *additional* mile driven. In economics this extra cost for an additional unit, $C'(m)$, is called *marginal cost*. In the case of $C(m)$, marginal cost is constant because $C(m)$ is a linear function. If the total cost function for q units, $C(q)$, is not linear, as in

$$C(q) = 50 + 2q + 0.005q^2,$$

then marginal cost, $C'(q)$, is not constant. Thus,

$$\text{Marginal cost} = C'(q) = 2 + 0.01q$$

so that at output $q = 100$, we have

$$C'(100) = \$3.$$

We interpret this by saying that at 100 units output, total cost is increasing at the *rate* of \$3 per additional unit produced.

Exercise.   Find marginal cost at $q = 50$ units output and write the interpretation of the result. Answer: $C'(50) = \$2.50$. At $q = 50$ units output, total cost is increasing at the rate of \$2.50 per additional unit produced.

We have seen that marginal cost, \$3, at $q = 100$ is higher than the value \$2.50 at $q = 50$. For this cost function, marginal cost increases as output increases. This follows from the fact that $C(q)$ is concave upward as shown in Figure 10–12 because its second derivative $C''(q) = 0.01$ is always positive. As the figure shows, the extra cost (the vertical rise in each triangle), per change of 1 in q, increases as we move to the right.

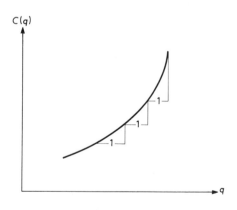

FIGURE 10–12

If we now turn to the selling aspect of the situation and find that selling price is constant at \$10 per unit, then total revenue, $R(q)$, from the sale of q units is

$$R(q) = \$10q,$$

so marginal revenue, defined as $R'(q)$, is constant at

$$R'(q) = 10.$$

Now recall that at $q = 100$, marginal cost is $C'(100) = \$3$, so total cost is increasing at the rate of $3 per additional unit, whereas revenue increases at $10 per additional unit. Marginal profit, the difference between marginal revenue and marginal cost, is $\$10 - \$3 = \$7$, meaning that total profit is increasing at the rate of $7 per additional unit made and sold. Clearly, total profit will increase if we make and sell more than 100 units. Suppose we consider $q = 1000$ units. Now marginal cost is

$$C'(1000) = 2 + (0.01)(1000) = \$12$$

so total cost is increasing at the rate of $12 per additional unit and revenue continues to increase at a rate of $10, so profit is *decreasing* at the rate of $2 per additional unit made and sold. If maximum profit is our goal, we should not make 1000 units, nor any number for which marginal cost exceeds marginal revenue. It follows that *maximum profit is achieved at the output for which marginal cost equals marginal revenue*. At outputs below this level, marginal cost is below marginal revenue (so profit can be increased by additional output).

In our example, marginal cost equals marginal revenue when

$$C'(q) = R'(q)$$
$$2 + 0.01q = 10 \quad \text{or} \quad q^* = 800 \text{ units.}$$

Writing profit $P(q)$ as revenue minus cost,

$$P(q) = R(q) - C(q)$$

we find the maximum profit to be

$$P(q^*) = R(q^*) - C(q^*)$$
$$= 10(800) - [50 + 2(800) + (0.005)(800)^2]$$
$$= 8000 - 4850 = \$3150.$$

Exercise. If unit price is constant at $30 per unit, whereas total cost is $C(q) = 20q + 0.001q^2$, find the output which maximizes profit and find the maximum profit. Answer: $q = 5000$ units, maximum profit $25,000.

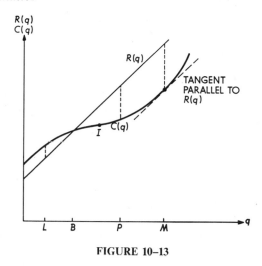

FIGURE 10–13

Figure 10–13 shows a linear revenue function, $R(q)$, and a cost function $C(q)$ which starts out concave downward, inflects at I, and becomes concave upward. At L, cost exceeds revenue and the vertical separation of the functions represents *loss*. At B, where the functions intersect, we have *break-even*, with cost equal to revenue. The vertical separation at P represents *profit*, and the maximum separation (profit) occurs at M, where the tangent to $C(q)$ is parallel to $R(q)$; that is, where $C'(q) = R'(q)$. Observe that to the left of the inflection at I, the downward concavity means marginal cost (the tangent slope) is decreasing to its minimum which occurs at the inflection point, I. To the right of I, the upward concavity means marginal cost is increasing but remains less than marginal revenue, the slope of $R(q)$, up to the point M where $C'(q)$ equals $R'(q)$. It is possible, of course, for a tangent to $C(q)$ to be parallel to $R(q)$, that is for $C'(q)$ to equal $R'(q)$, at a point to the *left* of I, but this would not provide maximum profit. The maximum occurs when $C'(q) = R'(q)$ *and* $C'(q)$ is increasing.

What we have found so far can be summarized simply by writing the profit function, $P(q)$, as

$$P(q) = R(q) - C(q).$$

Taking derivatives with respect to q, we have

$$P'(q) = R'(q) - C'(q).$$

If $P(q)$ is to be a maximum, $P'(q)$ must be zero. Hence,

$$0 = R'(q) - C'(q)$$

or

$$C'(q) = R'(q)$$

which means that marginal cost must equal marginal revenue if profit is to be maximized.

In economic analysis, demand q and price per unit, $p(q)$, are related by a demand function, $p(q)$, such as

$$p(q) = 152 - 0.5q.$$

We note that at 20 units demand, $p(20) = \$142$ per unit and at 80 units demand, $p(80) = \$112$ per unit, which is less than $p(20)$. This behavior is typical; that is, low unit price goes with high demand, and conversely, so that the demand curve $p(q)$ slopes downward to the right.

If q units are sold at $p(q)$ dollars per unit, total revenue is

$$R(q) = qp(q)$$

which in our case would be

$$R(q) = q(152 - 0.5q) = 152q - 0.5q^2.$$

Suppose now that the cost function is

$$C(q) = 25 + 2q + 0.25q^2.$$

We have that

$$R'(q) = 152 - q \quad \text{and} \quad C'(q) = 2 + 0.5q.$$

Exercise.   Using the last expressions, find the q which maximizes profit, and find the maximum. Answer: $q = 100$ units, maximum profit $= \$7475$.

10.16 AVERAGE VERSUS MARGINAL COST

Consider the linear cost function

$$C(q) = 100 + 5q.$$

The total cost of 10 units is $C(10) = \$150$ for an average cost per unit of

$$AC(10) = \frac{150}{10} = \$15.$$

Exercise.   Find average cost per unit at $q = 100$ units. Answer: $\$6$ per unit.

We observe first that average cost per unit declines in this case as output increases and, second, that average cost per unit is not the same as marginal cost. Marginal cost is

$$C'(q) = 5,$$

meaning that total cost increases at the rate of $\$5$ per *additional* unit produced, whereas average cost spreads total cost over *all* units produced. We have seen that marginal cost is the relevant concept in profit maximization. Average cost, as we shall now explain, is relevant when considering a firm's break-even level of operation. That is, breakeven occurs when average cost per unit equals price received per unit. If the firm's average unit cost is above the selling price per unit, the firm loses money, and conversely. Average unit cost, $AC(q)$, is total cost of q units divided by q. Thus,

$$AC(q) = \frac{C(q)}{q}.$$

For example, if

$$C(q) = 25 + 2q + 0.25q^2$$

then

$$AC(q) = \frac{25}{q} + 2 + 0.25q.$$

Average unit costs at $q = 4$, 5, and 50 are

$$AC(4) = \frac{25}{4} + 2 + 0.25(4) = \$9.25 \text{ per unit.}$$

$$AC(5) = \frac{25}{5} + 2 + 0.25(5) = \$8.25 \text{ per unit.}$$

$$AC(50) = \frac{25}{50} + 2 + 0.25(50) = \$15.00 \text{ per unit.}$$

Thus, $AC(q)$ first declines, then rises. Its minimum occurs when d/dq $[AC(q)] = 0$. We have

$$\frac{d}{dq}AC(q) = -\frac{25}{q^2} + 0.25.$$

Seeking the minimum,

$$\frac{-25}{q^2} + 0.25 = 0 \quad \text{or} \quad q^2 = 100$$

and $q = 10$ yields a minimum average unit cost. This minimum is

$$AC(10) = \frac{25}{10} + 2 + 0.25(10) = \$7 \text{ per unit.}$$

It follows that if a firm has the given cost function, it can determine what selling price per unit and output will just allow it to break even. In the case at hand, an output of 10 units selling at $7 per unit (which equals the firm's minimum average unit cost) would allow the firm just to break even.

The break-even output level is the q^* for which average unit cost is a minimum and equals the selling price per unit. It is useful for economic analysis to point out that at this q^*, marginal cost equals average cost per unit. To see why, recall that $AC(4)$ was \$9.25. We may verify that marginal cost, $C'(4)$, is \$4. Hence, if we add another unit, adding \$4 to cost, the average cost of the 5 units will be less than \$9.25.[11] That is, when marginal cost is below average cost, average cost is declining, and when marginal cost is above average cost, average cost is increasing. For average cost to be stationary (minimum), marginal cost must equal average cost. To illustrate the point, recall our example

$$C(q) = 25 + 2q + 0.25q^2$$

so that

$$C'(q) = 2 + 0.5q \quad \text{and} \quad AC(q) = \frac{25}{q} + 2 + 0.25q.$$

Equating the two, we have

$$2 + 0.5q = \frac{25}{q} + 2 + 0.25q$$

or

$$0.25q = \frac{25}{q}, \ 0.25q^2 = 25, \ q^2 = 100$$

so $q = 10$, which we found earlier was the value for which $AC(q)$ was a minimum, \$7 per unit.

Figure 10–14 shows the average cost function, and the *marginal* cost function $C'(q) = 2 + 0.5q$. Note that marginal cost intersects average cost at its lowest point.

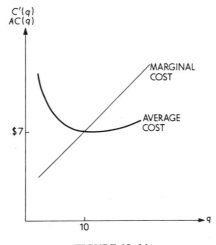

FIGURE 10–14

---

[11]What happens to a student's average of 80 if he receives a 60 on his next test?

To demonstrate in general that average cost must equal marginal cost if average cost is to be minimized, we express average cost as

$$AC(q) = \frac{C(q)}{q}.$$

Applying the quotient rule to find the derivative,

$$\frac{d}{dq}AC(q) = \frac{qC'(q) - C(q)(1)}{q^2}.$$

Setting the derivative equal to zero to obtain the condition for a minimum, we have

$$\frac{qC'(q) - C(q)}{q^2} = 0 \quad \text{or} \quad qC'(q) - C(q) = 0,$$

and

$$C'(q) = \frac{C(q)}{q},$$

which states that the minimum occurs when marginal cost equals average cost.

In summary, we can determine from $C(q)$ the output and selling price which will just allow the firm to break even either by finding the q^* which minimizes average unit cost, or the (same) q^* for which average cost equals marginal cost. Thus, if

$$C(q) = 512 + 5q + 0.08q^2,$$

we may apply the first method and write

$$\frac{d}{dq}AC(q) = \frac{d}{dq}\left(\frac{512}{q} + 5 + 0.08q\right) = -\frac{512}{q^2} + 0.08$$

which, when equated to zero, yields $q = 80$ and $AC(80) = \$17.80$, so at a selling price of $17.80 the firm can just break even at an output of 80 units.

Exercise. Verify the last result by the second method noted in the foregoing.

10.17 MARGINAL PROPENSITY TO CONSUME

If we think of the aggregate income, Y, received by families and available for their use as either being *spent on consumption* (food, clothing, and so on) *or saved,* and let $C(Y)$ be the consumption expenditures at income level Y billion dollars, then $C(Y)$ is the *propensity to consume function* or, more briefly, the consumption function. Suppose

$$C(Y) = 5 + 0.9Y.$$

The derivative

$$C'(Y) = 0.9$$

is called the *marginal propensity to consume* and means that $0.90 of an *extra* dollar of income would be spent on consumption, and $0.10 would be saved. Note carefully that this does *not* mean families spend 90% of their income on consumption. The *proportion* spent would be $p(Y)$ where

$$p(Y) = \frac{C(Y)}{Y} = \frac{5}{Y} + 0.9.$$

Exercise. Assuming $C(Y)$ holds for income levels of $Y = \$100$ and $500 billion, what proportion of income would be spent at each income level? Answer: 95% at $100 billion and 91% at $500 billion.

The aggregate propensity to consume is a combination of spending habits of many individual families. Looking at these individual families, we find that those with low incomes have high marginal propensities to consume, but high income families save substantial amounts and have lower marginal propensities to consume. As an example of such behavior, consider

$$C(Y) = 90 - 90e^{-Y/90}$$

where Y is in thousands of dollars. We find the marginal propensity to consume (henceforth MPC) to be

$$\text{MPC} = C'(Y) = -90e^{-Y/90}\left(-\frac{1}{90}\right)^* = e^{-Y/90},$$

where* calls attention to the application of the chain rule. At income levels $Y = \$20$ and $50 thousand, we find

$$C'(20) = e^{-20/90} = e^{-0.22} = 0.80,$$
$$C'(50) = e^{-50/90} = e^{-0.56} = 0.57.$$

Exercise. Interpret the MPC's 0.80 and 0.57. Answer: at $20 thousand income $0.80 of an extra dollar of income would be spent on consumption, whereas at $50 thousand income, only $0.57 of an extra dollar would be so spent.

10.18 THE MULTIPLIER

If at income Y, $C(Y)$ is spent on consumption, the amount saved is $Y - C(Y)$. We shall assume that the amount saved is invested and designate it as I, for investment. Hence,

$$I = Y - C(Y).$$

For example, if $C(Y) = 5 + 0.9Y$, then

$$I = Y - (5 + 0.9Y) = 0.1Y - 5.$$

The point of interest is to determine the rate at which income, Y, changes per \$1 change in investment, I. This is the derivative, dY/dI. Taking the derivative *with respect to I* of

$$I = 0.1Y - 5$$

we have

$$\frac{dI}{dI} = 0.1\frac{dY}{dI} \quad \text{or} \quad 1 = 0.1\frac{dY}{dI}$$

from which

$$\frac{dY}{dI} = \frac{1}{0.1} = 10.$$

The derivative, dY/dI, is called the *multiplier.* The value 10 just obtained means that income increases at a rate of \$10 per additional dollar of investment.

To see the relationship between the multiplier, dY/dI, and the MPC, $dC(Y)/dY$, we start with the definition

$$I = Y - C(Y)$$

and take the derivative with respect to I. Thus,

$$\frac{dI}{dI} = \frac{dY}{dI} - \frac{d}{dI}[C(Y)].$$

Applying the chain rule to the rightmost term, we find

$$1 = \frac{dY}{dI} - \frac{dC(Y)}{dY} \cdot \frac{dY}{dI}.$$

Factoring, we have

$$1 = \frac{dY}{dI}\left[1 - \frac{dC(Y)}{dY}\right]$$

or

$$\frac{dY}{dI} = \frac{1}{1 - \dfrac{dC(Y)}{dY}} = \frac{1}{1 - \text{MPC}}.$$

Thus, for the consumption function

$$C(Y) = 5 + 0.9Y$$

for which MPC, $dC(Y)/dY$, is 0.9,

$$\text{Multiplier} = \frac{dY}{dI} = \frac{1}{1 - 0.9} = \frac{1}{0.1} = 10$$

as we found earlier.

‖‖

Exercise. Find the multiplier if the propensity to consume function is $C(Y) = 6 + 0.75\,Y$ and interpet the result. Answer: The multiplier, 4, means that income increases at a rate $4 for each additional dollar of investment.

‖‖

MPC and its related multiplier play an important role in the economic analysis of income generation. We close this section by finding the multiplier first by taking dY/dI directly and second by using $1/(1 - \text{MPC})$. Starting with the function which defines propensity to consume,

$$C(Y) = 90 - 90e^{-Y/90}$$

we find $I = Y - C(Y)$ to be

$$I = Y - (90 - 90e^{-Y/90}).$$

The multiplier dY/dI is found by first taking the derivative with respect to I. Thus,

$$\frac{dI}{dI} = \frac{dY}{dI} - 0 + 90 \cdot \frac{d}{dI}e^{-Y/90}.$$

We must apply the chain rule twice to the rightmost term, as follows:

$$1 = \frac{dY}{dI} + 90\left[\frac{de^{-Y/90}}{dY} \cdot \frac{dY}{dI}\right]$$

$$1 = \frac{dY}{dI} + 90\left[(e^{-Y/90})\left(-\frac{1}{90}\right) \cdot \frac{dY}{dI}\right]$$

$$1 = \frac{dY}{dI} - e^{-Y/90}\frac{dY}{dI}$$

$$1 = \frac{dY}{dI}[1 - e^{-Y/90}].$$

Hence,

$$\text{Multiplier} = \frac{dY}{dI} = \frac{1}{1 - e^{-Y/90}}.$$

Equivalently, we find $\text{MPC} = dC(Y)/dY$ from

$$C(Y) = 90 - 90e^{-Y/90}$$

as

$$\text{MPC} = \frac{dC(Y)}{dY} = -90e^{-Y/90}\left(-\frac{1}{90}\right) = e^{-Y/90},$$

then write

$$\text{Multiplier} = \frac{1}{1 - \text{MPC}} = \frac{1}{1 - e^{-Y/90}},$$

as before.

10.19 PROBLEM SET 6

1. A firm sells all it produces at $8 per unit. Its total cost of making q units if $C(q)$ = $$(20 + 2q + 0.01q^2)$.
   a) What number of units yields maximum profit?
   b) What is the maximum profit?

2. A firm sells all it produces at $4 per unit. Its total cost of making q units is $C(q)$ = $$(50 + 1.3q + 0.001q^2)$.
   a) What number of units yields maximum profit?
   b) What is the maximum profit?

3. At demand q units, a firm receives a price per unit of $p(q) = \$(300 - 0.5q)$. The total cost of making q units is $C(q) = \$(2000 + 100q)$.
   a) What number of units yields maximum profit?
   b) What is this maximum profit?

4. At demand q units, a firm receives a price per unit of $p(q) = \$(40 - 0.05q)$. The total cost of making q units is $C(q) = (0.01/3)q^3 - 0.45q^2 + 47q + 10$.
   a) At what two values of q does marginal cost equal marginal revenue? (Use the quadratic formula.)
   b) Is marginal cost increasing or decreasing at $q = 10$? At $q = 70$? (Use the second derivative.)
   c) What value of q leads to maximum profit? Why?
   d) What value of q makes marginal cost a minimum, and what is the minimum marginal cost?

5. At demand q units, a firm receives a price per unit of $p(q) = 10 + 500/(q + 5)$. The cost of making q units is $C(q) = 5 + 11q$. Find the output which yields maximum profit. (This involves application of the quotient rule.)

6. Prove by derivatives that if profit, $P(q) = R(q) - C(q)$ is to be a maximum, marginal cost must equal marginal revenue.

7. If the total cost of producing q units is $C(q) = \$(400 + 5q + 0.01q^2)$, find the output level q^* and the unit selling price which would allow the firm just to break even:
   a) By finding the q^* which minimizes average unit cost.
   b) By finding the q^* for which average cost equals marginal cost.

8. The total cost of producing q units is $C(q) = \$(200 + 10q + 0.02q^2)$.
   a) Find the average cost per unit at outputs of $q = 20, 50, 80, 100, 200$, and 400.
   b) Find the q^* and the selling price which just allow the firm to break even by the minimal average cost procedure.
   c) Find marginal cost at $q = 50, 100$, and 200.
   d) Using results of a), b), and c), sketch a graph of average cost *and* marginal cost. Place appropriate labels on the graph.
   e) What does d) illustrate?
   f) If all units made could be sold at $20 per unit, what number of units would yield maximum profit?

9. The total cost of producing q units is $C(q) = 140 + 2q \ln q$. By equating average cost and marginal cost, find the q^* and selling price which allow the firm just to break even. (This involves the use of the product rule.)

10. Given any cost function $C(q)$, prove that if average cost is to be a minimum, then marginal cost must equal average cost.

11. If $C(Y)$ represents consumption expenditures when income is Y billion dollars and $C(Y) = 4 + 0.8Y$,
    a) What is the marginal propensity to consume?
    b) Interpret the answer to a).
    c) If income is \$200 billion, what *proportion* of income will be used for consumption expenditures?
    d) What is the multiplier?
    e) Interpret the answer to d).

12. If consumption expenditures at income Y thousand dollars are $C(Y) = 80 - 80e^{-Y/80}$,
    a) What is the marginal propensity to consume at income levels of \$20 and \$60 thousand?
    b) What proportion of income will be used for consumption at income of \$20 thousand?
    c) What is the multiplier for this function at income levels of \$20 and \$60 thousand?

13. Total income, Y, is distributed in part to investment, I, and the remainder to consumption, $C(Y)$, so that $Y = I + C(Y)$. If $C(Y) = 0.4Y + 1.28Y^{3/4}$, find the expression for the multiplier.

10.20 DIFFERENTIALS: RATE OF CHANGE: SENSITIVITY ANALYSIS

For any *linear* function, $f(x)$, the slope $df(x)/dx$ is the *rate* of change (vertically) in the function per unit change in x. It follows that one half of $[df(x)/dx]$ is the change in the linear function if x increases by $1/2$, and $[df(x)/dx](-1/4)$ is the change if x decreases by $1/4$, and so on. We shall assign the symbols $df(x)$ and dx, called respectively the *differential* of $f(x)$ and the *differential* of x, to represent, again respectively, the change in the linear function $f(x)$ for a change of dx in x.

||||||||||||||||||||||||||||||||||||||||||||||||||||||||||||||||||||||||||||||||||||||||||||||||||||||||||||||||||

Exercise. If $f(x) = 3 + 2x$, what is $df(x)$ if dx is 0.3? Answer: The rate of change in $f(x)$ per unit change in x is $df(x)/dx$ which is 2. If x changes by $dx = 0.3$, $f(x)$ changes by $df(x) = 2(0.3) = 0.6$.

||||||||||||||||||||||||||||||||||||||||||||||||||||||||||||||||||||||||||||||||||||||||||||||||||||||||||||||||||

The relationship, symbolically, for a straight line is

$$\text{Change in function} = (\text{slope})(\text{change in } x)$$

or

$$df(x) = \left[\frac{df(x)}{dx}\right]dx.$$

If the function $f(x)$ is not linear, its derivative $df(x)/dx$ varies from point to point and at any point represents the slope of a tangent line at that point, so $df(x)$ represents change on the tangent line, not on the curve. See Figure

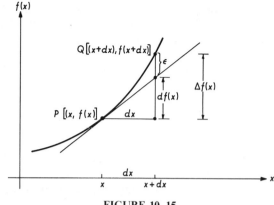

FIGURE 10–15

10–15 which shows x changing by dx and the corresponding change $df(x)$ on the tangent. The *actual change in the function* is represented as $\Delta f(x)$. Our interest in this section is to use $df(x)$ to approximate $\Delta f(x)$. Clearly, there will be an error, indicated by ϵ in Figure 10–15, and this error will be smaller the smaller dx and the less rapid the change in the curve.

To see how we can apply the foregoing, consider the positive square root function

$$f(x) = x^{1/2}.$$

The rate of change in the square root per unit change in x is

$$\frac{df(x)}{dx} = \frac{1}{2}x^{-1/2} = \frac{1}{2x^{1/2}}.$$

At $x = 16$, we have $f(x) = 4$ and the square root is changing at the rate of

$$\frac{1}{2(16)^{1/2}} = \frac{1}{8} = 0.125$$

per unit change in x. Now suppose we move from 16 to 16.1 so that $dx = 0.1$. The change, $df(x)$, on the tangent at $x = 16$ is

$$df(x) = \left[\frac{df(x)}{dx}\right] dx = (0.125)(0.1) = 0.0125.$$

Thus, we estimate the actual change, $\Delta f(x)$, as

$$\Delta f(x) \doteq df(x)$$

where $\doteq$ means *approximately equals*. We conclude that if x changes from 16 to 16.1 its square root increases from 4 to 4.0125, so

$$\sqrt{16.1} \doteq 4.01250.$$

For comparison, we state that the correct root is 4.01248 to five decimal places. This close estimate occurs because $f(x)$ is not changing rapidly over the interval 16 to 16.1.

We now repeat the procedure, approximating $\sqrt{0.042}$ from $\sqrt{0.040} = 0.2$, using $dx = 0.002$. We shall outline the method in the manner to be followed in the Problem Set.

Question: Given $f(x) = x^{1/2}$.

a) What is the rate of change of the function per unit change in x at $x = 0.04$? Answer: $df(x)/dx$ is $1/2x^{1/2}$ and is $(1/2\sqrt{0.04}) = [1/2(0.2)] = 2.5$ at $x = 0.04$.

b) Estimate the change in the function if x increases from 0.040 to 0.042. Answer: $dx = 0.002$, so $\Delta f(x) \doteq df(x) = [df(x)/dx]dx = 2.5(0.002) = 0.005$.

c) What is the approximate value of $\sqrt{0.042}$? Answer: $0.2 + 0.005 = 0.205$.

Exercise. Given the cube root function $f(x) = x^{1/3}$. a) What is the rate of change in the function per unit change in x at $x = 8$? b) Estimate the change in the cube root as x increases from 8 to 8.1. c) Estimate the cube root of 8.1. Answer: a) $df(x)/dx$ is $1/3x^{2/3}$ and equals $1/12$ or 0.0833 at $x = 8$. b) Here, $dx = 0.1$, so $\Delta f(x) \doteq (0.0833)(0.1) = 0.00833$. c) 2.00833. (Note: 2.00830 is the value correct to five decimals.)

Estimates by differentials can be used as shown to extend tabulated functions. As another example, Table X at the end of the book shows that $\ln 5 = 1.6094$. The function

$$f(x) = \ln x$$

has

$$\frac{df(x)}{dx} = \frac{1}{x}$$

which is $1/5 = 0.2$ at $x = 5$. Suppose we wish to estimate $\ln (5.03)$ from $\ln (5)$. Using $dx = 0.03$, we estimate

$$\Delta f(x) \doteq df(x) = \frac{df(x)}{dx}dx = (0.2)(0.03) = 0.006.$$

Hence,

$$\ln 5.03 \doteq \ln 5 + 0.006 = 1.6094 + 0.006 = 1.6154.$$

Sensitivity Analysis. The rate of change of a function per unit change in a variable shows how *sensitive* the function is to changes in the variable. In our inventory model earlier in the chapter (see Section 10.8), we determined that the lot size L which minimized cost was given by

$$L = \left[\frac{2NF}{i}\right]^{1/2},$$

where N was the number of units required in a year, F was fixed setup cost per lot, and i was the annual unit cost per item in average inventory. To find how sensitive the optimal solution is to the inventory cost parameter i, we write

$$L(i) = \left[\frac{2NF}{i}\right]^{1/2}$$

and find

$$\frac{dL(i)}{di} = \frac{1}{2}\left[\frac{2NF}{i}\right]^{-1/2}\left[\frac{-2NF}{i^2}\right]^{*}$$

where * calls attention to the use of the chain rule. The factor

$$\left[\frac{2NF}{i}\right]^{-1/2}$$

on the right is $1/L(i)$, or $1/L$, and

$$\frac{-2NF}{i^2} = -\frac{L^2}{i},$$

so we may write

$$\frac{dL(i)}{di} = \frac{1}{2L}\left[-\frac{L^2}{i}\right] = -\frac{L}{2i}.$$

Suppose $N = 500$, $F = 100$, $i = 10$, so that the optimum solution is $L = [2(500)(100)/10]^{1/2} = 100$ units. Then

$$\frac{dL(i)}{di} = -\frac{100}{2(10)} = -5,$$

which means that for the parameter values at hand, the optimum lot size decreases by 5 units per $1 increase in unit inventory cost. Thus, if the value $10 used for i was in error by $1 and should have been $11, the optimal lot size would be down about 5 units from 100 to 95.

||||||||||||||||||||||||||||||||||||||||||||||||||||||||||||||||||||||||||||||||||||||||||||||||||||||||||||||||||||||||||||||||||

Exercise. Treating the optimum lot size as a function of fixed cost, F, a) Find $dL(F)/dF$. b) For the values assumed in the foregoing example, what is the rate of change in optimal lot size per $1 change in F? c) If F were $90 instead of $100, by how much, approximately, would the optimal lot size change? Answer: a) $dL(F)/dF = N/Li$. b) 1/2 unit. c) Down 5 units.

||||||||||||||||||||||||||||||||||||||||||||||||||||||||||||||||||||||||||||||||||||||||||||||||||||||||||||||||||||||||||||||||||

10.21 PRICE ELASTICITY OF DEMAND

At q units demand, selling price per unit is

$$p(q) = 120 - 2q.$$

We find

$$p(10) = \$100 \text{ per unit}$$

and

$$p(11) = \$98 \text{ per unit.}$$

Demand, q, changed by 1 from 10 to 11 for a proportionate increase of

$$\frac{\text{change in } q}{q} = \frac{1}{10} = 0.1 \text{ or a } 10\% \text{ increase,}$$

whereas price per unit changed by $-\$2$ from $\$100$ to $\$98$ for a proportionate decrease of

$$\frac{\text{change in } p(q)}{p(q)} = \frac{-2}{100} = -0.02 \text{ or a } 2\% \text{ decrease.}$$

The ratio,

$$\frac{\text{percent change in } q}{\text{percent change in } p} = \frac{10\%}{-2\%} = \frac{-5\%}{1\%} = -5$$

may be interpreted by saying that a 1 percent *increase* in price is accompanied by a 5 percent *decrease* in demand. It is *conventional* in economics to change the sign of the last ratio with the understanding that demand changes are customarily in the direction opposite to price change.

The ratio of the percent changes, with sign changed, is called $e(q)$, the price elasticity of demand. An $e(q) = 2$ is interpreted to mean that a 1 percent price *increase* is accompanied by a 2 percent *decrease* in demand. We define $e(q)$ at a point as

$$e(q) = -\frac{\dfrac{\text{change in } q}{q}}{\dfrac{\text{change in } p}{p}} = -\frac{\dfrac{dq}{q}}{\dfrac{dp}{p}} = -\frac{p}{q} \cdot \frac{dq}{dp}.$$

Generally, $e(q)$ is not constant from point to point. Thus, for our earlier demand function

$$p(q) = 120 - 2q \quad \text{or} \quad p = 120 - 2q$$

we take the derivative *with respect to p* and find

$$\frac{dp}{dp} = -2q\frac{dq}{dp} \quad \text{or} \quad 1 = -2\frac{dq}{dp}.$$

Solving the last for dq/dp, we find

$$\frac{dq}{dp} = -\frac{1}{2}.$$

We then have

$$e(q) = -\frac{p}{q}\frac{dq}{dp} = -\frac{p}{q}\left[-\frac{1}{2}\right] = \frac{p}{2q}$$

or, replacing $p = p(q)$ by $120 - 2q$,

$$e(q) = \frac{120 - 2q}{2q} = \frac{60 - q}{q}.$$

We find

$$e(10) = \frac{50}{10} = 5$$

$$e(11) = 4.45$$

so that the elasticity 5 (or -5 as we first found it) at the beginning of the discussion holds at demand $q = 10$, and only approximately at $q = 11$.

As another example, let us find $e(q)$ for the demand function

$$p(q) = p = 400 - 0.2q^{1/2}.$$

Taking the derivative *with respect to p* and applying the chain rule (*), we find

$$\frac{dp}{dp} = -0.2\left(\frac{1}{2}\right)q^{-1/2} \cdot \frac{dq^*}{dp}$$

$$1 = \frac{-0.1}{q^{1/2}}\frac{dq}{dp}$$

or

$$-10q^{1/2} = \frac{dq}{dp}.$$

We have

$$e(q) = -\frac{p}{q} \cdot \frac{dq}{dp} = -\frac{p}{q}\left[-10q^{1/2}\right] = \frac{10p}{q^{1/2}}.$$

To find elasticity at $q = 1,000,000$ units, we first find

$$p = p(q) = 400 - 0.2(1,000,000)^{1/2} = 400 - 200 = 200.$$

Then,

$$e(q) = \frac{10(200)}{(1,000,000)^{1/2}} = \frac{2000}{1000} = 2,$$

so that at this demand level, an increase of 1 percent in price is accompanied by a decrease of *approximately* 2 percent in demand. The italicized word calls attention to the fact that $e(q)$ applies exactly at a point and only approximately if a change is made from that point.

‖‖

Exercise.   Given the demand function $p(q) = 500 - 0.001q^2$. *a)* Find the expression for $e(q)$. *b)* Compute elasticity at 400 units demand. Answer: *a)* $500p/q^2$. *b)* 17/16.

‖‖

A special property of $e(q)$ is that its value must be 1 if revenue is to be maximized. To show this, we note that revenue is

$$R(q) = qp(q) = qp.$$

For $R(q)$ to be a maximum, $R'(q)$ must be zero. Applying the product rule,

$$R'(q) - \frac{dR(q)}{dq} - q\frac{dp}{dq} + p \cdot \frac{dq}{dq} - q\frac{dp}{dq} + p.$$

$R'(q) = 0$ when

$$q\frac{dp}{dq} + p = 0 \quad \text{or} \quad q = -\frac{p}{\dfrac{dp}{dq}} = -p\left[\frac{1}{\dfrac{dp}{dq}}\right].$$

Now we may write[12]

$$\frac{1}{\dfrac{dp}{dq}} = \frac{dq}{dp}$$

and find that for $R(q)$ to be a maximum

$$q = -p\left[\frac{dq}{dp}\right].$$

Substituting this q into

$$e(q) = -\frac{p}{q}\left[\frac{dq}{dp}\right]$$

we have

$$e(q) = \frac{-p\left[\dfrac{dq}{dp}\right]}{-p\left[\dfrac{dq}{dp}\right]} = 1$$

which shows that $e(q)$ must be 1 if revenue is to be maximized.

---

[12]This assumes that if p is a function of q, then q is also a function of p, and this will be true over intervals where $p(q)$ never increases (as is assumed for demand functions), or never decreases (as is assumed for supply functions).

||||||||||||||||||||||||||||||||||||||||||||||||||||||||||||||||||||||||||||||||||||||||||||||||||||||||||||||||||||||||||||||

Exercise. If $p(q) = 100 - 2q$ so that $R(q) = q(100 - 2q)$,
a) Find the q^* which maximizes $R(q)$ and b) Find $e(q^*)$. Answer:
a) Maximum at $q^* = 25$. b) $e(25) = 1$.

||||||||||||||||||||||||||||||||||||||||||||||||||||||||||||||||||||||||||||||||||||||||||||||||||||||||||||||||||||||||||||||

10.22 PROBLEM SET 7

1. As shown in Table II at the end of the book, $1 compounded at the rate i per
   period will grow to $C(i)$ in n periods, where

$$C(i) = (1 + i)^n.$$

   a) Find the expression for the rate of change in the amount per change of 1 in
   the interest rate.
   b) Evaluate the rate in a) for $i = 0.04$, $n = 20$ periods.
   c) By how much, approximately, will the compound amount at $n = 20$ periods
   change if i is 4.1 percent rather than 4 percent?
   d) Use b) and c) to estimate the compound amount $C(0.041)$ at $n = 20$ periods.

2. Taking the compound amount in Problem 1 as a function of n, we have

$$C(n) = (1 + i)^n.$$

   a) What derivative rule must be applied to find $dC(n)/dn$?
   b) Find the rate of change in the compound amount at $i = 0.05$ and $n = 10$ years.
   (Note: ln $1.05 = 0.04879$.)
   c) By how much, approximately, will the compound amount change in 3 months;
   that is, between 10 and 10.25 years?
   d) Use b) and c) to estimate $C(10.25)$.

3. The optimal length of time, t days, to run an advertising campaign depends on a
   response parameter, r, as shown by

$$t(r) = \frac{\ln (500r)}{r}.$$

   a) Find the expression for the rate of change in $t(r)$ per unit change in r. (Quotient
   rule.)
   b) What is the rate in a) if $r = 0.2$? (Note: ln $100 = 4.60517$.)
   c) Find the optimal number of days to run the campaign if $r = 0.2$.
   d) Using b), by how much, approximately, would the optimal number of days
   in c) change if r were 0.21 rather than 0.20?

4. a) Find the expression for the rate of change of

$$f(x) = xe^{-1/x}$$

   per unit change in x.
   b) What is the rate in a) at $x = 1$?
   c) Find the approximate change in $f(x)$ if x increases from 1 to 1.01.
   d) Using b) and c) approximate $f(1.01)$.

5. Given

$$f(x) = x^{1/2},$$

a) What is the expression for the rate of change in the square root per unit change in x?
b) Evaluate the rate in a) if $x = 25$.
c) By how much, approximately, will the square root change if x decreases from 25 to 24.9?
d) Using b) and c), estimate $(24.9)^{1/2}$.

6. Given the demand function,

$$p(q) = p = 100 - 20 \ln q,$$

a) Find the expression for $e(q)$, the price elasticity of demand.
b) Evaluate $e(q)$ at demand 20 units.
c) Interpret the result in b).

7. Given the demand function

$$p(q) = p = (200 - 0.5q),$$

a) What is dq/dp?
b) What is the expression for elasticity of demand, $e(q)$?
c) Find elasticity at 100 units demand.
d) Interpret the result in c).
e) At what demand and price is elasticity equal to 1?
f) What will total revenue be at the price and quantity found in e?

8. Refer to Problem 7 and note that revenue $= R(q) = qp(q)$. Verify that the q which maximizes revenue is the q found in Problem 7e).

10.23 CALCULUS OF TWO INDEPENDENT VARIABLES

A substantial discussion of the calculus of two, three, or more independent variables is beyond the scope of this text. However, the existence of multivariate calculus should be mentioned, and we shall close this chapter with a brief introduction to the calculus of two independent variables. Consider, for example,

$$f(x, y) = 3x^2 + 2xy$$

where f is a function of the two variables, x and y. The *partial* derivative of $f(x, y)$ with respect to x may be symbolized as f_x, which means to take the derivative with respect to x, treating y as a constant. Thus,

$$f_x = 6x + 2y.$$

Similarly, to find f_y, the partial of f with respect to y, x is treated as a constant and

$$f_y = 2x.$$

The second derivatives, f_{xx} and f_{yy}, are found in like manner from the first derivatives. Thus,

$$f_{xx} = 6 \quad \text{and} \quad f_{yy} = 0.$$

The *mixed* partials $f_{xy} = f_{yx}$ mean to take the first partial with respect to x and the second with respect to y, or conversely. Thus, for $f(x, y) = 3x^2 + 2xy$

$$f_x = 6x + 2y \quad f_y = 2x$$
$$f_{xy} = 2 \quad f_{yx} = 2.^{13}$$

Exercise. For $f(x, y) = x^2y + 3xy^3$, find: *a)* f_x. *b)* f_y. *c)* f_{xx}. *d)* f_{yy}. *e)* f_{xy}. Answer: *a)* $2xy + 3y^3$. *b)* $x^2 + 9xy^2$. *c)* $2y$. *d)* $18xy$. *e)* $2x + 9y^2$.

The one-variable requirement that $f'(x)$ be zero if $f(x)$ is to be a stationary point (maximum or minimum), is matched in the case of two variables by the requirement that both partial derivatives be zero. Thus, to find candidates for maxima or minima, we set

$$f_x = 0$$
$$f_y = 0$$

and solve the two equations simultaneously for x and y. For example, if

$$f(x, y) = 475x - 4x^2 - 3xy + 300y - 3y^2,$$

then,

$$f_x = 475 - 8x - 3y$$
$$f_y = -3x + 300 - 6y.$$

Setting f_x and f_y to zero, we have

$$e_1: \quad -8x - 3y + 475 = 0$$
$$e_2: \quad -3x - 6y + 300 = 0.$$

We eliminate y by finding $e_2 - 2(e_1)$, which is

$$13x - 650 = 0$$

$$x = \frac{650}{13} = 50$$

and find by substitution that

$$y = 25.$$

Hence, (50, 25) is a candidate for maximum or minimum.

---

[13]There are several symbols for partial derivatives. One such employs the round delta, ∂, to signify *partial*. Thus, for example, $\partial f/\partial x$ means f_x and $\partial f/\partial y$ means f_y.

The next question, whether we have a maximum or minimum, or neither, can be answered easily if $(f_{xx})(f_{yy}) \neq (f_{xy})^2$ at the candidate point.

Maximum-Minimum Test.

1. If $(f_{xx})(f_{yy}) > (f_{xy})^2$, candidate marks a maximum (minimum) if f_{xx} and f_{yy} are both negative (positive).
2. If $(f_{xx})(f_{yy}) < (f_{xy})^2$, candidate yields neither a maximum nor minimum.
3. If $(f_{xx})(f_{yy}) = (f_{xy})^2$, the test fails. (We shall not consider this case.)

Returning to our example, we have

$$f_{xx} = -8, f_{yy} = -6, f_{xy} = -3$$

so $(f_{xx})(f_{yy}) = (-8)(-6) = 48$ is greater than $(f_{xy})^2$ which is 9. Both f_{xx} and f_{yy} are negative, so the point $x = 50$, $y = 25$ yields a maximum which is $f(50, 25) = 475(50) - 4(50)^2 - 3(50)(25) + 300(25) - 3(25)^2 = 15,625$.

We shall now apply partial derivatives to maximize a profit function. The illustration concerns two competing products; that is, products whose demands are related in a manner such that if the price of one increases, its demand decreases and the demand for the other product increases. Consider the following demand and cost functions for two products sold by a firm.

$$p_1(q_1, q_2) = 500 - 4q_1 - 2q_2$$
$$p_2(q_1, q_2) = 400 - q_1 - 3q_2.$$

The cost of making q_1 units of the first product and q_2 units of the second is given as

$$C(q_1, q_2) = 25q_1 + 100q_2.$$

The symbol $p_1(q_1, q_2)$ means the unit price of the *first* product at demands of q_1 units for the first product and q_2 units for the second product, and similarly for $p_2(q_1, q_2)$.

$P(q_1, q_2)$ is the profit at demands q_1 and q_2, and

$$\text{Profit} = (\text{Revenue}) - (\text{Cost})$$
$$P(q_1, q_2) = [q_1(500 - 4q_1 - 2q_2) + q_2(400 - q_1 - 3q_2)] - [25q_1 + 100q_2].$$

Expanding and combining terms, we find

$$P(q_1, q_2) = 475q_1 - 4(q_1)^2 - 3q_1q_2 + 300q_2 - 3(q_2)^2.$$

This function is the same as $f(x, y)$ in our previous example with $p_1 = x$ and $p_2 = y$, so a maximum occurs at $q_1 = 50$ and $q_2 = 25$, and the maximum profit is \$15,625. From the original demand functions we find that at $q_1 = 50$ and $q_2 = 25$, $p_1 = \$250$ per unit and $p_2 = \$275$ per unit. In summary,

$$\text{Maximum profit} = \$15,625 \text{ at}$$
$$q_1 = 50 \text{ units;} \qquad q_2 = 25 \text{ units}$$
$$p_1 = \$250 \text{ per unit;} \qquad p_2 = \$275 \text{ per unit.}$$

10.24 PROBLEM SET 8

1. Given $f(x, y) = x^2 + 2y^2 + 3xy$, find
   a) f_x.   b) f_y.   c) f_{xx}.   d) f_{yy}.   e) f_{xy}.   f) f_{yx}.
2. Repeat Problem 1 for $f(x, y) = 2xy^2 + 3yx^2 - 2xy$.
3. Does $f(x, y) = 200x - 0.1x^2 + 150y - 0.25y^2$ have a maximum or minimum? If so, which is it, where does it occur, and what is its value?
4. Answer Problem 3 for $f(x, y) = 1.5x^2 - 2xy - 2x + 1.5y^2 - 7y$.
5. The demand and cost function for two products are:

$$p_1(q_1, q_2) = 100 - 0.2q_1 - 0.1q_2$$
$$p_2(q_1, q_2) = 80 - 0.1q_1 - 0.2q_2$$
$$C(q_1, q_2) = 25q_1 + 38q_2.$$

   a) Find the quantities and unit prices which yield maximum profit.
   b) Find the maximum profit.

10.25 REVIEW PROBLEMS

Find $f'(x)$ if

1. $f(x) = 2x^4 - 3x^2 + 5x - 7$.   2. $f(x) = 7x + 10$.
3. $f(x) = 25$.   4. $f(x) = x^2 - 7$.
5. $f(x) = 2e^x$.   6. $f(x) = 6 \ln x$.
7. $f(x) = x^{1/2}$.   8. $f(x) = x^{-2/3}$.
9. $f(x) = \dfrac{1}{x^3}$.   10. $f(x) = 1 - \dfrac{1}{x}$.
11. $f(x) = 2x^{3/2} + \dfrac{1}{5}x^{5/3}$.   12. $f(x) = 5^x$.
13. $f(x) = 2(4)^x$.   14. $f(x) = e^x + 2^x + \ln x + x + 6$.

Write the coordinates of the points on the curve for the stated values of x and find the slope of the curve at that point:

15. $f(x) = 2x^3$ at $x = 2$.   16. $f(x) = 3x^{1/3}$ at $x = 27$.
17. $f(x) = 3e^x$ at $x = -1$.   18. $f(x) = 2 \ln x$ at $x = 2$.
19. $f(x) = (0.6)^x$ at $x = 2$.   20. $f(x) = x^{-1} - \dfrac{1}{x^2} + \ln x$ at $x = 1$.

21. The number of gallons of fuel consumed per mile traveled by a truck operated at a speed of x miles per hour is $g(x)$, where

$$g(x) = \frac{x}{80} + \frac{20}{x}.$$

   a) What speed provides the minimum fuel consumption per mile traveled?
   b) Show a) yields a minimum.
   c) What is the minimum?

Write the coordinates of every stationary point for each of the following, indicating which are maxima and which are minima:

22. $f(x) = \ln x$, $x > 0$.   23. $f(x) = 2x + 3$.
24. $f(x) = 2x^2 - x + 1$.   25. $f(x) = x^3 - (5/2)x^2 - 2x + 4$.

26. $f(x) = 3x^4 - 4x^3 - 12x^2 - 24.$ 27. $f(x) = 2x^4.$
28. $f(x) = x^2 + (54/x).$ 29. $f(x) = 2 \ln x - 2x.$

Write the coordinates of every stationary or inflection point, stating the type of point in each case:

30. $f(x) = 3x - x^3.$ 31. $f(x) = x^4 - 4x^3 + 35.$
32. $f(x) = x^2 + 32 \ln x.$

33. At any point, a polynomial is in one of the eight conditions shown in the tabular boxhead below:

		Rising			Falling		
Maxi-mum	Mini-mum	Concave Up	Concave Down	Inflec-tion	Concave Up	Concave Down	Inflec-tion

In which of these conditions is the curve

$$y = x^4 - 4x^3$$

at $x = \pm 4, \pm 3, \pm 2, \pm 1, 0$?

34. Interior and exterior walls of a rectangular 80,000-square-foot warehouse cost $90 per running foot. The warehouse is to be divided into ten rooms by four interior walls in the x direction and one interior wall in the y direction. What should be the warehouse dimensions if wall cost is to be minimized?

35. A rectangular area is to be enclosed, then divided into thirds by two fences across the area parallel to the sides. If the area to be enclosed is 1250 square feet, what dimensions will lead to the use of a minimum amount of fence?

36. (See Problem 35.) If the fence on the two ends costs $0.64 per running foot and the dividers and the sides cost $1 per running foot, what should be the dimensions if the cost of the fence is to be a minimum, and what is this minimum cost?

37. If a rectangular area is to be fenced in the manner of Problem 35, what is the maximum area that could be enclosed with 1000 feet of fence?

38. A rectangular manufacturing plant with a floor area of 5400 square feet is to be built in a location where zoning regulations require buffer strips 30 feet wide front and back, and 20 feet wide at either end. (A buffer strip is a grass and tree belt which must not be built upon.) What plot dimensions will lead to minimum total area for plant and buffer strips? What is this minimum total area?

39. A box with square bottom and no top is to contain 32 cubic inches. Find the dimensions which will lead to a box of minimum area. What is this minimum area?

40. If the bottom material for the box in Problem 39 costs eight cents per square inch, and the side material costs one cent per square inch, what dimensions will lead to minimum cost? What is this minimum cost?

41. A box with no top is to be made from an 8 by 15 inch piece of cardboard by cutting equal sized squares from the corners, then turning up the sides. What should be the dimensions of the squares if the box is to have maximum volume? (Note: The quadratic expression encountered in the solution is factorable.)

42. A parcel delivery service accepts cylindrical packages whose length, L, plus girth, $2\pi r$, does not exceed 120 inches. A shipper who uses cylindrical cartons, perforated, wishes to design a carton with maximum ventilation (area). What should be the length and radius of the carton?

43. An appliance service company is located centrally in a roughly square area x miles on a side. It charges $27 per call, less parts and labor, and travel cost is figured at $1.50 per mile. The average distance traveled per call is $1.2x$ miles. In a month, the average number of calls per square mile of service area is 30.
    a) What should x be if net travel income (which excludes parts and labor) is to be maximized?
    b) What is this maximum?

44. Following the inventory model of this chapter (Section 10.8), suppose it costs $250 to set up a plant to make a batch (lot) of a product and $2 for each unit made after set-up. Inventory cost is $1.20 per year per unit in average inventory. The annual requirement for the product is 9600 units.
    a) Write the expression for total annual cost as a function of the lot size, L.
    b) What lot size yields minimum cost?

Find $df(x)/dx$:

45. $f(x) = (x^2 - 2x)^3$.

46. $f(x) = (2x)^{3/2}$.

47. $f(x) = 2e^{-0.5x^2}$.

48. $f(x) = 0.5e^{2x+3}$.

49. $f(x) = 2^{3x}$.

50. $f(x) = 4^{0.5x}$.

51. $f(x) = 3 \ln 2x$.

52. $f(x) = (1/2) \ln (3x^2 + 2x)$.

53. a) Complete the statement of the chain rule:

$$\frac{dg(y)}{dx} = \qquad .$$

   b) In a) what is assumed about y?

54. a) Complete the statement of the chain rule:

$$\frac{dh(p)}{dq} = \qquad .$$

   b) In a) what is assumed about p?

Find the derivative with respect to x for each of the following:

55. $x^2 \ln x$.

56. $x(3x - 2)^6$.

57. $\dfrac{x^2 - 3}{2x}$.

58. xe^{2x}.

59. $e^x \ln x$.

60. $\dfrac{e^x}{x}$.

61. $\dfrac{e^{3x}}{x}$.

62. $\dfrac{x^2}{e^x}$.

63. $6xe^{-0.5x}$.

64. $\dfrac{x}{(x + 2)^{1/2}}$.

65. The potential audience of an advertising campaign is 2000 people. Revenue per response averages $5. The campaign has a fixed cost of $100, plus a variable cost of $105.40 per day. Find the number of days the campaign should continue to maximize profit if the proportion of the audience responding at time t is $1 - 0.9^t$.

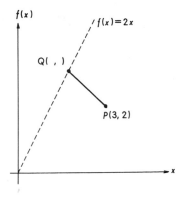

FIGURE A

66. See Figure A. A railway is on a line shown as $f(x) = 2x$, and a plant is at point $P(3, 2)$. A siding is to be placed on the railway at Q so that the distance $\overline{PQ}$ is minimized.
    a) Letting z be the x-coordinate of Q, what will be the y-coordinate?
    b) Write the expression for the distance $s = \overline{PQ}$.
    c) Find the value of z which minimizes s.
    d) Compute the minimal s.
    e) Prove, by slopes, that $\overline{PQ}$ is perpendicular to $y = 2x$.

67. See Problem 66. Let the straight line be $f(x) = mx$ and the point be $P(a, b)$. Letting z be the x-coordinate of Q, find the expression for z which leads to minimal s.

68. A firm makes q units at a total cost of

$$C(q) = \$(100 + 10q + 0.025q^2)$$

and sells them at $20 per unit.
    a) What number of units yields maximum profit?
    b) What is this maximum?

69. At demand q units, a firm receives for its product a unit price of $p(q) = \$(205 - 0.95q)$. The cost of making q units is $C(q) = \$(200 + 5q + 0.05q^2)$.
    a) What number of units yields maximum profit?
    b) What is the maximum profit?

70. At demand q units, a firm receives for its product a unit price of $p(q) = \$(200 - 0.1q)$. The cost of making q units is

$$C(q) = \$\left(\frac{0.01q^3}{6} - 0.7q^2 + 222q\right).$$

    a) What number of units yields maximum profit?
    b) At what number of units will marginal cost be at its minimum, and what is this minimum?

71. Given $C(q)$ as in Problem 70, find the quantity q^* and the unit selling price which will allow the firm just to break even by the following methods.
    a) Minimal average cost.
    b) Marginal cost equals average cost.

72. At $\$Y$ thousand income, consumption expenditures are

$$C(Y) = 0.7Y + 0.36Y^{2/3}.$$

a) Find the marginal propensities to consume at income levels of $8 and $64 thousand.
b) Interpret the answer to part a).
c) What is the multiplier at income level $64 thousand?
d) Interpret the answer to c).
e) What is the multiplier at income level $8 thousand?

73. Total income, Y, is distributed for consumption $C(Y)$ and investment, I, so that $Y = I + C(Y)$. Using derivative symbols,
a) What are the expressions for marginal propensity to consume, MPC, and for the multiplier?
b) Prove that the multiplier $= 1/(1 - \text{MPC})$. (See text discussion.)

74. Values of $f(x) = e^{-x}$ are shown in Table IX at the end of the book.
a) Find the expression for the rate of change of $f(x)$ per unit change in x.
b) Evaluate the rate in a) at $x = 1.5$.
c) If x changes from 1.50 to 1.52, by how much, approximately, will $f(x)$ change?
d) What is the approximate value of $f(1.52) = e^{-1.52}$?

75. Table IV at the end of the book shows the amount of $1 per period for n periods is

$$S(i) = \frac{(1 + i)^n - 1}{i}.$$

a) Find the expression for the rate of change $S(i)$ per unit change in i.
b) Evaluate the rate in a) for $i = 0.04$ and $n = 10$.
c) By how much, approximately, would the amount change if i were 0.041 rather than 0.040?
d) Estimate $S(0.041)$.

76. Given $p(x) = x(2x + 5)^{1/2}$,
a) Find the expression for the rate of change in $p(x)$ per unit change in x.
b) Evaluate this rate at $x = 2$.
c) Find $p(2)$.
d) Estimate $p(1.9)$ using b) and c).

77. Does $f(x, y) = 10xy - 5x^2 - 6y^2 + 10x$ have a maximum or minimum? If so, which is it, what is its value, and where does it occur?

78. The demand and cost functions for two products are

$$p_1(q_1, q_2) = 1000 - 3q_1 - q_2$$
$$p_2(q_1, q_2) = 500 - q_1 - 2q_2$$
$$C(q_1, q_2) = 20q_1 + 140q_2.$$

a) Find the quantities and unit prices which yield maximum profit.
b) Find the maximum profit.

11

Integral Calculus

11.1 INTRODUCTION

APPLICATIONS of integral calculus deal with problems which require determination of a function when the expression for its rate of change (derivative) is given, and with problems which involve the computation of an area which has the given function as part of its boundary. We shall see that the procedures for solving such problems are essentially the reverse of derivative procedures. That is, in the differential calculus, we are given a function and asked to find its derivative, whereas in the integral calculus we are given the derivative of a function and asked to find the function.

We start the chapter by deriving a few rules for integration, and present an example in which integration is applied to determine the total amount of oil that will be pumped by a well during its productive life. Next, we relieve the reader of the task of memorizing the numerous special rules for integration by inserting a table containing all the integral rules required in this book. Then we show that integration is a summing process, the result of which may be interpreted as an area, and present a number of applications of area interpretations, including some which involve areas having two curves as parts of their boundary and others which involve areas having an infinite boundary. Finally, we show how to solve some equations involving derivatives (*differential equations*), and offer applications of this subject matter.

11.2 ANTIDERIVATIVES–INTEGRALS

The expression

$$\text{antiderivative of } 3x^2$$

means the function which has $3x^2$ as its derivative. Clearly, x^3 is one proper answer, but $x^3 + 5$ or $x^3 - 6$, or x^3 plus any constant would also be correct because the derivative of x^3 plus a constant is $3x^2$. To include all possibilities, we write

$$\text{antiderivative } (3x^2) = x^3 + C.$$

||||||||||||||||||||||||||||||||||||||||||||||||||||||||||||||||||||||||||||||||||||||||||||||||||||||||||||||||||||

Exercise.   What is the antiderivative of $2x$? Answer: $x^2 + C$.

||||||||||||||||||||||||||||||||||||||||||||||||||||||||||||||||||||||||||||||||||||||||||||||||||||||||||||||||||||

For reasons which will become clear later when we consider integrating or *summing*, we shall adopt here the integral symbol, an elongated *S*, as the instruction to determine an antiderivative and write

$$\int 3x^2 dx = x^3 + C.$$

The *differential* symbol, dx, is appended to indicate that x is the variable of interest.[1] In the last expression, $3x^2$ is called the *integrand*, and the antiderivative, $x^3 + C$, is called the *indefinite integral*, or simply the *integral*. The adjective, indefinite, whether or not used, refers to the presence of the unknown constant C, which is called the *constant of integration*. To prove the correctness of an integral rule, we need only to show that the derivative of the integral equals the integrand. Thus,

$$\int x^2 dx \neq x^3 + C.$$

The integral is incorrect because its derivative is

$$\frac{d}{dx}(x^3 + C) = 3x^2$$

and the integrand is x^2, not $3x^2$. However,

$$\int x^2 dx = \frac{x^3}{3} + C$$

because

$$\frac{d}{dx}\left(\frac{x^3}{3} + C\right) = \frac{3x^2}{3} + 0 = x^2.$$

||||||||||||||||||||||||||||||||||||||||||||||||||||||||||||||||||||||||||||||||||||||||||||||||||||||||||||||||||||

Exercise.   Find $\int x^4 dx$ and prove it is correct. Answer: $(x^5/5)$ + C, because $d/dx[(x^5/5) + C]$ is $5x^4/5 = x^4$.

||||||||||||||||||||||||||||||||||||||||||||||||||||||||||||||||||||||||||||||||||||||||||||||||||||||||||||||||||||

We may formulate a general rule by noting that

$$\int x^6 dx = \frac{x^7}{7} + C; \quad \int x^7 dx = \frac{x^8}{8} + C$$

and it is true that if $x \neq -1$

$$\int x^n dx = \frac{x^{n+1}}{n+1} + C.$$

---

[1] The differential has more substance than implied here, as we saw in Chapter 10, and will see again later in this chapter.

The exceptional case is

$$\int x^{-1}dx = \int \frac{1}{x}dx = \ln x + C.$$

Exercise.   Prove the last statement. Answer: d/dx (ln x + C) = $1/x$.

We may state:

Power rule: $\int x^n dx = \frac{x^{n+1}}{n+1} + C$ if $n \neq -1,$

$$\int x^{-1}dx = \int \frac{1}{x}dx = \ln x + C.$$

For example, with $n = 10$, so $n + 1 = 11$,

$$\int x^{10}dx = \frac{x^{11}}{11} + C$$

and with $n = 1/2$, so that $n + 1 = 3/2$,

$$\int x^{1/2}dx = \frac{x^{3/2}}{\frac{3}{2}} + C = \frac{2}{3}x^{3/2} + C,$$

and

$$\int \frac{dx}{x^2} = \int x^{-2}dx = \frac{x^{-1}}{-1} + C = -x^{-1} + C.$$

However, as noted earlier,

$$\int \frac{dx}{x} = \ln x + C.$$

Note in particular that

$$\int dx = \int x^0 dx = \frac{x^1}{1} + C = x + C,$$

and

$$\int 2\,dx = 2x + C.$$

Exercise.   Find $\int 0.5\,dx$. Answer: $0.5x + C$.

Properties of Integration.   The integral of an expression may be taken term by term, and a constant factor may be placed outside the integral sign. For example,

$$\int (10x^4 + 2x - 5)dx = 10 \int x^4 dx + 2 \int x\,dx - 5 \int dx$$

$$= 10\left(\frac{x^5}{5}\right) + 2\left(\frac{x^2}{2}\right) - 5x + C$$

$$= 2x^5 + x^2 - 5x + C.$$

We do not need separate constants for the three foregoing integrals because the sum of three constants is another constant designated by C.

||||||||||||||||||||||||||||||||||||||||||||||||||||||||||||||||||||||||||||||||||||||||||||||||||||||||||||||||||||||||||||||

Exercise.   Find $\int (3x^{1/2} - 6x^2 + 2)dx$. Answer: $2x^{3/2} - 2x^3 + 2x + C$.

||||||||||||||||||||||||||||||||||||||||||||||||||||||||||||||||||||||||||||||||||||||||||||||||||||||||||||||||||||||||||||||

Recalling that the derivative of the exponential function, e^x, is e^x, it follows that

$$\int e^x dx = e^x + C.$$

Next, if we apply the chain rule, we find

$$\frac{d}{dx}e^{bx} = be^{bx},$$

where b is a constant. Dividing both sides by b we find

$$\frac{d}{dx}\frac{e^{bx}}{b} = e^{bx}.$$

Hence, since the derivative of e^{bx}/b is e^{bx}, it follows that the integral of e^{bx} is e^{bx}/b plus C.

Exponential Rule: $\int e^{bx}dx = \frac{e^{bx}}{b} + C.$

For example,

$$\int e^{0.4x}dx = \frac{e^{0.4x}}{0.4} + C = 2.5e^{0.4x} + C.$$

||||||||||||||||||||||||||||||||||||||||||||||||||||||||||||||||||||||||||||||||||||||||||||||||||||||||||||||||||||||||||||||

Exercise.   Find $\int e^{-2.5x}dx$. Answer: $-0.4e^{-2.5x} + C$.

||||||||||||||||||||||||||||||||||||||||||||||||||||||||||||||||||||||||||||||||||||||||||||||||||||||||||||||||||||||||||||||

11.3 DETERMINING THE CONSTANT OF INTEGRATION

Suppose the slope of a line is 2 and the line passes through the point $(3, 9)$. We seek the equation of the line which we shall designate as the linear function $f(x)$. The given slope, 2, is $f'(x)$, the derivative of $f(x)$. Thus,

$$f'(x) = 2.$$

To find $f(x)$ from its derivative, we integrate

$$f(x) = \int f'(x)dx = \int 2dx = 2x + C.$$

We now apply the condition that the line passes through (3, 9); that is, $f(3) = 9$. We have

$$f(x) = 2x + C$$
$$f(3) = 2(3) + C = 9$$

so that

$$6 + C = 9$$

and

$$C = 3.$$

It follows that the desired equation is

$$f(x) = 2x + 3.$$

The last example shows that if we are given the derivative of a function, and a condition the function must satisfy, the function itself can be found by integrating the derivative and using the condition to determine the value of the constant of integration.

As another example, suppose that in the manufacture of a product, fixed cost is \$100 and marginal cost at q units output is \$$(2 + 0.02q)$. We wish to find the total cost function, $C(q)$. Marginal cost is the derivative, $C'(q)$. Integrating, we have

$$C(q) = \int C'(q)dq = \int (2 + 0.02q)dq = 2q + 0.01q^2 + K,$$

where K is used here as the constant of integration. The given condition that fixed cost is \$100 means that $C(0) = 100$. Hence,

$$C(0) = 100 = 2(0) + (0.02)(0) + K$$

from which

$$K = 100$$

and

$$C(q) = 2q + 0.01q^2 + 100.$$

Exercise. If marginal cost at q units output is \$$(5 + 0.004q)$ and the total cost of producing 100 units is \$1520, find the total cost function, $C(q)$. Answer: $C(q) = 1000 + 5q + 0.002q^2$.

11.4 INTEGRATING RATES

The foregoing marginal interpretation of the derivative of a cost function is a special case of a more general class of applications where the derivative is interpreted as a rate. In this section we introduce more examples of the general class of rate applications.

Example 1. Suppose that the amount of oil produced by a well after it has been operating t years is $A(t)$ thousand barrels, but we do not have the expression for $A(t)$. Upon inquiry, we learn that the *rate* of output per year was 40 thousand barrels per year at the beginning ($t = 0$), and is expected to decrease steadily by two thousand barrels per year. That is, at any time t years the *rate* of production will be

$$40 - 2t$$

thousand barrels per year. We reason that if $A(t)$ is total production at time t years, then production will be at the rate of $A'(t)$ per year at time t years. Thus, the rate $40 - 2t$ is $A'(t)$ and we can obtain $A(t)$ by

$$A(t) = \int A'(t)dt = \int (40 - 2t)dt = 40t - t^2 + C.$$

At the instant $t = 0$, no oil has been produced, so $A(0) = 0$. Hence,

$$A(0) = 0 = 40(0) - (0)^2 + C$$

and $C = 0$ so that

$$A(t) = 40t - t^2.$$

At time 10 years, total output will be

$$A(10) = 40(10) - (10)^2 = 300 \text{ thousand barrels.}$$

Exercise. a) When will the rate of production drop to zero? b) What will be the total output of the well during its life? Answer: a) The rate, $40 - 2t$, will be zero when $t = 20$ years. b) $A(20) = 400$ thousand barrels.

Example 2. Ten months ($t = 10$) after introducing a product, total sales, $S(t)$, were \$64.148 thousand. By increased advertising, the company estimates it can increase sales to the *rate* of

$$12 - 12e^{-0.1t}$$

per month at time t months, $t > 10$. Find the expression for total sales at time t months.

We have given that

$$S'(t) = 12 - 12e^{-0.1t}, \ t > 10.$$

Hence,

$$S(t) = \int S'(t)dt = 12t - \frac{12e^{-0.1t}}{-0.1} + C$$

$$S(t) \qquad\qquad = 12t + 120e^{-0.1t} + C.$$

Using the condition that $S(t) = 64.148$ when $t = 10$, we have

$$64.148 = 12(10) + 120e^{-1} + C$$
$$= 120 + 44.148 + C$$

or

$$C = -100$$

so that

$$S(t) = 12t + 120e^{-0.1t} - 100, \; t > 10.$$

Exercise. Using the last expression, find total sales at time $t = 20$ months. Answer: $156.24 thousand.

Exercise. The amount of a bank account is increasing at the *rate* of $80e^{0.08t}$ per year at time t years, and the amount is $1000 at time $t = 0$. Find the expression for $A(t)$, the amount in the account at time t years. Answer: $A(t) = 1000e^{0.08t}$.

In passing, we note that the material presented in this section and the one preceding it generally are considered as part of the subject of *differential equations*, which we shall discuss later in the chapter. Our purpose for introducing the material at this point has been to emphasize the rate concept and to give meaning to the constant of integration.

11.5 TABLES OF INTEGRALS

The chain, product, and quotient rules, which greatly extended our ability to take derivatives, do not have generally applicable counterparts in integration. Indeed, small alterations in an expression can change it from a function which has an exact integral to one whose integral can only be approximated. For example,

$$\int e^x dx = e^x + C,$$

but there is no exact finite expression for

$$\int e^{x^2} dx.$$

Some integral rules can be developed by trial and error. Thus, the integral

$$\int (3 + 2x)^5 dx$$

must contain

$$(3 + 2x)^6.$$

However, the derivative of the latter, using the chain rule, is

$$\frac{d}{dx}(3 + 2x)^6 = 6(3 + 2x)^5(2)$$

and contains the factors 6 and 2 which are not present in the original integrand, so we remove them by division and find

$$\int (3 + 2x)^5 dx = \frac{(3 + 2x)^6}{2(6)} + C.$$

If we now replace the constants in the original integrand (3, 2, and the exponent 5) by a, b, and n, respectively, we have the general rule

$$\int (a + bx)^n dx = \frac{(a + bx)^{n+1}}{b(n + 1)} + C, \text{ if } n \neq -1.$$

The trial and error procedure does not work in most situations, and a number of methods can be applied to integrate functions having special forms. We shall not develop these methods but, instead, present a short Table of Integrals which includes the functions we shall need in our applications.[2] In a later section on numerical integration, we shall show how integrals for which formulas do not exist can be approximated.

For ready reference, the integral table appears just prior to Problem Set 1. To use the table, we must first identify the appropriate form, then assign the proper values to the constants appearing in that form. For example, if we seek

$$\int (3 - 2x)^5 dx$$

we identify the integrand as being the form of the integrand of Rule 5 which states

$$\int (a + bx)^n dx = \frac{(a + bx)^{n+1}}{b(n + 1)} + C.$$

Comparing

$$(3 - 2x)^5 \quad \text{with} \quad (a + bx)^n$$

we see that

$$a = 3, b = -2, n = 5.$$

Hence,

$$\int (3 - 2x)^5 dx = \frac{(3 - 2x)^6}{(-2)(5 + 1)} + C = -\frac{(3 - 2x)^6}{12} + C.$$

||||||||||||||||||||||||||||||||||||||||||||||||||||||||||||||||||||||||||||||||||||||||||||||||||||||||||||||||||||||||||||

Exercise. Find $\int \frac{dx}{3 + 2x}$. Answer: Apply Rule 6 with $a = 3$, $b = 2$ to obtain $(1/2)\ln (3 + 2x) + C$.

||||||||||||||||||||||||||||||||||||||||||||||||||||||||||||||||||||||||||||||||||||||||||||||||||||||||||||||||||||||||||||

[2] For a more extensive table of integrals, see a recent edition of *Standard Mathematical Tables*, published by the Chemical Rubber Company, Cleveland, Ohio.

As another example, we identify the integrand of the next expression with that of Rule 13, $b = -0.5$, treat the constant -2 as specified by Rule 1, and find

$$\int -2xe^{-0.5x}dx = -2\left[\frac{e^{-0.5x}(-0.5x - 1)}{0.25}\right] + C$$

$$= 8e^{-0.5x}(0.5x + 1) + C.$$

Exercise. Find $\int 4x(2x + 3)^{-1}dx$. Answer: Applying Rule 1 then Rule 7 with $a = 3, b = 2$, we have $2x - 3 \ln (2x + 3) + C$.

Table of Integrals*

1. $\int cf(x)dx = c \int f(x)dx.$

2. $\int [f(x) + g(x)]dx = \int f(x)dx + \int g(x)dx.$

3. $\int x^n dx = \frac{x^{n+1}}{n + 1} + C, n \neq -1.$

4. $\int x^{-1}dx = \int \frac{dx}{x} = \ln x + C.$

5. $\int (a + bx)^n dx = \frac{(a + bx)^{n+1}}{b(n + 1)} + C, n \neq -1.$

6. $\int (a + bx)^{-1}dx = \int \frac{dx}{a + bx} = \frac{\ln (a + bx)}{b} + C.$

7. $\int \frac{x \, dx}{a + bx} = \frac{x}{b} - \frac{a}{b^2} \ln (a + bx) + C.$

8. $\int \frac{dx}{x(a + bx)} = \frac{1}{a} \ln \left(\frac{x}{a + bx}\right) + C.$

9. $\int e^x dx = e^x + C.$

10. $\int a^x dx = \frac{a^x}{\ln a} + C.$

11. $\int e^{b+cx}dx = \frac{e^{b+cx}}{c} + C.$

12. $\int a^{b+cx}dx = \frac{a^{b+cx}}{c \ln a} + C.$

13. $\int xe^{bx} = \frac{e^{bx}(bx - 1)}{b^2} + C.$

14. $\int \ln x \, dx = x \ln x - x + C = x(\ln x - 1) + C.$

15. $\int \ln (a + bx)dx = \frac{(a + bx)[\ln (a + bx) - 1]}{b} + C.$

*See also Table XII at the end of the book.
Note: $a, b, c,$ and n are constants.

11.6 PROBLEM SET 1

Find the following integrals:

1. $\int x^3 dx.$

2. $\int x^{3/2} dx.$

3. $\int x^{-1/2} dx.$

4. $\int 3 dx.$

5. $\int dx.$

6. $\int \dfrac{dx}{x}.$

7. $\int (2x^{-1} + e^x) dx.$

8. $\int (\ln x + 1) dx.$

9. $\int 3^a dx.$

10. $\int [x^{-2} + 2(5^x)] dx.$

11. $\int \dfrac{3 dx}{x(1 + 6x)}.$

12. $\int e^{5 + 2x} dx.$

13. $\int (2 + 3x)^{1/2} dx.$

14. $\int e^{-2x} dx.$

15. $\int \dfrac{4x}{1 + 2x} dx.$

16. $\int 3xe^{2x} dx.$

17. $\int (2)^{-0.1x} dx.$

18. $\int 3 \ln (2 + 6x) dx.$

Prove by differentiation that the following rules are correct:

19. $\int xe^x dx = e^x(x - 1) + C.$

20. $\int \dfrac{dx}{x(a + bx)} = \dfrac{1}{a} \ln \left(\dfrac{x}{a + bx} \right) + C.$

21. If the fixed cost of manufacturing a product is $1500 and marginal cost at q units output is $(1.5 + 0.1q)$,
    a) Find the function for $C(q)$, the total cost of making q units.
    b) Find the total cost of 100 units.

22. A beverage has 35 quality units when first made and placed on a shelf for sale. At time t days, the number of quality units is changing at the rate, per day, of

$$-0.9e^{-0.06t}.$$

a) Find the function for $Q(t)$, the number of quality units at time t days.
b) Find the limit of $Q(t)$ as t approaches ∞.

23. A seller accepts orders for up to 1000 units of a product and applies a continuous discount to the unit price, so that unit price declines at the rate of $0.01 per unit at q units purchased. The price of $q = 100$ units is $p(q) = $19. Find the function for $p(q)$, the unit price at q units ordered.

24. If the amount of money in an account is increasing at the rate of

$$150e^{0.06t}$$

per year at time t years, and the amount is zero at $t = 0$, find the function for $A(t)$, the total amount in the account at time t years.

25. If marginal cost at q units output is $(3 + 2 \ln q)$, $q > 0$, and the total cost of 1 unit is $140, find the function for $C(q)$, the total cost of producing q units.

11.7 THE DEFINITE INTEGRAL

Integration rules are written as indefinite integrals which contain the constant of integration. The definite integral is written as

$$\int_a^b f(x) \, dx.$$

When a, the *lower limit*, and b, the *upper limit*, are constants, the definite integral is a constant which is evaluated by writing the integral, $F(x)$, without the constant of integration, and computing

$$F(b) - F(a).$$

For example, to evaluate

$$\int_2^6 x \, dx$$

we find the integral as $F(x) = x^2/2$ and evaluate

$$F(6) - F(2) = \frac{6^2}{2} - \frac{2^2}{2} = 16.$$

A conventional method of expressing the foregoing is

$$\int_2^6 x \, dx = \frac{x^2}{2}\Big|_2^6 = \frac{36}{2} - \frac{4}{2} = 16,$$

where the vertical line segment means to evaluate the expression to its left at the upper and lower limits and subtract, as shown.

As another example,

$$\int_1^5 \ln x \, dx = (x \ln x - x)\Big|_1^5 - (5 \ln 5 \quad 5) \quad (1 \ln 1 \quad 1)$$

$$= [5(1.6094) - 5] - (0 - 1)$$

$$= 3.047 + 1$$

$$= 4.047.$$

Again,

$$\int_0^5 3e^{-0.2x} \, dx = \frac{3e^{-0.2x}}{-0.2}\Big|_0^5$$

$$= -15e^{-0.2x}\Big|_0^5$$

$$= -15(e^{-1} - e^0)$$

$$= -15(0.3679 - 1)$$

$$= 9.4815.$$

Exercise. Evaluate $\int_1^8 x^{1/3} \, dx$. Answer: $\frac{45}{4}$.

11.8 THE DEFINITE INTEGRAL AS AN AREA

Figure 11–1 shows the function

$$f(x) = 5$$

with verticals at $x = 0$ and $x = 10$. The function and the verticals complete a rectangle whose area by geometry is

$$(\text{width})(\text{height}) = (10)(5) = 50.$$

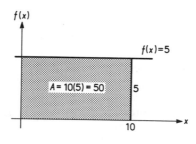

FIGURE 11–1

Next we find that *the same answer is obtained by evaluating*

$$\int_0^{10} f(x)\,dx = \int_0^{10} 5\,dx = 5x\,\Big|_0^{10} = 50.$$

This result is not a coincidence. That is, we can find an area bounded by a function, verticals, and the x-axis by simply finding the antiderivative of the function and evaluating it between limits to obtain the definite integral. The proof of this remarkable fact is called the *fundamental theorem of integral calculus*. We shall not show the proof here, but shall illustrate two more cases. Next we consider

$$f(x) = x$$

over the x-interval from 0 to b as shown in Figure 11–2. By geometry, the area of the right triangle is

$$\frac{1}{2}(\text{base})(\text{altitude}) = \frac{1}{2}(b)(b) = \frac{1}{2}b^2.$$

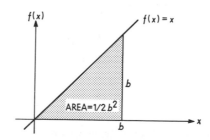

FIGURE 11–2

Applying the definite integral, we have

$$\int_0^b f(x)\,dx = \int_0^b x\,dx = \tfrac{1}{2}x^2 \Big|_0^b = \tfrac{1}{2}b^2,$$

the same result as that obtained from plane geometry.

Moving on to

$$f(x) = x^2$$

as shown in Figure 11–3, we no longer can determine the shaded area from

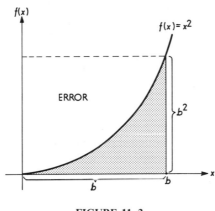

FIGURE 11–3
(not to scale)

plane geometry. It is clear, however, that if we draw the dashed horizontal line shown in the Figure to complete a rectangle, the area of the rectangle is

$$b(b^2) = b^3$$

which exceeds the desired shaded area by the error amount indicated in Figure 11–3. Obviously, the shaded area is a *fraction* of the rectangular area, b^3. According to the fundamental theorem, this fraction should be $\tfrac{1}{3}$ because

$$\int_0^b f(x)\,dx = \int_0^b x^2\,dx = \frac{x^3}{3}\Big|_0^b = \tfrac{1}{3}b^3.$$

We now wish to prove the last result because the method of proof discloses the fundamental *limit* nature of the integral and, incidentally, indicates the reason for the choice of an elongated S (for sum) as the integral symbol. We start by saying that the area, b^3, in Figure 11–3 is a *first* approximation of the shaded area, and the error is the unshaded part of the rectangle. Next consider Figure 11–4, wherein the interval from 0 to b has been divided into two segments and two rectangles constructed. If we use the *sum* of these two rectangular areas as a *second* approximation to the shaded area, the error will be less than that of the first approximation by the amount shown as

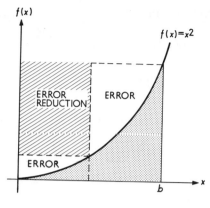

FIGURE 11–4
(not to scale)

error reduction in Figure 11–4. This suggests that we can continually decrease the error by increasing the number of subdivisions and summing the rectangular areas so obtained. Following this suggestion, we divide the interval 0 to b into n equal segments of width b/n as shown in Figure 11–5.

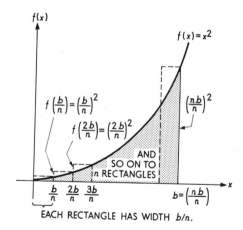

EACH RECTANGLE HAS WIDTH b/n.

FIGURE 11–5
(not to scale)

The first rectangle has

$$\text{width} = \frac{b}{n}.$$

And, since the function is

$$f(x) = x^2$$

we have that its

$$\text{height} = f\left(\frac{b}{n}\right) = \left(\frac{b}{n}\right)^2.$$

Consequently, the area of the first rectangle is

$$\text{width times height} = \left(\frac{b}{n}\right)\left(\frac{b}{n}\right)^2.$$

Similarly, the second rectangle has the same width, b/n, but

$$\text{height of second rectangle} = f\left(\frac{2b}{n}\right) = \left(\frac{2b}{n}\right)^2.$$

Consequently, the area of the second rectangle is

$$\left(\frac{b}{n}\right)\left(\frac{2b}{n}\right)^2.$$

Exercise. What is the area of the third rectangle? Answer: $(b/n)(3b/n)^2$.

Continuing, we note that the area of the last rectangle can be expressed as

$$\frac{b}{n}\left(\frac{nb}{n}\right)^2.$$

Letting S_n be the *sum* of the rectangular areas, we have

$$S_n = \frac{b}{n}\left(\frac{b}{n}\right)^2 + \left(\frac{b}{n}\right)\left(\frac{2b}{n}\right)^2 + \left(\frac{b}{n}\right)\left(\frac{3b}{n}\right)^2 + \cdots + \left(\frac{b}{n}\right)\left(\frac{nb}{n}\right)^2$$

$$= \frac{b^3}{n^3}(1^2 + 2^2 + 3^2 + \cdots + n^2).$$

There is a formula,[3] which we shall not derive, which states that $(1^2 + 2^2 + 3^2 + \cdots + n^2) = \frac{1}{6}[n(n + 1)(2n + 1)]$. Hence, substituting, we find

$$S_n = \frac{b^3}{n^3}\frac{[n(n + 1)(2n + 1)]}{6} = \frac{b^3}{n^2}\frac{[(n + 1)(2n + 1)]}{6}$$

$$= \frac{b^3}{6n^2}(2n^2 + 3n + 1)$$

$$S_n = \frac{b^3}{6}\left(2 + \frac{3}{n} + \frac{1}{n^2}\right).$$

Now S_n is an approximation to the shaded area under $f(x) = x^2$, and the error decreases as n, the number of rectangles, increases. Consequently,

---

[3]Evaluate the formula for $n = 5$, then verify that the result equals $1^2 + 2^2 + 3^2 + 4^2 + 5^2$.

we define the area under the curve to be the limit of S_n as n becomes infinite,

$$\lim_{n\to\infty} S_n = \lim_{n\to\infty} \frac{b^3}{6}\left(2 + \frac{3}{n} + \frac{1}{n^2}\right)$$

$$= \frac{b^3}{6}(2) = \frac{b^3}{3}$$

and we have the result predicted by the fundamental theorem that the area is

$$\int_0^b x^2\,dx = \frac{x^3}{3}\Big|_0^b = \frac{b^3}{3}.$$

Our purpose in the last demonstration was not only to show that the area under the curve is given by the definite integral, but also to call attention to the fundamental *sum* nature of the integral. Thus, integration means combining or summing elements of area and taking the limit of the sum as the number of elements become infinite. The elongated S, which is the integral symbol, represents this limiting sum.

We have used zero as the lower limit in our discussions thus far in this section. If we wish to calculate the area under $f(x) = x^2$ over the interval

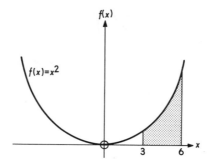

$f(x)$

$f(x)=x^2$

3   6 x

FIGURE 11–6
(not to scale)

3 to 6, as shown in Figure 11–6, we may find the area over 0 to 6 and subtract from this the area over 0 to 3, giving

$$\int_0^6 x^2\,dx - \int_0^3 x^2\,dx.$$

It is clear, however, that the last expression has the same value as

$$\int_3^6 x^2\,dx = \frac{x^3}{3}\Big|_3^6 = \frac{6^3 - 3^3}{3} = 63.$$

In general, the area bounded by $f(x)$, the x-axis, and verticals at $x = a$ and $x = b$ is given by

$$\int_a^b f(x)\, dx.^4$$

As another example, let us find the area under

$$f(x) = 3 + x^{1/2}$$

over the x-interval from 1 to 4, shown in Figure 11–7. We have

$$\int_1^4 (3 + x^{1/2})\, dx = \left(3x + \frac{2}{3}x^{3/2}\right)\Big|_1^4$$

$$= \left[3(4) + \frac{2}{3}(4)^{3/2}\right] - \left[3(1) + \frac{2}{3}\right]$$

$$= \left[12 + \frac{2}{3}(8)\right] - \left[3 + \frac{2}{3}\right]$$

$$= \frac{41}{3}.$$

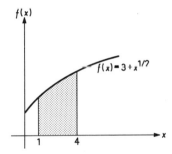

FIGURE 11–7
(not to scale)

Exercise.   Find the area under $f(x) = 5 + x^2$ over the interval $x = 1$ to $x = 4$. Answer: 36.

---

[4]If, for the interval at hand, part of the curve is above and part below the x-axis, the area above will be positive and that below will be negative. The integral will be the difference of the two areas, which could be meaningless. Such cases can be handled by evaluating the positive and negative parts separately and adding their absolute values. However, all applications we shall consider deal with intervals over which functions are positive, so we shall not be concerned with this procedure.

11.9 AREA—INTERPRETIVE APPLICATIONS

Example 1.   Suppose that at time *t* years, oil flows from a well at the *rate* of $F(t) = 40 - 2t$ thousand barrels per year. We wish to determine the total amount of oil the well will produce in 12 years.

First, note carefully that $F(t)$ is a rate at a *point* or instant (zero duration) in time and no oil flows *at* a point in time. However, if at time *t* we take a small interval of time, Δt years, and compute

(rate of flow per year)(Δt years) $= F(t)\Delta t$,

the result at time *t* will be the rectangular element of area shown in Figure 11–8, and this will be the *approximate* amount of oil produced in Δt years. If we were to proceed to divide the whole interval 0 to 12 into such rectangular elements, the sum of the resultant areas would *approximate* total oil production during the time interval. As we have already learned, the exact amount is the limit of this sum as the number of intervals increases ($\Delta t \rightarrow 0$), given by the definite integral

$$\int_0^{12} F(t)dt = \int_0^{12} (40 - 2t)dt$$

$$= 40t - t^2 \Big|_0^{12} = 336 \text{ thousand barrels.}$$

Thus we see that the area under the curve in Figure 11–8 can be interpreted as barrels of oil in the application at hand.

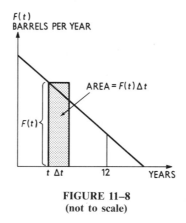

FIGURE 11–8
(not to scale)

Example 2.   Consider the function

$$p(q) = \frac{10}{q + 1} + 5$$

shown in Figure 11–9. We shall interpret this as a seller's price (or buyer's cost) function with a continuous discount. That is, each little bit more a

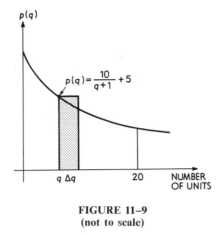

FIGURE 11–9
(not to scale)

customer orders has a slightly lower price per unit. Figure 11–9 shows a little bit more, Δq, and the price *per unit* of any such little bit is *approximately* $p(q)$. The cost to the buyer of the little bit is the area element

$$\text{(price per unit)}(\Delta q \text{ units}) - p(q)\Delta q.$$

Earlier little bits have a higher price per unit, and later bits a lower price per unit. If we sum the areas of the rectangles (the costs of the bits) over the interval 0 to 20, we have the approximate cost of the 20 units. Of course, the bits may be taken smaller and smaller and the exact cost (the limit as $\Delta q \rightarrow 0$) is the area

$$\int_0^{20} \left(\frac{10}{q+1} + 5 \right) dq = [10 \ln (q+1) + 5q]_0^{20}$$

$$= [10 \ln (21) + 5(20)] - [10 \ln 1 + 0]$$

$$= 10(3.0445) + 100 - 0$$

$$= \$130.445.$$

In this example, the area under the curve has been interpreted as the total dollar cost for an order of 20 units.

‖‖

Exercise. If the pricing function of the last example was

$$P(q) = 2 + e^{-0.02q},$$

what would a buyer pay for 100 units? Answer:

$$[2q - (e^{-0.02q})/(0.02)]_0^{100} = \$243.24.$$

‖‖

11.10 CONSUMERS' SURPLUS

If the function of the last example

$$p(q) = \frac{10}{q + 1} + 5$$

is considered as a demand function in the usual economic sense of a competitive market, the forces of supply and demand would lead to an *equilibrium* price which applies to all units demanded. Thus, if demand at equilibrium is $q_m = 49$ units, then all units would sell at the market price

$$p_m = \frac{10}{49 + 1} + 5 = \$5.20 \text{ per unit}$$

and total revenue would be

$$p_m q_m = 49(5.20) = \$254.80.$$

This amount is the area of the rectangle shown in Figure 11–10 and is the actual amount consumers pay for the 49 units demanded. However, the

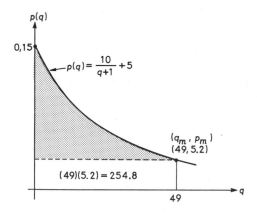

FIGURE 11–10
(not to scale)

demand curve implies that a demand would exist at prices *higher* than the equilibrium price of \$5.20 and that therefore the item is worth more than \$5.20 to some consumers. Consequently some consumers benefit from the competitive situation which results in the common market price of \$5.20 per unit, and the total of such benefits is called the consumers' surplus. As a method of measuring this surplus, we return to the idea of the last section and ask how much consumers would have paid if the first little bit was sold for the (high) price it is worth to some consumer, and the next little bit was sold for the (somewhat) lower price it is worth to another consumer, and so on. Thinking this way, we compute the amount consumers *would* have paid as the entire area under the curve from 0 to 49. The surplus would then be the difference between the total area and the rectangular area. Thus,

$$\text{Consumers' Surplus} = \int_0^{49} \left[\frac{10}{q+1} + 5 \right] dq - (49)(5.2)$$

$$= [10 \ln (q+1) + 5q] \Big|_0^{49} - 254.8$$

$$= 10 \ln 50 + 5(49) - 10 \ln 1 - 0 - 254.8$$

$$= 10(2.3026)(1.6990) + 245 - 254.8^5$$

$$= \$29.32.$$

In general terms, if we have a demand function, $p(q)$, with market equilibrium at (q_m, p_m), we compute

$$\text{Consumers' Surplus} = \int_0^{q_m} p(q)dq - q_m p_m.$$

||||||||||||||||||||||||||||||||||||||||||||||||||||||||||||||||||||||||||||||||||||||||||||||||||||||||||||||||||||||||||||||||||||||

Exercise.   Find consumers' surplus if market equilibrium occurs at demand 100 units and the demand function is $p(q) = 2 + e^{-0.02q}$. Answer: At $q_m = 100, p_m = 2 + e^{-2} = 2.1353$. We found in the last exercise that $\int_0^{100} (2 + e^{-0.02q}) dq = 243.24$. Hence, consumers' surplus is $243.24 - (100)(2.1353) = \$29.71$.

||||||||||||||||||||||||||||||||||||||||||||||||||||||||||||||||||||||||||||||||||||||||||||||||||||||||||||||||||||||||||||||||||||||

11.11 PRODUCERS' SURPLUS

Figure 11–11 shows a supply function which relates the amount producers will supply to the unit market price. In the competitive situation where

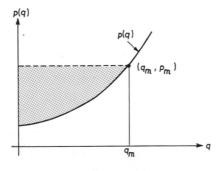

FIGURE 11–11

---

[5]Here, ln 50 has been computed as 2.3026 times the common logarithm of 50.

equilibrium exists at a point, (q_m, p_m), all units supplied command the same unit price, so the producers' total revenue is the rectangular area

$$q_m p_m.$$

However, the supply function implies that some suppliers would have offered product for sale at prices below the equilibrium and therefore benefit by the equilibrium price.

Exercise. What represents the total revenue to suppliers if product is offered for sale at all prices up to p_m? Answer: $\int_0^{q_m} p(q)\,dq$, which is the area under the supply curve over the interval from 0 to q_m.

The shaded area in Figure 11–11 represents the difference between what is obtained at equilibrium and the lesser amount which would have been received under the conditions of the last exercise. This is

$$\text{Producers' Surplus} = p_m q_m - \int_0^{q_m} p(q)\,dq.$$

For example, if the supply function is

$$p(q) = 100 + 0.1q$$

and equilibrium exists at $q_m = 1000$ units so that

$$p_m = 100 + 0.1(1000) = \$200, \text{ then}$$

$$\text{Producers' Surplus} = (1000)(200) - \int_0^{1000} (100 + 0.1q)\,dq$$

$$= 200{,}000 - [(100q + 0.05q^2)]\Big|_0^{1000}$$

$$= 200{,}000 - 150{,}000 = \$50{,}000.$$

Exercise. Find producers' surplus if equilibrium exists at 100 units supplied and the supply function is $p(q) = q^{1/2}$. Answer: $\$1000/3$.

11.12 NUMERICAL INTEGRATION—TRAPEZOIDAL RULE

Although extensive lists of integrals may be found in published tables, many functions of interest cannot be integrated exactly. For example, we know that

$$\int e^x \, dx = e^x + C,$$

but there are no comparable formulas for

$$\int e^{x^2} \, dx, \quad \int e^{x^3} \, dx, \quad e^{x^{1/2}} \, dx$$

or for e to any positive power of x except e^x itself. In such cases, approximate formulas for the indefinite integrals can be derived by advanced methods, and some of these results are given in tables of integrals. In the case of definite integrals, approximations may be made quite easily for even very complicated functions by various methods of *numerical integration*. We used one such method in the foregoing (Figure 11–5) when we approximated the area under a curve (the definite integral) by the sum of the areas of small rectangles. In this case, the upper horizontal side of each rectangle replaced the corresponding curved segment. If we instead take the same set of segments, but use a slant top to make trapezoids as in Figure 11–12, the

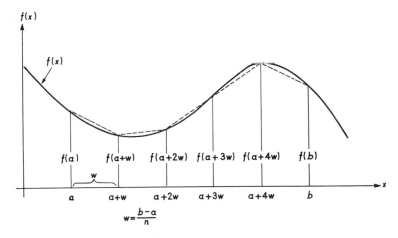

FIGURE 11–12

sum of the trapezoidal areas will generally give a closer approximation because the slant top approximates the curve segment better than a horizontal top. An even closer approximation, known as *Simpson's Rule*, uses parabolic top segments which will approximate the curve segment even more closely, and there are methods for securing even greater accuracy for a given number of subdivisions. We shall develop the *trapezoidal rule* here.

Consider Figure 11–12. We wish to compute

$$\int_a^b f(x)\,dx.$$

We interpret this definite integral as an area, and approximate the area by a series of n trapezoidal areas. To this end, the interval from a to b is divided into n equal segments of width w, so that

$$w = \frac{b-a}{n}.$$

The points on the x-axis marking out the segments are at

$$a, a+w, a+2w, a+3w, \text{ and so on to } b.$$

The vertical function values at these points are

$$f(a), f(a+w), f(a+2w), f(a+3w), \ldots, f(b).$$

The area of each of the trapezoids shown in Figure 11–12 is computed as one-half its base times the sum of the parallel sides, and the sum of the areas of all the trapezoids approximates the area under the curve which is the desired definite integral. Thus, using the tilde, $\sim$, to mean *approximately*,

$$\int_a^b f(x)\,dx \cong \frac{w}{2}[f(a) + f(a+w)]$$

$$+ \frac{w}{2}[f(a+w) + f(a+2w)]$$

$$+ \frac{w}{2}[f(a+2w) + f(a+3w)]$$

$$+ \frac{w}{2}[f(a+3w) + f(a+4w)]$$

$$+ \frac{w}{2}[f(a+4w) + f(b)].$$

Note that all the function values except the first and last, $f(a)$, and $f(b)$, occur twice in the sum, and that w is a common factor of all terms. Hence,

$$\int f(x)\,dx \cong w\left[\frac{f(a)}{2} + f(a+w) + f(a+2w) + f(a+3w)\right.$$

$$\left. + f(a+4w) + \frac{f(b)}{2}\right].$$

The procedure can be applied to any number, n, segments. Thus,

$$\int f(x)\,dx \cong \left\{\frac{f(a)}{2} + f(a+w) + f(a+2w)\right.$$

$$\left. + f(a+3w) + \cdots + f[a+(n-1)w] + \frac{f(b)}{2}\right\}w.$$

The last may be expressed more compactly in summation symbols as

$$\int f(x)\, dx \cong \left\{ \frac{f(a)}{2} + \sum_{i=1}^{n-1} [f(a + iw)] + \frac{f(b)}{2} \right\} w, \text{ where } w = \frac{b - a}{n}.$$

To apply the rule, we first find w, then establish the points a, $a + w$, $a + 2w$, and so on to b. We next find the function value at each point, compute half the first and last values and add these to the sum of all the other function values, then multiply the overall sum by w.

As an example, let us approximate

$$\int_1^3 x^2\, dx$$

by the trapezoidal rule with $n = 5$ subdivisions. We have

$$w = \frac{b - a}{n} = \frac{3 - 1}{5} = 0.4.$$

The computations are organized in the table.

Point	Function Value	Numbers to be Summed
a $= 1$	1	0.50
$a + w$ $= 1.4$	1.96	1.96
$a + 2w$ $= 1.8$	3.24	3.24
$a + 3w$ $= 2.2$	4.84	4.84
$a + 4w$ $= 2.6$	6.76	6.76
b $= 3$	9	4.50
		21.80 (0.4) $= 8.72$

We have

$$\int_1^3 x^2\, dx \cong 8.72.$$

The correct value for this integral is

$$\frac{x^3}{3} \Big|_1^3 = \frac{26}{3} = 8.67,$$

to two decimal places. The approximation, 8.72, is not especially good in this computation. Better approximations are obtained by using larger values of n, and the limit of the approximations as $n \to \infty$ is the exact value of the integral. A modern high-speed computer can provide highly accurate approximations by the trapezoidal method in a few seconds for even very complicated functions.

Next, let us return to

$$\int_0^1 e^{x^2}\, dx$$

for which no simple integral rule exists. We shall approximate this integral by the area of $n = 5$ trapezoids. We have,

$$w = \frac{1 - 0}{5} = 0.2.$$

The work is organized in the table.

	Value of		
x	x^2	e^{x^2}	Numbers to be Summed
0.0	0.00	1.0000	0.5000
0.2	0.04	1.0408	1.0408
0.4	0.16	1.1735	1.1735
0.6	0.36	1.4333	1.4333
0.8	0.64	1.8965	1.8965
1.0	1.00	2.7183	1.3592
			7.4033 (0.2) = 1.48

We find that

$$\int_0^1 e^{x^2} \, dx \cong 1.48.$$

Exercise. *a*) Approximate $\int_1^5 \frac{1}{x + 1} \, dx$ with $n = 4$ trapezoids.

b) Find the correct value for this definite integral. Answer: *a*) $(0.2500 + 0.3333 + 0.2500 + 0.2000 + 0.0833)(1) = 1.1166$. *b*) $\ln 6 - \ln 2 = \ln 3 = 1.0986$.

11.13 PROBLEM SET 2

(If necessary, refer to Table XII at the end of the book for integration rules.)
Evaluate the following:

1. $\int_1^3 2x \, dx.$

2. $\int_5^{10} dx.$

3. $\int_1^2 (x^{-1} + 1) \, dx.$

4. $\int_0^2 (0.8)^x \, dx.$

5. $\int_1^{10} x^{-2} \, dx.$

6. $\int_1^e \ln x \, dx.$

7. $\int_0^2 e^{0.5x} \, dx.$

8. $\int_1^2 (2x - 1) \, dx.$

Find the area under the curve over the stated interval:

9. $f(x) = x^{1/2}$, $x = 1$ to $x = 9$.

10. $f(x) = \ln x$, $x = 1$ to $x = 2$.

11. $f(x) = 2^{-x}$, $x = -1$ to $x = 4$.

12. $f(x) = 5$, $x = 0$ to $x = 1$.

13. $f(x) = \dfrac{1}{x+1}$, $x = 0$ to $x = 1$.

14. *a)* Sketch the graph of $f(x) = 25 - x^2$.
    b) Find the first quadrant area bounded by the curve and the axes.

15. *a)* Sketch the graph of $f(x) = x - 3$.
    b) Find the first quadrant area bounded by the line, the *x*-axis, and a vertical at $x = 6$.

16. A product sells at the rate of

$$R'(t) = 10e^{-0.2t}$$

thousand gallons per year at time *t* years, as illustrated in Figure A.

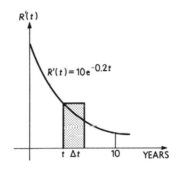

FIGURE A

a) Write the expression for the area of the rectangular element in Figure A.

b) Interpret the area in *a)* in relation to the function at hand.

c) How can the area under the curve over the interval 0 to 10 be approximated using rectangular elements?

d) How can the approximation in *c)* be improved?

e) Using limit terminology, state how the exact area is obtained from the rectangular areas.

f) Compute the area over the interval 0 to 10.

g) Interpret the result in *f)* in relation to the function at hand.

17. Maintenance costs are at the rate of

$$M'(t) = 2 + 0.1t$$

thousand dollars per year at time *t* years. Graph $M'(t)$ over the interval 0 to 10 years, put in an area element, then answer question *a)*–*g)* of Problem 16, using $M'(t)$ in place of $R'(t)$.

18. Given the demand function

$$p(q) = 5 + 20e^{-0.01q}.$$

Competitive market conditions have established demand at 100 units. Find consumers' surplus.

19. Compute consumers' surplus if market equilibrium exists at demand level 999 and the demand function is

$$p(q) = 50 - 5 \ln (q + 1).$$

20. Demand curves typically are concave upward and have negative slope at all points. Sketch a typical demand curve. Select point (q_m, p_m) for equilibrium and show, by shading, the area representing consumers' surplus.

21. Market equilibrium occurs at 300 units supplied. Find producers' surplus given the supply function

$$p(q) = 5 + e^{0.01q}.$$

22. Market equilibrium exists at 50 units supplied. Find producers' surplus if the supply function is

$$p(q) = 4 \ln (2q + 1).$$

23. Supply curves typically have positive slope and are concave upward. Show that the curve of Problem 22 is not typical with respect to concavity.

24. Given the demand function $p_d(q)$ and the supply function $p_s(q)$ for a product, as

$$p_d(q) = 100 - 0.2q$$
$$p_s(q) = 10 + 0.1q,$$

a) Graph the functions.
b) Find the equilibrium demand and price, (q_m, p_m).
c) Find consumers' surplus.
d) Find producers' surplus.

Evaluate approximately by the trapezoidal rule.

25. $\int_0^3 e^x dx$. Use $n = 6$.

26. $\int_0^1 e^{-x^2} dx$. Use $n = 5$.

27. $\int_1^3 \ln x\ dx$. Use $n = 5$.

28. $\int_0^5 2^x dx$. Use $n = 5$.

11.14 AREA BETWEEN CURVES

In Figures 11–10 and 11–11, the area of interest turned out to be one whose boundaries were not simply the curve, the x-axis, and verticals. We refer loosely to areas such as those in these figures as area *between curves* to distinguish them from what we have defined as the area under a curve. Area between curves can be evaluated from area under curves by appropriate addition or subtraction if we know the intersection points. As an example, suppose that we are asked to find the area bounded by the y-axis and the functions

$$f(x) = 8 - x \quad \text{and} \quad g(x) = 5 + \frac{x}{2}.$$

Figure 11–13 shows this area as A_1. We need the coordinates of the intersection point, P, in order to compute this area. At P, $f(x) = g(x)$, so that

$$8 - x = 5 + \frac{x}{2},$$

from which we find $x = 2$, and $f(x) = g(x) = 6$. We now find A_1 as the area

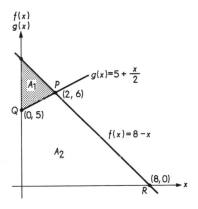

FIGURE 11–13

under $f(x)$ minus the area under $g(x)$. Thus,

$$A_1 = \int_0^2 f(x)\,dx - \int_0^2 g(x)\,dx = \int_0^2 [f(x) - g(x)]\,dx$$

$$= \int_0^2 \left(3 - \frac{3x}{2}\right) dx$$

$$= \left[3x - \frac{3x^2}{4}\right] \Big|_0^2$$

$$= 6 - 3 = 3.$$

To find the area shown as A_2 in Figure 11–13, we need the points Q and R. Q is the vertical intercept of $g(x) = 5 + x/2$, which is at $(0, 5)$, and R is the x-intercept of $f(x) = 8 - x$ which is at $(8, 0)$.

Exercise. *a*) Write the integral expression for the area A_2 in terms of the functional symbols $f(x)$ and $g(x)$. *b*) Evaluate A_2. Answer: *a*) $A_2 = \int_0^2 g(x)dx + \int_2^8 f(x)dx$. *b*) 29.

As another example, let us find the first quadrant area bounded by the y-axis and the curves

$$f(x) = x^2 \quad \text{and} \quad g(x) = \frac{x^2}{2} + 2.$$

See Figure 11–14. The intersection occurs when $f(x) = g(x)$; that is, when

$$x^2 = \frac{x^2}{2} + 2$$

$$x^2 = 4$$

which yields $x = 2$ for the first quadrant intersection point in the figure. The shaded area is

$$\int_0^2 g(x)dx - \int_0^2 f(x)dx = \int_0^2 [g(x) - f(x)]dx = \int_0^2 \left(2 - \frac{x^2}{2}\right)dx$$

$$= \left[2x - \frac{x^3}{6}\right]\Big|_0^2 = 4 - \frac{8}{6} = \frac{8}{3}.$$

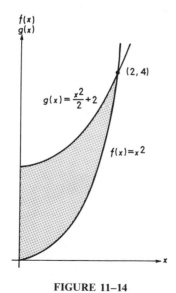

FIGURE 11–14

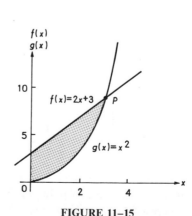

FIGURE 11–15

Exercise. See Figure 11–15. Find the area bounded by the curves and the y-axis, given $f(x) = 2x + 3$ and $g(x) = x^2$. Answer: Intersection occurs where $x^2 = 2x + 3$ or $(x - 3)(x + 1) = 0$, so P has coordinates (3, 9). The area is 9.

As a final example, we find the area bounded by

$$g(x) = 25 - x^2 \quad \text{and} \quad f(x) = 19 - x.$$

At the intersections

$$19 - x = 25 - x^2$$
$$x^2 - x - 6 = 0$$
$$(x - 3)(x + 2) = 0$$

so

$$x = 3 \quad \text{and} \quad x = -2.$$

The intersections, shown in Figure 11–16, are (3, 16) and (−2, 21). The shaded area is

$$\int_{-2}^{3} [g(x) - f(x)] dx = \int_{-2}^{3} [25 - x^2 - (19 - x)] dx$$

$$= \int_{-2}^{3} (6 + x - x^2) dx$$

$$= \left(6x + \frac{x^2}{2} - \frac{x^3}{3} \right) \Big|_{-2}^{3}$$

$$= \left[18 + \frac{9}{2} - 9 \right] - \left[-12 + 2 + \frac{8}{3} \right]$$

$$= 9 + \frac{9}{2} + 10 - \frac{8}{3}$$

$$= \frac{125}{6}.$$

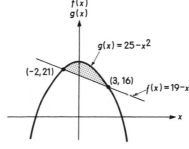

FIGURE 11–16
(not to scale)

11.15 PROFIT OVER TIME

Here we consider an undertaking for which the expenditure rate per day at time t days is

$$E'(t) = \$100,$$

where the prime calls attention to the fact that $E'(t)$ is a rate. See Figure

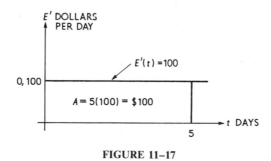

FIGURE 11–17

11–17. Total expenditure after five days is the area of the rectangle, and in integral notation we have

$$\text{Total expenditure} = \int_0^5 100 \, dt = 100t \Big|_0^5 = 500.$$

More generally, if $E'(t)$ is the expenditure rate per day, then total expenditure after t days is

$$\int_0^t E'(t) \, dt.$$

Note the *units* involved in

$$E'(t) \, dt.$$

$E'(t)$ is in *dollars per day*, and dt is in *days*, so that the product is

$$\left(\frac{\text{dollar}}{\text{day}}\right)(\text{days}) = \text{dollars}.$$

The integral over the interval from 0 to t days is the total dollars expended in t days.

Before proceeding, we should remark that if the undertaking incurred a *fixed* expense, then total expenditure would be the amount we compute by integration plus fixed expense. It will not be necessary for our purposes in this section to consider fixed expense, so we shall simply assume that it is zero and refer to the integral as total expenditure.

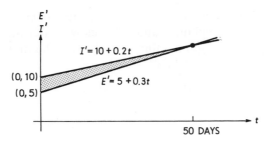

FIGURE 11–18

Now consider Figure 11–18 which shows the rate of expenditure per day as

$$E'(t) = 5 + 0.3t.$$

The second line,

$$I'(t) = 10 + 0.2t,$$

represents the rate of income per day after t days. The two lines intersect when $E'(t) = I'(t)$; that is, when

$$5 + 0.3t = 10 + 0.2t$$
$$0.1t = 5$$
$$t = 50 \text{ days.}$$

From a profit-maximizing point of view, the undertaking should be terminated at $t = 50$ days because beyond this point expenditure per day exceeds income per day. Hence, $t = 50$ days yields maximum profit, and this will be the shaded area between the two lines. We have

$$\text{Profit} = \text{Total Income} - \text{Total Expenditure}$$

$$= \int_0^{50} (10 + 0.2t)\,dt - \int_0^{50} (5 + 0.3t)\,dt$$

$$= \int_0^{50} (5 - 0.1t)\,dt$$

$$= 5t - 0.05t^2 \Big|_0^{50}$$

$$= 250 - 125 = \$125.$$

As another example, suppose that expenditure and income rates in thousands of dollars per day are given by, respectively,

$$E'(t) = \frac{120}{t + 10} + 8 \quad \text{and} \quad I'(t) = \frac{200}{t + 5} + 4.$$

Here, both rates decline as time goes on, but the income rate exceeds the expenditure rate at the beginning and falls below it after the time when the two rates are equal. The optimal time to terminate the undertaking occurs when $E'(t) = I'(t)$; that is, when

$$\frac{120}{t + 10} + 8 = \frac{200}{t + 5} + 4$$

or, dividing both sides by 4 and combining constants,

$$\frac{30}{t + 10} + 1 = \frac{50}{t + 5}.$$

Multiplying both sides by $(t + 10)(t + 5)$, we have

$$30(t + 5) + (t + 5)(t + 10) = 50(t + 10)$$

which, after expanding and combining terms yields,

$$t^2 - 5t - 300 = 0$$

or, factoring,

$$(t - 20)(t + 15) = 0.$$

We select the positive value, $t = 20$, for this real-world problem and disregard the value $t = -15$ which also satisfies the last equation. The profit maximum will be

$$\text{Profit} = \int_0^{20} \left[\frac{200}{t + 5} + 4\right] dt - \int_0^{20} \left[\frac{120}{t + 10} + 8\right] dt$$

$$= [200 \ln (t + 5) - 120 \ln (t + 10) - 4t] \Big|_0^{20}$$

$$= (200 \ln 25 - 120 \ln 30 - 80) - (200 \ln 5 - 120 \ln 10)$$

$$= 200(\ln 25 - \ln 5) - 120(\ln 30 - \ln 10) - 80$$

$$= 200 \ln 5 - 120 \ln 3 - 80$$

$$= 200(1.6094) - 120(1.0986) - 80$$

$$= 321.88 - 131.832 - 80$$

$$= \$110.05 \text{ thousand.}$$

Exercise. Following the foregoing example, find *a*) the optimum time to terminate the undertaking and *b*) the profit at this time if expenditures are at the rate of $4 + 15/(t + 1)$ per day and income is at the rate of $66/(t + 3)$ per day. Answer: *a*) $t = 9$. *b*) Profit $= 66 \ln 4 - 15 \ln 10 - 36 = \$20.9568$.

11.16 AREAS WITH INFINITE BOUNDARIES

In a number of applications it is of interest to inquire whether

$$\lim_{b \to \infty} \int_a^b f(x)\, dx$$

exists. For example, if we have

$$\lim_{b \to \infty} \int_1^b \frac{1}{x}\, dx = \lim_{b \to \infty} (\ln x)\Big|_1^b,$$

the lower evaluation is $\ln 1 = 0$, and the upper evaluation is

$$\lim_{b \to \infty} (\ln b).$$

Inasmuch as the $\ln b$ increases without limit as b increases, the definite integral does not exist in this case. However, if we consider

$$\lim_{b \to \infty} \int_1^b \frac{1}{x^2}\, dx = \lim_{b \to \infty} \left(-\frac{1}{x}\right)\Big|_1^b$$

$$= \lim_{b \to \infty} \left(-\frac{1}{b}\right) - (-1)$$

$$= 1,$$

the limit exists, and is 1, because $-1/b$ has zero as a limit as $b \longrightarrow \infty$.

Conventionally, integrals of the type under discussion are written with the ∞ symbol as a limit. In the last case, for example, we would write

$$\int_1^\infty \frac{1}{x^2}\, dx = 1,$$

where it is understood that this means to evaluate the expression for the integral as the upper limit increases without bound. As another example, using Table XII, we find

$$\int_1^\infty \frac{6\, dx}{x(3 + 2x)} = 6\left[\frac{1}{3}\ln\left(\frac{x}{3 + 2x}\right)\right]_1^\infty = 2\left[\ln\left(\frac{x}{3 + 2x}\right)\right]_1^\infty$$

$$= 2\left[\lim_{x \to \infty} \ln\left(\frac{x}{3 + 2x}\right) - \ln\left(\frac{1}{5}\right)\right]$$

$$= 2\left[\ln\left(\frac{1}{2}\right) - \ln\left(\frac{1}{5}\right)\right] \qquad {}^6$$

$$= 2[-0.6931 - (-1.6094)] = 2(0.9163)$$

$$= 1.8326.$$

---

[6] Note that as x increases and becomes very large, $x/(3 + 2x)$ approaches $\frac{1}{2}$ as a limit.

||||||||||||||||||||||||||||||||||||||||||||||||||||||||||||||||||||||||||||||||||||||||||||||||||||||||||||||||||||||||||||||||||||||||||||||||

Exercise.   Evaluate $\int_0^\infty 2^{-x} \, dx$. Answer: $1/\ln 2$.

||||||||||||||||||||||||||||||||||||||||||||||||||||||||||||||||||||||||||||||||||||||||||||||||||||||||||||||||||||||||||||||||||||||||||||||||

If we interpret

$$\int_0^\infty e^{-x} \, dx = 1$$

as an area, we may describe this area loosely as having an infinite boundary. See Figure 11–19. We understand, of course, that this description means only that the area expression has a limit as x increases without bound.

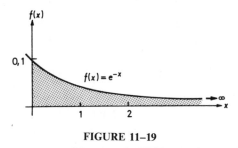

FIGURE 11–19

11.17 EXPECTED VALUE

If we wish to compute the weighted average of a group of numbers, we multiply each number by its weight, sum the products, and divide by the sum of the weights. For example, if the numbers 5 and 10 are weighted by 2 and 3, respectively, we have

Number x	Weight w	Number times Weight xw
5	2	10
10	3	30
	5	40

and the weighted average is $40/5 = 8$. The same weighted average will be obtained by the use of weights in proportion to 2 and 3. In particular, if we use the sum of the weights, 5, for proportioning, we have

Weight	Proportionate Weight
2	$2/5 = 0.4$
3	$3/5 = 0.6$
5	1.0

||||||||||||||||||||||||||||||||||||||||||||||||||||||||||||||||||||||||||||||||||||||||||||||||||||||||||||||||||||||||||||||||||||||||||||||||||||||||||||||||

Exercise. Compute the weighted average of 5 and 10 using 0.4 and 0.6, respectively, as weights. Answer: $8/1 = 8$, as before.

||||||||||||||||||||||||||||||||||||||||||||||||||||||||||||||||||||||||||||||||||||||||||||||||||||||||||||||||||||||||||||||||||||||||||||||||||||||||||||||||

Proportionate weights which sum to 1 often can be interpreted as probabilities. For example, there are 13 spades in a deck of 52 cards so that the proportion of spades is $^{13}/_{52} = 0.25$. The probability of obtaining a spade in a random drawing from a deck is $\frac{1}{4}$ or 0.25. Similarly, the probability is 0.25 for hearts, diamonds, and clubs, and the sum of the four probabilities is 1. Thus, like proportions of a total, probabilities have 1 as a sum if the probabilities of all the different events which might occur are added.

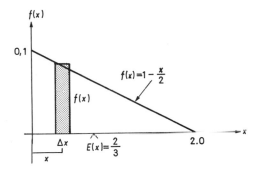

FIGURE 11–20

Refer now to Figure 11–20. The function

$$f(x) = 1 - \frac{x}{2}$$

has been designed so that the first quadrant area equals 1. That is,

$$\int_0^2 f(x)\, dx = \int_0^2 \left(1 - \frac{x}{2}\right) dx = \left(x - \frac{x^2}{4}\right)_0^2 = 2 - 1 = 1.$$

If we divide the interval into rectangles of area $f(x)\Delta x$ and use these areas as *weights* for the adjoining values of x, each weighted x value is

$$x f(x)\Delta x$$

and the weighted average of the x's is approximately the sum of the $x f(x)\Delta x$ values for the x's in the intervals selected, divided by the sum of the weights (the area elements), which is approximately 1. The exact weighted average, called $E(x)$, the *expected value* of x is

$$E(x) = \frac{\displaystyle\int_0^2 x f(x)\, dx}{1} = \int_0^2 x f(x)\, dx.$$

Exercise. Why is the sum of the weights in the denominator of the last expression equal to 1? Answer: The exact sum of the weights is $\int f(x)\,dx$, the total area, which equals 1 as shown in the foregoing.

For the function

$$f(x) = \left(1 - \frac{x}{2}\right), \quad 0 \le x \le 2$$

we find

$$E(x) = \int_0^2 x\left(1 - \frac{x}{2}\right) dx = \frac{x^2}{2} - \frac{x^3}{6}\Big|_0^2 = \frac{2}{3}.$$

Physically, the expected value of x is the x-coordinate of the center of gravity of the area under $f(x)$. That is, the triangular figure would balance on a fulcrum placed at $x = \frac{2}{3}$.

Functions which are positive and contain a total area of 1 over the entire interval of interest are called *density functions* in probability discussions. Thus, given that $f(x)$ is a density function over the x-interval from a to b, we define the expected value of x as

$$E(x) = \int_a^b xf(x)\,dx,$$

if this integral exists.

Exercise. *a)* Verify that $f(x) = 1/x^2$ qualifies as a density function over the interval 1 to ∞. *b)* Show that the expected value of x does not exist. Answer: *a)* $\int_1^\infty (1/x^2)\,dx = -1/x\Big|_1^\infty = 1$. *b)* $\int_1^\infty x(1/x^2)\,dx = \ln x\Big|_1^\infty$ and the limit specified in the upper evaluation does not exist.

Expected values, or averages, find many applications. For example, if

$$f(x) = 1 - \frac{x}{2}$$

in the foregoing example is the density function for the amount, x, spent by customers, then

$$E(x) = \frac{2}{3}$$

is the average amount spent per customer. This average serves as a basis for estimating the total amount a number of customers would spend. Thus, 600 customers would be expected to spend

(average amount per customer)(number of customers) $= \frac{2}{3}(600) = \$400.$

As another example, suppose that the density function for the amount spent by customers is

$$f(x) = \frac{3(x-5)^2}{125}$$

for amounts running from $0 to $5. How much would we expect 1000 customers to spend? The expected amount per customer is

$$E(x) = \int_0^5 xf(x)\, dx = \int_0^5 \frac{3x(x-5)^2}{125}\, dx$$

$$= \frac{3}{125} \int_0^5 (x^3 - 10x^2 + 25x)$$

$$= \frac{3}{125} \left(\frac{x^4}{4} - \frac{10x^3}{3} + \frac{25x^2}{2} \right) \Big|_0^5$$

$$= \frac{3}{125} \left(\frac{625}{4} - \frac{1250}{3} + \frac{625}{2} \right) - 0$$

$$= \frac{3}{125} \left(\frac{625}{12} \right)$$

$$= \frac{5}{4}.$$

We have $E(x) = \$1.25$ per customer, so expenditure by 1000 customers would be estimated as

$$1000E(x) = 1000(1.25) = \$1250.$$

‖‖‖

Exercise.   Given that the density function for amount spent by customers is $f(x) = 0.2$ over the x-interval from $0 to $5, find $E(x)$. Answer: $2.50.

‖‖‖

In the exercise, the density function is the horizontal line, $f(x) = 0.2$, so that all values of x receive a uniform weight and the expected value of x is the simple average of the endpoints of the interval, $(5 + 0)/2 = 2.5$.
As a final example, consider $f(x) = e^{-x}$ over the x-interval from 0 to ∞. This will serve as a density function because we have already verified that

$$\int_0^\infty e^{-x}\, dx = 1.$$

We seek the expected value of x,

$$E(x) = \int_0^\infty x e^{-x}\, dx.$$

From Table XII we have

$$\int_0^\infty x e^{-x}\, dx = -e^{-x}(x+1)\Big|_0^\infty$$

$$= -\frac{(x+1)}{e^x}\Big|_0^\infty$$

$$= -\lim_{x \to \infty}\left(\frac{x+1}{e^x}\right) - \left(-\frac{1}{1}\right)$$

$$= -0 + 1$$

$$= 1.$$

In the last, it may not be apparent immediately that

$$\lim_{x \to \infty} \frac{x+1}{e^x}$$

is zero, but if we think of substituting increasingly large values of x into the expression, it becomes clear that the denominator grows at an explosive rate compared with the numerator and the ratio rapidly approaches 0.

Forms of the negative exponential, e^{-x}, are used in a number of probability applications. Here, we have found that the expected value is 1 if the exponent is $-x$. More generally, if the density function is

$$f(x) = k e^{-kx},$$

the expected value of x is

$$E(x) = \frac{1}{k}.$$

Exercise. *a*) Show that $f(x) = 2x^{-3}$ can be used as a density function over the x-interval from 1 to ∞. *b*) Find $E(x)$. Answer:
a) $\int_1^\infty 2x^{-3}\, dx = 2(-x^{-2}/2)\big|_1^\infty = 1.$ *b*) $\int_1^\infty x(2x^{-3})\, dx = E(x) = 2.$

11.18 PROBLEM SET 3

1. See Figure A which shows the areas A_1 and A_2. Write the integral expressions for
   a) A_1.   *b*) A_2.

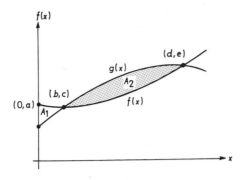

FIGURE A

2. *a*) Make a sketch of $f(x) = 3x$ and $g(x) = x^2$.
   b) Find the points of intersection.
   c) Find the area bounded by the functions.

3. *a*) Make a sketch showing $f(x) = 12/x$ and $g(x) = 13 - x$.
   b) Find the points of intersection.
   c) Find the area bounded by the functions.

4. Find the first quadrant area bounded by $f(x) = x^2$ and $g(x) = x^{1/2}$.

5. *a*) Make a sketch showing the area bounded by $f(x) = e^x$, $g(x) = e^{-x}$, and the vertical line $x = 2$.
   b) Find the area in (*a*).

6. *a*) Make a sketch showing $f(x) = 8 - x$ and $g(x) = 2x + 2$.
   b) Find the area bounded by the lines and the *y*-axis.

7. See Problem 6. Find the area bounded by the lines and the axes.

8. *a*) Make a sketch of $f(x) = x^2 + 2$ and $g(x) = x + 4$.
   b) Find the points of intersection.
   c) Find the area bounded by the functions.

9. Income per day from an undertaking is $20 - 0.3t$ (thousand) dollars after *t* days and expenditure per day at this time is $10 - 0.1t$ (thousand) dollars.
   a) What is the optimal time to terminate the undertaking?
   b) What will profit be at this time?

10. Income per day from an undertaking is $30 + 20e^{-0.01t}$ and expenditure is 34 per day at time *t* days.
    a) What is the optimal time to terminate the undertaking?
    b) What will profit be at this time?

11. Given that daily income and expenditure rates, in thousands of dollars, are $I'(t) = 20 - t^{1/2}$ and $E'(t) = 2 + 2t^{1/2}$, respectively,
    a) Make a sketch showing the area which represents maximum profit.
    b) Compute maximum profit.

Evaluate the following.

12. $\displaystyle\int_0^\infty e^{-2x}\, dx.$

13. $\displaystyle\int_0^\infty 2xe^{-x^2}\, dx.$

14. $\int_{-\infty}^{1} 2^x \, dx.$

15. $\int_{0}^{\infty} \frac{30}{(3x + 5)^2} \, dx.$

16. Given the density function $f(x) = 0.2 - 0.02x$ over the x-interval 0 to 10, find the expected value of x.

17. *a*) Show that

$$f(x) = \frac{30x - 3x^2}{500}$$

is appropriate as a density function over the x-interval 0 to 10.
b) Find the expected value of x.

18. Given that

$$f(x) = \frac{6(10 + 3x - x^2)}{275}$$

is a density function over the x-interval 0 to 5 for the amount, x, spent by customers, find how much 440 customers would be expected to spend.

19. Find $E(x)$ if $0 \leq x < \infty$, and

$$f(x) = \frac{1}{2}e^{-0.5x}.$$

11.19 DIFFERENTIAL EQUATIONS

If we have

$$y = x^2 + 10 \tag{1}$$

so that

$$\frac{dy}{dx} = 2x, \tag{2}$$

we define the differential of y, dy, to be $(dy/dx) \, dx$, so that from (2),

$$dy = 2x \, dx. \tag{3}$$

Equations (2) and (3) are examples of *differential equations;* that is, equations which contain derivatives or differentials. To solve the differential equation, we integrate to obtain the *general solution*

$$y = x^2 + C$$

if we have no information which makes it possible to assign a value to the constant C, or we find a *particular solution* if we are given *initial conditions* which permit us to assign a value to C.

Suppose that we are given the differential equation

$$dy = x^{1/2} \, dx$$

together with the initial conditions that $y = 20$ when $x = 9$. We integrate to obtain the general solution

$$y = \frac{2}{3}x^{3/2} + C,$$

then substitute the initial conditions to obtain

$$20 = \frac{2}{3}(9)^{3/2} + C$$

$$20 = 18 + C$$

$$C = 2.$$

Replacing C by 2 in the general solution, we have the particular solution,

$$y = \frac{2}{3}x^{3/2} + 2.$$

||||||||||||||||||||||||||||||||||||||||||||||||||||||||||||||||||||||||||||||||||||||||||||||||||||||||||||||||||||||||||||||||||||||||

Exercise.   *a*) Find the general solution of $dy = x\, dx$. *b*) Find the particular solution if $y = 15$ when $x = 4$. Answer: *a*) $y = (x^2/2) + C$. *b*) $y = (x^2/2) + 7$.

||||||||||||||||||||||||||||||||||||||||||||||||||||||||||||||||||||||||||||||||||||||||||||||||||||||||||||||||||||||||||||||||||||||||

11.20 SEPARABLE DIFFERENTIAL EQUATIONS

A differential equation which can be put in the form

$$g(y)\, dy = f(x)\, dx$$

where one side is a function of y and its differential and the other side is a function of x and its differential is called a *separable* differential equation. For example,

$$\frac{dy}{dx} = \frac{y}{x}$$

can be separated to yield

$$\frac{dy}{y} = \frac{dx}{x}.$$

The general solution is

$$\int \frac{dy}{y} = \int \frac{dx}{x}$$

$$\ln y = \ln x + C.$$

As another example,

$$x\frac{dy}{dx} - 1 = 0$$

becomes

$$x \, dy - dx = 0$$

$$dy = \frac{dx}{x}$$

$$\int dy = \int \frac{dx}{x}$$

$$y = \ln x + C.$$

Exercise.   *a*) Separate $dy/dx = xy$. *b*) Write the general solution. Answer: *a*) $dy/y = x \, dx$. *b*) $\ln y = (x^2/2) + C$.

11.21 FORMS OF THE CONSTANT

The definition and rules of logarithms are used often to obtain alternate forms of a solution. For example, the solution

$$\ln y = \ln x + C$$

can be written as

$$\ln y - \ln x = C$$

$$\ln \frac{y}{x} = C.$$

From the definition of logarithms,

$$\frac{y}{x} = e^C.$$

Now, e^C is itself a constant which we may call k. Hence, we have

$$\frac{y}{x} = k$$

or

$$y = kx$$

as an alternate, and simpler, statement of the solution.

Exercise.   Given the solution $\ln y = -\ln x + C$, write an alternate form in the manner of the last exercise. Answer: $xy = k$.

As a final example, if we have the solution

$$\ln y - 2 \ln x = C$$

we may write

$$\ln y - \ln x^2 = C$$

$$\ln \frac{y}{x^2} = C$$

$$\frac{y}{x^2} = e^C = k$$

$$y = kx^2.$$

11.22 APPLICATIONS OF DIFFERENTIAL EQUATIONS

If we deposit $100 in an account which bears interest at 6 percent per year, compounded *continuously* and let $S(t)$ represent the amount in the account at time t years, then the rate of change of the $S(t)$ *per year* at time t is 6 percent of $S(t)$. Thus,

$$\frac{d\,S(t)}{dt} = 0.06\,S(t).$$

Separating the variables, we have

$$\frac{d\,S(t)}{S(t)} = 0.06\,dt$$

and

$$\int \frac{d\,S(t)}{S(t)} = \int 0.06\,dt$$

so

$$\ln S(t) = 0.06t + C.$$

Inasmuch as $S(t) = 100$ at $t = 0$,

$$\ln (100) = C$$

and

$$\ln S(t) - 0.06t + \ln 100$$

$$\ln S(t) - \ln (100) = 0.06t$$

$$\ln \frac{S(t)}{100} = 0.06t.$$

From the definition of logarithms,

$$\frac{S(t)}{100} = e^{0.06t}$$

or

$$S(t) = 100e^{0.06t}.$$

‖‖‖

Exercise. If A dollars is deposited at rate i per year compounded continuously, what will be the amount in the account at time t years? Answer: $S(t) = Ae^{it}$.

‖‖‖

Observe that the last example leads to the same formula we deduced by rather cumbersome methods in Chapter 8. The efficiency of the differential equation approach is due to the fact that the given information is an instantaneous rate (a rate at a point in time), so the natural starting point in finding the expression for $S(t)$ is the known information, $d\,S(t)/dt$. To pursue this approach further, suppose that in addition to continuous compounding at rate i, money flows into an account continuously at the *rate* of \$100 per year at time t years, and that amount $A(t)$ is zero at $t = 0$.[7] It follows that the amount in the account, $A(t)$, is growing at the rate of $iA(t)$ per year at time t years, as in the previous example, and this rate is augmented by the inflow at the rate of \$100 per year. Hence,

$$\frac{d\,A(t)}{dt} = iA(t) + 100.$$

We have the differential equation

$$d\,A(t) = [iA(t) + 100]\,dt$$

which can be separated to yield

$$\frac{d\,A(t)}{[iA(t) + 100]} = dt$$

so that

$$\int \frac{d\,A(t)}{[iA(t) + 100]} = \int dt.$$

Using Table XII, we have

$$\frac{\ln\,[i\,A(t) + 100]}{i} = t + C.$$

---

[7]Note again that the limit concept underlies this illustration. No money flows into the account at a point (zero duration) in time, but *approximately* \$100Δt flows in during the time bit Δt years; $100(0.01) = \$1$ in $\Delta t = 0.01$ years; $100(0.001) = \$0.10$ in $\Delta t = 0.001$ years; $100(0.0001) = \$0.01$ in $\Delta t = 0.0001$ years, and so on.

It will be helpful to determine the constant of integration before solving for $A(t)$. Using the initial condition that $A(0) = 0$,

$$\frac{\ln (100)}{i} = C$$

so that

$$\frac{\ln [i\, A(t) + 100]}{i} = t + \frac{\ln 100}{i}$$

or, multiplying by i,

$$\ln [i\, A(t) + 100] = it + \ln 100$$

and

$$\ln [i\, A(t) + 100] - \ln 100 = it.$$

By a rule of logarithms,

$$\ln \frac{[i\, A(t) + 100]}{100} = it$$

and by the definition of the natural logarithm,

$$\frac{i\, A(t) + 100}{100} = e^{it}$$

so

$$i\, A(t) + 100 = 100 e^{it}$$

and

$$A(t) = \frac{100 e^{it} - 100}{i} = \frac{100(e^{it} - 1)}{i}.$$

If money flowed into an account at the rate of R dollars per year at time t years, rather than \$100, the last result would have been

$$A(t) = \frac{R(e^{it} - 1)}{i}.$$

The last is the formula for the amount of a *continuous flow compounded continuously.*

As another example, suppose that an oil company and a farmer upon whose land a well has been drilled agree that the farmer will receive a royalty of \$0.50 per barrel of oil produced. The company wishes to pay for a year's production in a lump sum at the end of the year. The farmer points out that a year's production does not occur in a lump sum at the year's end, but is spread out uniformly during the year, so that his royalty income is generated uniformly but will be held by the company until the end of the year. The farmer insists that he should receive interest on this continuous flow of income, and the company agrees to compute his royalties on this basis at 6 percent compounded continuously.

||||||||||||||||||||||||||||||||||||||||||||||||||||||||||||||||||||||||||||||||||||||||||||||||||||||||||||||||||||||||||||||||||||||||||||||||||||||||||||||||||||||||||||||||||||||||||||||

Exercise. Suppose the well produces 10,000 barrels in a year. What would the royalty amount be *a*) without and *b*) with the agreement? Answer: *a*) 10,000(0.50) = $5000. *b*) $5150, using Table IX. (More accurately, the figure is $5153.05).

||||||||||||||||||||||||||||||||||||||||||||||||||||||||||||||||||||||||||||||||||||||||||||||||||||||||||||||||||||||||||||||||||||||||||||||||||||||||||||||||||||||||||||||||||||||||||||||

We now consider a problem requiring more analysis. Suppose that a nation has $200(billion) of paper currency in circulation, and that an average of $1 (billion) passes through the banking system each banking day. Starting today, the old currency is to be replaced by new bills and, to that end, old bills appearing at banks are destroyed and replaced by new bills. We seek to find the expression for $A(t)$, the amount of new currency in circulation at time t. We have the initial conditions that $A(0) = 0$; that is, the amount of new currency at time $t = 0$ is zero. Further,

Amount of new currency at time $t = A(t)$
Amount of old currency at time $t = 200 - A(t)$.

We observe that even though $1 billion of currency passes through banks at time t days, on the average, only part of this will be old currency which will be replaced. Thus, if at a point in time only $1/4$ of the currency in circulation is old, then only $1/4$ of $1 billion is replaced by new currency. The *fraction* of old currency in circulation at time t will be

$$\frac{\text{Amount of old currency}}{\text{Total amount of currency}} = \frac{200 - A(t)}{200}.$$

At time t, then, the rate of increase in new currency, *per day*, will be

$$\frac{dA(t)}{dt} = \left[\frac{200 - A(t)}{200}\right] (\$1 \text{ billion}),$$

which is the differential equation we wish to solve. For brevity, let us omit the argument t for a while and write

$$dA = \left[\frac{200 - A}{200}\right] dt.$$

We may separate this to obtain

$$\frac{dA}{200 - A} = \frac{dt}{200}.$$

Integrating, we find

$$-\ln(200 - A) = 0.005t + C.$$

Because of the initial condition $A(0) = 0$,

$$-\ln 200 = C$$

and the particular solution is

$$-\ln (200 - A) = 0.005t - \ln 200.$$

The following steps lead to an explicit statement for $A(t)$:

$$-0.005t = \ln (200 - A) - \ln 200$$

$$-0.005t = \ln \left[\frac{200 - A}{200} \right]$$

$$e^{-0.005t} = \frac{200 - A}{200}$$

$$200e^{-0.005t} = 200 - A$$

$$A = 200 - 200e^{-0.005t}$$

$$A(t) = 200(1 - e^{-0.005t}).$$

We now have the desired equation and we can examine its consequences. For example, let us find when there will be $120 billion of new currency in circulation. We have

$$120 = 200(1 - e^{-0.005t})$$
$$0.6 = 1 - e^{-0.005t}$$
$$0.4 = e^{-0.005t}.$$

Taking natural logarithms and reversing the equation yields

$$\ln e^{-0.005t} = \ln 0.4$$

$$-0.005t \ln e = \ln 0.4$$

$$-0.005t = -0.9163$$

$$t = \frac{0.9163}{0.005} = 183.3 \text{ days.}$$

Exercise. Working from the foregoing model, find how much new currency will be in circulation after 50 days. Answer: $200(1 - e^{-0.25}) = 200(1 - 0.7788) = \$44.24$ billion.

11.23 PROBLEM SET 4

Find the general solution of the following differential equations:

1. $dy = dx$.
2. $y \, dy = x^2 \, dx$.
3. $y^2 \, dx = x \, dy$.
4. $xy \, dy = dx$.

5. $dy = x \, dx + e^{-x} \, dx.$

6. $xy + (1 - x^2)\dfrac{dy}{dx} = 0.$

7. $y\dfrac{dy}{dx} = x - 1.$

8. $\dfrac{1}{2}\dfrac{dy}{dx} + xy = 0.$

9. Find the particular solution of

$$dP = P \, dt$$

if $P = 10$ when $t = 0$.

10. Find the particular solution of

$$x \, dy - y \, dx - dx = 0$$

if $y = 0$ when $x = 1$.

11. If population at time t changes at an annual rate of 3 percent of population at time t, and population is 100 at $t = 0$, find population at time $t = 10$.

12. Total output, $P(h)$, is a function of man-hours worked. We shall assume that $P(h)$ is measured in thousands of dollars and h, man-hours, is in thousands also. At $h = 100$, $P(h)$ is 2500. However, output per man-hour (productivity) decreases as the number of man-hours worked increases. Output per man-hour at h man-hours worked is given by

$$25 + 5e^{-0.01h+1}.$$

Letting $dP(h)$ and dh be, respectively, the differential of total output and man-hours,

a) Write the differential equation relating $dP(h)$ and dh.

b) Find the particular solution of (a) using the initial conditions.

c) Find total output when h is 200.

13. Suppose that money is added continuously to a fund at the rate of $500 per year. The fund earns interest at the continuous rate of 6.2 percent per year. The amount in the fund at time t is $A(t)$, and $A(0) = 0$.

a) Write the differential equation.

b) Find the particular solution for the given conditions.

c) Find the amount in the fund at time 10 years.

11.24 REVIEW PROBLEMS

Integrate the following, using Table XII where necessary:

1. $\int (2x + 3)dx.$

2. $\int (3x^{1/2} - 6x^2 + 1)dx.$

3. $\int 3 \ln x \, dx.$

4. $\int 3(5)^x \, dx.$

5. $\int 2e^x dx.$

6. $\int \left(x + \dfrac{3}{x} \right) dx.$

7. $\int (4x + 2)^{3/2} dx.$

8. $\int 3^{2x} dx.$

9. $\int e^{2-0.1x} dx.$

10. $\int \dfrac{x}{3x + 2} \, dx.$

11. If marginal cost at q units output is $\$(10 + 0.04q)$ and fixed cost is $\$200$, find the total cost function, $C(q)$.

12. A firm is using fuel now at the rate of 1,000 tons per year and expects usage at time t years will be at the rate of

$$1000e^{0.05t}$$

tons per year. Find the function $U(t)$ for the total amount of fuel to be used during the next t years, taking $U(0) = 0$.

Evaluate the following:

13. $\displaystyle\int_1^4 2\,dx.$ 14. $\displaystyle\int_1^5 \left(x + \frac{2}{x}\right)dx.$

15. $\displaystyle\int_{0.2}^{0.4} 3e^{1.5x}\,dx.$ 16. $\displaystyle\int_1^5 (1 + \ln x)dx.$

Sketch the first quadrant section of the function, then find the first quadrant area bounded by the function and the axes:

17. $f(x) = 24 - 3x.$ 18. $f(x) = 9 - x^2.$

19. Sketch a graph of $f(x) = \ln x$ and find the first quadrant area bounded by the function, the x-axis, and a vertical at $x = 10$.

20. If at time t years from now, the rate of world petroleum consumption will be

$$C'(t) = 20e^{0.05t}$$

trillion barrels per year, as sketched in Figure A,

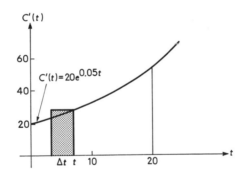

FIGURE A

a) Write the expression for the rectangular area element in Figure A.
b) Interpret *a)* in relation to petroleum consumption.
c) How can the area under the curve over the interval 0 to 10 be approximated using rectangular elements?
d) How can the approximation in *c)* be improved?
e) Using limit terminology, state how the exact area is obtained from the rectangular areas.
f) Compute the exact area over the interval 0 to 20 years.
g) Interpret the result in *f)* in terms of petroleum consumption.

21. Market equilibrium exists at demand $q = 100$ and the demand function is

$$p(q) = \frac{100}{2q + 1} + 50.$$

Find consumers' surplus.

22. Market equilibrium occurs at a supply level of $q = 100$. Find producers' surplus if the supply function is

$$p(q) = 5 + e^{0.01q}.$$

23. By means of derivatives, show that the demand function

$$p(q) = 500 - 5 \ln(q + 1)$$

has negative slope and is concave upward at every point.

24. Given the demand function $p_d(q)$ and the supply function $p_s(q)$ for a product as

$$p_d(q) = 220 - 0.1q$$
$$p_s(q) = 20 + 0.1q,$$

a) Find the equilibrium price and demand (q_m, p_m).
b) Find consumers' surplus.
c) Find producers' surplus.

Approximate the following by the trapezoidal rule.

25. $\displaystyle\int_1^5 x^3 dx.$ Use $n = 4$.

26. $\displaystyle\int_0^5 (0.5)^x dx.$ Use $n = 5$.

27. $\displaystyle\int_2^5 \ln(x^2 - 3) dx.$ Use $n = 6$.

28. Find the first quadrant area bounded by the y-axis and the curves

$$f(x) = 18 - x^2, g(x) = x^2.$$

29. Find the area bounded by the x-axis and the curves

$$f(x) = 18 - x^2, g(x) = x^2.$$

30. Find the area bounded by the line $x = 1$ and the curves

$$f(x) = x^{1/2}, g(x) = \frac{8}{x}.$$

31. Find the area bounded by the y-axis and the curves

$$f(x) = 2^{3-x}, \quad g(x) = 2^{x-1}.$$

32. Income from an undertaking is $35 thousand per week. Expenses per week, in thousands, at time t are

$$E'(t) = 15 + e^{0.1t}.$$

a) Find the optimal time to terminate the undertaking.
b) Find profit at this time.

33. Evaluate

$$\int_0^\infty \frac{dx}{(3x + 7)^2}.$$

34. Evaluate

$$\int_0^\infty xe^{-0.5x} \, dx.$$

35. Evaluate

$$\int_1^\infty \frac{25 \, dx}{x(5 + 5x)}.$$

36. Show that the following can be used as a density function over the *x*-interval from 0 to ∞.

$$f(x) = \frac{(2^{1-x}) \ln 2}{2}.$$

37. Find the expected value of *x* if the density function over the *x*-interval 1 to ∞ is

$$f(x) = \frac{3}{2x^{5/2}}.$$

38. Find the amount 1000 customers would be expected to spend if the density function for amounts spent by customers over the interval 0 to ∞ is

$$f(x) = \frac{e^{-0.1x}}{10}.$$

Find the general solution of the following differential equations.

39. $(x + 1) \, dy - dx = 0.$
40. $2xy \, dy = dx - y \, dy.$
41. $e^{-x} \, dy = y \, dx.$
42. $y^2 \, dy + x \, dx + 3 \, dy = 0.$
43. Find the particular solution of the following differential equation if $y = 1$ when $x = 0$.

$$e^{-x} \, dy = y \, dx.$$

44. Find the particular solution of the following differential equation if $y = 5$ when $x = 1$.

$$x \, dy = x(\ln x) \, dx - dx.$$

45. Responses, $R(t)$, to an advertising campaign are zero at time $t = 0$. Total potential responses are 10,000. At time *t*, responses are at the daily rate of 2 percent of those who have not yet responded. Letting $R(t)$ be total responses at time *t*,
   a) How many will not have responded at time *t*?
   b) At time *t*, what will be the rate of response per day?
   c) Find the particular solution of the differential equation in (*b*).
   d) How many will have responded after 100 days?
46. In Problem 45, let *P* be the total potential audience and *r* be the response rate per day. Write the solution in terms of the parameters *P* and *r*.

12

Probability

12.1 INTRODUCTION

ALMOST ALL STUDENTS of management and economics study statistics at some point during their undergraduate years. Typically, statistics courses devote only a small amount of time to probability, even though probability forms the foundation upon which statistical inference rests. Consequently, one objective of this chapter is to lay the foundation for the study of statistical inference. A second objective is to encourage readers to think in the probabilistic mode which is characteristic of decision-making problems. Finally, we should add, probability has its own interesting and useful applications and is worthy of study in its own right.

12.2 PROBABILITY AND ODDS

A weather forecaster may state in his radio broadcast that the probability of rain tomorrow is 60% and later, while playing poker, tell friends that the odds on rain tomorrow are 3 to 2. Other methods exist for expressing the degree of certainty, or uncertainty. In mathematics, we express probabilities as numbers on a scale from 0 to 1, inclusive. Thus, letting $P(R)$ represent the probability of rain, we convert 60% to its decimal equivalent and write

$$P(R) = 0.60.$$

The odds statement, 3 to 2, is viewed as meaning there are $3 + 2 = 5$ "chances" on tomorrow's weather, of which 3 are for rain, so

$$P(R) = \frac{3}{3 + 2} = \frac{3}{5} = 0.60$$

as before.

Exercise. If the odds are 13 to 7 that the stock market will rise tomorrow, what is the decimal value of the probability of a rise? Answer: 0.65.

Odds are in common use, but suffer lack of comparability because they do not have a fixed base. Thus, if the odds for a stock price rise are 11 to 7 while those for a rise in bond yields are 3 to 2, a conversion is needed to express the likelihood of rises in comparable form, and probabilities in decimal form are an appropriate conversion.

Exercise. Convert the last two odd statements as decimal probabilities. Answer: 0.611 and 0.600.

12.3 SOURCES OF PROBABILITIES

If we plan to draw a card from a well-shuffled standard deck of 52 cards, we would find it natural to assess the probability that the card will be an ace as

$$P(A) = (4 \text{ aces})/(52 \text{ cards}) = \frac{4}{52} = \frac{1}{4} = 0.25.$$

What this implies is that if we repeat the experiment of drawing a card from a well-shuffled deck of 52 cards many times, we would expect the proportion or *relative* frequency of drawing an ace to approach 0.25 as a limit as the number of drawings increases. *Relative frequency over the long run* under constant conditions (*e.g.,* drawing from a well-shuffled deck) is one source of probability measures. Most people feel confident in such probability assignments and refer to them as *objective* probabilities which can be verified by experiment. A novice in a poker game may feel that a pair of kings and a pair of aces should beat three deuces, but such is not the case. In poker, the more unusual (lower probability) hand wins, and three of a kind is more unusual than two pairs—not because of intuition or any subjective factor, but because of mathematical calculation which can be verified by experiment.

We should point out at once that objective probabilities, which arise as relative frequencies over the long run under constant conditions, are not as prevalent in management decision-making as they are in games of chance. To see why, consider a firm which is deciding on whether or not to bid on a $5 million project, at a point in time, and in the competitive environment of that time. The firm knows it will cost about $100,000 to prepare the bid so, naturally, it will not prepare a bid unless the probability of winning the contract is sufficiently high to justify the cost. How does the firm assess the probability of winning? Past experience will be of some help, but it is unlikely that the circumstances surrounding this particular situation have occurred very often, if at all, in the past, so that a relative frequency objective probability is not at hand. Nevertheless, a decision based upon a probability assessment must be made. We may assume that the appropriate officials of the firm, drawing upon both quantified and nonquantified experience and projection, will make the assessment. We refer to the result as a *subjective* probability.

In practice, probability assessments range from highly objective to highly subjective. In this chapter, we shall not distinguish between the two because probability theory is not concerned with the source of the probabilities we assign. The theory requires only that we consider *all* of the *different* events which may occur in a given situation and assign to each event a *nonnegative* number in a manner such that the sum of all the probabilities is one. For example, in drawing a card from a deck, the events red card, spade, and club are one way to specify *all* the *different* events that can occur, and we may assign probabilities 0.5, 0.25, and 0.25, respectively, to these events.

Certain and Impossible Events.   We assign a probability of 1 to an event which is certain to occur, such as the event of getting a red or a black card when drawing from a deck. The probability zero is assigned to an impossible event, such as drawing a red spade from the deck. Again, a weatherman may assign a probability of 0 for rain tomorrow if he believes rain is impossible, or 1 if he believes rain is certain. In summary, if $P(E)$ is the probability of event E,

$$0 \leq P(E) \leq 1$$
$$P(\text{impossible event}) = 0$$
$$P(\text{certain event}) = 1.$$

12.4 PROBABILITY SYMBOLS AND DEFINITIONS

We shall introduce probability terminology by reference to the data in Table 12–1. These data show the results of a poll of *all* 500 employees in a firm on the question of whether to change the work week from five 8-hour days to four 10-hour days.

Table 12–1

	Favor (F)	*Disfavor (D)*	*Neutral (N)*	*Totals*
Men (M)	70	10	20	100
Women (W)	170	150	80	400
Totals	240	160	100	500

The right margin shows there are 100 men in the group of 500 workers. If we were to draw an individual at random[1] from the 500, it would be natural to assign the probability of selecting a man, $P(M)$, as

$$P(M) = \frac{100}{500} = 0.20.$$

Similarly, the probability that the individual disfavors, $P(D)$, is

$$P(D) = \frac{160}{500} = 0.32.$$

Exercise. From Table 12–1, find *a*) $P(W)$. *b*) $P(F)$. Answer: *a*) 0.80. *b*) 0.48.

Joint Events: Intersection. In Table 12–1, sketch a horizontal line across the M row and a vertical line down the F column. At the *intersection* of the two lines there are 70 people who are both men *and* in favor. In set terminology, the symbol ∩ represents intersection, so the event at hand is M intersection F, or

$$M \cap F.[2]$$

An alternative description refers to the intersection as the *joint* event, M and F. Clearly,

$$P(M \cap F) = \frac{70}{500} = 0.14.$$

Similarly,

$$P(W \cap N) = \frac{80}{500} = 0.16,$$

and is the probability that the selected individual is a woman *and* is neutral. Remember always that the intersection means *both* occur and is characterized by the word *and.*

Exercise. Write the symbols for the probability that a man who disfavors is selected, and compute this probability. Answer: $P(M \cap D) = {}^{10}\!/_{500} = 0.02.$

---

[1]The concept of random selection underlies all probability rules. In the case of card playing, randomness is sought by shuffling the cards before dealing a hand. In the case of Table 12–1, we could shuffle the 500 employee time cards or use other methods to insure random selection.
[2]Later in the chapter we will, for brevity, omit ∩ and express the intersection as MF. See also Appendix 1, Section A1.7.

Mutually Exclusive (Disjoint) Events. In Table 12–1 the intersection

$$F \cap D$$

is empty (has no elements) because an individual cannot both favor and dis-favor the proposal. In set terminology

$$F \cap D = \phi$$

where ϕ is the *empty* or *null* set. We shall say F and D are *disjoint* or, more commonly, that they are *mutually exclusive.* Clearly,

$$P(F \cap D) = 0$$

and, in general,

A and B are mutually exclusive if P(A ∩ B) = 0

Exercise. If a card is drawn from a deck, what is the proba-bility that it will be red and be a spade? Why? Answer: $P(R \cap S)$ = 0 because red and spade are mutually exclusive.

Conditional Probability. Suppose next that an individual has been selected. We are told that the individual is a man, but not told his attitude toward the proposal. What now is the probability that this individual is in favor? Here it is *given* that we have a man, so in Table 12–1 we consider only the 100 men and the 70 of these who are in favor. We find that

$$P(\text{in favor given man}) = P(F|M) = \frac{70}{100} = 0.70,$$

where the vertical line segment in $P(F \mid M)$ is read as *given,* and the whole symbol as the probability of F given M.

Similarly, if we are told the person selected is neutral and asked to find the probability that the person is a man, we compute

$$P(M \mid N) = \frac{20}{100} = 0.20.$$

Exercise. If the person selected is in favor, write the symbols for, and compute the probability that, the person is a woman. Answer: $P(W \mid F) = {}^{170}\!/_{240} = 0.708$.

Dependent and Independent Events.   **Table** 12–1 shows

$$P(M) = \frac{100}{500} = 0.200 \quad \text{but} \quad P(M \mid F) = \frac{70}{240} = 0.292.$$

We note that

$$P(M \mid F) \neq P(M)$$

and say that M and F are *dependent* in the probability sense. The idea is that the probability that the selected person is a male is related to, or depends on, whether the person is in favor. In this case, the *unconditional* probability that the selected person is a man is 0.200, but it is more likely (0.292) that the person is a man if we are given that the person is in favor.

On the other hand, we note that

$$P(M) = \frac{100}{500} = 0.20 \quad \text{and} \quad P(M \mid N) = \frac{20}{100} = 0.20$$

so that

$$P(M \mid N) = P(M)$$

which we translate by saying that the probability that the selected person is a male does not depend on (is independent of) whether the person is neutral. In general, we state that *in the probability sense,*

> *If $P(A \mid B) = P(A)$,   A and B are independent.*
> *If $P(A \mid B) \neq P(A)$,   A and B are dependent.*

‖‖

Exercise.   See Table 12–1. Are the following independent or dependent? Why? *a*) W and D. *b*) W and N. Answer: *a*) $P(W \mid D)$ = $^{150}/_{160}$ which does not equal $P(W) = 0.8$, so W and D are dependent. *b*) $P(W \mid N) = 0.80 = P(W)$, so W and N are independent.

‖‖

The concepts of independence and mutual exclusiveness often are a source of confusion to beginners in probability. It will help to remember that independence is concerned with $P(A \mid B)$ which may, but usually is not, equal to zero, whereas mutual exclusiveness is concerned with the joint event $A \cap B$, and requires that $P(A \cap B) = 0$. For example, if we toss a coin twice we usually assume that the probability of heads on the second toss is not affected by (is independent of) what happened on the first toss. That is, the probability of heads on the second toss, H_2, is 0.5, and $P(H_2 \mid H_1)$ = $P(H_2) = 0.5$, so H_2 and H_1 are independent. On the other hand, it is possible to get heads on both tosses. $P(H_1 \cap H_2)$ is not zero, so H_1 and H_2 are not mutually exclusive.

‖‖

Exercise. If we draw two cards from a deck and let A_1 and A_2 represent first card ace and second card ace, respectively, are A_1 and A_2 mutually exclusive? Independent? Explain. Answer: The events A_1 and A_2 (e.g., first card ace of spades, second card ace of clubs) both can occur, so $P(A_1 \cap A_2) \neq 0$ and the events are not mutually exclusive. However, the probability that the second card is an ace *does* depend on whether the first card was an ace, so the events are *not* independent.

‖‖

Union of Events. Refer to Table 12–1. The expression

$$M \cup F,$$

where $\cup$ is read as *union* and $M \cup F$ as *M union F*, means the set of individuals who are either men *or* in favor *or* both. The table shows there are 100 men and an additional uncounted 170 in favor, for a total of 270, so

$$P(M \cup F) = \frac{270}{500} = 0.54.$$

Another counting method we shall find useful soon is to add the number of men to the number in favor, then subtract the *doubly* counted number of men in favor. This yields $100 + 240 - 70$, so

$$P(M \cup F) = \frac{100 + 240 - 70}{500} = \frac{270}{500} = 0.54,$$

as before.

On the other hand, F and D are mutually exclusive (the intersection is empty), so

$$P(F \cup D) = \frac{240 + 160 - 0}{500} = 0.80.$$

‖‖

Exercise. From Table 12–1, find: *a*) $P(M \cup W)$. *b*) $P(F \cup N)$. *c*) $P(W \cup D)$. *d*) $P(W \cup F)$. Answer: *a*) 1. *b*) 0.68. *c*) 0.82. *d*) 0.94.

‖‖

12.5 PROBLEM SET 1

1. The table shows, for example, that area A has 30 large stores and area C has 150 small stores.

Geographic Area	Store Size ($ volume)		
	Large L	*Medium* M	*Small* S
A............	30	45	75
B............	150	125	275
C............	20	130	150

Find the following probabilities:

a) $P(M)$.   b) $P(B)$.   c) $P(M \cap S)$.

d) $P(B \cap M)$.   e) $P(A \cap C)$.   f) $P(A \cap L)$.

g) $P(L \cap A)$.   h) $P(A \mid L)$.   i) $P(L \mid A)$.

j) $P(A \mid M)$.   k) $P(S \mid B)$.   l) $P(L \cup M)$.

m) $P(B \cup S)$.   n) $P(M \cup C)$.   o) $P(B \cup C)$.

2. See table of Problem 1.
   a) Does $P(M \mid A) = P(M)$?
   b) What does the answer to a) mean?
   c) Are A and L independent? Explain.
   d) Are B and L independent? Explain.
   e) What is $P(A \cap B)$? What does this mean?

3. A box contains two defective and three good items, *DDGGG*.
   a) Suppose an item is drawn at random, replaced, and a second drawing is made. Let *DG* mean the first is defective and the second is good. Are *D* and *G* independent? Explain.
   b) In a), are *D* and *G* mutually exclusive? Explain.

4. See Problem 3. Suppose the two items are drawn at random without replacement of the first drawn.
   a) Are *D* and *G* independent? Explain.
   b) Are *D* and *G* mutually exclusive? Explain.

5. Among the various events that may occur in Problem 4 are *DG* and *GD*. Are these two events mutually exclusive? Explain.

12.6 PROBABILITY RULES

Table 12–1 showed the numbers of people in each category, the marginal totals, and the grand total, 500. We now divide each number in Table 12–1 by 500 to yield the *probabilities* shown in Table 12–2.

TABLE 12–2

	Favor (F)	*Disfavor* (D)	*Neutral* (N)	*Total*
Men (M)	0.14	0.02	0.04	0.20
Women (W)	0.34	0.30	0.16	0.80
Total.........	0.48	0.32	0.20	1.00

Note: Intersections, such as $M \cap F$, are very often of interest, and for brevity we shall henceforth omit the intersection symbol and express a joint event such as $M \cap F$ as simply MF. We shall refer to MF as M *and* F, keeping in mind that the word *and* means *both* M and F. The union symbol, as in $M \cup F$ shall be retained, and we shall refer to $M \cup F$ as M *or* F, keeping in mind that this means M or F, or MF.

From Table 12–1 we computed

$$P(M \cup F) = \frac{100 + 240 - 70}{500}$$

$$= \frac{100}{500} + \frac{240}{500} - \frac{70}{500}$$

$$= 0.20 + 0.48 - 0.14 = 0.54.$$

The corresponding calculation can be made directly from Table 12–2 as

$$P(M \cup F) = P(M) + P(F) - P(MF)$$
$$= 0.20 + 0.48 - 0.14 = 0.54.$$

On the other hand, from Table 12–2,

$$P(F \cup D) = P(F) + P(D) - P(FD)$$
$$= 0.48 + 0.32 - 0$$
$$= 0.80$$

where, here, $P(FD) = 0$ because F and D are mutually exclusive. We have the following rule.

Addition Rule. The probability that A or B occurs is the probability of A plus the probability of B, minus the probability of A and B. Thus,

$$P(A \cup B) = P(A) + P(B) - P(AB).$$

If A and B are mutually exclusive, then

$$P(A \cup B) = P(A) + P(B).$$

||||||||||||||||||||||||||||||||||||||||||||||||||||||||||||||||||||||||||||||||||||||||||||||||||||||||||||||||||||||||||||||

Exercise.   From Table 12–2 find: *a*) $P(M \cup N)$. *b*) $P(M \cup W)$. *c*) $P(D \cup N)$. Answer: *a*) $0.20 + 0.20 - 0.04 = 0.36$. *b*) $0.20 + 0.80 - 0 = 1$. *c*) $0.32 + 0.20 - 0 = 0.52$.

||||||||||||||||||||||||||||||||||||||||||||||||||||||||||||||||||||||||||||||||||||||||||||||||||||||||||||||||||||||||||||||

As another example, suppose a political candidate runs for two offices, A and B. She assesses her probabilities of winning at 0.30 and 0.20 for A and B, respectively, and thinks she has an outside chance, probability 0.05, of winning both offices. The probability of winning A or B would then be

$$P(A \cup B) = P(A) + P(B) - P(AB)$$
$$= 0.30 + 0.20 - 0.05$$
$$= 0.45.$$

||||||||||||||||||||||||||||||||||||||||||||||||||||||||||||||||||||||||||||||||||||||||||||||||||||||||||||||||||||||||||||||||||||||||||||||||||||||||||||||||||||||||||||||||||||||||||

Exercise. An investor thinks the probability that stock P will rise tomorrow is 0.70, and that Q will rise is 0.80. He thinks there is a 50–50 chance that both will rise. What is the probability that P or Q will rise? Answer: $0.70 + 0.80 - 0.50 = 1.00$.

||||||||||||||||||||||||||||||||||||||||||||||||||||||||||||||||||||||||||||||||||||||||||||||||||||||||||||||||||||||||||||||||||||||||||||||||||||||||||||||||||||||||||||||||||||||||||

It is worth noting in the last exercise that the investor's subjective probability assignments lead logically to the conclusion that he is certain (probability 1) that P or Q will rise. Note also that if the investor had assessed the probability that both will rise at 0.4 (rather than 0.5), then $0.7 + 0.8 - 0.4 = 1.1$. This would lead the investor to reassess probabilities because the probability of an event cannot exceed 1. The last sentence illustrates a fundamental reason for understanding probability rules; namely, to monitor probability assessments and insure the internal logical consistency of such assessments.

Returning to Table 12–1, recall that

$$P(F \mid M) = \frac{70}{100} = 0.70.$$

If we express the latter in the equivalent manner,

$$P(F \mid M) = \frac{\frac{70}{500}}{\frac{100}{500}} = \frac{0.14}{0.20} = 0.70$$

and relate the 0.14 and 0.20 to Table 12–2, we observe that

$$P(MF) = 0.14 \quad \text{and} \quad P(M) = 0.20.$$

It follows that

$$P(F \mid M) = \frac{P(FM)}{P(M)}.$$

Thus, the probability of F given M is the probability of the joint event, FM, divided by the probability of M. Correspondingly,

$$P(M \mid F) = \frac{P(MF)}{P(F)}$$

which, from Table 12–2, is

$$P(M \mid F) = \frac{0.14}{0.48} = 0.292.$$

Exercise.   Complete the following: *a*) $P(N \mid W) =$.
b) $P(W \mid N) =$. Compute *a*) and *b*) from Table 12–2. Answer:
a) $P(N \mid W) = P(NW)/P(W)$. *b*) $P(W \mid N) = P(WN)/P(N)$. The
probabilities are *a*) 0.20 and *b*) 0.80.

Conditional Probability Rule.   The probability that B will occur given
that A has occurred is the probability of AB divided by the probability of
A. Thus,

$$P(B \mid A) = \frac{P(AB)}{P(A)}.$$

The rule just stated can be solved for $P(AB)$ after multiplying both sides
by $P(A)$. The result is:

Joint Probability Rule.   The probability of AB is the probability of A
times the probability of B given A. Thus,

$$P(AB) = P(A)P(B \mid A).$$

As an example, suppose a box contains two defective and three good
items, *DDGGG*. Two items are to be selected and we seek the probability,
$P(GD)$, that the first is good and the second defective.

$$P(GD) = P(G)P(D \mid G).$$

The probability, $P(G)$, that the first is good is $\frac{3}{5}$. We reason that if the first
selected is G, the four remaining are *DDGG*, so the probability, $P(D \mid G)$,
that the second is defective given the first is good is $\frac{2}{4}$. Hence,

$$P(GD) = \frac{3}{5}\left(\frac{2}{4}\right) = 0.30.$$

Observe that *order* is significant in the context of the last example. Thus,
GD means *first* good, *second* defective. Similarly, *GGD* means first good,
second good, third defective. Using a continuation of the joint probability
rule,

$$P(GGD) = P(G)P(G \mid G)P(D \mid GG)$$

which means that $P(GGD)$ is the probability that the first is good times the
probability that the second is good given that the first is good, times the
probability that the third is defective given that the first two are good.
Arithmetically,

$$P(GGD) = \frac{3}{5} \cdot \frac{2}{4} \cdot \frac{2}{3} = 0.20.$$

Exercise.   For the foregoing example, compute: *a*) $P(DD)$.
b) $P(DDG)$. *c*) $P(DDD)$. Answer: *a*) 0.1. *b*) 0.1. *c*) 0.

Recalling that independence means

$$P(B \mid A) = P(B),$$

we may substitute $P(B)$ for $P(B \mid A)$ in the joint probability rule and find:

Joint Probability Rule, Independent Events.   If A and B are independent in the probability sense, then the probability of the joint event AB is the probability of A times the probability of B. Thus,

$$P(AB) = P(A)P(B).$$

We may use this rule as a test for independence. For example, in Table 12–2, we note that

$$P(M) = 0.20, \quad P(N) = 0.20, \quad \text{and} \quad P(MN) = 0.04.$$

Hence,

$$P(MN) = P(M)P(N) = 0.04$$

so M and N are independent.

Exercise.   Refer to Table 12–2. Are W and D independent? Explain. Answer: $P(W) = 0.80$, $P(D) = 0.32$, whereas $P(WD) = 0.30$. $P(WD) \neq P(W)P(D)$, so W and D are not independent.

The assumption of independence underlies a number of probability applications. For example, suppose that, by test, a fire alarm functions successfully, S, in the presence of fire 90 percent of the time and fails, F, 10 percent of the time. For added protection, a store installs two alarms which operate independently. If there is a fire, what is the probability that both will fail? Here $P(F) = 0.1$ and because of independence,

$$P(FF) = P(F)P(F) = 0.1(0.1) = 0.01.$$

Exercise.   For the foregoing example, find the probability that: a) Both will function successfully. b) The first will function and the second fail. Answer: a) $P(SS) = 0.9(0.9) = 0.81$. b) $(0.9)(0.1) = 0.09$.

12.7 EXPERIMENT, EVENT, SAMPLE SPACE

The word *experiment* usually is associated with the natural sciences. However, it is also used in a very broad sense in probability to specify what we plan to do. For example, one experiment may be to select 10 items from a continuous production line, examine them, and record the number of defectives. Another experiment would be to select a sample of 100 families and record the date at which they last purchased an automobile.

An outcome of an experiment is called an *event*, and a totality of all possible mutually exclusive events is called a *sample space* for the experiment. For example, consider the experiment of testing two fire alarms by subjecting them to fire and recording whether they worked successfully or failed. *One* way to construct the sample space is to use ordered pairs such as *SF*, which means the first was successful and the second failed. We have

Events, Sample Space 1
SS
SF
FS
FF

Observe that the occurrence of one of the events (say *SS*) excludes the possibility of occurrence of any of the others, so the events are mutually exclusive. Moreover, every possible event is listed, so this is a proper sample space.

Usually, more than one sample space can be constructed for an experiment. In the case at hand, we could describe all possible events by listing the three mutually exclusive events 0 fail, 1 fails, 2 fail. Thus,

Events, Sample Space 2 *Number Failing*
0
1
2

Returning to Sample Space 1 and recalling the assumption of independence with $P(S) = 0.9$ and $P(F) = 0.1$, we can compute $P(SS) = (0.9)(0.9) = 0.81$, and so on leading to the following listing of the sample space and probabilities.

Sample Space 1, with Probabilities	
Event	*Probability*
SS	0.81
SF	0.09
FS	0.09
FF	0.01
	1.00

The sum of the probabilities for the entire sample space (which includes all possible mutually exclusive events) must be 1, as shown.

Turning now to probabilities for Sample Space 2, the event 0 fail is the event *SS*, so $P(0) = 0.81$. Similarly, the event 2 fail is the event *FF*, so $P(2) = 0.01$. However, the event 1 fails occurs if *FS* or *SF* occurs. Keeping in mind that the latter two are mutually exclusive, we have

$$P(FS \cup SF) = P(FS) + P(SF) = 0.09 + 0.09 = 0.18.$$

We thus have:

Sample Space 2, with Probabilities	
Number Failing	*Probability*
0	0.81
1	0.18
2	0.01
	1.00

A special way of constructing a sample space is to place each event in one or the other of two categories. In the last table, for example, we could say E is the event zero fail and E' is the event one or more fail. We have

$$P(E) = 0.81$$
$$P(E') = 0.19$$
$$P(E) + P(E') = 1.00.$$

The two events, E and E', are called *complementary* events because they constitute a sample space. We have:

Rule for Complementary Events. If E and E' are complementary events, then

$$P(E) + P(E') = 1 \quad \text{or} \quad P(E') = 1 - P(E).$$

The last rule finds frequent application. Returning to the fire alarm example, the most important consideration is the probability that at least one of the two alarms functions in case of fire. This could be computed as

$$P(SS) + P(SF) + P(FS),$$

but the computation is simplified if we note that the event FF is the complement of the three just mentioned. Hence, the desired probability is

$$1 - P(FF) = 1 - 0.01 = 0.99.$$

||||||||||||||||||||||||||||||||||||||||||||||||||||||||||||||||||||||||||||||||||||||||||||||||||||||||||||||||||||||||||||||||||||||||||||||

Exercise.   If the store had three independent alarms with $P(S)$ = 0.9 and $P(F)$ = 0.1, what is the probability that at least one functions successfully? Answer: $1 - P(0) = 1 - P(FFF) = 1 - (0.1)(0.1)(0.1) = 0.999.$

||||||||||||||||||||||||||||||||||||||||||||||||||||||||||||||||||||||||||||||||||||||||||||||||||||||||||||||||||||||||||||||||||||||||||||||

12.8 PROBLEM SET 2

1. Convert the following to a probability table.

	Store Size ($ Volume)		
Geographic Area	Large L	Medium M	Small S
A	30	45	75
B	150	125	275
C	20	130	150

2. *a*) Complete the following probability table. (Note C' and F' are complements of C and F, respectively.)

	F	F'	Total
C	0.24		0.60
C'			
Total	0.40		

Find the following probabilities:

b) $P(CF')$.	c) $P(C \cup F)$.	d) $P(C \cap C')$.
e) $P(F\|C)$.	f) $P(C\|F)$.	g) $P(F\|F')$.

h) Are C and F independent? Why?

i) Why is $P(F \cup F') = 1$?

3. A political candidate runs for two offices, A and B. She assesses her probabilities of winning at 0.6 and 0.2, respectively, and thinks she has only a probability of 0.01 of winning both.

   a) What is the probability that she wins one office *or* the other?

   b) If the candidate sets $P(A) = 0.7$, $P(B) = 0.4$, and $P(AB) = 0.01$, what advice should she be given? Why?

4. A box contains seven good and three defective items. If a sample of two items is selected, what is the probability that:

   a) Both will be good?

   b) Both will be defective?

   c) *Exactly* one will be defective. (Note: Two events are involved.)

   d) If a sample of 3 items is selected, what is the probability that *at least one* will be defective?

   (Hint: What is the complement of at least one defective?)

5. A weatherman states that if it is colder tomorrow, the probability of snow is 0.7. He also states the probability that it will be colder is 0.5. What is the probability it will be colder *and* snow tomorrow? Why?

6. The probability that a student will graduate with honors and get a good job is 0.09, whereas the probability that a student will graduate with honors is 0.1. What is the probability that a student will get a good job if he graduates with honors?

7. In the manufacture of an expensive product, each item is inspected independently by two women. The probability that either woman will correctly classify an item as defective is 0.95. If a defective item is inspected, what is the probability that:

   a) Both will correctly classify the item?

   b) Neither will correctly classify the item?

   c) At least one will correctly classify the item?

   d) *Exactly* one will correctly classify the item? (Two events.)

   e) The second will correctly classify the item if the first did not? Why?

8. Three coins are to be tossed. Write a sample space

   a) In terms of *HHT*, and so on, where *HHT* means first coin heads, second heads, and third tails. (There are 8 events.)

   b) In terms of the number of heads.

   c) Construct a probability table for *a*).

   d) Construct a probability table for *b*).

   What is the probability of:

   e) At least one head?

   f) At least two heads?

   g) Exactly one head?

   h) Exactly two heads?

9. An election results in a tie among two women (W_1 and W_2) and one man (M). It is decided to select two people at random from the three and appoint the first selected as president and the second selected as vice president.

   a) Construct a sample space using, for example, $W_1 M$ to mean woman one is president, and the man is vice president. Assign probabilities to each event in the sample space.

   Using the results of a), what is the probability that:

   b) The man will be president?

   c) The man will be either president or vice president?

   d) What is the probability women will occupy both offices?

10. See Problem 9, but do not refer to the sample space.

   a) Using W to represent either woman, write the symbol for the joint event which specifies the man is vice president.

   b) Compute the probability for the event in a) using the conditional probability rule.

   c) Write the symbols for the event that a man is selected for neither office.

   d) Compute the probability for the event in c) using the conditional probability rule.

12.9 PRACTICE WITH PROBABILITY RULES

In this section we present a series of examples to provide further practice with probability rules.

Example 1. A company has two employees who work and answer the telephone on Saturday and two different employees who cover the telephone on Sundays. On a particular week-end, all four request a day off, but the company decides it can grant only two requests. To be fair, the two are selected at random. What is the probability that the telephone will be covered on Saturday and Sunday?

The first step in this and other problems is to describe the event(s) whose probability we seek. Let us use S for Saturday and D (standing for domingo, or Sunday in Spanish) for Sunday. Then, a worker will be on hand to cover the telephone if from the S, S, D, D either of the two mutually exclusive events SD or DS occurs, where, for example, SD means the first selected for a day off is a Saturday worker and the second selected is a Sunday worker. Hence,

$$
\begin{aligned}
P(SD \quad \text{or} \quad DS) &= P(SD \cup DS) = P(SD) + P(DS) - 0 \\
&= P(S)P(D \mid S) + P(D)P(S \mid D) \\
&= \frac{2}{4} \cdot \frac{2}{3} + \frac{2}{4} \cdot \frac{2}{3} = \frac{2}{3} = 0.667.
\end{aligned}
$$

In this, as in many problems, we can arrive at the solution using more than one method of approach. For example, we note that the telephone will *not* be covered both days if either both Saturday or both Sunday workers get a day off. That is, SS or DD is the complement of the event whose probability we seek. Then, using $P(S|S)$ to mean the probability that the second is S given that the first is an S, and similarly for $P(D|D)$, we find

$$P(SD \cup DS) = 1 - P(SS \cup DD)$$
$$= 1 - [P(SS) + P(DD)]$$
$$= 1 - [P(S)P(S|S) + P(D)P(D|D)]$$
$$= 1 - \left[\frac{1}{2} \cdot \frac{1}{3} + \frac{1}{2} \cdot \frac{1}{3}\right]$$
$$= 1 - \frac{2}{6} = \frac{2}{3},$$

as before.

Another way to view the problem is to note that the telephone will be covered if, no matter which of S or D is chosen first, the second is *different*. In this approach, no matter which is first chosen, there are two different ones in the remaining three, so the probability that the second chosen is different is $\frac{2}{3}$.

Finally, the sample space approach can be applied in a simple problem such as the one at hand. If we specify events as S_1, S_2, D_1 and D_2, where S_1 refers to one Saturday worker and S_2 the other Saturday worker, and similarly for D_1 and D_2, we may write the sample space as:

$$S_1 \quad S_2*$$
$$S_1 \quad D_1$$
$$S_1 \quad D_2$$
$$S_2 \quad D_1$$
$$S_2 \quad D_2$$
$$D_1 \quad D_2*.$$

Each of the six events listed is *equally likely*, probability $\frac{1}{6}$, and mutually exclusive. Four of the six (those not marked by*) are events in which a worker is available on both days to answer the telephone, and the sum of the probabilities is $\frac{4}{6}$ or $\frac{2}{3}$, as before.

The foregoing shows that some probability questions can be answered by constructing a sample space of mutually exclusive events in which each event is *equally likely* and summing the relevant probabilities. The exercise is instructive, but the number of events in sample spaces for all but simple problems is unmanageable, so probability rules are the preferred technique for solving most of the problems we shall encounter.

Example 2. As a variation on Example 1, suppose that only one of the Saturday workers and one of the Sunday workers speaks English and can answer the telephone. Now what is the probability that the telephone is covered both days?

Here we must distinguish workers, which we shall do by subscripts, writing S_1, S_2, D_1, D_2. Let S_1 and D_1 be English-speaking. We seek

$$P(S_1 D_1 \cup D_1 S_1) = P(S_1 D_1) + P(D_1 S_1) - 0$$

$$= P(S_1)P(D_1 \mid S_1) + P(D_1)P(S_1 \mid D_1)$$

$$= \frac{1}{4} \cdot \frac{1}{3} + \frac{1}{4} \cdot \frac{1}{3}$$

$$= \frac{1}{6} = 0.167.$$

Example 3. A job applicant assigns probabilities as follows: The probability, $P(A)$, of being offered a job at company A is 0.6; the probability, $P(B)$, of being offered a job at company B is 0.5; the probability of being offered a job at both companies is 0.4. What, consequently, is the probability of being offered a job with at least one of the two companies?

Here we shall apply a tabular approach to the problem. The given probabilities are entered in the table at the left, and the entries in the table at the right follow as logical consequences.

Given:

	B	B'	*Total*
A	0.4	. .	0.6
A'.	. .	. .	. .
Total	0.5	. .	. .

Completed Table:

	B	B'	*Total*
A	0.4	0.2	0.6
A'.	0.1	0.3	0.4
Total	0.5	0.5	1.0

The event in question consists of the mutually exclusive events AB'; $A'B$; and AB, the sum of whose probabilities is 0.7. The complement of the event in question is $A'B'$, so that the desired probability could have been found as $1 - 0.3 = 0.7$.

Example 4. Probabilities often are assigned in a manner which assures independence. For example, suppose that a test consists of two four-choice multiple-choice questions. A student guesses at the first and assigns $1/4$ as the probability of getting the first correct. He is almost certain of the answer to the second and assigns probability 0.9 of getting this one correct. Moreover, he assumes that whether or not he is correct on one of the questions does not affect the probability of being correct on the other, and translates this assumption into the requirement that the events in each of the ordered pairs CC; CW; WC; WW are independent, where, for example, CW means first correct, second wrong. The given probabilities are 0.25 and 0.9; the assumption of independence leads to

$$P(CC) = (0.25)(0.9) = 0.225.$$

After entering 0.225 in the table, the remaining probabilities can be computed. One may verify that

$$P(CW) = (0.25)(0.1) = P(C)P(W) = 0.025$$

and similarly for the other joint events in the table.

	Second Question		
First Question	*C*	*W*	*Total*
C........	.0.225	0.025	0.25
W	.0.675	0.075	0.75
Total.....	.0.900	0.100	1.00

When more than two events are at hand, it can happen that the events in each pair are independent, but the events as a whole are not independent. We shall not consider this and similar circumstances. Instead, when dealing with more than two events, we shall restrict our discussion of independence to situations where the events in any combination are independent. For example, if A, B, C, D, and E are independent, then

$$P(ABCDE) = P(A)P(B)P(C)P(D)P(E)$$
$$P(ACE) = P(A)P(C)P(E)$$

and similarly for other combinations.

Example 5. A multiple-choice examination has three four-choice questions, and a student guesses the answer to each question, assigning 0.25 as the probability of guessing the correct answer. Assuming independence:
a) What is the probability of getting all three correct?

$$P(CCC) = P(C)P(C)P(C) = (1/4)(1/4)(1/4) = 1/64.$$

b) What is the probability of getting exactly one correct? The joint events (CWW), (WCW), and (WWC) constitute the event *exactly one correct*. These three events are mutually exclusive. Hence:

$$P(CWW) + P(WCW) + P(WWC) = P \text{ (exactly one correct)}$$

$$
\begin{aligned}
P(CWW) &= (1/4)(3/4)(3/4) = 9/64 \\
P(WCW) &= (3/4)(1/4)(3/4) = 9/64 \\
P(WWC) &= (3/4)(3/4)(1/4) = 9/64 \\
& 27/64.
\end{aligned}
$$

Observe that each of the three events has the same probability, so that the answer is $3(9/64) = 27/64$. The multiplier, 3, is simply the number of ways of arranging the two W's and one C. We shall explore this matter a little later.

c) What is the probability of getting at least one correct? At least one correct means one correct, two correct, or three correct, and could be found as the sum of the probabilities of these three mutually exclusive events. However, it is more direct to compute the probability of the complementary event, *zero correct,* and subtract from one. Thus:

$$P \text{ (at least one correct)} = 1 - P(\text{zero correct})$$
$$= 1 - P(WWW)$$
$$= 1 - (3/4)(3/4)(3/4) = 37/64.$$

For practice the reader should verify that

$$3P(CWW) + 3P(CCW) + P(CCC) = 37/64.$$

The multiplier, 3, is the number of ways of arranging two W's and one C, or two C's and one W.

Exercise. In the last example, what would constitute the event "at most one correct," and what would be the probability of this event? Answer: The event is made up of the events in part b) of the previous example, plus the event WWW which adds another 27/64 to make the total probability 54/64.

Example 6. If a student passes the final examination, the probability of passing the course is 0.9; if he does not pass the final, the probability of passing the course is 0.2. The student assigns 0.7 as the probability of passing the final. What is the probability of passing the course?

Let C and C' represent passing and not passing the course, and F and F' represent passing and not passing the final examination. We have

$$P(F) = 0.7$$
$$P(C \mid F) = 0.9$$
$$P(C \mid F') = 0.2.$$

From

$$P(C \mid F) = \frac{P(CF)}{P(F)}$$

we have

$$0.9 = \frac{P(CF)}{0.7}$$

so that

$$P(CF) = (0.9)(0.7) = 0.63.$$

Similarly, $P(CF')$ can be found from $P(C \mid F')$ and

$$P(F') = 1 - P(F)$$

to be

$$(0.2)(0.3) = 0.06.$$

	F	F'	Total
C	0.63	0.06	0.69
C'	0.07	0.24	0.31
Total	0.70	0.30	1.00

The desired probability is

$$P(C) = P(CF) + P(CF') = 0.69.$$

Exercise. The probability of good weather, G, is 0.6 and the probability of accident, A, is 0.014. The probability of the joint event, accident and good weather, is 0.006. Find the probability of accident if the weather is not good. Answer: We may complete a table, as follows:

	G	G'	Total
A	0.006	0.008	0.014
A'	0.594	0.392	0.986
Total	0.600	0.400	1.000

Hence:

$$P(A \mid G') = 0.008/0.4 = 0.02.$$

Example 7. The first time an insurance salesman calls on a new client, he has a probability of 0.1 of selling a policy. *If* he does not sell the policy on the first call, he makes a second call and has a probability of 0.3 of making the sale. *If* he does not make the sale on either of the first two calls, he makes one final call and has a probability of 0.05 of making the sale. What is the probability that the insurance man will sell the policy to a client?

We shall use S_1 and S_1' to mean sale or no sale, respectively, on the first call, and similarly for S_2, S_2', and S_3, S_3'. The events of interest are S_1 which is sale on the first call; $S_1'S_2$, which is no sale at first call and sale at the second call; $S_1'S_2'S_3$, which is no sale on call 1, no sale on call 2, and sale on call 3. These are mutually exclusive, so that

$$\begin{aligned} P(\text{Sale}) &= P(S_1) + P(S_1'S_2) + P(S_1'S_2'S_3) \\ &= P(S_1) + P(S_1')P(S_2 \mid S_1') + P(S_1'S_2')P(S_3 \mid S_1'S_2') \\ &= P(S_1) + P(S_1')P(S_2 \mid S_1') + P(S_1')P(S_2' \mid S_1')P(S_3 \mid S_1'S_2') \\ &= 0.1 + 0.9(0.3) + 0.9(0.7)(0.05) \\ &= 0.1 + 0.27 + 0.0315 \\ &= 0.4015. \end{aligned}$$

Exercise. As we moved from the third to the fourth line in the preceding problem, we replaced $P(S_2' \mid S_1')$ by 0.7. Explain how the 0.7 is derived. Answer: The problem states that the probability of a sale on the second call, given no sale on the first, is 0.3. Thus, $P(S_2 \mid S_1') = 0.3$. Now, given no sale on the first, the second call either results in a sale or in no sale, so $P(S_2 \mid S_1') + P(S_2' \mid S_1') = 1$ and, therefore, $P(S_2' \mid S_1') = 1 - P(S_2 \mid S_1') = 1 - 0.3 = 0.7$.

In general, we may write

$$P(B' \mid A') = 1 - P(B \mid A')$$

but it is important to note that, in general

$$P(A \mid B) \neq 1 - P(A \mid B').$$

For example, the probability of snow if it is colder tomorrow may be $P(S \mid C) = 0.7$, but this does not require that the probability of snow if it is not colder be 0.3. Indeed, the latter probability might well be zero, or it might be 0.5, or it might be 0.3. We see that $P(S \mid C) = 0.7$ does not require that $P(S \mid C')$ be 0.3.

12.10 BAYES' RULE

As a result of past hiring procedure, a company finds that 60% of its employees are good workers, G, and 40% are poor workers, which we shall designate by the complement, G'. The woman in charge of hiring believes that the proportion of good workers can be increased by designing a test to be administered to job applicants, and hiring only those who pass, P, the test. A consulting firm supplies the test and offers to administer it for a fee to applicants. Because of the cost, it is decided to determine first how well the test discriminates between good and poor workers by trying it on current employees. It is found that 80% of the good workers and 40% of the poor workers pass the test, which leads to the conclusion that the probability of passing the test if the employee is a good worker is $P(P|G) = 0.80$ and similarly, $P(P|G') = 0.40$.[3]

At first glance it may appear that the test functions well because it is twice as likely (0.80 vs 0.40) that an employee will pass if he is a good worker as it is that he will pass if he is a poor worker. The important point to note, however, is that the real question at hand is whether the test should be used in selecting employees from job applicants. Thus, the issue is not whether an employee will pass if he is a good worker, $P(P|G)$, but rather $P(G|P)$ which is the probability that an employee will be a good worker if he passes the test. Thus, we know $P(P|G)$ and we seek the *inverse* probability. $P(G|P)$. We have given:

$$P(G) = 0.60 \qquad P(G') = 1 - P(G) = 0.40$$
$$P(P|G) = 0.80 \qquad P(P|G') = 0.40.$$

From the foregoing, we can compute:

$$P(PG) = P(G)P(P|G) = (0.60)(0.80) = 0.48$$
$$P(PG') = P(G')P(P|G') = (0.40)(0.40) = 0.16.$$

We now have

	P	P'	Total
G.	0.48	. .	0.6
G'	0.16	. .	. .
Total.	0.64	. .	. .

It follows from the table that

$$P(G|P) = \frac{0.48}{0.64} = 0.75.$$

To see the significance of the last result, recall that 60% of current employees are good workers. This means that past or prior employment procedures, without the test, had a probability of $P(G) = 0.6$, which we shall call the *prior* probability, of selecting a good worker. If, now, we change the selection procedure and employ only those who pass the test, we *revise* the probability of hiring a good worker to $P(G|P) = 0.75$, and this revised probability is called the *posterior* probability. Management must now decide whether the additional information provided by a test which results in an increase in the probability of hiring good workers from 0.60 to 0.75, is worth the cost of administering the test.

---

[3]Note, again, that $P(P|G') \neq 1 - P(P|G)$.

To develop Bayes' Rule, refer to the foregoing table where we computed $P(G) = 0.75$ in the following manner:

$$0.75 = \frac{0.48}{0.48 + 0.16}.$$

In symbols, the last is

$$P(G|P) = \frac{P(GP)}{P(GP) + P(G'P)}.$$

Using the rule for joint probabilities, the last may be written as:

Bayes' Rule:

$$P(G|P) = \frac{P(P|G)P(G)}{P(P|G)P(G) + P(P|G')P(G')}.$$

The rule can be remembered easily if we note that in the numerator on the right we start with $P(P|G)$ which is simply the left side, $P(G|P)$, inverted. The numerator is completed in the manner of any joint probability by multiplying by $P(G)$ to give $P(P|G)P(G)$. The first term in the denominator on the left repeats the numerator, and the second term follows by replacing the G in the first term by G'. As another example,

$$P(B|A) = \frac{P(A|B)P(B)}{P(A|B)P(B) + P(A|B')P(B')}.$$

‖‖

Exercise. Write Bayes' Rule for $P(X|Y)$. Answer: $P(X|Y) = [P(Y|X)P(X)]/[P(Y|X)P(X) + P(Y|X')P(X')]$.

‖‖

As another example, suppose that the probability that a person has disease X is

$$P(X) = 0.09.$$

The probability that medical examination will indicate the disease if a person has it is

$$P(I|X) = 0.6.$$

The probability that medical examination will indicate the disease if a person does not have it is

$$P(I|X') = 0.05.$$

What is the probability that a person has the disease if medical examination so indicates?

By formula:

$$P(X|I) = \frac{P(I|X)P(X)}{P(I|X)P(X) + P(I|X')P(X')}$$

$$= \frac{(0.6)(0.09)}{(0.6)(0.09) + (0.05)(0.91)}$$

$$= \frac{(0.054)}{(0.0995)} = 0.543.$$

The same result could be obtained by applying the tabular analysis used earlier in this section. We see that the prior probability of the person having the disease, 0.09, has been revised to the posterior probability 0.543 as a consequence of the additional information that medical examination indicated the disease.

12.11 PROBLEM SET 3

1. An arena has scheduled two events on Thursday and three on Friday of a particular week. To obtain time to carry out repairs, two of the events, selected at random, are to be postponed.
   a) Using T_1 for the first Thursday event, F_2 for the second Friday event, and so on, write a sample space for this experiment in which events are equally likely.
   b) From a) find the probability that there will be an event on Thursday and an event on Friday.
   c) Calculate the answer to b) by probability rules.

2. See Problem 1. Suppose there are three events scheduled on each day. Using probability rules find:
   a) The probability that both postponed events will be on the same day.
   b) The probability that two events will remain on Thursday and two on Friday.

3. Suppose an arena has two events, hockey and basketball, scheduled on Thursday and another two, hockey and track, on Friday. If two are selected to be played and two cancelled, calculate by probability rules the probability that there will be a hockey game on Thursday or on Friday.

4. Given $P(X) = 0.6$, $P(Y) = 0.4$, and $P(XY) = 0.1$. Find: $P(X|Y)$ and $P(Y|X)$.

5. If $P(Y|X) = 0.4$, $P(X|Y) = 0.5$, and $P(XY) = 0.2$, find $P(X)$ and $P(Y)$.

6. If $P(Y|X) = 0.72$, $P(XY) = 0.18$, and $P(Y') = 0.4$, find $P(X \cup Y)$.

7. A candidate runs for two political offices, A and B. He assigns 0.4 as the probability of being elected to both, 0.7 as the probability of being elected to A if he is elected to B, and 0.8 as the probability of being elected to B if he is elected to A.
   a) What is the probability of being elected to A?
   b) What is the probability of being elected to B?
   c) What is the joint probability of being elected to neither?
   d) What is the probability of being elected to at least one of the offices?

8. The probability of snow is 0.4, of colder 0.5; and the conditional probability of snow if colder is 0.7. Find the probability of:
   a) Colder and snow.
   b) No snow.
   c) Either colder or snow.
   d) Neither colder nor snow.

9. Job candidates are screened by means of a preliminary interview. The probability is 0.6 that a screened candidate will be a good worker. Screened candidates are given a test. If the candidate is one who will prove to be a good worker, the probability of his passing the test is 0.8. If the candidate is one who will prove to be a poor worker, the probability of his passing the test is 0.4.
   a) What is the probability that a screened candidate will be a good worker and pass the test?
   b) What is the probability that a screened candidate will not be a good worker and will pass the test?

10. A multiple-choice examination has four three-choice questions, and a student guesses each answer, assigning the same probability to each choice. Find the probability that the student
    a) Gets all four correct.
    b) Gets at least three correct.
    c) Gets at least one correct.
    d) Gets exactly one correct.

11. If snow, colder are the events with $P(S) = 0.8$, $P(C) = 0.6$, and $P(S'C) = 0.1$:
    a) Show that the events are not independent in the probability sense.
    b) What probability would $(S'C)$ have if the events were to be independent in the probability sense?

12. The probability that machine A will break down on a particular day is

    $$P(A) = 1/50.$$

    Similarly, for machine B:

    $$P(B) = 1/80.$$

    Assuming independence, on a particular day:
    a) What is the probability that both will break down?
    b) What is the probability that neither will break down?
    c) What is the probability that one *or* the other will break down?
    d) What is the probability that exactly one machine will break down?

13. A pair of dice is rolled, and a coin is tossed. Assuming independence, what is the probability of
    a) Heads on the coin and a seven on the dice?
    b) Heads on the coin or a seven on the dice?
    c) Heads on the coin and an even number on the dice?
    d) Heads on the coin or a number greater than eight on the dice?

14. Five parts go into the assembly of item X. The assembly is defective if any one of the parts is defective, and each part has a probability of 0.03 of being defective. Assuming independence, what is the probability that an assembly is defective?

15. In the table, *H, M, L* stand for "high absenteeism," "medium absenteeism," and "low absenteeism," respectively. *S* stands for "salaried" and *HP* for "hourly paid." The table shows, for example, that there were 40 salaried workers who had high absentee records. Assigning probabilities as relative frequencies, show that method of pay and absenteeism (for these data) are independent in the probability sense.

	H	*M*	*L*
S	40	60	100
HP	60	90	150

16. A plant has 300 workers, 200 of whom are females. Classified by *H, M, L* absenteeism, there are 30 workers in the *H* class and 30 workers in the *L* class. Assuming independence of sex and absenteeism, how many workers should there be in each of the six possible classes?

17. Company *A* plans to bid on a contract. It does not know whether or not a competitor, Company *B*, will bid, but assesses the probability that it will bid at 0.6. *A* judges that it has a probability of 0.8 of winning if *B* does not bid, but only a probability of 0.4 if *B* does bid. What is the probability that *A* wins?

18. A company assesses the probability that its product will fail during the first month after sale as 0.01. The probability it will fail during the next 11 months if it did not fail during the first month is 0.001. The company guarantees the product for the first year. What is the probability that the product will fail in the first year?

19. See Problem 18. Suppose the probability of failure during the first month is 0.05, the probability of failure during the next 5 months (if the product did not fail in the first month) is 0.02, and the probability of failure for the remainder of the year if failure did not occur during the first noted time intervals is 0.01. What is the probability of failure during the first year?

Note: Problems 20–23 involve Bayes' rule.

20. The probability that a customer will be a bad debt is 0.01. The probability that he will make a large down payment if he is a bed debt is 0.20, and the probability that he will make a large down payment if he is not a bad debt is 0.60. Suppose that a customer makes a large down payment. Find the posterior probability that he will be a bad debt.

21. Complete the tabular analysis for Problem 20, and find the probability that a customer will not be a bad debt if he does not make a large down payment.

22. You note that your girl friend is happy on 60 percent of your visits, so you assign $P(H) = 0.6$. You have noticed also that if she is happy, she makes cookies for you with probability 0.4, whereas if she is not happy, she makes cookies with probability of 0.1. You call one day and find her making cookies. What is the probability that she is happy?

23. The probability that a machine is running properly is 0.95. From time to time, samples of output are selected and measured, and the sample average is computed. If the machine is running properly, the probability that the sample average will be in a certain range is 0.9. If the machine is not running properly, the probability that the sample average will be in this range is 0.04. A sample is selected, and its average is outside the range. What is the probability that the machine is running correctly?

12.12 THE BINOMIAL PROBABILITY DISTRIBUTION

Let us assume that a basketball player is successful (hits) on 40% of his free-throw trials. Letting H represent *hit* and M represent *miss*, we assign

$$P(\text{Hit}) = P(H) = p = 0.40$$
$$P(\text{Miss}) = P(M) = q = 1 - q = 0.60.$$

We assume further that the outcome of one trial is *independent* of the outcomes of other trials so that p and q are the same on every trial.[4] If we consider the probability of five hits in a row, we find

$$P(S) = P(HHHHH) = (0.4)(0.4)(0.4)(0.4)(0.4) = (0.4)^5$$
$$= 0.01024.$$

Next, consider $P(4)$, the probability of four hits in the next five trials. One way this can be done is $HHHHM;$ that is, hit on the first four trials and miss on the fifth. Another way is $HHHMH$. Writing all the events which constitute four hits and a miss, together with their probabilities, we find:

Components of Event Four Hits	Probability
$HHHHM$	$(0.4)(0.4)(0.4)(0.4)(0.6) = (0.4)^4(0.6)^1$
$HHHMH$	$(0.4)(0.4)(0.4)(0.6)(0.4) = (0.4)^4(0.6)^1$
$HHMHH$	$(0.4)(0.4)(0.6)(0.4)(0.4) = (0.4)^4(0.6)^1$
$HMHHH$	$(0.4)(0.6)(0.4)(0.4)(0.4) = (0.4)^4(0.6)^1$
$MHHHH$	$(0.6)(0.4)(0.4)(0.4)(0.4) = (0.4)^4(0.6)^1$

The five events just written are mutually exclusive, so the total probability, $P(4)$, is the sum of the five probabilities. Moreover, the five probabilities are equal, so that

$$P(4) = 5(0.4)^4(0.6)^1 = 5(0.01536) = 0.0768.$$

In the last statement, note that

$$P(4) = (\text{number of ways of getting four hits and a miss})p^4q^1.$$

---

[4]This assumption, which is *required* in the development of the binomial distribution, may be subject to question in this case because players do have their ups and downs.

If we next ask for the probability of three hits in the next five trials, we would have

$$P(3) = \text{(number of ways of getting three hits and two misses)}p^3q^2.$$

One way to get three hits is *HHHTT,* another way is *HHTHT,* and if we persist we will find there are 10 different sequences in the event (three hits, two misses) in five trials. Fortunately, we can compute this number by the formula

$$C_3^5 = \frac{5!}{3!(5-3)!} = \frac{5!}{3!(2!)} = \frac{5(4)3!}{3!(2)(1)} = \frac{20}{2} = 10.$$

The symbol at the left in the foregoing is read as *the number of combinations of five taken three at a time.* In the present application, it is the number of different ways of arranging five things of which three are of one kind and the remaining $5 - 3 = 2$ are of another kind.

We now have

$$P(3) = C_3^5 p^3 q^2$$
$$= 10(0.4)^3(0.6)^2 = 10(0.02304) = 0.2304.$$

Exercise.   Find the probability of two hits and three misses in the next five trials. Answer: $C_2^5 p^2 q^3 = 10(0.4)^2(0.6)^3 = 10(0.03456) = 0.3456.$

We may now generalize our demonstrations and state that if n independent trials are made, where p is the probability of outcome A and $q = 1 - p$ is the probability of the complementary outcome A' on any trial, then the probability that A will occur x times in the n trials is:

Binomial Probability Rule:

$$P(x) = C_x^n p^x q^{n-x}.$$

As an example, let us compute the probability that the basketball player hits exactly three times in his next 10 tries. We have

$$n = 10, \quad p = 0.4, \quad q = 0.6, \quad x = 3,$$

and

$$P(3) = C_3^{10} p^3 q^{10-3} = \frac{10!}{3!7!} = (0.4)^3(0.6)^7$$

$$= \frac{10(9)(8)}{3!}(0.4)^3(0.6)^7$$

$$= 120(0.064)(0.0279936)$$

$$= 0.215.$$

||||||||||||||||||||||||||||||||||||||||||||||||||||||||||||||||||||||||||||||||||||||||||||||||||||||||||||||||||||||||||||||

Exercise. Compute the probability that the player will hit exactly twice in his next six trials. Answer: $15(0.16)(0.1296) = 0.31104$.

||||||||||||||||||||||||||||||||||||||||||||||||||||||||||||||||||||||||||||||||||||||||||||||||||||||||||||||||||||||||||||||

If we seek the probability that $x = 0$ in $n = 7$ trials, it is clear that there is only one way, $MMMMMMM$, that this can occur. The binomial rule states

$$P(0) = C_0^7(0.4)^0(0.6)^7$$

$$= \frac{7!}{0!7!}(0.4)^0(0.6)^7.$$

The count,

$$\frac{7!}{0!7!},$$

must be one so, for consistency, we state:

Definition: $0! = 1$.

Then,

$$P(0) = \frac{7!}{0!7!}(0.4)^0(0.6)^7 = 1(1)(0.6)^7$$

$$= 0.0280.$$

Binomial situations arise in many applications. For example, we may record an inspected item as good or defective, check male or female on a questionnaire, classify accounts receivable as active or bad debts, and so on. In some situations we apply the binomial rule even though the requisite constancy of probability from trial to trial is only approximately correct. For example, if we draw items from a batch of 100,000 items, of which 100 are defective, we would have

$$P(\text{first is defective}) = P(D_1) = \frac{100}{100,000} = 0.001.$$

The probability that the second is also defective, correctly computed, is not 0.001 but is

$$P(D_2 \mid D_1) = \frac{99}{99,999} = 0.00099.$$

The difference noted between 0.001 and 0.00099 would be lessened if we were drawing from a batch larger than 100,000. To call attention to this approximate use of the binomial, we shall indicate that selection is made from a (very) *large* group and *not* state the number in the group. For example, suppose that 35 percent of the people in a *large* area are independent voters. We select 10 people and compute the probability that exactly six are independents as, approximately,

$$P(6) = C_6^{10}(0.35)^6(0.65)^4$$

$$= \frac{10(9)(8)(7)}{(4)(3)(2)(1)}(0.35)^6(0.65)^4$$

$$= 210(0.001838)(0.1785) = 0.0689.$$

Exercise. In a large batch of items, 10 percent are defective. If a sample of five is selected, what is the probability that exactly three will be defective? Answer: $C_3^5(0.1)^3(0.9)^2 = 10(0.001)(0.81) = 0.0081$.

At Least One Occurrence. Some applications require the computation of the probability of at least one occurrence. For example, suppose the assembly of a machine requires five independent operations, and the probability that any operation will result in a defect is 0.01. Further, the assembly is defective if one or more operations is defective; that is, if the number of defective operations is 1, 2, 3, 4, or 5. Hence, we seek the probability of at least one defective operation, which is the complement of zero defective.

$$P(\text{at least one defective}) = 1 - P(0 \text{ defective})$$
$$= 1 - C_0^5(0.01)^0(0.99)^5$$
$$= 1 - 1(1)(0.99)^5$$
$$= 1 - 0.951$$
$$= 0.049.$$

Exercise. A town has three ambulances for emergency transportation to the hospital. The probability that any one of the ambulances will be available at a point in time is 0.90. If a person calls for an ambulance, what is the probability that at least one will be available? Answer: $1 - (0.1)^3 = 0.999$.

As a variation on the last example and exercise, consider a car salesman who makes telephone calls during the day to obtain prospective car buyers. He assesses the probability of obtaining a prospect on a given call as 0.10.

How many calls should he make if his goal is to have a probability of 0.90 of obtaining at least one prospect? Here we have:

$$P(\text{at least one prospect}) = 1 - P(0 \text{ prospects}) = 0.90,$$

or

$$1 - C_0^n(0.10)^0(0.90)^n = 0.90$$
$$1 - (0.90)^n = 0.90$$
$$-(0.90)^n = -0.10$$
$$(0.90)^n = 0.10.$$

We solve for n by taking logarithms of both sides, writing $\text{Log}(0.90)^n = n \text{ Log}(0.90)$. Thus,

$$n \text{ Log}(0.90) = \text{Log}(0.10)$$

$$n = \frac{\text{Log}(0.10)}{\text{Log}(0.90)}$$

$$= \frac{-1}{0.9542 - 1}$$

$$= \frac{-1}{-0.0458} = 21.8,$$

or about 22 calls.

> **Exercise.**   A complicated computer program has a flaw or "bug" in it, and will not execute properly. The program is to be sent to n experts, each of whom has a probability of 0.4 of finding the bug. What should n be if the probability that at least one expert will find the flaw is to be 0.99? Answer: $2/0.2218 = 9$.

12.13 CUMULATIVE BINOMIAL PROBABILITIES

The binomial rule as applied in the foregoing computes the probability of *exactly* x occurrences in n trials. Thus, in examples where we drew, say, five items from a large lot and recorded the number of defectives, the binomial rule calculates $P(2)$, the probability of exactly two defectives, or $P(1)$, and so on. In its real-world application, the purpose of drawing the sample is to make a decision on whether or not to *accept the whole lot* on the basis of the sample evidence. Thus, the quality control department may have a *sampling plan* which specifies that five items are to be drawn and inspected, and the *whole lot* is to be accepted if no more than one defective is found in the sample. That is, the lot is accepted if the sample contains zero or one defective; otherwise, the lot is rejected. Hence, if the *lot* is 10 percent defective

$$P(\text{Acceptance}) = P(0 \text{ defective}) + P(1 \text{ defective})$$
$$= P(0) + P(1)$$
$$= C_0^5(0.1)^0(0.9)^5 + C_1^5(0.1)^1(0.9)^4$$
$$= (0.9)^5 + 5(0.1)(0.9)^4$$
$$= 0.59049 + 0.32805$$
$$= 0.91854.$$

The probability of acceptance is about 0.92, so if lots 10 percent defective are submitted to this sampling plan, 92 percent will be accepted and only 8 percent rejected.

Exercise. If, in the foregoing, lots submitted are one percent defective, what is the probability of acceptance and rejection? Answer: $(0.99)^5 + 5(0.01)(0.99)^4 = 0.951 + 0.048 = 0.999$ as the probability of acceptance and 0.001 as the probability of rejection.

If we consider 10 percent defective as poor quality and one percent defective as good quality, then the sampling plan at hand, with a sample of $n = 5$, has a high probability (0.999) of accepting good quality, but also an undesirably high probability (0.92) of accepting poor quality. We can easily see that the probability of accepting poor quality would be reduced by selecting a larger sample, say 20, and accepting the lot only if *none* of the 20 is defective. Now, if lots are 10 percent defective,

$$P(\text{Acceptance}) = P(0 \text{ defective}) = C_0^{20}(0.1)^0(0.9)^{20} = 0.12.$$

This plan offers better protection against the acceptance of poor lots. On the other hand, a good lot (say one percent defective) now has

$$P(\text{Acceptance}) = P(0 \text{ defective}) = C_0^{20}(0.01)^0(0.99)^{20} = 0.82,$$

so in reducing the probability of accepting poor lots (from 0.92 to 0.12) we also lower the probability of accepting good lots (from 0.999 to 0.82). Clearly, the choice of sampling plan (number in the sample to be inspected and the number of defectives permitted in the sample) can be adjusted to achieve a balance between the chances of accepting good and poor lots. For example, we may select 25, and accept the lot if it has no more than three defectives. If p is the proportion defective in a lot, then

$$P(\text{Acceptance}) = P(0, 1, 2, \text{ or } 3 \text{ defectives})$$
$$= P(0) + P(1) + P(2) + P(3)$$
$$= C_0^{25}p^0q^{25} + C_1^{25}p^1q^{24} + C_2^{25}p^2q^{23} + C_3^{25}p^3q^{22}$$
$$= \sum_{x=0}^{3} C_x^{25} p^x q^{25-x}.$$

We refer to the last expression as a *cumulative* binomial probability, and hasten to add that tables are available for these cumulative probabilities. A single such table for $n = 25$ is presented as Table XIII at the end of the book.[5] For the problem at hand, we find from Table XIII that for $p = 0.01$

$$\sum_{x=0}^{3} C_x^{25}(0.01)^x(0.99)^{25-x} = 1.000.$$

The tabular entries have been rounded to three decimals and the last-written number, 1.000, does not mean exactly 1, but a number less than 1 which rounds to 1.000. We interpret this result by saying that if $n = 25$ items are selected from a lot which is one percent defective and three defectives are allowed in the sample, it is almost certain that the lot will be accepted.

‖‖‖

Exercise.   For the foregoing find the probability of acceptance if the lot is 10 percent defective. Answer: 0.764.

‖‖‖

To demonstrate an understanding of cumulative probabilities, we should be able to write summation expressions similar to the one we wrote in the foregoing. Thus, if we toss a coin 50 times and seek the probability of getting *at most* 10 heads, we have $n = 50, p = 0.5$. The probability of 0, 1, 2, $\cdots$, 10 heads would be expressed as

$$\sum_{x=0}^{10} C_x^{50}(0.5)^x(0.5)^{50-x}.$$

‖‖‖

Exercise.   If a salesperson's probability of making a sale on any contact is 0.08, express in summation symbols the probability of making at most 15 sales in 100 contacts. Answer: $\sum_{x=0}^{15} C_x^{100}(0.08)^x(0.92)^{100-x}$.

‖‖‖

Now, for brevity, we shall express the cumulative probability for, say, 0, 1, 2, 3, 4, 5, as

$$\sum_{0}^{5}.$$

---

[5] Cumulative binomial probabilities to five decimal places for p in intervals of 0.01 and for selected values of n up to 1,000 may be found in *Tables of the Cumulative Binomial Probability Distribution,* Harvard University Press, 1955.

Thus, from Table XIII, with $n = 25$, $p = 0.4$

$$\sum_0^5 = 0.029$$

is the probability of *at most* five occurrences. If we seek the probability of *at least* five occurrences, which would be 5, 6, 7, ..., 25, it would be found as the complementary probability

$$1 - \sum_0^4 = 1 - 0.009 = 0.991.$$

Again, the probability of six to 10 occurrences, inclusive, would be

$$\sum_0^{10} - \sum_0^5 = 0.586 - 0.029 = 0.557.$$

Finally, if we want the probability of *exactly* five occurrences, we compute

$$\sum_0^5 - \sum_0^4 = 0.029 - 0.009 = 0.020.$$

Exercise.   For $n = 25$, $p = 0.3$, find the probability of: *a*) Less than eight occurrences. *b*) At most five occurrences. *c*) At least 10 occurrences. *d*) Between five and 10 occurrences, inclusive. *e*) Exactly eight occurrences. Answer: *a*) 0.512. *b*) 0.193. *c*) 0.189. *d*) 0.812. *e*) 0.165.

As an application of Table XIII, suppose a student takes a 25-question true-false test and determines the answer to each by flipping a coin, so that the probability of getting the correct answer is $p = 0.5$ for each question. If 60 percent or more correct is passing, what is the probability of passing? We note that 60 percent or more is $0.6(25) = 15$ or more correct. Hence,

$$P(\text{Pass}) = \sum_{15}^{25} = 1 - \sum_0^{14} = 1 - 0.788 = 0.212.$$

The student's probability of not passing is 0.788, or about 0.8. In other terminology, the odds are about 4 to 1 against passing.

Exercise.   If a student takes a 25-question four-choice multiple choice examination and judges that his probability of getting any question correct is about 0.6, what is the probability that he will get 80 percent or more correct? Answer: 0.029.

12.14 PROBLEM SET 4

Compute the answers to Problems 1–8 using the binomial rule.

1. A basketball player has a probability of 0.3 of hitting on any shot from the foul line. What is the probability of:
   a) Exactly one hit in three trials?
   b) At least one hit in three trials?
   c) Exactly two hits in five trials?
   d) More than two hits in four trials?

2. The probability that an inspector will properly classify an item is 0.8. If each item is inspected independently by three inspectors, what is the probability that at least one will properly classify the item?

3. A company has bid on five projects, assessing the probability of winning a contract at 0.6. To have a successful year, it must win at least two of the contracts. What is the probability for a successful year?

4. A sample of six items is selected from a large lot. The *lot* is accepted if the *sample* contains no more than one defective item. Find the probabilities of accepting and rejecting a lot if the proportion defective in the lot is: a) 0.1. b) 0.2.

5. A department store employs four women who take orders over the telephone. Each woman is busy taking an order 70 percent of the time. What is the probability that an operator will be free to take an order at the time of a call:
   a) If one customer calls at a point in time?
   b) If three customers call at the same time?

6. See Problem 5. How many telephone operators should the store have if the probability that an operator will be free when a customer calls is to be 0.83?

7. A true-false examination has five questions, and a student guesses the answer for each question, assigning probability of 0.5 of being correct. Assuming independence, what is the probability that he gets
   a) All five correct?   b) At least four correct?
   c) Exactly three correct?   d) At least three correct?
   e) At least four incorrect?

8. A test has five four-choice questions, and a student guesses the answer to each question. Assuming independence, what is the probability of
   a) All five correct?   b) At least four correct?
   c) Exactly three correct?   d) At least three correct?

9. The probability that an item in a large group is defective is 0.05. Express the following by use of the summation symbol, but do not try to calculate the answer.
   a) The probability that at most 15 of 100 items purchased are defective.
   b) The probability that more than 10 of 200 items purchased are defective.

Use Table XIII to answer Problems 10–13.

10. A sample of 25 items is selected from a large lot and the lot is rejected if the sample contains more than four defectives. Find the probability of acceptance if the proportion defective in a lot is: a) 0.05. b) 0.20.

11. Repeat Problem 10 assuming the lot is rejected if more than one defective is found in the sample.

12. A test has 25 five-choice questions. A student gives answers at random. What is the probability that he gets:
   a) Less than 40 percent correct?
   b) More than 40 percent correct?
   c) At least 20 percent correct?
   d) Exactly eight correct?
   e) Six to 10, inclusive, correct?
   f) At most five correct?
13. Repeat Problem 12 if the student assesses his probability of getting a correct answer at 0.5.

12.15 EXPECTED MONETARY VALUE (EMV)

If we bet $1 that heads will appear on the toss of a coin for which we have assigned

$$P(H) = P(T) = 1/2$$

we win $1 if heads appears, and win $-$1 (lose $1) if tails appears. We compute the *expected monetary value* of the *act* of tossing the coin as

EMV = (payoff if *event H* occurs)$P(H)$ + (payoff if *event T* occurs)$P(T)$
   = $1.00(1/2) + (-$1.00)(1/2) = 0.

More generally, the expected monetary value of an *act* is the sum of the products formed by multiplying the dollar payoff of each *event* by the probability of the event. It is assumed that the events constitute a sample space. For example, if we consider act A as having three events, E_1, E_2, and E_3, with probabilities 0.4, 0.5, and 0.1, respectively, and payoffs $10, $-$8, and $2, then

EMV of act A = 0.4(10) + 0.5($-$8) + 0.1(2) = $0.20.

Example 8. An urn contains five red, one white, and four green balls, and we assign probabilities

$$P(R) = 0.5$$
$$P(W) = 0.1$$
$$P(G) = 0.4.$$

A ball is to be drawn, and the payoffs are red ball, lose $1; white ball, win $3; green ball, win nothing. Compute the EMV of the act of drawing a ball.

P	Event	Payoff
0.5	R	$-$1
0.1	W	3
0.4	G	0

EMV = 0.5($-$1) + 0.1(3) + 0.4(0)
   = $-$$0.20.

512   *Mathematics*

Exercise. What is the expected monetary value of the act of tossing two coins if the payoffs are $0 for zero heads, $1 for one head, and $-\$1$ (loss of $1) for two heads? Answer: $(0)(1/4) + (\$1)(1/2) + (-\$1)(1/4) = \$0.25$.

Expected monetary value has been advanced as one criterion to aid decision making. The notion is that we list the various events which might arise in a certain situation and assign a probability to each event. In addition, we consider the payoffs which would occur for each event for each decision we might make, the decisions being the choice of act 1, act 2, and so on. Suppose that we use EMV to choose between act 1 and act 2 in Table 12–3. According to the EMV criterion, the decision maker would choose act 1 rather than act 2 because act 1 has the higher expected monetary value.

TABLE 12–3
Payoff Table

		Act	
P	Event	A_1	A_2
0.3	E_1	$2.00	$2.00
0.4	E_2	1.00	3.00
0.3	E_3	8.00	3.00
1.0			

EMV, act 1 = 0.3(2) + 0.4(1) + 0.3(8) = 3.4.
EMV, act 2 = 0.3(2) + 0.4(3) + 0.3(3) = 2.7.

Example 9. Items are manufactured for sale. Each unit made and sold yields a profit of $3; each unit made but not sold yields a loss of $1. It is believed that zero, one, two, or three units might be demanded by customers, but the event *four or more* units demanded is considered impossible and assigned probability zero. Other probabilities are assigned by experience and judgment (Table 12–4). Use the expected monetary value to decide

TABLE 12–4

Events (Number of Units Demanded)	Probability of Number of Units Being Demanded
0	0.2
1	0.4
2	0.3
3	0.1
4 or more	0.0
	1.0

whether to make zero units (act 1), one unit (act 2), or two units (act 3).

The payoff table can be filled in from the given information. For example, if two units are made and one unit is demanded, one of the two would yield a profit of $3 and the other a loss of $1, for a payoff of $2 net. Again, if one unit is made and two are demanded, the payoff is $3 on the single unit made. These and the remaining payoffs are shown in Table 12–5.

TABLE 12–5
Payoff Table

		Acts			
P	*Events* *(Units Demanded)*	A_1 *(Make 0)*	A_2 *(Make 1)*	A_3 *(Make 2)*	A_4 *(Make 3)*
0.2 0		0	−1	−2	−3
0.4 1		0	3	2	1
0.3 2		0	3	6	5
0.1 3		0	3	6	9
0.0 . . . . 4 or more		0	3	6	9
1.0					

EMV of $A_1 = 0.0$
EMV of $A_2 = 2.2$
EMV of $A_3 = 2.8$
EMV of $A_4 = 2.2$

The decision would be to choose A_3, the act with the highest EMV, and so make two units.

EMV can be a useful criterion in some decisions. However, it is easy to illustrate that this criterion does not have general applicability. For example, the EMV of A_1 in the following table is $2500, compared to an EMV of $200 for A_2, and yet some persons would prefer A_2 to A_1. The point here is that

P	*Event*	A_1	A_2
0.5 . . . . E_1		$10,000	$400
0.5 . . . . E_2		− 5000	0

EMV of $A_1 = $2500
EMV of $A_2 = 200$

even though the EMV of A_1 is much larger than that of A_2, some people would not feel they could afford a loss of $5000 which would arise if they chose A_1 and event E_2 occurred. Others would prefer A_2 on the ground that they cannot lose if they choose A_2, and have a 0.5 probability of gaining $400. Of course, a person possessing a large amount of money might well choose A_1 because he can afford to lose $5000 and thinks a 50–50 gamble of winning $10,000 or losing $5000 is sensible.

Hence, we see that the act chosen depends upon the person making the decision and the amounts involved. EMV may or may not be a proper guide for action. A criterion applicable when EMV is not appropriate is *expected utility value* (EUV), which allows a person to inject his own circumstances and inclinations into the analysis. Exploration of the EUV criterion would carry us beyond our immediate goals.[6]

12.16 PROBLEM SET 5

1. A pair of dice is to be rolled. If the number appearing is even, you win that even number of dollars; if the number appearing is odd, you lose that odd number of dollars. Compute the EMV of the act "rolling the pair of dice."

2. An act is accompanied by three possible events with probabilities 0.2, 0.3, and 0.5, and payoffs $2, $3, and −$1, respectively. Compute the expected monetary value of the act.

3. Urn number one contains four red, nine white, and seven green balls with payoffs $2, −$4, and $2, respectively. Urn number two contains four red and six black balls with payoffs $3 and −$1.80, respectively. If act 1 is selecting a ball from urn number one and act 2 is selecting a ball from urn number two, which act should be chosen according to the criterion of expected monetary value?

4. Which act should be chosen according to EMV?

Payoff Table

P	Event	A_1	A_2	A_3	A_4
0.2 . . . . E_1		$2	$1	$0	$0
0.1 . . . . E_2		2	2	−1	−3
0.4 . . . . E_3		2	3	3	3
0.3 . . . . E_4		2	2	4	5

5. If you make a unit of product and it is sold (demanded), you gain $5; if you make a unit which is not sold, you lose $2. You assign probabilities as follows:

Number of Units Demanded	Probability of Number of Units Demanded
0	0.10
1	0.20
2	0.25
3	0.40
4	0.05
5 or more	0.00

[6]The reader is encouraged to investigate Robert Schlaifer, *Introduction to Statistics for Business Decisions* (New York: McGraw-Hill Book Co., 1961) as his next step toward achieving a fuller understanding of the role of probability in business decisions.

According to the EMV criterion, how many units should you make?

6. In setting premiums to charge for protection against various hazards, insurance companies must start with a base figure (exclusive of overhead and profit) which represents their expected loss. A building is to be insured in the amount of $60,000 for fire damage. The probabilities of total, 75 percent, 50 percent, and 25 percent losses in a year are, respectively, 0.0001, 0.00015, 0.0005, and 0.001. Assuming these are the only losses to be considered,

   a) What base figure should be used in computing the annual premium?

   b) Why do the probabilities given not add up to 1?

7. Think seriously about your present circumstances, and then decide in each case whether you would choose act 1 or act 2. For example, in part (*a*), would you prefer a 0.6 probability of gaining $3, 0.4 of losing $1, to a gamble which has a 0.6 probability of gaining $1? (There are no correct answers to this question.)

a) *P*	A_1	A_2		*b*) *P*	A_1	A_2
0.6	$3	$1		0.6	$30	$10
0.4	− 1	0.		0.4	− 10	0.

c) *P*	A_1	A_2
0.5	$3000	$500
0.5	− 1000	0.

12.17 REVIEW PROBLEMS

1. The table shows, for example, that 40 cars of make Y had gear train malfunctions.

	Malfunction		
Make of Car	*Electrical E*	*Gear Train G*	*Carburetor C*
X	17	60	23
Y	20	40	60
Z	15	48	117

Find the following probabilities:

a) $P(Y)$.	*b*) $P(E)$.	*c*) $P(C)$.	*d*) $P(X \cap G)$.
e) $P(G \cap X)$.	*f*) $P(E \cap G)$.	*g*) $P(Z \cap X)$.	*h*) $P(Y \cap C)$.
i) $P(X \mid C)$.	*j*) $P(C \mid X)$.	*k*) $P(E \mid Z)$.	*l*) $P(Z \mid G)$.
m) $P(X \cup Y)$.	*n*) $P(X \cup C)$.	*o*) $P(G \cup C)$.	*p*) $P(Y \cup G)$.

2. See the table of Problem 1.

   a) Are Z and C independent? Explain.

   b) Are Y and C independent? Explain.

   c) What is $P(X \cap Y)$? What does this mean?

3. *a*) In some areas, one may often predict tomorrow's weather correctly by stating it will be the same as today's weather. In such areas, is tomorrow's weather independent of today's weather? Explain.

   b) A card is to be drawn from a deck. Let B represent black card and D represent diamond. Are B and D mutually exclusive? Independent? Explain.

c) Consider the physical traits of brown eyes, *B*, and dark hair, *D*. Are *B* and *D* mutually exclusive? Do *you* think *B* and *D* are independent?

4. *a)* Complete the following table if all event pairs such as AX are independent.

	X	Y	Z	Total
A				0.40
B				0.60
Total	0.30	0.20	0.50	

Find the following probabilities from the completed table.

b) $P(A \mid X)$.   *c)* $P(X \mid A)$.   *d)* $P(X)$.   *e)* $P(A)$.

f) $P(A \cup X)$.   *g)* $P(X \cup Y)$.

5. *a)* Complete the following probability tables. (Note: A' and B' are complements of *A* and *B*, respectively.)

	B	B'	Total
A			0.70*
A'		0.12*	
Total		0.34*	

b) Are *A* and *B* independent? Explain.

Compute:

c) $P(B' \mid A')$.   *d)* $P(A \cup B')$.

6. An investor assesses the probability that the Dow-Jones stock market average will rise tomorrow as 0.65, and the probability that the price of stock *X* will rise if the Dow-Jones rises as 0.90. What is the probability that the Dow-Jones will rise and *X* will rise? Why?

7. A candidate runs for offices *A* and *B*, assessing the probability of winning both at 0.10, and the probability of winning *B* at 0.25. What is the probability of winning *A* if he wins *B*? Why?

8. An investor has funds in banks *A* and *B*. He assesses the probability that *A* will fail at 0.0001, and assigns the same failure probability to *B*. Further, he thinks the probability that both banks will fail is 0.00001.

a) What is the probability that *A* or *B* will fail?

b) Are the events *A* fails, *B* fails independent? Why?

c) What is the probability that *B* fails if *A* fails?

9. A box of eight items contains six good and two defective items. If a sample of two items is selected, what is the probability that:

a) Both will be good?

b) Both will be defective?

c) Exactly one will be defective? (Note: Two events are involved.)

d) If three items are drawn, what is the probability that at least one will be defective?

10. A box contains four items, of which two are good and two defective. A sample of two items is selected.

    a) Letting *G* and *D* stand for good and defective, write a sample space for the experiment using four events and enter the probabilities for each event.

    b) Write a sample space using as events the count of the number of defectives in the sample, and enter the probabilities for each event.

    c) Write a sample space using equally likely events. (Hint: Let G_1D_2 mean good number one and defective number two.)

11. Graduation exercises are to be held outdoors on Friday if it does not rain. If it rains on Friday, the exercises will be postponed until Saturday and held outdoors if it does not rain, and indoors if it rains. The probability that it will rain on Friday is 0.3, and the probability it will not rain on Saturday if it rains on Friday is 0.4. Find the probability that the exercises will be held outdoors.

12. Two women and three men are equally qualified for two positions in a firm. The firm decides to select two of the five at random.

    a) Write a sample space for the experiment using equally likely events.

    b) From *a)* determine the probability that a man and a woman are selected.

    c) Calculate the answer to *b)* by probability rules.

13. A job applicant assigns probabilities as follows: The probability, $P(A)$, of being offered a job at company *A* is 0.4; the probability of being offered a job at *B* is $P(B) = 0.3$, and the probability of being offered a job at both companies is 0.12. What is the probability of being offered a job at at least one of the two companies? At exactly one of the two companies?

14. Probability of colder is assigned at 0.7, probability of snow at 0.4 and probability of neither colder nor snow at 0.2. What is the probability

    a) That it will get colder and snow?

    b) That it will get colder but not snow?

    c) That it will snow but not get colder?

15. A test has two questions. A student assigns probability 0.6 of getting the first correct, 0.3 of getting the second correct, and 0.25 of getting both wrong.

    a) Show that with this assignment of probabilities, the outcome of the second question is not independent of the outcome of the first in the probability sense.

    b) What would the probability of getting both wrong be if the outcomes are to be independent in the probability sense?

16. The claim is made that whether or not an employee's attendance record is good depends upon the sex of the employee. On the basis of the table, using probability terminology, refute the claim.

| | Number of Employees with | |
| | | |
Sex	*Good Attendance Records*	*Poor Attendance Records*
Male 40		10
Female 80		20

17. The probability that machine A will break down on a particular day is $P(A) = 1/100$; similarly, for machine B, $P(B) = 1/200$. Assuming independence, on a particular day,
   a) What is the probability that both will break down?
   b) What is the probability that neither will break down?
   c) What is the probability that one or the other will break down?
   d) What is the probability that exactly one will break down?

18. Five parts go into the assembly of item X. The assembly is defective if any one of the parts is defective, and each part has probability of 0.01 of being defective. Assuming independence, what is the probability that an assembly is defective?

19. There are five intersections between two cities where a driver can bear left or bear right. If, and only if, the proper turn is made at each intersection will a driver starting from one city arrive at the second city. Suppose that the driver flips a coin to choose each turn. What is the probability that he will arrive at the second city?

20. Given $P(X) = 0.5$, $P(Y) = 0.7$, and $P(XY) = 0.30$, find $P(X \mid Y)$ and $P(Y \mid X)$.

21. If $P(Y \mid X) = 0.80$, $P(X \mid Y) = 0.75$, and $P(XY) = 0.60$, find $P(X)$ and $P(Y)$.

22. If $P(Y \mid X) = 0.80$, $P(XY) = 0.20$, $P(Y') = 0.30$, find $P(X \cup Y)$.

23. Job candidates are screened by means of a preliminary interview. The probability is 0.6 that a screened candidate will be a good worker. Screened candidates are given a test. If the candidate is one who will prove to be a good worker, the probability of his passing the test is 0.90. If the candidate is one who will prove to be a poor worker, the probability of his passing the test is 0.40.
   a) What is the probability that a screened candidate will be a good worker and pass the test?
   b) What is the probability that a screened candidate will not be a good worker and will pass the test?

24. A candidate runs for two political offices, A and B. He assigns 0.27 as the probability of being elected to both, 0.50 as the probability of being elected to A if he is elected to B, and 0.90 as the probability of being elected to B if he is elected to A.
   a) What is the probability of being elected to A?
   b) What is the probability of being elected to B?
   c) What is the probability of being elected to neither?
   d) What is the probability of being elected to at least one of the offices?

25. Mr. M thinks his probability of winning an election is 0.9 if Ms. W does not run, but only 0.3 if Ms. W does run. Mr. M judges the probability that Ms. W will run to be 0.4. What is the probability that Mr. M wins the election?

26. Complete the Bayes' Rule formulation which starts with $P(A' \mid B) = \qquad$.

27. The probability that a customer will be a bad debt is 0.01. The probability that he will make a large down payment if he is a bad debt is 0.10, and the probability that he will make a large down payment if he is not a bad debt is 0.50. Suppose that a customer makes a large down payment. Find the posterior probability that he will be a bad debt.

28. Complete the tabular analysis for Problem 27, and find the probability that a customer will not be a bad debt if he does not make a large down payment.

29. The probability that a person has disease X is $P(X) = 0.008$. The probability that medical examination will indicate the disease if a person has it is $P(I \mid X) = 0.75$, and the probability that examination will indicate the disease if a person does not have it is $P(I \mid X') = 0.01$. What is the probability that a person has the disease if medical examination so indicates?

30. The probability that a machine is running properly is 0.80. From time to time, samples of output are selected and measured, and the sample average is computed. If the machine is running properly, the probability that the sample average will be in a certain range is 0.95. If the machine is not running properly, the probability that the sample average will be in this range is 0.05. A sample is selected, and its average is outside the range. What is the probability that the machine is running properly?

Do Problems 31–41 using the binomial rule.

31. A baseball player assesses his probability of getting on base each time at bat at 0.20. What is the probability that he will get on base:
    a) Exactly once in his next three times at bat?
    b) At least once in his next three trips?
    c) At least twice in his next three trips?
    d) Exactly twice in six trips?

32. The four engines of an airplane operate independently and on an overseas flight the probability that an engine will fail is 0.001. On such a flight, what is the probability that:
    a) Exactly one will fail?
    b) More than one will fail?

33. A true-false examination has five questions, and a student assigns probability of 0.7 of being correct on each of the questions. Assuming independence, what is the probability that he gets:
    a) All five correct?   *b*) At least four correct?
    c) Exactly three correct?   *d*) At least three correct?
    e) At least four incorrect?

34. A test has five five-choice questions and a student guesses the answer to each question. Assuming independence, what is the probability of:
    a) All five correct?   *b*) At least four correct?
    c) Exactly three correct?   *d*) At least three correct?

35. A test has five four-choice questions and a student assigns probability 0.8 of getting each one correct. What is the probability of getting:
    a) All five correct?   *b*) At least three correct?
    c) At least one correct?

36. The probability of success on a single trial of an event is 0.3. In 100 independent trials, write, but do not evaluate, the expression for the probability of:
    a) At least one success.   *b*) At most one success.
    c) Exactly five successes.   *d*) At most three successes.

37. A test has 50 five-choice questions and a student guesses the answer to each question. Write, but do not evaluate, the expression for each of the following probabilities:
    a) Ninety percent or more correct.   *b*) At least one correct.

38. The probability that any particular item is defective is 0.10. Using binomial probabilities, find:
    a) The probability of at most one defective in 10 items.
    b) How many items would have to be selected to have a probability of 0.99 that the group would contain at least one good item.

39. A binomial experiment consists of m trials. Write the summation expression using m, p, q, r, and x for:
    a) The probability of at most r occurrences in the m trials.
    b) The probability of more than r occurrences in the m trials.

40. A department store employs three people who take orders over the telephone. Each person is busy taking an order 80 percent of the time. What is the probability that an operator will be free to take an order at the time of the call:
    a) If one customer calls at a point in time?
    b) If two customers call at a point in time?

41. See Problem 40. How many telephone operators should the store have if the probability that an operator will be free when a customer calls is to be approximately 0.6?

   Do Problems 42–44 using Table XIII.

42. A sample of 25 items is selected from a large lot, and the lot is rejected if the sample contains any defective items. Find the probability of acceptance if the proportion defective in the lot is:
    a) 0.01.   b) 0.10.   c) 0.20.

43. Repeat Problem 42 assuming the lot is rejected if more than two defectives are found in the sample.

44. From past experience, a market research firm knows that 60 percent of the people contacted by telephone will agree to a telephone interview. If the firm calls 25 people, what is the probability of completing:
    a) At most 10 interviews?
    b) At least 10 interviews?
    c) More than 15 interviews?
    d) From 15 to 20 interviews, inclusive?
    e) Exactly 10 interviews?

45. The act is to draw a card. If the card is a face card you win $20; if it is not a face card, you lose $20. Compute the expected monetary value of the act. (Face cards are jack, queen, king, and ace.)

46. The act is rolling a pair of dice. If the dice come up 2, 7, or 11, you win $100; otherwise you lose $16. Compute the expected monetary value of the act.

47. Which act should be chosen according to the EMV criterion?

		Payoffs		
Probability	Event	A_1	A_2	A_3
0.5 . . . . . . E_1		$5	$4	$4
0.3 . . . . . . E_2		5	7	7
0.2 . . . . . . E_3		5	3	5

48. If you make a unit of a product and it is sold (demanded), you gain $10; if you make a unit and it is not sold, you lose $5. You assign probabilities as follows:

Number of Units Demanded	Probability of Number of Units Demanded
0	0.1
1	0.2
2	0.4
3	0.3
4 or more	0.0

According to the EMV criterion, how many units should you make?

49. One million tickets are sold at $1 each for a lottery. There is a first prize of $100,000, two second prizes of $50,000, 10 third prizes of $1000, and 20 fourth prizes of $500. If a person buys a ticket, what is the (expected) value of the ticket?

50. A company offers insurance covering damage of 20 percent, 40 percent, 60 percent, 80 percent, or 100 percent. The owner of a particular property wishes to insure it for $200,000. The company assesses yearly damage probabilities (for the various respective damage percentages) at 0.0010, 0.0008, 0.0006, 0.0004, and 0.0002. What base figure (before overhead and profit) should the company use in establishing the annual premium to charge for insuring the property?

13

Probability as an Area

13.1 INTRODUCTION

MODELS which employ areas under a curve in the assignment of probabilities play an important role in mathematics. In this chapter, we shall discuss such probability assignments, turning to the integral calculus for area measurement, and then introduce the normal probability model and show some of its applications.

13.2 PROBABILITY AND INFINITE SETS

If an urn contains 10 balls, only 1 of which is red, and we mix the balls thoroughly, then select 1 at random, we claim each ball has equal probability of being selected and therefore assign 1/10 as the probability of selecting the red ball. Now, suppose that we retain the 1 red ball, and think of the total number of balls being first 100, then 1000, then 10,000, and so on. The probability of a red ball would change from 0.1 to 0.01 to 0.001 to 0.0001, and so on, continually approaching zero. In conceptual terms, we say that as the total number of balls increases without limit, the probability for any single ball, the red one in this case, approaches zero. We express the concept of the last sentence by saying that if we have an infinite set of balls, the probability for a single ball is zero.

It is true, of course, that the sets we encounter in the real world do not have an unlimited number of elements. However, the concept of an infinite set is not altogether unfamiliar. Consider the set of *all* points on the line segment from 6 to 8 in Figure 13–1. This geometric collection of points is not limited in number. For example, we could divide the segment in half, obtaining the point 7, then divide each half in half, yielding the points $6\frac{1}{2}$ and $7\frac{1}{2}$. The division process has no end. It can be repeated *ad infinitum.* The totality of points on the line segment is an infinite set.

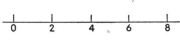

FIGURE 13–1

If, now, we conceive of a random selection process in which all points on the segment from 6 to 8 are equally likely, the probability of selection for any single point is zero. One might object to this statement on the ground that the number 7, say, exists on the line, so it seems reasonable that it be *possible* for the number 7 to be selected. We must agree that this objection is well taken and face the question: "Can it be possible for a number to occur if its probability of occurrence is zero?" The answer is yes if we are to use probability models involving infinite sets.

We call attention here to the discussion of probability in Chapter 12, wherein none of the problems involved infinite sets, and events judged impossible were assigned probability zero. Our attitude here does not contradict such assignments. Rather, our attitude here is that we do not accept the converse; that is, we assign probability zero to an impossible event, but we do not assert that an event with probability zero is impossible. This attitude is dictated by our desire to achieve the advantages of exceedingly useful probability models which involve infinite sets. If we wish to use such models, we must accept the stated attitude as a necessary consequence. The reader may find it instructive to think about the following questions. Suppose that a part is to be made by a machine, and the diameter of the part is to be 2 inches. First question: Is it *possible* that a part will have a diameter of exactly 2 inches? Second question: How *likely* is it that a part will have a diameter of *exactly* 2 inches? Keep in mind here that the concept of *exactly* does not refer to 2.0 inches, nor 2.00, nor 2.00000000000. It refers to 2.000 . . . , the three dots signifying "and so on without limit."

Returning to Figure 13–1, suppose we seek the probability that a point selected at random from the 6 to 8 interval will be in the subinterval from 6 to 7. Now, of course, the numbers in the interval from 6 to 7 form an infinite set, as do those in the interval from 6 to 8. However, we find it quite natural to assign the sought-for probability as 1/2. In so doing, we have measured the sets by the lengths of the line segments. The measure of the set from 6 to 7 is 1, that of the whole set is 2, and the probability assignment is the ratio of the measures, 1/2. Here, then, we have a probability assignment involving infinite sets.

13.3 AREA ASSIGNMENT OF PROBABILITIES

Refer to Figure 13 – 2 whereon is depicted the graph of

$$f(x) = c$$

where c is a constant, over the interval from 6 to 8. We shall now define the probability that x lies in a given *interval* as the area over that interval under the curve (which here is a straight line) divided by the total area from 6 to 8. This means, of course, that the probability that x lies in 6 to 8 is one, and

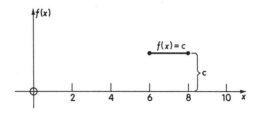

FIGURE 13–2

any interval outside 6 to 8 has a probability of zero. Because we wish to extend our thinking to areas with a curved upper boundary, we shall express areas in terms of integrals. The probability that x lies in the interval from 6 to 7 is

$$\frac{\int_6^7 c\,dx}{\int_6^8 c\,dx} = \frac{cx\,\Big|_6^7}{cx\,\Big|_6^8} = \frac{7c - 6c}{8c - 6c} = \frac{1}{2}.$$

The probability that x will be exactly 7 is

$$\frac{\int_7^7 c\,dx}{\int_6^8 c\,dx} = \frac{0}{2c} = 0,$$

illustrating again that the probability assigned to a point is zero; nonzero probabilities are associated with *intervals.*

The function $f(x) = c$, in a probability sense, represents the case of *equal likelihood.* Again, we must pay attention to concepts. We do not here mean equal likelihood in the sense that every point has zero probability, for this is true for any infinite set. What we mean is that every *interval* of the *same length* has equal probability. Consider next the function

$$f(x) = 1 + 2x$$

shown in Figure 13–3. Equal likelihood is not presumed. We define the probability that x lies outside the interval from 2 to 6 to be zero; that is, the probability that x lies in the interval 2 to 6 is one. We would like to connote the idea that the "amount" of probability over an interval of *fixed* length depends upon the position of the interval. To do so, we refer to the *density* of the probability. In the case of Figure 13–3, for example, the probability is greater over the interval from 4 to 6 than over the *same* length interval from 2 to 4, so it is more dense in the former interval.

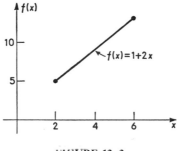

FIGURE 13–3

For the function at hand, the probability that a randomly selected x will lie in the interval from 2 to 4 is

$$\frac{\text{Area over interval from 2 to 4}}{\text{Total area over interval from 2 to 6}}$$

$$= \frac{\int_2^4 (1 + 2x)\,dx}{\int_2^6 (1 + 2x)\,dx} = \frac{x + x^2 \Big|_2^4}{x + x^2 \Big|_2^6} = \frac{7}{18}.$$

Exercise. For the function under discussion, $f(x) = 1 + 2x$, find the probability that a randomly selected x will lie in the interval from 2 to 5 using integral calculus notation. Answer: 2/3.

13.4 PROBABILITY DENSITY FUNCTIONS (p.d.f.'s)

In applied probability models, functions used in the probability sense are adjusted so that the total area under the curve, over the entire range of permissible values of x, is unity, and the functions are then called *probability density functions* or, in abbreviation, p.d.f.'s. We shall use the letter p to represent such functions.

Consider the function just discussed, $f(x) = 1 + 2x$. The permissible range of x was from 2 to 6. The total area under the line over the interval from 2 to 6 is found to be 36. The p.d.f. therefore is

$$p(x) = \frac{1 + 2x}{36}.$$

Inasmuch as total area for any p.d.f. is one, the probability ratio of areas has a denominator of one, which need not be written. Hence the probability that x will lie in the interval from a to b can be expressed simply as

$$\int_a^b p(x)\, dx.$$

Given the probability density function in the immediately foregoing, we express the probability that x will lie in the interval from 2 to 4 as

$$\int_2^4 p(x)\, dx = \int_2^4 \frac{1+2x}{36}\, dx = \frac{7}{18}.$$

Exercise. Consider $f(x) = 6 - 2x$ in the probability sense over the x-interval from 0 to 3. Convert $f(x)$ to a p.d.f., $p(x)$. Answer: $p(x) = (6 - 2x)/9$.

As another example, we shall convert

$$f(x) = 4x - x^2 - 3$$

to a p.d.f. over the interval from 1 to 3:

$$\int_1^3 (4x - x^2 - 3)\, dx = 2x^2 - \frac{x^3}{3} - 3x \Big|_1^3 = \frac{4}{3}$$

Hence:

$$p(x) = \frac{3}{4}(4x - x^2 - 3).$$

Figure 13–4 is the graph of $p(x)$. The probability that a randomly selected

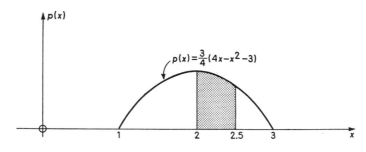

FIGURE 13–4

x will lie in the interval from 2 to 2.5 is shown as a shaded area on the figure. Computing this probability, we find:

$$\int_2^{2.5} \frac{3}{4}(4x - x^2 - 3)\, dx = \frac{3}{4}\left(2x^2 - \frac{x^3}{3} - 3x\right)\Big|_2^{2.5} = 0.344.$$

Exercise. Using the p.d.f. of Figure 13–4, find the probability that a randomly selected x will lie in the interval from 1 to 2.5. What is the relationship of this answer to the shaded area, 0:344, in the figure? Answer: The probability is 0.844, which is 0.5 plus the shaded area in the figure; that is, half the total area, plus the shaded area.

In passing, we note that probability density functions often are referred to as probability *distribution functions,* or distribution functions, or simply distributions.

13.5 PROBLEM SET 1

1. Consider $f(x) = x$ in the probability sense over the interval from 0 to 10.
   a) Integrate to show that the distribution function (the p.d.f.) is

   $$p(x) = \frac{x}{50}.$$

   b) Using the p.d.f., find the probability that a randomly selected x will lie in the intervals 0 to 3, 5 to 7, 4 to 5.
2. Consider $f(x) = x^{1/2}$ in the probability sense over the interval 0 to 9.
   a) Find the p.d.f.
   b) Using the p.d.f., compute the probability that x lies between 1 and 4.
3. Given $p(x) = 1.5x^{1/2}$ over the interval 0 to 1, find the probability that a randomly selected x will lie between 1/9 and 1/4.
4. If kx^2 is a p.d.f. on the interval 1 to 5, what must be the numerical value of k?
5. Interpreting $f(x) = 2 - x^{-2}$ over the interval from 1 to 20 in the probability sense, what is the probability that a randomly selected x will lie between 1 and 10?

13.6 SOME GRAPHIC OBSERVATIONS

Before introducing the normal probability distribution, we shall examine simpler curves of a similar nature and point out some important characteristics. The type of function to be discussed is exponential, the exponent containing the negative of the square of the variable. We choose

$$f(x) = 10^{-x^2}$$

because the problem of finding points for plotting is simplified by selecting 10 as the base of the exponential. For instance, to evaluate the function when x is 0.4, we have

$$10^{-(0.4)^2} = 10^{-0.16}.$$

The number we seek is the number whose logarithm to the base 10 is

$$-0.16 = 0.8400 - 1.$$

Hence:

$$\text{Antilog}(0.8400 - 1) = 0.6918.$$

When x is 0.4, the function is 0.6918.

Proceeding in this manner, points have been determined and graphs drawn for

$$f(x) = 10^{-x^2}$$
$$g(x) = 10^{-(x-1)^2}$$
$$h(x) = 10^{-(x/2)^2}.$$

The graphs are shown on Figure 13–5. Each of the curves is symmetrical with respect to a vertical line drawn to its peak. Each extends indefinitely to the left and to the right, so we say the variable ranges from minus to plus infinity. The x-axis is an asymptote to each curve to the left and to the right; the functions are everywhere positive.

So much for the similarity of the curves. Observe that the central point of

$$g(x) = 10^{-(x-1)^2}$$

is one unit to the right of the central point of

$$f(x) = 10^{-x^2}.$$

It is clear, in general, that the number in the position b of

$$\text{Function of } x = 10^{-(x-b)^2}$$

will determine the central value of x. In a *symmetrical* distribution, **this** central value of x is the *mean* of the distribution. Thus, for the three last-written functions, the means are, respectively, 1, 0, and b.

Comparing the curves for

$$f(x) = 10^{-x^2}$$

and

$$h(x) = 10^{-(x/2)^2}$$

we observe that the latter spreads out more than the former. We use the term *variability* to describe this characteristic, and state that $h(x)$ is more variable, or has greater variability, than $f(x)$. The point of particular interest here is that the difference in variability for the two last-written functions is traceable to the divisor in the exponent. It is clear that the number in the position c of

$$\text{Function of } x = 10^{-(x/c)^2}$$

controls variability. The larger c, **the greater is the variability.**

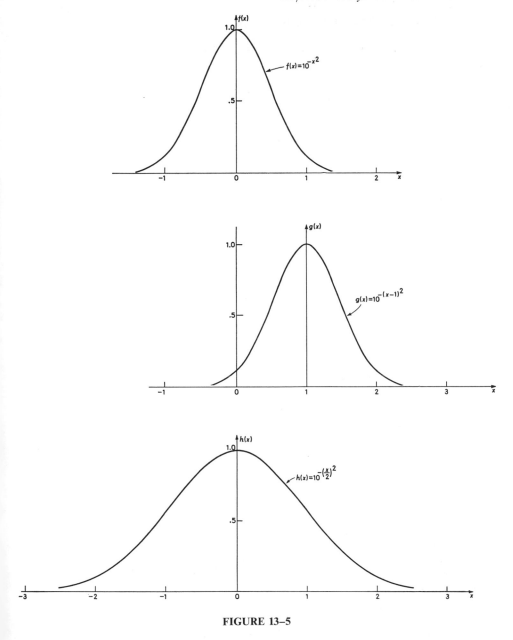

FIGURE 13–5

None of the last three functions qualifies as a p.d.f. because these functions do not have unit total area. However, they serve well to introduce the ideas of the mean and variability.

13.7 THE NORMAL DISTRIBUTION

The exponential giving rise to the so-called "normal" distribution is

$$f(x) = e^{-\frac{1}{2}\left(\frac{x-\mu}{\sigma}\right)^2}.$$

To qualify as a p.d.f., it is necessary to determine k so that

$$k\int_{-\infty}^{+\infty} f(x)\, dx = 1.$$

By advanced methods, it can be proved that

$$k = \frac{1}{\sigma\sqrt{2\pi}}.$$

Hence the equation of the normal probability density function is

$$p(x) = \frac{1}{\sigma\sqrt{2\pi}} e^{-\frac{1}{2}\left(\frac{x-\mu}{\sigma}\right)^2}.$$

The letter μ is the Greek mu; σ is the small Greek sigma. Following the discussion of the previous section, the position of μ shows it to be the *mean* of the distribution. The position of σ relates this quantity to the variability of the distribution. The quantity σ is called the *standard deviation* of the distribution; its square, σ^2, is called the *variance* of the distribution. The larger the standard deviation (variance) of a normal distribution, the more the distribution spreads out. Often, we specify a normal distribution by $N(\mu, \sigma)$, meaning "normal, with mean mu and standard deviation sigma." For example, $N(100, 5)$ specifies a normal distribution with mean 100, standard deviation 5 (variance 25). This specific distribution function is

$$p(x) = \frac{1}{5\sqrt{2\pi}} e^{-\frac{1}{2}\left(\frac{x-100}{5}\right)^2}.$$

Normal curves are similar to (but not the same as) the exponentials plotted on Figure 13–5. They are described as being bell-shaped, symmetrical about the mean, approaching the x-axis asymptotically in both directions.

Exercise. Write the equation of the normal distribution $N(40, 8)$. Geometrically, how would this distribution compare with $N(30, 2)$? Answer: $p(x) = [1/8\sqrt{2\pi}]\left[e^{-\frac{1}{2}\left(\frac{x-40}{8}\right)^2}\right]$. This distribution would have its mean to the right of $N(30, 2)$ and would spread out more than $N(30, 2)$.

13.8 NORMAL PROBABILITY TABLE

Starting from first principles to find the probability that a randomly selected x will lie in a specified interval, we would try to integrate the normal p.d.f. between the interval limits. As it happens, this p.d.f. cannot be integrated exactly, but it is possible to make approximations. It seems natural to suggest that the approximations for various sets of limits be made and the outcomes tabulated, but this raises a problem. There is not just one normal distribution. There is a normal distribution for each of the limitless combinations of mu and sigma. Fortunately, this latter problem can be handled by changing every normal distribution to a *standard* form. To see how this is done, consider the problem of evaluating

$$\frac{1}{\sigma\sqrt{2\pi}} \int_{\mu}^{a} e^{-\frac{1}{2}\left(\frac{x-\mu}{\sigma}\right)^2} dx.$$

(See Figure 13–6.) We introduce a new variable, z, called the standardized normal deviate, where

$$z = \frac{x-\mu}{\sigma}.$$

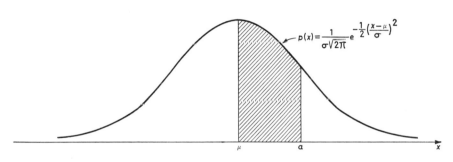

$$p(x) = \frac{1}{\sigma\sqrt{2\pi}} e^{-\frac{1}{2}\left(\frac{x-\mu}{\sigma}\right)^2}$$

FIGURE 13–6

We must now make the following conversions:
If

$$z = \frac{x-\mu}{\sigma}$$

then

$$\frac{dz}{dx} = \frac{d}{dx}\left(\frac{x-\mu}{\sigma}\right) = \frac{1}{\sigma}$$

$$\sigma\, dz = dx.$$

The last statement says we must replace dx in the integral by $\sigma\, dz$. Attention turns next to the limits on the integral. The limit $x = \mu$ becomes

$$z = \frac{x-\mu}{\sigma} = \frac{\mu-\mu}{\sigma} = 0.$$

Similarly, the limit $x = a$ becomes

$$z = \frac{a - \mu}{\sigma}.$$

Substituting, we now have the expression

$$\frac{1}{\sqrt{2\pi}} \int_0^{\frac{a-\mu}{\sigma}} e^{-\frac{1}{2}z^2}\, dz$$

in place of the original integral. See Figure 13–7.

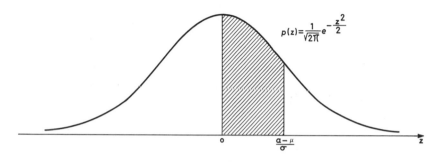

FIGURE 13–7

To see what has been accomplished, consider the two normal distributions $N(50, 5)$ and $N(78, 2)$. Asked to integrate $N(50, 5)$ between the limits 50, which is μ, and 55, which is a, we find

$$z = \frac{55 - 50}{5} = 1$$

for the z corresponding to 55. The z corresponding to μ is, of course, always zero. Our problem here is to integrate the standardized normal distribution from $z = 0$ to $z = 1$.

Suppose, next, we are asked to integrate the second distribution, $N(78, 2)$, between the x limits 78, which is μ, and 80, which is a. We find again that the z limits are 0 and 1. Both problems lead to the evaluation of

$$\frac{1}{\sqrt{2\pi}} \int_0^1 e^{-z^2/2}\, dz.$$

Clearly, the same evaluation arises if the transformation to z yields the interval from $z = 0$ to $z = 1$, no matter what normal distribution is at hand.

Exercise. Given $N(20, 3)$, what z-values correspond to $x = 26$ and $x = 18.5$? Answer: 2, $-1/2$.

Values of the definite integral

$$\frac{1}{\sqrt{2\pi}} \int_0^z e^{-z^2/2} \, dz$$

are given in Table VIII at the end of the book for various values of z. We find, for example, when $z = 1$, the tabulated probability is 0.3413.

13.9 USING THE NORMAL PROBABILITY TABLE

The particular manner in which the normal probabilities are given in Table VIII must be kept in mind so that the problem at hand will be matched properly with the table. Frequently, a sketch will serve to lessen the chance of improper use of the table.

Example. Given a normal distribution with mean 50 and standard deviation 10, what is the probability that a randomly selected number will lie in the interval from 50 to 65?

The desired probability is shown in Figure 13–8 as the shaded area over the interval from 50 to 65. At the point 65:

$$z = \frac{65 - 50}{10} = 1.5.$$

From Table VIII, with $z = 1.5$, we find the desired probability to be 0.4332.

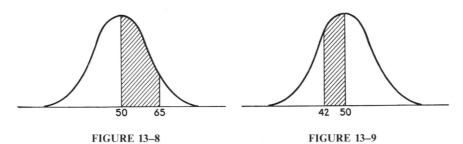

FIGURE 13–8   FIGURE 13–9

Example. Given $N(50, 10)$, as in the last example, what is the probability that x will fall in the interval 42 to 50? As Figure 13–9 shows, 42 lies to the left of the mean. However, the curve is symmetrical. The probability is the same for a given value of z, whether it be positive or negative. Here

$$z = \frac{42 - 50}{10} = -0.8.$$

The desired probability is found by entering Table VIII with z equal 0.8. It is 0.2881.

||||||||||||||||||||||||||||||||||||||||||||||||||||||||||||||||||||||||||||||||||||||||||||||||||||||||||||||||

Exercise. Given $N(50, 10)$ as in the last example, what is the probability that (*a*) x will fall in the interval 56–60? and (*b*) what is the probability that x will be less than 40? Answer: 0.1156 and 0.1587.

||||||||||||||||||||||||||||||||||||||||||||||||||||||||||||||||||||||||||||||||||||||||||||||||||||||||||||||||

Example. Given $N(100, 20)$, what is the probability that x will be greater than 145?

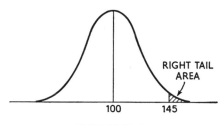

FIGURE 13–10

The pertinent area is shown in Figure 13–10. It is a *right tail* area. Table VIII provides areas only for intervals which start at the mean (that is, at $z = 0$). We can find the desired area by applying the table, then subtracting from 0.5, inasmuch as the entire area to the right of the mean is 0.5.

$$z = \frac{145 - 100}{20} = 2.25.$$

The tabulated entry for $z = 2.25$ is 0.4878. Hence the desired area is

$$0.5000 - 0.4878 = 0.0122.$$

Example. Given $N(2, 0.1)$, what is the probability that x will lie in the interval 1.95 to 2.1?

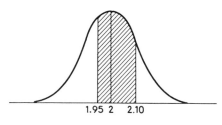

FIGURE 13–11

Figure 13–11 shows that we must add two tabulated entries.

At 1.95, z is -0.5, and the area is 0.1915.
At 2.1, z is 1, and the area is 0.3413.

Adding, we find the desired probability as

$$0.1915 + 0.3413 = 0.5328.$$

Example. Given $N(15, 2)$, find the probability that x will lie in the interval from 16 to 17.

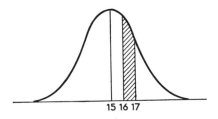

15 16 17

FIGURE 13–12

Analysis of Figure 13–12 shows that the tabulated areas at $z = 1$ and $z = 0.5$ must be subtracted. Hence:

$$0.3413 - 0.1915 = 0.1498.$$

13.10 PROBLEM SET 2

1. Given a normal distribution with mean 150 and standard deviation 20, find the area over each of the following intervals:
   a) 150 to 160, inclusive.
   b) 150 up to but not including 160.
   c) Beyond 170.
   d) 160 to 170.
   e) Below 135.
   f) 145 to 165.
   g) 142 to 150.
   h) 138 to 146.

2. Given $N(50, 5)$ find the probability that a randomly selected x will lie in each of the following intervals:
   a) 45 to 65.
   b) 40 to 60.
   c) 35 to 65.
   d) Outside the interval from 40 to 60.
   e) Above 46.
   f) 44 to 48.
   g) Below 55.
   h) Above 56.

3. Given $N(10, 2)$, find the probability that a randomly selected x will lie in each of the following intervals:
   a) Above 6.
   b) 11 to 15.
   c) Below 9.
   d) 8 to 10.
   e) 8 to 9, excluding the end points.
   f) 8 to 12.
   g) Above 11.5.
   h) 9 to 13.
   i) Above 16.

13.11 ESTIMATING THE MEAN AND THE STANDARD DEVIATION

The normal probability density function as a mathematical model can be related approximately to numerous real-world situations. Suppose, for example, that an automatic machine is adjusted to produce shafts which have a diameter of 2 inches. As shaft after shaft comes off the machine, it is not expected that each shaft will have a diameter of precisely 2 inches. Shaft diameters will vary somewhat, the amount of variability being dependent upon the precision of the machine. Quality control engineers often assume that the probability of various departures from the nominal figure, 2 inches, can be estimated by application of the normal distribution.

If we are to apply a normal distribution to a real-world problem, we must have numbers for the mean and standard deviation of the distribution. In the mathematical model, of course, these quantities represent the central point and the degree of spread of an infinite set of values of the variate x. In the real world, we work with a finite set of numbers and think of them as a *sample* of x's drawn at random from the infinite set. The numbers in the sample reflect the central tendency and variability of the infinite set. We estimate the mean and the standard deviation for the p.d.f. of the infinite set from the numbers in the sample. In the remainder of this section, we shall show how the estimates are made.

Suppose that we select a sample of 10 shafts from the output of a machine and measure the diameter of each shaft, obtaining the numbers in the next table:

Diameters of 10 *Shafts, in Inches*				
2.00	2.01	1.97	2.02	1.97
1.99	2.03	1.99	2.01	2.01

The mean of the p.d.f. model we shall apply to this machine process is estimated by computing the mean (the ordinary average) of the sample. The variance of the p.d.f. is estimated by a procedure which is not intuitively obvious, but which is required by our model. The procedure is as follows:

1. Find the average of the sample data.
2. Subtract the average from a sample number, and square the difference.
3. Repeat Step 2 for each number in the sample, and then sum all the squares.
4. Divide the sum of squares in Step 3 by $n - 1$, where n is the number of values in the sample. The result of this step is the desired estimate of the variance.

The steps are illustrated in Table 13–1. The difference between each value of x and the sample average, 2, is shown in the second column whose heading is $x - 2$, that is, sample value minus sample average. The squares of the differences are shown in the third column, whose heading is descriptive of the method for obtaining its entries. The sum of squares is 0.0036; dividing this sum of squares by one less than the number of values in the sample, that is, by $n - 1$ where n is 10, we obtain the variance estimate, 0.0004. The square root of the variance estimate, 0.02, is the estimated standard deviation. On the evidence of this small sample, we establish the model $N(2, 0.02)$ as the p.d.f. for the shaft diameters. In practice, we prefer to use much larger samples for estimating purposes, but to work with larger samples here would only add to the computational burden.

TABLE 13–1

Sample Value x	x − 2	(x − 2)²
2.00	0.00	0.0000
1.99	−0.01	0.0001
2.01	0.01	0.0001
2.03	0.03	0.0009
1.97	−0.03	0.0009
1.99	−0.01	0.0001
2.02	0.02	0.0004
2.01	0.01	0.0001
1.97	−0.03	0.0009
2.01	0.01	0.0001
20.00	0.00	0.0036

Average $= \dfrac{20}{10} = 2$; $n = 10$.

Variance estimate $= \dfrac{0.0036}{9} = 0.0004$.

Standard deviation estimate $= \sqrt{0.0004} = 0.02$.

The summation symbol, Σ, large sigma, can be put to good use in summarizing instructions for computing estimates of mean and standard deviation. Conventionally, if the variate at hand is x and we wish to indicate the average of a sample of x's, we write the variate with a bar over it, thus: $\bar{x}$, which is read as "x bar" or "bar x." If we now let Σx represent the sum of a set of values of x, and n represent the number of values in the set, the average of the set is

$$\bar{x} = \frac{\Sigma x}{n}.$$

Returning to Table 13–1, n is 10, Σx is 20; hence:

$$\bar{x} = \frac{20}{10} = 2.$$

Turning to the second column of Table 13 − 1, we express the operation of subtracting the average from each value of x by the instruction $x - \bar{x}$. In the third column, we square each difference, an operation denoted by $(x - \bar{x})^2$. The sum of the third column is expressed as $\Sigma (x - \bar{x})^2$. Finally, dividing the last expression by $n - 1$ provides the variance estimate which we now label as s^2:

$$s^2 = \frac{\Sigma (x - \bar{x})^2}{n - 1}.$$

The standard deviation estimate, s, is

$$s = \sqrt{\frac{\Sigma (x - \bar{x})^2}{n - 1}}.$$

Example.   Compute $\bar{x}$ and s for the values of x given in the first column of Table 13–2.

TABLE 13–2

x	$x - \bar{x}$	$(x - \bar{x})^2$
5	−2.2	4.84
8	0.8	0.64
8	0.8	0.64
9	1.8	3.24
6	−1.2	1.44
36		10.80

$$\bar{x} = \frac{\Sigma x}{n} = \frac{36}{5} = 7.2.$$

$$s^2 = \frac{\Sigma (x - \bar{x})^2}{n - 1} = \frac{10.80}{4} = 2.7.$$

$$s = \sqrt{\frac{\Sigma (x - \bar{x})^2}{n - 1}} = \sqrt{\frac{10.80}{4}} = \sqrt{2.7} = 1.64.$$

13.12 PROBLEM SET 3

Compute $\bar{x}$ and s for each given set of x's:

1. 8, 8, 9, 6, 4.
2. 17, 15, 14, 20, 18, 16, 10, 12, 15, 13.
3. 8, 13, 12, 13, 10.
4. 2, 2, 6, 5, 1, 5.
5. 0.64, 0.58, 0.57, 0.57, 0.52, 0.58, 0.60, 0.58.

13.13 $N(\mu, \sigma)$ IN THE REAL WORLD

In applications the areas for the normal distribution are interpreted as percentages and proportions as well as probability, as we shall see in the examples which follow. In each example the mean and the standard deviation are presumed to have been estimated from sample data by the methods of the previous section.

Example.   An automatic filling machine can be set to pour a certain number of ounces of fluid into containers. Data collected from runs of the machine show that the *average* amount of fill per container is, for practical purposes, equal to the machine setting. However, the amount of fill per can varies, and the standard deviation of amount of fill per can has been esti-mated from sample data to be 0.4 ounces.

1. If the machine is set to pour 32 ounces, what is the probability that a container will receive less than 31.5 ounces?

We interpret this problem as one of finding the probability that x will be less than 31.5 in a normal distribution whose mean is 32 and whose

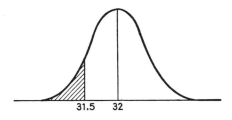

FIGURE 13–13

standard deviation is 0.4. Figure 13–13 illustrates the probability sought.

$$\text{At } 31.5, \; z = \frac{31.5 - 32}{0.4} = -1.25.$$

From Table VIII, with $z = 1.25$, the tabulated area is 0.3944. The shaded area in the sketch is

$$0.5000 - 0.3944 = 0.1056.$$

We estimate the probability that a can will receive less than 31.5 ounces to be 0.1056.

2. Given the circumstances in part 1, what percent of the containers will receive less than 31.5 ounces?

The probability, 0.1056, here is interpreted as the percent equivalent, 10.56 percent; that is, over the long run, about 10.56 percent of the containers will receive less than 31.5 ounces.

3. Given the circumstances in part 1, if 1000 containers are filled, approximately how many will contain less than 31.5 ounces?

Following the interpretation of part 2, we find 10.56 percent of 1000. This is

$$0.1056(1000) = 105.6$$

or about 106 containers.

Example. Given the same machine as in the last example, suppose that we wish to set the machine so that not more than 5 percent of the containers will receive less than 32 ounces. What should be the setting?

The nature of this problem is seen in Figure 13–14. We wish to find the mean (the setting) so that the probability below 32 is 0.05 (that is, the given

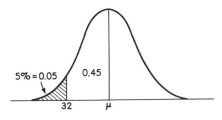

FIGURE 13–14

5 percent). This means, by subtraction, that the area over the interval from 32 to μ must be

$$0.5 - 0.05 = 0.45.$$

We can find, by inverse use of Table VIII, what value of z corresponds to the area 0.45.

Turning to Table VIII, we see in the field that 0.4500 is between the tabulated values 0.4495 and 0.4505. The corresponding marginal values, z, are 1.64 and 1.65, respectively. By interpolation, $z = 1.645$.

Now, it is important to understand that in terms of the units of the original problem, *a value of z means z times the standard deviation*. In the present case, $z = 1.645$ means

$$1.645(0.4 \text{ ounces}) = 0.658 \text{ ounces}.$$

Hence the interval from 32 up to μ is 0.658 ounces, which tells us the machine should be set at

$$\mu = 32 + 0.658 = 32.658 \text{ ounces}$$

if not more than 5 percent of the containers are to receive less than 32 ounces.

The italicized words in the immediately foregoing refer to the definition

$$z = \frac{x - \mu}{\sigma}$$

from which it follows that

$$z\sigma = x - \mu.$$

The interval $x - \mu$ is the distance from x to the mean. It equals z (for the x in question) times the standard deviation. In passing, we note that the question whether to add $z\sigma$ to x or to subtract $z\sigma$ from x to locate the mean depends upon which side of the mean our sketch shows x to be. In the present case the mean is to the right of 32, so we added $z\sigma$.

Example.   If light bulbs of a certain type have an average burning life of 500 hours and a standard deviation of 40 hours, as estimated from experience, within what limits symmetrically located above and below 500 will the burning lives of half of such bulbs lie?

The desired limits are shown by question mark in Figure 13–15. If we

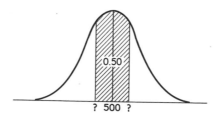

FIGURE 13–15

want half (50 percent) of the numbers to fall between these limits which are symmetrically located with respect to the mean, then 25 percent will be in each of the intervals 500 to ?. Entering Table VIII, we find that the z corresponding to an area of 0.25 is 0.675. Hence the length of the interval from 500 to ? is 0.675 times the standard deviation:

$$0.675(40) = 27.$$

The desired limits are

$$500 \pm 27 \quad \text{or} \quad 473 \text{ to } 527 \text{ hours.}$$

Example. A testing service reports the average score on a test as 130 points, and the standard deviation of test scores as 20 points. Jones takes the test and scores 118 points. What is Jones's percentile standing? See Figure 13–16.

We interpret percentile standing as the percent scoring less than Jones, and estimate it as being the percent below 118 in $N(130, 20)$:

$$z = \frac{130 - 118}{20} = 0.6.$$

From Table VIII, with $z = 0.6$, we obtain the area 0.2257. Figure 13–16

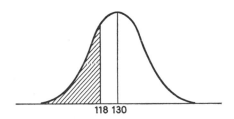

118 130

FIGURE 13–16

shows that the area *below* Jones's 118 points is found by the subtraction

$$0.5000 - 0.2257 = 0.2743 = 27.43\%.$$

Hence, we estimate Jones's score as being at the 27th percentile.

Example. Study of records of the number of units of item K3 in inventory day by day shows average inventory to be 50 units, and the standard deviation of the day-by-day numbers to be 10 units. Suppose that during a day, 25 units are requested. What is the probability that there will not be enough units in inventory to meet this day's requests?

We interpret this problem as one of finding the probability that x is less than 25 in $N(50, 10)$.

$$z = \frac{25 - 50}{10} = -2.5.$$

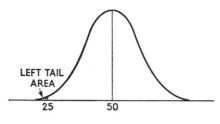

FIGURE 13–17

From Table VIII, with $z = 2.5$, we obtain the area 0.4938. According to Figure 13–17, the desired area is

$$0.5000 - 0.4938 = 0.0062.$$

The probability of not being able to meet the requests is the small figure 0.0062; that is, the chances are only six in a thousand that such requests could not be satisfied.

13.14 PROBLEM SET 4

All problems in this set are to be solved under the assumption that the normal p.d.f. model is applicable.

1. Given $N(50, 5)$:
   a) The probability is 0.17 that x will lie in the interval from 50 up to what number?
   b) The probability is 0.10 that x will exceed what number?
   c) The probability is 0.05 that x will be less than what number?
   d) The probability is 0.50 that x will lie in what interval symmetrically located above and below 50?
   e) The probability is 0.95 that x will be less than what number?

2. Assuming that a machine will turn out parts whose average diameter is the figure at which the machine is set, and that the standard deviation has been estimated from sample data to be 0.001 inches:
   a) If the machine is set at 2.00 inches, what percentage of parts made will have diameters exceeding 2.001 inches?
   b) If the machine is set at 0.100 inches, what percentage of parts made will have diameters exceeding 0.098 inches?
   c) If the machine is set at 1.500 inches, and 1000 parts are made, approximately how many parts will have diameters less than 1.498 inches?
   d) Specifications state that parts are to have 2.000-inch diameter, tolerance plus or minus 0.002; that is, parts whose diameters differ by more than 0.002 from 2.000 are to be classified as defectives. What percent of the parts will be defective?
   e) Within what limits symmetrically located about a setting of 1.575 inches will the diameters of 95 percent of the parts lie?

f) If the machine is set at 2.250 inches, 5 percent of the parts will have diameters less than what number of inches?

g) A loss of 40 cents is incurred for every part which is scrapped because its diameter is too *small*. Specifications state that the part is to have diameter 1.500 inches, tolerance ± 0.0015 inches. The machine is set at 1.500 inches. If 2000 parts are made, what will be the cost of the scrapped undersized parts?

3. The amounts poured by an automatic can-filling machine average out at the machine setting. Collected data indicate that the standard deviation of amounts of fill per can is 0.2 ounces.
   a) If the machine is set to pour 32 ounces, what is the probability that a can will contain less than 31.9 ounces?
   b) If the machine is set to pour 32 ounces, what percent of the cans will contain at least 31.9 ounces?
   c) If the machine is set to pour 32 ounces and 2000 cans are filled, approximately how many cans will contain less than 31.7 ounces?
   d) If the can label states the contents to be 32 ounces and the machine is set to pour 32.25 ounces, what percent of the cans will be underfilled?
   e) If the can labels state contents to be 16 ounces, at what level should the machine be set if 99 percent of the cans are to contain at least 16 ounces?

4. The average breaking strength of certain connectors is 2000 pounds, $s = 80$ pounds.
   a) Within what limits symmetrically located about 2000 will the breaking strength of 75 percent of the connectors lie?
   b) What percent of the connectors will have a breaking strength of more than 2100 pounds?
   c) The breaking strength of 95 percent of the connectors will exceed what number of pounds?
   d) If a connector with breaking strength less than 1800 pounds is classified as defective, what percent of connectors will be defective?

5. A testing service reports the average score on a certain test to be 200 points. $s = 25$ points.
   a) What percent of those taking the test score more than 275?
   b) What percent of those taking the test score less than 160?
   c) Half of those taking the test make scores in what interval symmetrically located above and below 200?
   d) Ninety-five percent of those taking the test score above what number of points?
   e) What would be the percentile standing of a person who scored 260 points on the test?

13.15 REVIEW PROBLEMS

1. Consider $f(x) - x \mid 1$ in the probability sense over the interval from $x = 0$ to $x = 10$:
   a) Integrate to find the probability density function.
   b) Using the probability density function, find the probability that a randomly selected x will lie in the intervals 0 to 1; 5 to 6.

2. Given $f(x) = x^{1/3} - 1$ in the probability sense over the interval 1 to 8:
   a) Find the probability density function.

b) Find the probability that a randomly selected x will lie in the x-interval 1 to 27/8.

3. Given that

$$p(x) = \frac{12x - 3x^2}{32}$$

is a probability density function over the interval $x = 0$ to $x = 4$, find the probability that x will lie in the intervals 0 to 1; 0 to 2; 2 to 3.

4. If $p(x) = kx^{3/2}$ is a probability density function on the interval 0 to 100, what must be the value for k?

5. Interpreting $f(x) = 5\sqrt{x} - x$ over the interval from 0 to 25 in the probability sense, what is the probability that a randomly selected x will lie between 0 and 9?

6. Given a normal distribution with mean 30 and standard deviation 6, find the area over each of the following intervals:

   a) 30 to 33.   *b*) 36 to 42.   *c*) Less than 27.
   d) More than 48.   *e*) 24 to 33.   *f*) More than 21.

7. Given $N(100, 5)$, find the probability that a randomly selected x will lie in each of the following intervals:

   a) 90 to 105.   *b*) 105 to 115.   *c*) 92 to 96.
   d) Less than 103.   *e*) More than 88.   *f*) 112 to 115.

8. Compute the mean, $\bar{x}$, and the standard deviation, s, for each of the following:

   a) 3, 7, 7, 9, 14.   *b*) 3, 5, 7, 7, 8.
   c) 15, 15, 15.   *d*) 1.2, 0.7, 1.2, 1.3.

9. Given $N(20, 4)$:

   a) The probability is 0.20 that x will lie in the interval from 20 up to what number?
   b) The probability is 0.16 that x will exceed what number?
   c) The probability is 0.025 that x will be less than what number?
   d) The probability is 0.50 that x will lie in what interval symmetrically located above and below 20?
   e) The probability is 0.95 that x will be less than what number?

10. Assuming that a machine will turn out parts whose average diameter is the figure at which the machine is set, and that the standard deviation has been estimated from sample data to be 0.002 inches:

   a) If the machine is set at 2.500 inches, what percentage of parts made will have diameters exceeding 2.503 inches?
   b) If the machine is set at 2.000 inches, what percentage of parts made will have diameters exceeding 1.999 inches?
   c) If the machine is set at 2.250 inches, and 1000 parts are made, approximately how many parts will have diameters less than 2.245 inches?
   d) Specifications state that parts are to have 2.500-inch diameter, plus or minus 0.003; that is, parts whose diameters differ by more than 0.003 from 2.500 are to be classified as defectives. If the machine is set at 2.500 inches, what percent of parts will be defective?
   e) Within what limits symmetrically located about a setting of 3.000 inches will the diameters of 95 percent of the parts lie?
   f) If the machine is set at 2.500 inches, 10 percent of the parts will have diameters less than what number of inches?

g) A loss of 50 cents is incurred for every part which is scrapped because its diameter is too small. Specifications state that the part is to have diameter 2.000 inches, tolerance ± 0.004 inches. If the machine is set at 2.000 inches and 5000 parts are made, what will be the cost of the scrapped undersized parts?

11. If 100 coins are tossed and the number of heads recorded, the number of heads, *x*, will be *approximately* normally distributed with mean 50 and standard deviation 5. Assuming $N(50, 5)$,
    a) What percent of the time would more than 65 heads appear in tossing 100 coins?
    b) What is the probability that the number of heads will differ by more than 15, one way or the orther, from 50?
    c) What percent of the time would the number of heads be between 40 and 60?

12. If light bulbs of a certain type have an average burning life of 1000 hours with a standard deviation of 100 hours,
    a) Within what limits symmetrically located above and below 1000 will burning lives of 50 percent of the bulbs fall?
    b) If a large number of these bulbs burn continuously, how long will it be before 40 percent have burned out?
    c) How many bulbs out of 2000 will have burning lives exceeding 1250 hours?

13. The amounts poured by an automatic can-filling machine average out at the machine setting. Collected data indicate that the standard deviation of the amounts of fill per can is 1 percent of the machine setting; that is, for example, if the machine is set at 50 ounces, the standard deviation is 1 percent of 50 = 0.50 ounces.
    a) If the machine is set to pour 20 ounces, what is the probability that a can will contain less than 19.7 ounces?
    b) If the machine is set to pour 50 ounces, how many cans out of 1000 will contain between 49 and 51 ounces?
    c) If can labels state contents to be 48 ounces and the machine is set to pour 49 ounces, what percent of the cans will be underfilled?
    d) If specifications call for cans to contain at least 100 ounces, and the machine is set to pour 101 ounces, what percent of cans will meet specifications?
    e) In (*d*), if cans cannot hold any more than 102.3 ounces, what percent of cans will overflow?
    f) If the maximum amount cans hold is 50.8 ounces and the machine is set at 50 ounces, some cans will overflow. Overflows are defectives and are removed from the batch under production. How many cans will have to be filled if the number left after removing defectives is to be 5000?

14. A testing service reports the average score on a test to be 500 points with a standard deviation of 100 points.
    a) What percent of those taking the test score above 600? Above 700?
    b) Seventy-five percent of those taking the test make scores in what interval symmetrically located about the average?

15. Inventory on hand for item *X* is brought up to 150 at the end of each day. Demand for *X* averages 100 units a day, with a standard deviation of 20 units. On a particular day, what is the probability that demand will exceed inventory on hand?

Appendix 1

Sets

A1.1 INTRODUCTION

IT IS NOT NECESSARY to cover all of this appendix systematically before starting the text. The material contained here is referred to at various points in the text, and it will be sufficient to review the relevant sections when the referral is made.

A1.2 SET TERMINOLOGY

Some simple notions about groups or collections or *sets* are core ideas in mathematics. We use braces to indicate a set, and specify the *members* or *elements* of the set within the braces. Thus,

$$\{\text{Boston, Wellesley, Newton}\}$$

is a set whose members (elements) are the cities Boston, Wellesley, and Newton. Again,

$$\{3, 4, 7, 8\}$$

is a set of numbers whose elements (members) are 3, 4, 7, and 8. The symbols $\in$ and $\notin$ are membership symbols, the first being read as "is a member of," the second as "is not a member of." For example,

$$3 \in \{3, 4, 7, 8\}$$
$$5 \notin \{3, 4, 7, 8\}$$

say that 3 is a member (or element) of the set $\{3, 4, 7, 8\}$ but 5 is not an element of this set.

Each element of a set must be unique. For example, in set terminology, the numbers 1, 3, 4, 4 would be a set whose elements are 1, 3, and 4. A set may have no elements, a finite number of elements, or an unlimited number of elements. The set with no elements is called the *empty* (or *null*) set and is

546

symbolized by ϕ, without braces. A set with an unlimited number of elements is said to be an *infinite* set. Two sets are *equal* if, and only if, they contain exactly the same elements. The expression

$$\{\text{Integers between 5 and 6}\} = \phi$$

says the set of integers between 5 and 6 has no elements; it is the empty set.

Exercise. Read the following aloud:

$$\{\text{Red spades in a bridge deck}\} = \phi.$$

The expression

$$\{\text{Integers between 5 and 7}\} = \{6\}$$

says that the set of integers between 5 and 7 is the set with the single element, 6. The next expression states that the set of integers between 5 and 10 is the set whose elements are 6, 7, 8, 9

$$\{\text{Integers between 5 and 10}\} = \{6, 7, 8, 9\}.$$

Exercise. Express in set symbols that the numbers whose square is 25 are 5 and -5.

The set specified by

$$\{\text{Integers}\}$$

is an infinite set, so we cannot list all its members.

Exercise. Using braces and the symbols for set membership express the facts that $\frac{1}{2}$ is not an integer and 13 is an integer. Answer: $\frac{1}{2} \notin \{\text{Integers}\}$; $13 \in \{\text{Integers}\}$.

It is conventional to use a capital letter, without braces, when an entire set is to be named by one symbol. For example, we might let E represent the (infinite) set of even integers; thus

$$E = \{\text{Even integers}\}.$$

We may then refer to E by statements such as

$$3 \notin E \quad \text{and} \quad 4 \in E$$

which say 3 is not a member of E and 4 is a member of E.

Set terminology is not limited to collections of numbers. We may talk, for example, about the set of residents of New York, classifying residents as being members of the set and nonresidents as not being members of the set.

‖‖‖

Exercise. In set symbols, write the relationship of $ to L, and Q to L, if $L = \{$Capital English letters$\}$.

‖‖‖

A1.3 SET SPECIFICATION

A set is specified by identifying its elements. The *roster* method of specification *lists* each member of the set. The *descriptive* method states the rule or condition which distinguishes members of the set from nonmembers. Thus,

$$S = \{1, 3, 5, 7, 9\}$$

specifies a set by the roster method. The same set could be specified by the condition that its members be odd numbers between 0 and 10; a descriptive specification therefore would be

$$S = \{\text{Odd numbers between 0 and 10}\}.$$

In either specification it is clear whether an object is or is not a member of the set, for example,

$$3 \in S \quad \text{and} \quad 4 \notin S.$$

‖‖‖

Exercise. Specify $W = \{$Days in the week$\}$ by the roster method.

‖‖‖

The set of positive integers, like any infinite set, cannot be listed completely. We may specify the set by the descriptive method as $I = \{$Positive integers$\}$ or by

$$I = \{1, 2, 3, \ldots\}$$

where the three dots are read "and so on." We shall classify this last expression as being a roster specification, although in fact it is a roster-like descriptive specification.

‖‖‖

Exercise. Specify the set of positive odd integers by the "three dot" convention.

‖‖‖

A1.4 SOLUTION SETS FOR EQUATIONS[1]

The statement, "two plus three equals five," is a *sentence* which can be written as

$$2 + 3 = 5.$$

Similarly, the sentence, "*y* plus three equals five," can be expressed as

$$y + 3 = 5.$$

Sentences which use the equality symbol $=$ are called equations. If we replace the symbol *y* by a number which makes the equation become a *true* sentence, the replacement number is a member of the *solution set* of the equation. In the case at hand, the solution set has only one element, 2. We may write

$$\{y : y + 3 = 5\} = \{2\}$$

where, at the left, the symbol $y :$ is read as *the set of y's such that,* and the whole statement says the set of *y*'s such that $y + 3 = 5$ is the set with the element 2.

‖‖

Exercise. *a*) Write the set symbols for $y + 1 = 6$. *b*) Translate the symbols $\{y : y - 2 = 10\} = \{12\}$. Answer: *a*) $\{y : y + 1 = 6\} = \{5\}$. *b*) The set of *y*'s such that *y* minus two is ten is the set with the element 12.

‖‖

A1.5 RELATIONS AND FUNCTIONS[2]

Mrs. Smith is related to her three children, and each of the children is related to Mrs. Smith. However, daughter Mary Smith's relation to her mother is unique because for each child there is exactly one mother. In mathematics we refer to a relation in which for each given first element of a pair there is one and only one second element as a *function*. Thus, a function is a special type of relation. It is helpful to think of a function as a *rule* which states how the second number of the pair is determined when the first number is given. For example, in the function

$$y = 2x,$$

the rule is "multiply by 2." Given 1 as the first number of the pair, the second number is 2 and we write this number of the solution set as (1, 2).

---

[1]Extension of this material is used in Section 1.27 of the text.
[2]This material is referenced in Section 9.2.

Other pairs which are members of the infinite set of pairs in the solution set are (2, 4); (3, 6); (4, 8); and so on.

|||||||||||||||||||||||||||||||||||||||||||||||||||||||||||||||||||||||||||||||||||||||||||||||||||||||||||||||||||||||||||||||||||||||||

> ***Exercise.***   *a*) What is the function rule in $y = 3x + 2$? *b*) How many number pairs are there in the solution set? *c*) Write the pairs in the solution set corresponding to $x = 1, 2,$ and 3. Answer: *a*) Multiply by three and add two. *b*) This is an infinite set. *c*) (1, 5); (2, 8); (3, 11).

|||||||||||||||||||||||||||||||||||||||||||||||||||||||||||||||||||||||||||||||||||||||||||||||||||||||||||||||||||||||||||||||||||||||||

If we define the symbol $<$ to mean is *less than*, then

$$y < x$$

is a relation, but not a function, because if, for example, $x = 10$, y can be any number less than 10 so there is not a unique second number for a given first number.

|||||||||||||||||||||||||||||||||||||||||||||||||||||||||||||||||||||||||||||||||||||||||||||||||||||||||||||||||||||||||||||||||||||||||

> ***Exercise.***   Recall that the product of two numbers of the same sign is positive. Thus, $(+3)(+3) = 9$ and $(-3)(-3) = 9$. Is the relation y times y equals x (that is, $y^2 = x$) a function? Explain. Answer: The relation is not a function because, for example, if x is 16, then y can be either $+4$ or -4.

|||||||||||||||||||||||||||||||||||||||||||||||||||||||||||||||||||||||||||||||||||||||||||||||||||||||||||||||||||||||||||||||||||||||||

A1.6 PROBLEM SET 1

1. Read the following aloud: for example, $A = \{5, 10, 15, \ldots, 100\}$ is read, "A is the set whose elements are 5, 10, 15, and so on, through 100."
   a) $B = \{2, 4, 6\}$.   *b*) $C = \{0\}$.
   c) $Q = \{1, 3, 5, \ldots, 29\}$.   *d*) $F = \{a, e, i, o, u\}$.
   e) $L = \{\text{George, Charles}\}$.   *f*) $K = \{2, 4, 6, \ldots\}$.

2. Specify each of the following sets by the descriptive method:
   a) $F = \{a, e, i, o, u\}$.   *b*) $S = \{3, 6, 9, \ldots\}$.
   c) $P = \{1, 2, 3, 4, 5\}$.   *d*) $G = \{A, B, C, D, E\}$.
   e) $R = \{100, 101, 102, \ldots\}$.   *f*) $J = \{-1, -3, -5, \ldots, -101\}$.

3. Specify each of the following by roster:
   a) {Odd numbers less than 9}.
   b) {First four months of the year}.
   c) {Positive even multiples of 7}.
   d) $\left\{ \begin{array}{l} \text{Positive numbers, less than 1000,} \\ \text{which are exactly divisible by 5} \end{array} \right\}$.

   e) $\left\{ \begin{array}{l} \text{Numbers representing the ratios} \\ \text{formed by dividing integers by 0} \end{array} \right\}$.
   f) {Squares of the integers from 1 through 10}.

4. Why is {0} not equal to ϕ?

5. Using set membership symbols and the roster specification of the sets, write the following:

   a) 1/2 is not a positive integer.   *b*) 64 is a multiple of 4.

   c) *a* is a lower case vowel.   *d*) 4 is not a positive odd number.

   e) \$ is not a letter in the lower case English alphabet.

   f) This problem does not have a part *g*,

6. Given $A = \{1, 3, 4, 7\}$; $B = \{3, 7, 12\}$; $C = \{1, 5, 8\}$, write the following sets:

   a) The set containing all elements which are members of *A*, or members of *B*, or members of both *A* and *B*.

   b) The set of elements that are members of both *A* and *B*.

   c) The set of elements that are members of both *B* and *C*.

   d) The set of elements that are members of *A* but not members of *B*.

   e) The set of elements that are members of both *A* and *C*.

   f) The set of elements that are members of all three sets.

7. *a*) Write the set symbols for $y + 2 = 10$.

   b) Translate the symbols $\{y : y + 5 = 8\} = \{3\}$.

8. *a*) Write the set symbols for $m - 6 = 4$ and its solution set.

   b) Translate the symbols $\{q : 2q = 6\} = \{3\}$.

9. Given the relation $y = 5x + 4$:

   a) Is the relation a function? Explain.

   b) What is the function rule?

   c) What is meant by "the solution set" for the equation?

   d) How many elements are there in the solution set?

   e) What are the elements of the solution set corresponding to $x = 1, 2, 3, 4$?

10. What is the solution set for $\{z : 2z + 5 = 2z + 7\}$?

11. Take any point within a circle as the first point in a relation. Erect a vertical line through that point and take any intersection with the circle as the second point in the relation. Is this relation a function? Explain.

A1.7 SUBSETS, UNIONS, AND INTERSECTIONS[3]

If every element of a set *B* is also an element of a set *A*, then *B* is called a *subset* of *A*. For example, if

$$A = \{1, 3, 6, 9\} \quad \text{and} \quad B = \{3, 6\}$$

then *B* is a subset of *A*. Moreover, in this example, *B* is a *proper* subset of *A* because it does not contain all the elements of *A*. If

$$C = \{8, 9, 10\} \quad \text{and} \quad D = \{8, 9, 10\}$$

then *D* is an *improper* subset of *C*, and vice versa, because all the elements of one are elements of the other. It follows that if two sets are equal, they are improper subsets of each other. The empty set, ϕ, is by definition a

---

[3]This material is referenced in Section 12.4.

proper subset of every set except itself. If $S = \{a, b, c\}$, then the **proper subsets** of S are

$$\{a\}, \{b\}, \{c\}, \{a, b\}, \{a, c\}, \{b, c\}, \text{ and } \phi.$$

Exercise.   List all eight subsets of $K = \{2, 4, 6\}$.[4]

The *intersection* of two sets A and B is a set whose elements are elements of *both* A and B. The intersection symbol is like an inverted U. Thus

$$A \cap B$$

is read as "A intersection B," or "the intersection of A and B." If we have

$$A = \{4, 8, 9, 11\} \quad \text{and} \quad B = \{8, 11, 14\} \quad \text{and} \quad C = \{3, 5\}$$

then

$$A \cap B = \{8, 11\} \quad \text{and} \quad A \cap C = \phi.$$

The last expression says that the intersection of A and C is the empty set because the two sets have no common members. Sets with no common members also are said to be *disjoint* sets.

Exercise.   Given $A = \{a, b, c, d, e\}$ and $B = \{e, f, g, \ldots, z\}$, write $A \cap B$.

The *union* of sets A and B is a set containing those elements which are members of A or members of B, or members of both A and B. A U-like symbol represents the union; thus,

$$A \cup B$$

is read "A union B" or "the union of A and B." If

$$A = \{a, b, c, d, e\} \quad \text{and} \quad B = \{c, e, f, k\}$$

then

$$A \cup B = \{a, b, c, e, f, k\}.$$

Exercise.   Find $A \cup B$ if $A = \{2, 7, 8, 9, 13\}$ and $B = \{1, 8, 13, 22\}$.

---

[4]If a set has n elements, then the set will have 2^n subsets, counting the null set and the improper subset.

It is helpful to illustrate the ideas of intersection and union diagrammatically as in Figure A1–1. If we think of *A* as a set containing all the points in

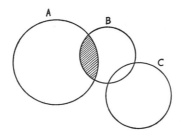

FIGURE A1–1

circle *A*, and similarly for *B* and *C*, then : *A* ∩ *B* is the set of points indicated by the shaded area; *A* ∪ *B* is the set of points containing all the points in both circles *A* and *B*; *A* ∪ *C* contains all the points in circles *A* and *C*; and *A* ∩ *C* = ϕ because *A* and *C* have no points in common, so are disjoint.

A1.8 PROBLEM SET 2

1. Figure A shows set *R* with points *a*, *b*, *c*, *d*, and *e* and similarly for sets *S* and *T*. Write the following sets by the roster method:
   a) *R* ∪ *S*.   b) *R* ∪ *T*.   c) *R* ∩ *S*.
   d) *R* ∩ *T*.   e) *S* ∪ *T*.   f) *S* ∩ *T*.

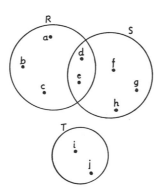

FIGURE A

2. If *A* = {1, 3, 5, ...}, *B* = {0, 2, 4, 6, ...}, and *C* = {5, 7, 9} write the following sets by the roster method:
   a) *A* ∪ *B*.   b) *A* ∩ *B*.   c) *A* ∩ *C*.
   d) *B* ∩ *C*.   e) *A* ∪ *C*.

3. If *A* is the set of face cards in a standard 52-card deck and *B* is the set of queens, what will be the elements of *A* ∪ *B* and *A* ∩ *B*?

4. If *A* is the set of face cards in a standard 52-card deck, *B* is the set of red cards, and *C* is the set of nines, what will be the elements in *A* ∪ *B*, *A* ∩ *B*, *A* ∩ *C*, *B* ∩ *C*?

5. If M represents the set of all points on one line in a plane, and N the points on a different line in the plane, what is $M \cap N$.
   a) If the lines are parallel?
   b) If the lines are not parallel?

6. If R means rain tomorrow and W means warmer tomorrow, what do the following mean?
   a) $R \cup W$.
   b) $R \cap W$.

A1.9 REVIEW PROBLEMS

1. Read the following aloud: for example, $A = \{5, 10, 15, \ldots, 100\}$ is read, "A is the set whose elements are 5, 10, 15, and so on, through 100."
   a) $\{5, 6, 7, 9\}$.
   b) $B = \{$Amherst, Babson, Colgate, Dartmouth$\}$.
   c) $C = \{1, 3, 5, 7, \ldots\}$.
   d) $D = \{A, B, C, \ldots, Z\}$.
   e) $E = \left\{ 1, \dfrac{1}{2}, \dfrac{1}{4}, \dfrac{1}{8}, \ldots \right\}$.
   f) $F = \{$Tom, Dick, Harry$\}$.

2. Specify each of the following sets by the descriptive method:
   a) $A = \{1, 3, 5, 7, 9, \ldots\}$.
   b) $B = \{5, 10, 15, \ldots\}$.
   c) $C = \{101, 102, 103, \ldots, 999\}$.
   d) $D = \{M, A, R, Y\}$.

3. Specify each of the following by the roster method:
   a) $A = \{$Letters in the word HARVARD$\}$.
   b) $B = \{$First nine positive prime numbers$\}$.
   c) $C = \{$Positive odd multiples of 4$\}$.
   d) $D = \{$The first five positive even multiples of 3$\}$.

4. Using set membership symbols and the roster method of specification of the sets, write the following:
   a) The letter S is among the letters in the word BABSON.
   b) 7 is not a positive integral multiple of 3.
   c) 3 is not a positive integral power of 2.
   d) 5 is not a lower case English letter.

5. a) Write the set symbols for $t + 1 = 20$ and its solution set.
   b) Translate the symbols $\{ y : 3y + 5 = 11 \} = \{2\}$.

6. Given the relation $y = x$:
   a) Is the relation a function? Explain.
   b) What is the function rule?
   c) What is meant by the "solution set" for the equation?
   d) How many elements are there in the solution set?
   e) What are the elements in the solution set corresponding to $x = 1, 2, 3, 4$?

7. What is the solution set for $\{ t : t = t + 1 \}$?

8. Suppose x is any positive whole number and y is any positive whole number which is an exact divisor of x. Is the relation between y and x a function? Explain.

9. Given $A = \{5, 10, 12, 13, 15\}$; $B = \{2, 10, 13, 14\}$; $C = \{12, 16, 17\}$; write the following sets:
   a) The set containing all elements which are members of A or members of B, or members of both A and B.
   b) The set of elements that are members of both A and B.
   c) The set of elements that are members of both B and C.
   d) The set of elements that are members of A but not members of B.
   e) The set of elements that are members of both A and C.
   f) The set of elements that are members of all three sets.

10. Figure A shows set R with points p, q, r, s, and t, and similarly for sets S and T. Write the following sets by the roster method:
   a) $R \cup S$.   b) $R \cup T$.   c) $R \cap S$.
   d) $R \cap T$.   e) $S \cup T$.   f) $S \cap T$.

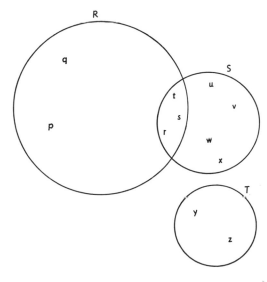

FIGURE A

11. If $A = \{5, 10, 12, 15, 19\}$, $B = \{3, 10, 15\}$, and $C = \{7, 12\}$, write the following sets by the roster method:
   a) $A \cup B$.   b) $A \cap B$.   c) $A \cap C$.
   d) $B \cap C$.   e) $A \cup C$.   f) $B \cup C$.

12. Make a diagram showing sets R, S, and T (like the diagram in Problem 10) corresponding to the following requirements:
   a) $R \cap T = \{x\}$.   b) $R \cap S = \{z\}$.
   c) $T \cap S = \emptyset$.   d) $R \cup S = \{v, x, y, z\}$.
   e) $R \cup T = \{w, x, y, z\}$.

13. If sets M and N are, respectively, the sets of points on the perimeters of two concentric circles, what is $M \cap N$?

14. If P is the set of all cars with air conditioners, Q is the set with automatic shifting, and R is the set with manual (or stick) shifting, what are the following?
   a) $P \cup Q$.   b) $P \cap Q$.
   c) $Q \cap R$.   d) $P \cap R$.

Appendix 2

Elements of Algebra

A2.1 INTRODUCTION

THIS APPENDIX presents the basic definitions, conventions, and rules of algebra, together with discussions and illustrations of the fundamental properties of numbers. Each topic is accompanied by examples which will meet the needs of those who use isolated parts of the appendix for reference purposes. However, topics are presented in logical order so that the needs of those seeking a substantial review of elements will be met by starting at the beginning and moving in an orderly fashion through the material.

A2.2 THE REAL NUMBERS

The set of positive and negative whole numbers, with zero, form the *integers*. Combining the integers with the fractions, we have the set of *rational* numbers. Thus, a rational number (ratio number) is a number which can be expressed in the form of a fraction which has integers as numerator and denominator. For reasons to be discussed soon, fractions with denominator zero are excluded from our number system.

||||||||||||||||||||||||||||||||||||||||||||||||||||||||||||||||||||||||||||||||||||||||||||||||||||||||||||||||||||||||||||||||||||||||||

Exercise. How would the number 1.15 be classified? Answer: This is a rational number because it can be expressed as $115/100$ which is the ratio of the two integers 115 and 100.

||||||||||||||||||||||||||||||||||||||||||||||||||||||||||||||||||||||||||||||||||||||||||||||||||||||||||||||||||||||||||||||||||||||||||

Numbers such as the square root of 2 or the cube root of 6 cannot be expressed as a ratio of two integers, and serve as examples of *irrational* numbers. The set of all rational and irrational numbers is called the set of *real* numbers. Only real numbers will be used in this book. Numbers with an imaginary component (that is, a component involving the square root of −1) are not in the set of reals and play no role in our discussions.

A2.3 JUSTIFICATION FOR MATHEMATICAL STATEMENTS

Justifications for mathematical statements fall into three rough categories: first, definitions and conventions which represent common agreement about the meanings of words and symbols; second, fundamental properties of numbers; and third, rules which are derived from fundamental properties. Definitions and conventions are, of course, a part of statements of fundamental properties and rules. As illustrations of the categories, observe that it is *conventional* to write the product of the numbers *a* and *b* without a multiplication sign as *ab*. It is true by *definition* that the sum of a number and its negative is zero. Thus:

$$+a + (-a) = 0.$$

Exercise. Why is *a* the same as 1*a*? Answer: *Conventionally,* if a quantity is written without a coefficient, the coefficient is assumed to be one.

The statement that the sum of two numbers is the same whether found by adding the first to the second or by adding the second to the first is a *fundamental property* of numbers. Finally, the idea that "minus a minus is plus" is a *rule* which can be derived using fundamental properties, definitions, and conventions.

Exercise. Of the operations *multiply, divide, subtract,* which has the property just illustrated for *add*? Answer: *Multiply.*

A2.4 RULES OF SIGN

The numbers +2 and −2 are opposites in the sense that each is the negative of the other. The statement that −2 is the negative of +2 is true by definition; the statement that the negative of −2 is +2 is a rule which can be derived from definitions and axioms.[1] Thus:

$$-(+2) = -2$$

and

$$-(-2) = +2.$$

---

[1] The rule is not derived here.

Another derived rule justifies the statement

$$+(-2) = -(+2).$$

That is, the positive of the negative of a number is the same as the negative of the positive of the number. As a consequence of this rule, the sign of a number is the same no matter what the *order* is of the signs immediately preceding the number. For example:

$$+[-(-5)] = -[+(-5)] = -[-(+5)].$$

The ultimate sign is shown by working with the last expression to be

$$-[-(+5)] = -[-5] = +5.$$

Exercise. Write $-[-(-5)]$ with one sign. Answer: -5.

A2.5 SIGN SEQUENCE RULE

Any sequence of signs immediately preceding a number can be converted to a single ultimate sign, the ultimate sign being *minus* if the sequence has an *odd* number of *minus* signs, and being *plus* if the sequence has an *even* number of minus signs. For example:

$$-[+(-3)] = +3$$

and

$$-\{-[+(-3)]\} = -3.$$

A2.6 ADDITION AND SUBTRACTION OF SIGNED NUMBERS

If two numbers have the same sign, their sum is found by adding the two numbers and affixing the common sign. Thus:

$$(+4) + (+3) = +7$$

and

$$(-3) + (-4) = -7.$$

Exercise. Add: $(-10) + (-4)$. Answer: -14.

The sum of a positive and a negative number is found by taking the difference between the two numbers, disregarding sign, and affixing the sign of the larger to the difference. For example:

$$(+7) + (-3) = +4$$

and

$$(-7) + (+3) = -4.$$

Exercise.   Add: $(+5) + (-3)$. Answer: $+2$.

In terms of fundamentals, subtraction is developed from addition. Subtraction can be converted to addition by following the rule which states that subtracting a number is the same as adding its negative. This rule, together with a rule already stated, permits us to write

$$+3 - (+4) = +3 + (-4) = -1.$$

As another example:

$$-3 - (-4) = -3 + (+4) = +1.$$

Exercise.   Write as an addition problem, and find the sum: $+5$ $(+11)$. Answer: $+5 + (-11) = -6$.

Observe that the minus symbol is used both to designate negative and to indicate the operation of subtraction. Again, the plus symbol designates both positive and the operation of addition. For example, we write $- (+2)$ to designate the *negative* of $+2$, but in the expression $+3 - (+2)$, the same symbol, $-(+2)$, means to *subtract* $+2$. Now, let us adopt two conventions. First, we shall assume that an unsigned number is positive; second, we shall indicate the subtraction of a number by a minus sign, rather than by addition of its negative. Thus, instead of $+3 - (+2)$ we shall write the condensed statement, $3 - 2$. The conventions just stated, taken together with the rule for sign sequence, contribute to simplicity of expression, as illustrated next:

$$+1 + (+2) - (+3) + (-4) - (-5) - [+(-4)]$$

becomes

$$1 + 2 - 3 - 4 + 5 + 4 = 12 - 7 = 5.$$

Exercise.   Simplify and evaluate $+(-1) + (-3) - (+2) + (+4) + [-(-6)]$. Answer: 4.

We usually seek to reduce additions and subtractions of signed numbers to simplified form, as just shown. However, in the interest of keeping track of fundamentals, we should be able to convert from the simplified form to an expression in terms of addition of signed numbers. Thus:

$$4 - 3 + 2$$

means

$$+4 + (-3) + (+2).$$

Exercise.   What does $2 + 5 - 8$ mean in terms of addition of signed numbers? Answer: $+2 + (+5) + (-8)$.

A2.7 NEGATIVE OF A SUM

The negatives of the sum of two numbers is the sum of the negatives of the numbers. Thus:

$$-(7 + 5) = (-7) + (-5) = -7 - 5 = -12.$$

It will be observed that as we go from $-(7 + 5)$ to $-7 - 5$, the effect is to remove the parentheses and change the signs of the numbers inside the parentheses. The methodology applies generally; that is, parentheses preceded by a minus sign alone can be removed by changing the sign of each term[2] inside the parentheses. As examples:

$$-(4 - 6) = -4 + 6$$

and

$$-(4 + 8) = -4 - 8.$$

Parentheses preceded by a plus sign alone (or no sign) serve no necessary purpose and may be omitted. Thus:

$$+(8 - 6) = 8 - 6$$

and

$$(2 - 4 + 3) = 2 - 4 + 3.$$

Exercise.   Remove grouping symbols: $(2 + 3 - 5) - (5 - 2 + 4)$. Answer: $2 + 3 - 5 - 5 + 2 - 4$.

---

[2]See "Definitions: Expression, Term, Factor," Section A2.18 of this Appendix.

A2.8 MULTIPLICATION AND DIVISION OF SIGNED NUMBERS

Multiplication is a fundamental operation with numbers, and division is defined in terms of multiplication. One rule of sign suffices for both operations—namely, if two numbers have the same sign, their product (or quotient) is positive; if two numbers are of opposite sign, their product (or quotient) is negative. In multiplication, for example:

$$(-4)(-2) = 8$$

and

$$2(-4) = -8.$$

Observe the use of parentheses in the last written expression. In the first, the absence of a symbol between the two sets of parentheses means that the numbers inside the parentheses are to be multiplied. We could have written $-4(-2)$ instead of $(-4)(-2)$ without ambiguity, but multiplication would not be in order if the expression read $(-4) - 2$. The last written number is a difference, not a product.

The division symbol, $\div$, is not used often in algebraic statements. The fractional form of expression is preferred. Thus, 8 divided by 4 is written as 8/4. Moreover, the general practice in mathematics is to speak of the last number as "8 over 4," not "4 into 8," and the student is advised to follow this practice.

According to the rule for division of signed numbers:

$$\frac{8}{-4} = -2, \quad \frac{-8}{4} = -2, \quad \text{and} \quad \frac{-8}{-4} = 2.$$

Exercise. Evaluate: $-2(4)$; $3(6)$; $-(-3)(-5)$; $-4/-2$; $6/-3$; $-(6/-3)$; $-(-6/3)$; $-(-6/-3)$. Answer: -8; 18; -15; 2; -2; 2; 2; -2.

The sign of a term which involves *only* multiplications and divisions of numbers can be determined by counting the number of minus signs; if the count is odd, the ultimate sign of the term is minus; if the count is even, the ultimate sign of the term is positive. For example:

$$\frac{-4(2)(-3)}{-2(-1)(4)} = 3$$

and

$$\frac{-(-4)(-5)}{2(-1)(-10)} = -1.$$

||||||||||||||||||||||||||||||||||||||||||||||||||||||||||||||||||||||||||||||||||||||||||||||||||||||||||||||||||||||||||||||||||||||||||||||||

Exercise. Write as a single signed number: $\dfrac{-2(3)(6)(-10)}{12(-5)}$.

Answer: -6.

||||||||||||||||||||||||||||||||||||||||||||||||||||||||||||||||||||||||||||||||||||||||||||||||||||||||||||||||||||||||||||||||||||||||||||||||

A2.9 PROBLEM SET 1

Reduce to a single signed number, or zero:
1. $-2 + (-3) - [-(-4)] + (-5)$.
2. $-(-8) + 8 - (+3)$.
3. $-(+4) + (-4) - [-(-4)]$.
4. $6 - (+3) + (+5) - (-7) + (-2) - (+10)$.
5. $3.15 - (+1.08) - (-3.27)$.
6. $-4 - (3)(-2) + (-5)(-4)(-1) - (3)(2)$.
7. $-4(3) + (-1)(2)(-6)(-1) - (-1)(-1)(-1)$.
8. $-(-4 + 3) + [-1(-2)]$.
9. $14(-2) - (-13) + (-3)(-2) - (-5)$.
10. $(3 - 4) - (5 - 2)$.
11. $\dfrac{-3(20)}{-4}$.
12. $\dfrac{5(-2)(3)}{-4(-2.5)}$.
13. $\dfrac{-(-7)}{-2}$.
14. $\dfrac{6(-2)(-1)(-3)}{-(-12)(-2)}$.
15. $\dfrac{-(-26)(16)}{-[-(-13)]}$.
16. $\dfrac{4(-1)(-2)(3)(-7)}{(-5)(-3)(2)}$.

Rewrite as the sum of signed numbers:
17. $3 - 4 + 5 - 2 + 6$.
18. $3 - 2 + 4 - 5$.
19. $-1.5 - 3 - 2.5$.
20. $-3 + 7 - 4 + 2$.

Reduce to a single signed number:
21. $-3(2) + (3 - 7) - (2 + 4)$.
22. $-[-2 + (-3)] - 2(-3) + (-3)(-4)$.
23. $-(3 - 5) + (2 - 6)$.
24. $-[+(-3) + (-5)] + [-2 + (-3)]$.
25. $-[+(-5)] + [-(-3)] + (8 - 4) - (2 - 6)$.

A2.10 FUNDAMENTALS OF ORDER AND GROUPING

Mathematics beyond arithmetic is characterized by its use of letter representation of numbers. The sum, difference, and quotient of two numbers in literal form, such as a and b, are written in the usual manner as, respectively, $a + b$, $a - b$, and a/b. The product conventionally is written as ab, the multiplication symbol being omitted. This convention is, of course, in conflict with the place system of decimal notation. For example, 34 means thirty-four, not 3 times 4. We would interpret 34*a* as thirty-four times *a*, not 3 times 4 times *a*. It follows that care must be exercised when writing numbers which have literal and numerical factors. Sometimes a dot is employed to indicate multiplication, as in 3 · 4*ab*. We shall generally follow the more common practice, already introduced, and use parentheses; thus, 3(4)*ab* or (3)(4)*ab*. The first of the last two expressions is the simpler. The second is equivalent to the first, but uses two sets of parentheses where only one is required.

The first advantage of literal representation of numbers is their utility in stating generalizations. For example, we may illustrate the point that the sum of any two real numbers is a real number by mentioning that the sum of 3 and 4 is 7, and 7 is a real number. To generalize the point and state it as a property of all real numbers, we need a statement which is not restricted to a particular pair of numbers. The statement desired is: "*If a and b are any real numbers, then a + b is a real number.*" In this statement the word *any* means that the property applies to *all* real numbers. If someone asks if the statement applies to *x* and *y*, we inquire whether *x* and *y* are real numbers; if they are, the statement applies, and *x* + *y* is a real number. The student will find it helpful to keep in mind the *all* implication of the word *any* in mathematical statements.

The fundamental properties we list for real numbers are formal statements of observations of arithmetic. The very fact that they seem obvious is a primary reason for stating them as fundamental properties. The fact that they seem obvious is not, however, justification for thinking time spent on them is wasted. Many questions which perplex beginning students arise simply because of failure to have in mind the fundamental properties of numbers.

A2.11 CLOSURE

Countless illustrations can be written showing that when two real numbers are added, subtracted, multiplied, or divided (division by zero excluded), the outcome itself is a real number. In other words, the "answer" to problems in addition, subtraction, multiplication, and division of real numbers are part of the real number system. Indeed, our algebra with literal numbers would falter if we did not assume *closure* to be a fundamental property. The closure property is this: *If a and b are any real numbers, then a + b, a − b, ab, and a/b are real numbers, except that a/b is not defined if b is zero.*

The property is sometimes described by stating that the real numbers are closed with respect to addition, subtraction, multiplication, and division, except division by zero. The property extends to more than two numbers according to the logic which says, for example, that if *a*, *b*, and *c* are real numbers, then *a* + *b* is a real number, and this last real number when added to *c* yields a real number, so that *a* + *b* + *c* is a real number.

The concept of closure can be understood easily by thinking of addition of clock numbers. If we start at 2 and add 3, we go to 5. If we start at 10 and add 4 we go to 2. True, the sums in such additions are contrary to those obtained for real numbers, but the point is that the addition of any two clock numbers has an answer which is a clock number, so that we may say the clock numbers are closed with respect to addition.

As a simple example of lack of closure, it is easy to see that the set of integers is not closed with respect to division; for example, 3 and 7 are integers, but 3/7 is not an integer. Again, the real numbers are not closed with respect to the operation "square root," because negative numbers are real, but the square roots of negatives are not real.

||||||||||||||||||||||||||||||||||||||||||||||||||||||||||||||||||||||||||||||||||||||||||||||||||||||||||||||||||||||||||||||||||

Exercise.   Is the set of odd numbers closed with respect to addition? Explain. Answer: No. The sum of two odd numbers is not an odd number.

||||||||||||||||||||||||||||||||||||||||||||||||||||||||||||||||||||||||||||||||||||||||||||||||||||||||||||||||||||||||||||||||||

A2.12 DISTRIBUTIVE PROPERTY

If we wish to write an expression in which the *sum* of the numbers a and b is to be multiplied by 2, a convention must be established to ensure that the expression is interpreted as 2 times the sum, and not 2 times one of the numbers. Parentheses accomplish the purpose. We write $2(a + b)$ and interpret this to mean that the quantity inside the parentheses is a number to be multiplied by 2, the number inside being the sum of a and b. It may be helpful to practice a bit here by reading the following parallel statements:

In Words	*In Symbols*
Multiply 2 by a	$2a$
Multiply 2 by a, and then add b to the product	$2a + b$
Multiply 3 by the sum of a and c	$3(a + c)$
To a, add b times the sum of c and d	$a + b(c + d)$
Multiply the sum of a and b by the sum of c and d	$(a + b)(c + d)$

The distributive property is illustrated by the observation that

$$2(3 + 4) = 2(7) = 14$$

can be evaluated in the alternative manner:

$$2(3) + 2(4) = 6 + 8 = 14.$$

As another example:

$$-7[+5 + (-6)] = -7[-1] = 7$$

can be evaluated as

$$-7[+5 + (-6)] = -7(+5) + (-7)(-6) = -35 + 42 = 7.$$

These and the countless similar observations one could devise are generalized by stating the distributive property of the real numbers. In brief, we say that multiplication may be distributed over addition (or subtraction). More formally, we state that *if a, b, and c are any real numbers, then*

$$a(b + c) = ab + ac.$$

This property finds extensive application in algebra. As examples:

$$2a(b + c + 3d) = 2ab + 2ac + 6ad$$
$$-3(a - 2b) = -3a + 6b.$$

In the last example, it is instructive to match the procedure with the formal statement of the property. To do so, we remember that $-3(a - 2b)$ is shorthand for $-3[+a + (-2b)]$. Applying the distributive property to the right-most statement, we obtain

$$-3(+a) + (-3)(-2b)$$

which, reverting to shorthand, is $-3a + 6b$.

Exercise. Apply the distributive property to $-3(-2 + a)$. Answer: $6 - 3a$.

As another example, if c does not equal zero,

$$\frac{a}{c}(x + 2y) = \frac{a}{c}(x) + \frac{a}{c}(2y).$$

The point here is that the distributive property holds for any real numbers, and a/c is a real number if $c \neq 0$.

Exercise. Why is a/c a real number? Answer: In the foregoing it was stated that a and c are real numbers. Inasmuch as the real numbers are closed with respect to division if $c \neq 0$, a/c must be a real number.

A2.13 COMMUTATIVE PROPERTIES

The observation that we may find the sum of 3 and 4 by starting with 4 and then adding 3, or by starting with 3 and then adding 4, is one of countless illustrations that the sum of two numbers is the same in whatever *order* the numbers are added. Thus:

$$3 + 4 = 4 + 3$$

and

$$-2 + (-7) = -7 + (-2).$$

In shorthand, the last statement would have been

$$-2 - 7 = -7 - 2.$$

Formally, the commutative property for addition states that *if a and b are any real numbers, then*

$$a + b = b + a.$$

The property is not limited to the sum of two numbers, of course. It extends by closure to any series of numbers in addition. For example, if we have $a + b + c$, then $a + b$ is a real number, by closure, and we may change the order of this real number and c to obtain $c + a + b$. Again, the commutative property for addition justifies the following rearrangement:

$$2a + 3b + 6 + b + 7 = 2a + 3b + b + 6 + 7.$$

Finally, we note that

$$a - b = -b + a$$

because $a - b$ is shorthand for $a + (-b)$, which, by the commutative property, is $-b + a$.

Exercise.   Rearrange $c - a + b$ so that the letters are in alphabetical order. Answer: $-a + b + c$.

The commutative property for multiplication states that *if a and b are any real numbers, then $ab = ba$.* That is, the *order* of the factors in a product may be changed. As numerical examples, we have:

$$3(4) = 4(3); \ 2(-5) = -5(2); \ -3(-2) = -2(-3).$$

The property extends to more than two numbers by closure. For example, in abc, ab is a real number, by closure, and the commutative property says we may change the order of this real number and c; hence:

$$abc = cab.$$

Similarly:

$$abcd = cdab = bcda.$$

A2.14 ASSOCIATIVE PROPERTIES

Associative properties for addition and multiplication have to do with the manner of *grouping* numbers. If we have the addition problem

$$2 + 3 + 4$$

we can find the sum as

$$(2 + 3) + 4$$

or as

$$2 + (3 + 4)$$

that is, as $5 + 4$ or as $2 + 7$. The same sum results whether we group the first two and then add the third number or whether we group the second two and add to the first number. Formally, the associative property for addition states that *if a, b, and c are any real numbers, then*

$$(a + b) + c = a + (b + c).$$

The parentheses in the last statement are not necessary. They serve only to indicate the different groupings. The number in question is $a + b + c$, no matter what order or grouping is used.

Exercise.   According to the associative property, $5 + 2 + 8$ may be evaluated in what ways? Answer: $7 + 8$, $5 + 10$, and (if the commutative property is applied first), $13 + 2$.

The associative property for multiplication asserts that *if a, b, and c are any real numbers, then*

$$a(bc) = (ab)(c).$$

Again, unnecessary parentheses are used to show that the property has to do with the manner of grouping the factors. To illustrate the property, we observe that $(3)(4)(2)$ can be evaluated as $(12)(2)$ or as $(3)(8)$. That is, we can multiply the product of the first pair of numbers by the third number, or multiply the first number by the product of the second pair of numbers.

Exercise.   Illustrate the associative property for multiplication by reference to the product $(3)(5)(2)$. Answer: The product may be evaluated as $(15)(2)$ or as $(3)(10)$.

A2.15 IMPORTANCE OF FUNDAMENTAL PROPERTIES

The distributive, associative, commutative, and closure properties of the real numbers are important because they are the fundamental justifications for many of the steps taken in algebraic procedures. Asked if $(a + b)c$ is the same as $c(a + b)$, we answer yes, and cite the commutative law for multiplication as justification. Now:

$$c(a + b) = ca + cb$$

by the distributive property, and the commutative property assures us that the right side can be altered to obtain

$$c(a + b) = ac + bc.$$

From what has been said, it follows that

$$(a + b)c = ac + bc.$$

The point here is that the distributive law does not tell us directly how to expand an expression such as $(2 + a)b$, but we can easily justify by fundamentals that the expansion may be written as

$$(2 + a)b = 2b + ab.$$

Exercise. In $(2)(a)(3) = (2a)(3) = 2(3)(a) = 6a$, what justifies each of the three expressions following the $=$ sign? Answer: Associative law for multiplication; commutative law for multiplication; associative law for multiplication.

Carrying on one step further, let us learn how to expand

$$(a + b)(c + d).$$

We know by closure that $(a + b)$ is a real number, and the distributive law justifies writing

$$(a + b)(c + d) = (a + b)c + (a + b)d.$$

The right member of the last expression can be expanded, leading to

$$(a + b)(c + d) = (a + b)c + (a + b)d = ac + bc + ad + bd.$$

The distributive law works both ways. Writing it in reverse, we have

$$ab + ac = a(b + c).$$

Converting in this manner from the sum of ab and ac to the product of a by $(b + c)$ is referred to as *factoring*. Hence the distributive property is the fundamental underlying factoring. When applied to obtain

$$2a + 3a = (2 + 3)a = 5a$$

the procedure is called *combining like terms*. As other examples:

$$2ab - 3ab + 5ab - 6 = 4ab - 6$$
$$4a - 2a + 5b - 2b = 2a + 3b.$$

In the last two examples the associative law for addition justifies the separate combinations of like terms. We may say in general that terms having the same literal factor can be added (subtracted) by adding (subtracting) their numerical coefficients. Thus, in

$$3ab + 4ab - 2ab$$

each term has the same literal factor, ab. Combining the numerical coefficients, we have

$$(3 + 4 - 2)ab = 5ab.$$

Exercise. Remove symbols of grouping and combine like terms: $a(2 - b) - 3(a - 2b)$. Answer: $a(2) - ab - 3a + 6b = 6b - ab - a$.

The following example shows how the fundamental properties come into play. Starting with the expression at the top left, we change to the expression on the next line, indicating on that next line, at the right, the justification for the step:

$$-2a(3b) + ab$$
$$= -2(3b)a + ab \qquad \text{Commutative law, multiplication}$$
$$= -6ba \quad + ab \qquad \text{Associative law, multiplication}$$
$$= -6ab \quad + ab \qquad \text{Commutative law, multiplication}$$
$$= -5ab. \qquad \text{Distributive law (combining like terms, or factoring)}$$

A similar example follows:

$$2 + ba + 3$$
$$= 2 + \ 3 + ba \qquad \text{Commutative law, addition}$$
$$= \qquad 5 + ba \qquad \text{Associative law, addition}$$
$$= \qquad 5 + ab. \qquad \text{Commutative law, multiplication}$$

We do not mean to suggest by these examples that the reader cite the fundamental reason for every algebraic step he takes. We do mean to suggest, and strongly, that he become so thoroughly familiar with these fundamentals that he never has to ask his instructor questions such as:

"Is $(2 + 3b)(-c)$ the same as $-c(2 + 3b)$?"

Exercise. What fundamental law specifies that the two last written expressions are equal? Answer: The commutative law for multiplication.

A2.16 PROBLEM SET 2

1. When we write *ab* or *cxy,* in what way does algebraic convention differ from the usual decimal notation?

2. What would *abc* mean in decimal notation?

3. What is the connotation of the word *any* in mathematical statements?

4. If we add the odd number 3 to the odd number 5, the sum is an even number, 8. State this odd-even relationship in a manner which shows it is a fundamental property of integers.

5. Explain the property of closure by using addition of clock numbers as an example.

6. Why does $3a + c$ not mean three times the sum of *a* and *c?*

7. How many specific numerical illustrations of the distributive property do you think you could devise?

8. Explain what is meant by the statement that the distributive property equates a product involving a sum to a sum of products.

What fundamental property or convention justifies each of the following?

9. $a(2) = 2a.$

10. $2 - a = -a + 2.$

11. $a(2 + 3) = a2 + a3.$

12. $a + 2 + b = a + b + 2.$

13. $+a = a.$

14. $a(b + c) = (b + c)a.$

15. $a + (-3) = a - 3.$

16. $a2 + a3 = 2a + 3a.$

17. $2 + 3 + b = 5 + b.$

18. $3a2 = 3(2)a.$

19. $3(2)a = 6a.$

20. $+b + (-a) = b - a.$

21. $5 - 4 + b = 1 + b.$

22. $2a + 2b = 2(a + b).$

23. $a + 2 + b = a + b + 2.$

24. $a + (b + c) = (b + c) + a.$

25. $a(2)(4) = a(8).$

Combine like terms:

26. $2b + (-3b).$

27. $2a - 3b + 5a - 2b.$

28. $3abc - 2d - (-2abc).$

29. $5 - (-3x) + 2 - (+2x).$

30. $2abc - (-3abc) + 3abc.$

31. $3 + 2a - 3b + 5 - (-2b).$

Name the fundamental property justifying each step:

32. $-b + a$
   $= \ a - b.$

33. $2b(3a)$
   $= 2b3(a)$
   $= 2(3)ba$
   $= 6ba$
   $= 6ab.$

34. $3 + xy + \ 5 + 3(2ab)$
   $= 3 + 5 \ + xy + 3(2ab)$
   $= \ \ \ \ 8 \ + xy + 3(2ab)$
   $= \ \ \ \ 8 \ + xy + 6ab.$

35. $3 + a + 2$
   $= 3 + 2 + a$
   $= \ \ \ \ 5 + a.$

36. $a + 3 + c + 2$
   $= a + c + 3 + 2$
   $= a + c + 5.$

37. $acdb$
   $= adcb$
   $= adbc.$

Write in algebraic form:

38. The sum of a and b.
39. To a, add the sum of b and c.
40. To the sum of a and b, add twice the product of c and d.
41. From twice the sum of a and b, subtract three times the sum of c and $2d$.
42. Multiply the sum of a, b, and c by the product of 2 and d.

A2.17 REMOVING GROUPING SYMBOLS

Instructions to remove grouping symbols are carried out by systematic application of the distributive property. Thus:

$$a[b - c(d + 2)] = a[b - cd - 2c] = ab - acd - 2ac.$$

Observe that the innermost grouping symbols, the parentheses, were removed first. This order of attack lessens the chance of errors of omission. As another example:

$$\begin{aligned}
a - \{-2 - 3[-4 + (5 - a)]\} &= a - \{-2 - 3[-4 + 5 - a]\} \\
&= a - \{-2 + 12 - 15 + 3a\} \\
&= a + 2 - 12 + 15 - 3a \\
&= -2a + 5.
\end{aligned}$$

Exercise. Remove symbols of grouping and combine like terms: $3x - 2[y - 4(x - 3y)]$. Answer: $11x - 26y$.

Care must be exercised in the interpretation of grouping symbols. Thus:

$$5 - 3(b + c) = 5 - 3b - 3c.$$

If we had wished to indicate that the difference of 5 and 3 was to be multiplied by the sum of b and c, the expression would have been

$$(5 - 3)(b + c) = 2b + 2c.$$

Extension of the distributive property justifies the expansions

$$(3 - a)(2b + 5) = 6b + 15 - 2ab - 5a$$

and

$$(a - b + 2)(c + d) = ac - bc + 2c + ad - bd + 2d.$$

A2.18 DEFINITIONS: EXPRESSION, TERM, FACTOR

Any statement involving mathematical symbols may be referred to as a mathematical *expression*. When an expression consists of parts separated by plus or minus signs, or by an equal sign, the parts, together with their signs, are called *terms* of the expression. Thus, in the expression

$$a + 2b - 3ac + 4$$

the terms are $+a$, $+2b$, $-3ac$, and $+4$.

Each term in an expression consists of one or more *factors*, a factor being one of the separate multipliers in a *product*. Thus, in the expression $2a - 3bc$, the term $+2a$ consists of the two factors, 2 and a; the term $-3bc$ consists of the three factors, -3, b, and c.

The words we have just defined make possible clear descriptions of mathematical statements. We shall provide illustrations for practice purposes. Consider the expression

$$(2a + b)(x + 2y).$$

As written, the expression is a single term consisting of the factors $(2a + b)$ and $(x + 2y)$. We may go on to say that the first factor is an expression containing the two terms, $+2a$ and $+b$; the second factor is also an expression of two terms, $+x$ and $+2y$.

Exercise.   The expression $3ab + 4a - 3$ has how many terms? What is the composition of the first term? Do all terms have a common factor other than 1? Answer: The expression has three terms; the first term is composed of the factors 3, a, and b; the terms do not have a common factor other than 1.

The distributive property assures us that

$$3(a - 2b + c) = 3a - 6b + 3c.$$

If we think of this last statement as multiplying the parenthetical expression by 3, we come to the general statement that to multiply an expression by a number, we must multiply each term of the expression by the number. On the other hand, if we think of

$$3(2a) = 6a$$

as multiplying the term $2a$ by 3, we come to the general statement that to multiply a term by a number, we multiply one factor of the term by that number.

According to a rule of sign:

$$-(a - b + 2c) = -a + b - 2c.$$

If we think of this as changing the sign of the parenthetical expression, we see that to change the sign of an expression, we change the sign of every term of the expression. On the other hand, the sign of a term is changed by changing the sign of one of its factors. For instance, thinking of $-[a(-b)(c)(-2)]$ as changing the sign of the bracketed expression, the outcome of the sign change can be written as $-a(-b)(c)(-2)$ or $a(b)(c)(-2)$ or $a(-b)(-c)(-2)$ or $a(-b)(c)(2)$.

Finally, we may state that the sign of a term is unchanged if the signs of an even number of its factors are changed. For example:

$$-ab(-c) = a(-b)(-c) = abc$$

and

$$-3(4 - a) = 3(a - 4).$$

A2.19 ELEMENTARY FACTORING

We are assured by the distributive property that

$$ab + ac = a(b + c).$$

Thinking of this expression as changing from the form on the left to that on the right, we see that the sum of two terms has been converted to the product of two factors. Conversion from the sums and differences of terms to a single term with two or more factors is called *factoring*. Observe that an expression in completely factored form has but one term.

When terms have a factor in common, factoring is carried out by writing the product of the common factor times an expression (in grouping symbols) whose terms are the remaining factors of each of the original terms, as shown in the following examples:

$$2xy + axy = xy(2 + a)$$
$$ax - bx = x(a - b)$$
$$6x + 2y - 4a + 8b = 2(3x + y - 2a + 4b).$$

It is conventional to omit the coefficient one when writing a term, and this convention must be kept in mind when factoring. Thus:

$$ab + b = b(a + 1)$$
$$abc - ab + abd = ab(c - 1 + d).$$

Exercise.   Factor $xy + ax + x$. Answer: $x(y + a + 1)$.

Thus far, we have illustrated *monomial* factoring, that is, cases where the common factor has a single term. On rare occasions in this book the need for *binomial* factoring arises. As examples:

$$a(b + 2) - 3(b + 2) = (b + 2)(a - 3)$$
$$1 + i + i(1 + i) = (1 + i)(1 + i).$$

Exercise. Factor $bx + by - x - y$. Answer: $(x + y)(b - 1)$.

Finally, consider the expression

$$(3x + 4)(2x - 1)$$
$$= 3x(2x - 1) + 4(2x - 1)$$
$$= 6xx - 3x + 8x - 4$$
$$= 6x^2 + 5x - 4$$

where, in the last line, xx is written as x^2 (read as x *squared*). In a number of places in the text, we must do problems like the last one, but in reverse; that is, start with a trinomial which contains a term in x^2, a term in x, and a constant, such as

$$6x^2 + 5x - 4$$

and obtain the equivalent pair of binomial factors,

$$(3x + 4)(2x - 1).$$

We may do this by trial and error. As an example, let us factor

$$12x^2 + 7x - 10. \tag{1}$$

First we write

$$(\quad)(\quad),$$

where each set of parentheses contains two terms, a *first* term and a *second* term. The product of the first terms must be $12x^2$, so the first terms could be x and $12x$, $2x$ and $6x$, $3x$ and $4x$, $6x$ and $2x$, $12x$ and x, or any of the last pairs with signs changed. The product of the second terms must be -10, so the second terms could be 1 and -10, 2 and -5, 5 and -2, 10 and -1, or any of the last four pairs with the sign of both numbers changed. Let us try x and $12x$ as first terms, 1 and -10 as second terms. We fill in the parentheses as follows:

$$(x + 1)(12x - 10).$$

The product of the first terms is $12x^2$, and the product of the second terms is -10, as required. The *test* of our trial is the product of the inner terms $1(12x)$ plus the product of the outer terms $x(-10)$, which is

$$12x - 10x = 2x.$$

Hence,

$$(x + 1)(12x - 10) = 12x^2 + 2x - 10$$

and this does not have the same middle term as the original expression, (1),

$$12x^2 + 7x - 10,$$

so we try another set of first and second terms from our list. For example,

$$(3x + 2)(4x - 5).$$

All we need do at each trial is apply the test mentioned several lines back, and determine if the middle term in the expansion of the trial is the required $7x$. The test here yields

$$2(4x) + (3x)(-5) = 8x - 15x = -7x$$

which is the negative of the desired $7x$ so we need only change the signs of the second (or first) terms. Thus, instead of $(3x + 2)(4x - 5)$ we write $(3x - 2)(4x + 5)$. The test term is now

$$-2(4x) + (3x)5 = -8x + 15x = 7x,$$

so we have the desired factors of the original expression,

$$12x^2 + 7x - 10 = (3x - 2)(4x + 5).$$

With a little practice, the proper pairs of terms usually can be found quite rapidly if we take hints from the original expression and the results of a trial. Thus, in

$$2x^2 - 13x + 20,$$

the second terms have the *positive* product 20, so both must be positive or both must be negative. Inasmuch as the middle term is negative, it follows that both second terms are negative. As a trial, we write

$$(2x - 4)(x - 5)$$

and the test term is $-14x$ which is not the desired $-13x$ in the original. However, we will get the desired middle term if we interchange the second terms of the first trial. Thus,

$$2x^2 - 13x + 20 = (2x - 5)(x - 4).$$

Exercise.   Factor $10x^2 + 26x + 12$. Answer: $(5x + 3)(2x + 4)$.

The expression

$$x^2 - 9$$

has a square term and a constant, but no middle term. The first term is the square of x and the second term is the square of 3 (that is, 3 times 3 $= 3^2$ $= 9$), so the expression is called the *difference of two squares*. By trial and error we find

$$x^2 - 9 = (x + 3)(x - 3).$$

Thus, the difference of the squares of two numbers is the product of the sum of the numbers times the difference of the numbers. As another example, noting that $4a^2 = (2a)(2a)$ and $16b^2 = (4b)(4b)$, we have

$$4a^2 - 16b^2 = (2a + 4b)(2a - 4b).$$

Exercise. Factor: *a)* $y^2 - 25$. *b)* $y^2 - 25x^2$. Answer: *a)* $(y + 5)(y - 5)$. *b)* $(y - 5x)(y + 5x)$.

A2.20 PROBLEM SET 3

Remove grouping symbols and combine like terms, if any:

1. $2ab(c - 2)$.
2. $(a - 2)(b + 1)$.
3. $2 - 3[1 - (+4)]$.
4. $(c + 2)(a - b + 3)$.
5. $-(-2) + 3[a - (1 - b)]$.
6. $10 - 3[4 - 5(-4 + a)]$.
7. $a - 2\{-3 - 2[5a - 2(a - 6)]\}$.
8. $(3x - 2)[a - 2(b + 3)]$.
9. $-(a - x - 2b)$.
10. $(2x + 3y)$.
11. $(a - 2b) - b$.
12. $a + (3x + 2)$.
13. $b - 2b(a - 3)$.
14. $ab[c - 2(x - 5)]$.
15. $ax - 2b(a - 1)$.
16. $(a - bx)(3 - c)$.
17. $(a + b + 1)(x + y)$.
18. $(a - 1)(b + 1)$.

Reduce to a single signed number:

19. $-3(2) - 2(1 - 3)$.

20. $\dfrac{(-3)(2) - (-2)(4)}{-3(5) - 2.5(-4)}$.

21. $\dfrac{0.017(5 - 1.08) + 2.3[0.5 - 6.2(3.1)]}{-4(0.0025)}$.

22. $\dfrac{12 - (6 - 2)(3 - 1)}{(8 - 3)(-1 - 2) + 6}$.

23. $\dfrac{1.7 - [3 - 17(15.4 - 1.6)]}{-0.8(0.245 - 0.37)}$.

24. Define *expression, term,* and *factor,* giving an illustration in each instance.

Factor:

25. $ab - 2b$.
26. $3a + 5a$.
27. $4abc - 2ab + 6a$.
28. $ax - bx + x$.
29. $3ad - 5ac + a$.
30. $4uv - 2xv + 2$.
31. $abx + aby - ab$.
32. $2ax - 6ay + 4az$.
33. $2x + ax + bx$.
34. $-ab - 3ac - a$.
35. $a(x + 1) + b(x + 1)$.
36. $x + 1 + y(x + 1)$.
37. $2(x + y) - a(x + y)$.
38. $ax + bx + ay + by$.
39. $x^2 - x - 2$.
40. $2x^2 - 9x - 5$.
41. $12x^2 - 25x + 12$.
42. $10x^2 + 13x - 3$.
43. $x^2 - x - 6$.
44. $x^2 - 9$.
45. $x^2 - y^2$.
46. $4x^2 - 9y^2$.

Mark (T) for true or (F) for false:

47. ( ) The expression ab has only one factor.

48. ( ) The expression $a + 2b - c$ has three terms.

49. ( ) The expression $a(b - c)$, as written, is a single term.

50. ( ) Referring to $a(b - c)$, it would be proper to say that the expression has two factors.

51. ( ) If ab is to be doubled, both a and b must be doubled.

52. ( ) $(a - b)(-c) = c(b - a)$.

53. ( ) It is correct to state that "to multiply an expression by a number, every factor of each term in the expression must be multiplied by the number."

54. ( ) $-a(-b - c)(-d) = ad(b + c)$.

55. ( ) The sign of a term is changed if the sign of any one of its factors is changed.

56. ( ) The parentheses in $a + (b + c)$ are unnecessary.

57. ( ) $3 + 2(a + b) = 5(a + b)$.

58. ( ) $-a - b = +(-b) + (-a)$.

59. ( ) $(a + b) - c = -ac - bc$.

60. ( ) To change the sign of a term, it is sufficient to change the sign of one factor of the term.

A2.21 PROPERTIES OF THE NUMBERS ZERO AND ONE

The number *zero* is unique in several respects. First, for any number a

$$0(a) = 0;$$

that is, zero times any number is zero. Second,

$$a + (-a) = 0;$$

that is, the sum of any number and its negative is zero, and this property defines what we mean by the negative of any number. We also say that in addition, a number and its negative *cancel,* meaning their sum is zero. Third,

$$a + 0 = a - 0 = a;$$

that is, a number is not affected by adding zero to it, or subtracting zero from it. Finally,

$$\frac{a}{0} \text{ is not defined.}$$

We shall find the last statement to be of basic importance in our development of calculus. To understand why $a/0$ is not defined, we consider cases such as $6/0$ where $a \neq 0$, and $0/0$ where a is zero. First recall that $6/3$ means to find a number which when multiplied by 3 yields 6. This is the *definition* of division in terms of multiplication, and we prove $6/3$ is 2 by stating $6 = (3)(2)$. Applying the definition to

$$\frac{6}{0},$$

we would seek a number which when multiplied by 0 yields 6. There is no such number, because 0 times any number is zero. Hence, expressions such as $6/0$, $-5/0$, and so on are not defined.

Next consider

$$\frac{0}{0}.$$

The temptation is to say that this expression is 1 because $0(1) = 0$ which satisfies the definition of division. However, we could say also the expression is 2, or 3.17, 0, or any number because

$$0(\text{any number}) = 0.$$

It follows that if $0/0$ were permitted, the results of mathematical operations could be ambiguous or contradictory. To demonstrate the last statement, suppose that a and b both equal 1. Then

$$a = b,$$

and it is also true if $a = b = 1$ that

$$a^2 = ab$$

and the equality remains if we subtract $b^2 = 1$ from a^2 and from ab, obtaining

$$a^2 - b^2 = ab - b^2.$$

Factoring shows that

$$(a + b)(a - b) = b(a - b).$$

In general, *except for a division of 0*, if two numbers are equal and we divide them by the same number, the results are equal. If we here neglect the exception of division by zero and divide both numbers in the last equality by $a - b$ (which is zero), we have

$$\frac{(a + b)(a - b)}{(a - b)} = \frac{b(a - b)}{(a - b)}.$$

If we again forget the exception and cancel the $(a - b)$'s, we have

$$a + b = b.$$

Now recall that at the beginning we had $a = b = 1$. The last statement then would be

$$1 + 1 = 1 \quad \text{or} \quad 2 = 1,$$

which is the contradictory result we sought to demonstrate.

Remember: Expressions such as $5/0$, $0/0$, or any number divided by zero are not defined. We shall say alternatively that division by zero is impossible, or that it is not permitted.

Summarizing the properties, we state that *for any real number, a:*

$$(0)a = 0$$
$$a + 0 = a - 0 = a$$
$$a + (-a) = 0$$

$a/0$ is not defined.

The unique properties of one exist in reference to multiplication and its inverse, division. Thus a number is unchanged if it is multiplied or divided by one. We have

$$a = (1)a = \frac{a}{1}.$$

The equivalence of a and $1a$ is assumed conventionally in the writing of various algebraic expressions. Any term may be assumed to have a factor of one. Recall the factoring of

$$ab - a = a(b - 1)$$

as an illustration of the point of the last two sentences.

Exercise. A rational number is the quotient of two integers, yet the single number, 4, is rational. Explain. Answer: 4 is the rational number $4/1$.

Any number, zero excepted, divided by itself yields a quotient of one. This fact is employed often, as when we write

$$\frac{6}{6} = 1, \text{ or } \frac{ab}{ab} = 1, \text{ or } \frac{x - y}{x - y} = 1.$$

Circumstances arise also where we may wish to multiply an expression by

$$\frac{6}{6} \quad \text{or} \quad \frac{ab}{ab}$$

and this can be done without changing the expression because it is equivalent to multiplication by one.

The words *cancel* and *cancellation* are used with reference to zero and one. In addition, a number and its negative cancel each other, meaning their sum is zero. In multiplication, a number and its reciprocal cancel, meaning their product is one. Thus:

$$a\left(\frac{1}{a}\right) = 1$$

where $1/a$ is called the reciprocal of a. Often, the latter type of cancellation is thought of in terms of division rather than multiplication of reciprocals. Thus, in ab/a, we think of a over a as being one, so that

$$\frac{ab}{a} = 1(b) = b$$

and we say that the a's cancel. An important rule to keep in mind when working with fractions is that cancellation (replacement by one) can be performed *only for factors common to numerator and denominator.* We cannot cancel the a's in

$$\frac{a + 2}{a}$$

because a is not a factor of the numerator. On the other hand, the numerator of

$$\frac{ax + 2x}{x(b - 1)}$$

can be factored to permit cancellation; thus:

$$\frac{x(a + 2)}{x(b - 1)} = \frac{a + 2}{b - 1}.$$

Exercise. x is a factor of what parts of the expression

$$\frac{ax + 2}{x(b + c)}.$$

Answer: x is a factor of the denominator, and of the single term ax in the numerator; x is not a factor of the numerator.

A2.22 PRODUCT OF FRACTIONS

The product of two fractions is the product of their numerators over (divided by) the product of their denominators. For example:

$$\left(\frac{2}{5}\right)\left(\frac{3}{7}\right) = \frac{6}{35}$$

$$\left(\frac{a}{2}\right)\left(\frac{3}{b}\right) = \frac{3a}{2b}$$

$$\left[\frac{a(b + 2)}{3}\right]\left(\frac{2}{b}\right) = \frac{2a(b + 2)}{3b}.$$

Exercise. Express as a single fraction without grouping symbols:

$$\left(\frac{3}{x+y}\right)\left(\frac{x-y}{2}\right).$$

Answer: $\dfrac{3x-3y}{2x+2y}.$

Generally, cancellation should be performed where it is possible to do so. For example:

$$\left(\frac{6ab}{5c}\right)\left(\frac{c}{3a}\right)=\frac{2b}{5}$$

$$\frac{a(b+2)}{3}\left(\frac{2}{a}\right)=\frac{2(b+2)}{3}$$

$$\left(\frac{2a-2}{b}\right)\left(\frac{1}{2}\right)=\frac{2(a-1)}{b}\left(\frac{1}{2}\right)=\frac{a-1}{b}$$

$$(x+y)\left[3+\frac{a}{x+y}\right]=3(x+y)+a.$$

In the last example the factor $(x+y)$ can be assumed to have a denominator of one for multiplication purposes. Again:

$$3\left(\frac{a}{b}\right)=\left(\frac{3}{1}\right)\left(\frac{a}{b}\right)=\frac{3a}{b}$$

$$2\frac{(a-3)}{b}=\left(\frac{2}{1}\right)\frac{(a-3)}{b}=\frac{2(a-3)}{b}.$$

Exercise. Carry out the multiplication, leaving the result as the sum of two fractions:

$$2\frac{a}{b}\left[3+\frac{x+2}{ax}\right].$$

Answer: $\dfrac{6a}{b}+\dfrac{2x+4}{bx}.$

An equivalent expression is obtained if a given expression is multiplied or divided by -1 an even number of times because the net effect is multiplication or division by $+1$. For example, in the change

$$\frac{-(a-b)(c-d)}{-3}=\frac{(b-a)(d-c)}{3}$$

four multiplications by -1 were performed.

Keeping in mind that three signs are associated with a fraction (the signs of numerator and denominator, and the sign of the fraction itself), it is helpful to remember that an equivalent fraction results if any two of these signs are changed. For instance:

$$-\frac{2}{-3} = \frac{2}{3}$$

$$\frac{-b-2}{2(1-a)} = \frac{b+2}{2(a-1)}$$

$$\frac{a-1}{-2} = \frac{1-a}{2}.$$

In the first of the last three examples, the sign of the denominator and the sign preceding the fraction were changed. In the other two examples, signs of numerator and denominator were changed. Remember that numerator and denominator are expressions, and to change the sign of an expression, it is necessary to change the sign of every term in the expression, the change of a term's sign being accomplished by changing the sign of one (or an odd number) of its factors. In the middle example of the foregoing, the numerator is an expression with two terms, $-b$ and -2. The expression in the denominator is a single term with factors 2 and $1 - a$. The sign of the numerator is changed by altering the signs of both terms to obtain $b + 2$. The sign change in the denominator is made by altering the sign of one factor, $1 - a$, to obtain $a - 1$.

Exercise.   Write

$$\frac{a-b}{x-c}$$

in three equivalent forms by altering the signs in the manner just discussed.

Answer: $\dfrac{b-a}{c-x}$; $-\dfrac{a-b}{c-x}$; $-\dfrac{b-a}{x-c}$.

A final point worthy of note in the multiplication of fractions is the use of the word *of* to designate multiplication. Thus, two thirds *of* one half means

$$\left(\frac{2}{3}\right)\left(\frac{1}{2}\right) = \frac{1}{3}.$$

A2.23 ADDITION AND SUBTRACTION OF FRACTIONS

Addition and subtraction of fractions is accomplished by changing each fraction to the same (common) denominator and then placing the sums (differences) of the resultant numerators over the common denominator. A common denominator can always be found by forming the term which has each of the separate denominators as a factor. On the other hand, if all the factors of each denominator are set down and a term is constructed which contains each factor the maximum number of times it appears in any one denominator, this term is called the *lowest common denominator.* Consider

$$\frac{3}{5} + \frac{4}{15} - \frac{2}{3} + \frac{5}{18}.$$

Factors of 5 are 5, 1.
Factors of 15 are 5, 3, 1.
Factors of 3 are 3, 1.
Factors of 18 are 3, 3, 2, 1.
Lowest common denominator is $(5)(3)(3)(2) = 90$.

The mechanical procedure for changing each fraction to the common denominator is illustrated by reference to the fraction 3/5. We divide the lowest common denominator by 5 to obtain the conversion factor 90/5, which is 18, and then multiply the numerator, 3, by the conversion factor to obtain 54. By this procedure, 3/5 is changed to 54/90. We have

$$\frac{3}{5} + \frac{4}{15} - \frac{2}{3} + \frac{5}{18} = \frac{54}{90} + \frac{24}{90} - \frac{60}{90} + \frac{25}{90} = \frac{43}{90}.$$

The mechanical procedure is efficient, but in the interest of emphasizing fundamentals, it should be made clear that the process derives from a fundamental property of the number one; that is, a number is unchanged if it is multiplied by one. For example, when converting 3/5 to a denominator of 90, we observe that 5 must be multiplied by 18 to yield 90, so we multiply 3/5 by 18/18; that is, in this instance the unit multiplier is 18/18. In the case of 4/15 the unit multiplier is 6/6. It follows that the mechanical procedure is a consequence of the more lengthy, but also more fundamental, process shown next:

$$\frac{3}{5} + \frac{4}{15} - \frac{2}{3} + \frac{5}{18} = \frac{3(18)}{5(18)} + \frac{4(6)}{15(6)} - \frac{2(30)}{3(30)} + \frac{5(5)}{18(5)} = \frac{43}{90}.$$

As another example, follow the conversion of each fraction in the next expression to the lowest common denominator, $a(2)(3)$:

$$\frac{2}{a} + \frac{b}{2} + \frac{2c}{3} + \frac{1}{6} = \frac{2}{a}\left(\frac{6}{6}\right) + \frac{b}{2}\left(\frac{3a}{3a}\right) + \frac{2c}{3}\left(\frac{2a}{2a}\right) + \frac{1}{6}\left(\frac{a}{a}\right)$$

$$= \frac{12 + 3ab + 4ac + a}{6a}.$$

||||||||||||||||||||||||||||||||||||||||||||||||||||||||||||||||||||||||||||||||||||||||||||||||||||||||||||||||||||||||

Exercise. Add $5 + \frac{2}{y} + \frac{3}{4} + \frac{1}{2x}$.

Answer: $\dfrac{23xy + 8x + 2y}{4xy}$.

||||||||||||||||||||||||||||||||||||||||||||||||||||||||||||||||||||||||||||||||||||||||||||||||||||||||||||||||||||||||

As a final example, consider

$$\frac{5}{3} + \frac{c}{2a} - \frac{c}{a(b+2)}.$$

The factors of the denominator are, in turn:

$$3, 1$$
$$2, a, 1$$
$$a, (b+2), 1.$$

The lowest common denominator is $3(2)a(b+2)$. Hence:

$$\frac{5}{3}\frac{(2a)(b+2)}{(2a)(b+2)} + \frac{c}{2a}\frac{(3)(b+2)}{(3)(b+2)} - \frac{c}{a(b+2)}\frac{(3)(2)}{(3)(2)}$$

is equivalent to the original set of fractions. Multiplication followed by combination of like terms leads to the final expression

$$\frac{10ab + 20a + 3bc}{6a(b+2)}.$$

‖‖‖

Exercise.   Add: $\dfrac{1}{x} + \dfrac{2}{y+1} - \dfrac{2}{3}.$

Answer: $\dfrac{4x + 3y - 2xy + 3}{3x(y+1)}.$

‖‖‖

A2.24 DIVISION OF FRACTIONS

The division of the fraction a/b by the fraction c/d can be written in the form of a third fraction:

$$\frac{\dfrac{a}{b}}{\dfrac{c}{d}}.$$

This last expression can be simplified if it is multiplied by a suitably chosen one. The objective is to convert from the *complex* fraction (that is, a fraction whose numerator or denominator contains a fraction) to a *simple* fraction (which does not have a fraction in its numerator or denominator). Clearly, we can cancel the c/d of the denominator if we multiply it by d/c. We must then also multiply the numerator by d/c, so that the net effect is multiplication by one. Thus:

$$\frac{\dfrac{a}{b}}{\dfrac{c}{d}} = \frac{\dfrac{a}{b}\left(\dfrac{d}{c}\right)}{\dfrac{c}{d}\left(\dfrac{d}{c}\right)} = \frac{ad}{bc}.$$

The process often is described as "inverting the denominator and multiplying"; that is, invert c/d to give d/c, then multiply the numerator by d/c. This description is adequate when numerator and denominator are in completely factored form, but multiplication by a suitably chosen one not only is a more fundamental description, but also is somewhat more direct when numerator and denominator are not in factored form. Consider the problem of reducing the following to a simple fraction:

$$\frac{\frac{a}{2} + \frac{1}{3}}{\frac{1}{2} + b}.$$

We observe that the lowest common denominator of the terms in the numerator and denominator is 6, so we multiply the fraction by 6/6. Thus:

$$\frac{6\left(\frac{a}{2} + \frac{1}{3}\right)}{6\left(\frac{1}{2} + b\right)} = \frac{3a + 2}{3 + 6b}.$$

||||||||||||||||||||||||||||||||||||||||||||||||||||||||||||||||||||||||||||||||||||||||||||||||||||||

Exercise. Convert to a simple fraction by multiplying numerator and denominator by 12: $\dfrac{\frac{1}{2} + \frac{1}{3}}{\frac{3}{4} + \frac{1}{3}}$. Answer: 10/13.

||||||||||||||||||||||||||||||||||||||||||||||||||||||||||||||||||||||||||||||||||||||||||||||||||||||

In the next example the lowest common denominator of all terms in numerator and denominator is $3ab$. Hence, we choose our one to be $3ab/3ab$:

$$\frac{\frac{2}{a} + \frac{1}{b}}{\frac{1}{3} + \frac{2}{b}} = \frac{\left(\frac{2}{a} + \frac{1}{b}\right)(3ab)}{\left(\frac{1}{3} + \frac{2}{b}\right)(3ab)} = \frac{6b + 3a}{ab + 6a}.$$

A2.25 PROBLEM SET 4

Mark (T) for true or (F) for false:

1. (   ) 0/0 equals one.
2. (   ) 0/0 equals zero.
3. (   ) No matter what number a is, $a/0$ is meaningless.
4. (   ) If a is not zero, then $0/a$ equals zero.
5. (   ) The product of any number and zero is zero.
6. (   ) The reciprocal of 4 equals 0.25.
7. (   ) The reciprocal of 3 equals 0.3.
8. (   ) In addition, it is said that a number and its reciprocal cancel.

9. ( ) In division, cancellation is the equivalent of substituting the factor one in place of the product of a number and its reciprocal.

10. ( ) Multiplying a number by its reciprocal gives the same result as dividing the number by itself.

11. ( ) "Inverting and multiplying" is equivalent to multiplying by a reciprocal.

12. What does it mean to say a number and its reciprocal cancel?

Simplify by cancellation where possible:

13. $\dfrac{-3a(-6)}{12c}$.

14. $\dfrac{2a-3}{3}$.

15. $\dfrac{2a-3a}{-a}$.

16. $\dfrac{2a+6}{2}$.

17. $\dfrac{24acd}{4ad}$.

18. $\dfrac{x+y}{y}$.

19. $\dfrac{2xy+6ax+x}{4xy}$.

20. $4(a+2)\left(\dfrac{2x}{a+2}\right)$.

21. State the rule for multiplication of fractions.

Multiply, leaving no grouping symbols in the answer:

22. $\dfrac{ab}{2}\left(\dfrac{3}{4}\right)$.

23. $\dfrac{a+b}{3}\left(\dfrac{2}{5}\right)$.

24. $\dfrac{-2}{3}\left(\dfrac{9a}{8}\right)$.

25. $(-2)\left(\dfrac{a}{3}\right)\left(\dfrac{b+2}{-7}\right)$.

26. $(2)\left(-\dfrac{1}{3}\right)\left(\dfrac{1}{a+b}\right)$.

27. $3(a+2)\left(\dfrac{1}{3}+\dfrac{2b}{a+2}\right)$.

28. $6ab\left(\dfrac{2}{3b}-\dfrac{1}{a}\right)$.

Express with a single minus sign:

29. $\dfrac{-b(c-d)}{-2}$.

30. $-\dfrac{b-2}{-3-a}$.

31. $\dfrac{-2+(b-c)}{-2a}$.

32. $-\dfrac{-2+(b-c)-a}{2x(a+b)}$.

33. $-\dfrac{x+y}{x-y}$.

Reduce to one simple fraction:

34. $\dfrac{2}{3}-\dfrac{1}{2}+\dfrac{1}{6}$.

35. $\dfrac{a}{2}-\dfrac{3}{5}$.

36. $\dfrac{3}{2a}-\dfrac{1}{6}+\dfrac{2}{5b}$.

37. $\dfrac{2}{5}-\dfrac{2(a-10)}{5a}+\dfrac{1}{6}$.

38. $3\frac{1}{2}-2\frac{1}{3}$.

39. $\dfrac{x}{a-2}+\dfrac{1}{b}-2$.

40. $\dfrac{x}{2a}-b+\dfrac{3}{a}$.

41. $3x-\dfrac{1}{2}+\dfrac{2}{12ab}$.

42. $\dfrac{2a}{3(b-1)}-\dfrac{a-1}{4}+\dfrac{1}{6}$.

43. $\dfrac{7}{2(x+3)}-3+\dfrac{5}{4(x+3)}$.

44. Multiply $2\frac{1}{3}$ by $3\frac{1}{4}$, stating the product as a simple fraction.

45. Divide $1\frac{1}{8}$ by $7\frac{1}{3}$, stating the quotient as a simple fraction.

Reduce the following complex fractions to simple fractions by multiplying by a suitably chosen one:

46. $\dfrac{\dfrac{1}{2} + \dfrac{1}{3} - \dfrac{1}{4}}{\dfrac{2}{3} - \dfrac{1}{6}}.$

47. $\dfrac{\dfrac{6}{a} + 2}{-\dfrac{3}{b} + \dfrac{5}{a}}.$

48. $\dfrac{\dfrac{ab}{2} - \dfrac{b-3}{c}}{\dfrac{b}{3} - 1}.$

49. $\dfrac{\dfrac{2a}{3b} - \dfrac{1}{c} + 2}{\dfrac{1}{6} - \dfrac{2}{bc}}.$

50. $\dfrac{\dfrac{a}{2} + \dfrac{b}{3} - \dfrac{c}{6}}{b - \dfrac{a}{4}}.$

A2.26 EXPONENTS

The product $(a)(a)(a)(a)(a)$ is denoted by writing a with a superscript of 5; thus, a^5. It is called the fifth power of a. The number a is the *base*, and 5 is the *exponent* of the power. More generally, if n is a positive integer, a^n is read as "a to the nth" and means the term which has a as a factor n times. By convention, we interpret absence of an exponent to mean the exponent is one. We have

$$(a)(a) = a^2$$
$$(a)(a)(a) = a^3$$
$$a = a^1.$$

It is clear that

$$(a^2)(a^3) = (a)(a)[(a)(a)(a)] = a^5.$$

We see that

$$(a^2)(a^3) = a^{2+3} = a^5$$

and it follows that if two powers have the same base, their product is found by writing the common base with the sum of the exponents as its power. For example:

$$x^5(x)x^2 = x^{5+1+2} = x^8$$
$$3a^2(2a^3) = 6a^5$$
$$3^2(3) = 3^3 = 27$$
$$(-2)^3(-2) = (-2)^4 = 16$$
$$(-3)(-3)^2 = (-3)^3 = -27.$$

Exercise. Write $2x^2(3x^5)$ with a single exponent. Answer: $6x^7$.

Turning to division, we have, for example:

$$\frac{a^5}{a^2} = \frac{(a)(a)(a)(a)(a)}{(a)(a)} = (a)(a)(a) = a^3$$

by cancellation. Alternatively, the final exponent, 3, could have been obtained by the subtraction, $5 - 2$, that is, the numerator exponent minus the denominator exponent. In general, if two powers have the same base, their quotient is the common base with an exponent found by the subtraction procedure just mentioned. As examples:

$$\frac{a^4}{a^2} = a^{4-2} = a^2$$

$$\frac{a^2b^3}{ab} = ab^2$$

$$\frac{(a + b)^3}{a + b} = (a + b)^2$$

$$\frac{5^{12}}{5^{10}} = 5^2 = 25$$

$$\frac{3^4(2a^6)}{3a^4} = 54a^2$$

$$\frac{(-2)^5}{(-2)^2} = (-2)^3 = -8$$

$$\frac{(-3)^4(2x^5)}{-3x^2} = (-3)^3(2x^3) = -54x^3.$$

Exercise. Write $(5x^5)/3x^2$ with a single exponent. Answer: $5x^3/3$.

A2.27 ZERO EXPONENT

Following the procedure of the last discussion, we see that for any number a, not zero:

$$\frac{a}{a} = a^{1-1} = a^0.$$

Inasmuch as the beginning expression, a/a, equals one, we conclude that any nonzero number to the zero power equals one. Thus:

$$1^0 = 1, \qquad (ab^3c^2)^0 = 1, \qquad (14.6)^0 = 1, \qquad (-x)^0 = 1.$$

Exercise. Evaluate $3^0 + (x + 2y)^0$. Answer: 2.

A2.28 NEGATIVE EXPONENTS

When exponents are subtracted, the difference may be negative. For example:

$$\frac{2^3}{2^6} = 2^{3-6} = 2^{-3}.$$

Alternatively, we may evaluate the expression as

$$\frac{2^3}{2^6} = \frac{(2)(2)(2)}{(2)(2)(2)(2)(2)(2)} = \frac{1}{(2)(2)(2)} = \frac{1}{2^3}.$$

We see that

$$(2)^{-3} = \frac{1}{2^3}$$

The general definition applying to a negative exponent is

$$a^{-n} = \frac{1}{a^n}.$$

As illustrations of the definition, we see that

$$2^{-1} = \frac{1}{2}$$

$$3^{-2} = \frac{1}{3^2} = \frac{1}{9}$$

$$ax^{-2} = \frac{a}{x^2}$$

$$\left(\frac{2}{3}\right)^{-1} = \frac{1}{\frac{2}{3}} = \frac{3}{2}$$

$$\left(\frac{1}{3}\right)^{-2} = 9.$$

Exercise. Evaluate $(1/2)^{-2} + 5(2^{-3})$. Answer: $37/8$.

The following examples are self-explanatory and show how in some expressions we may avoid negative exponents by choice of procedure.

$$\frac{x^3}{x^5} = \frac{1}{x^{5-3}} = \frac{1}{x^2}$$

$$\frac{2x}{x^7} = \frac{2}{x^6}$$

$$\frac{a^2 b^3}{a^4 b} = \frac{b^2}{a^2}.$$

Exercise.   Apply the laws of exponents to simplify the following expression and write the result with positive exponents.

$$\frac{3x(x^{-2})y^5}{4x^4y^2}.$$

Answer: $\frac{3y^3}{4x^5}.$

Finally, we note that the expression

$$\frac{1 + x^{-n}}{2 + a}$$

is, in effect, a complex fraction because of the fractional nature of x^{-n}. Remembering that ·

$$x^{-n}(x^n) = x^0 = 1,$$

we may obtain a simple fraction by multiplying by x^n/x^n, as follows:

$$\frac{1 + x^{-n}}{2 + a} = \frac{(1 + x^{-n})x^n}{(2 + a)x^n} = \frac{x^n + 1}{(2 + a)x^n}.$$

As another example of the same procedure:

$$\frac{2 + 3^{-2}}{1 + 3^{-1}} = \left(\frac{2 + 3^{-2}}{1 + 3^{-1}}\right)\left(\frac{3^2}{3^2}\right) = \frac{18 + 1}{9 + 3} = \frac{19}{12}.$$

Exercise.   Remove negative exponents by multiplying numerator and denominator by x^n: $\frac{1 + x^{-n}}{x^{-n} + 2}.$ Answer: $\frac{x^n + 1}{1 + 2x^n}.$

A2.29 POWER TO A POWER

The expression $(a^2)^3$ is an example of a power raised to a power; that is, the second power of a is indicated as being raised to the third power. According to definition:

$$(a^2)^3 = (a^2)(a^2)(a^2)$$

which is a^6 or $a^{(3)(2)}$. The procedure is generalized by stating that in raising a power to a power, exponents are multiplied. As other examples:

$$(2^2)^4 = 2^8 = 256$$
$$(x^3)^2 = x^6$$
$$(x^{-1})^4 = x^{-4}$$
$$(5^{-2})^{-1} = 5^2 = 25.$$

Exercise. Evaluate $[(2)^{-2}]^{-3}$. Answer: 64.

A2.30 FRACTIONAL EXPONENTS

The number

$$8^{1/3}$$

may be read as *eight to the one-third power*. If we apply the rules discussed earlier for integral exponents to this rational fractional exponent, it follows that

$$(8^{1/3})(8^{1/3})(8^{1/3}) = 8^1 = 8,$$

so that the number symbolized as $(8^{1/3})$ must be 2. That is,

$$8^{1/3} = 2.$$

Moreover,

$$8^{2/3} = (8^{1/3})(8^{1/3}) = (2)(2) = 4.$$

Exercise. Express $8^{4/3}$ as an integer. Answer: 16.

The number $8^{1/3}$ is also called the *cube root* of 8 and expressed by the *radical* symbol,

$$\sqrt[3]{8}.$$

The number appearing in the opening of the radical symbol is called the *index* of the root, and the number under the symbol, here 8, is called the *radicand*. We note that the index of the root is the denominator of the fractional exponent. In similar fashion,

$$16^{1/2} = \sqrt[2]{16} = \sqrt{16}$$

is called the square root of 16 and, conventionally, the index 2 is not written. That is, if no index number appears on the radical, the index is assumed to be 2.

Both 4 and -4 are square roots of 16 because

$$(4)(4) = (-4)(-4) = 16.$$

Thus,

$$\sqrt{16} = \pm 4.$$

Exercise. What are the fourth roots of 16? Answer ± 2.

Even roots of positive numbers have both a positive and a negative value. However, *we shall generally follow the practice of indicating only the positive value for even roots of positive numbers.* Even roots of negative numbers are not real numbers. For example, in

$$(-4)^{1/2}(-4)^{1/2} = (-4)^1 = -4,$$

there is no real number for $(-4)^{1/2}$ which will make the statement true. Clearly, $(0)(0) \neq -4$ and moreover the product of numbers of like sign cannot be negative, which would be required if $(-4)^{1/2}$ was a real number as required by the statement.

Odd roots of negative numbers can be found, as in

$$(-8)^{1/3} = -2,$$

but we shall have no occasion for their use. Thus, when we use a fractional power of a number, x, it will be assumed that x is not negative.

If we write radical expressions at random, the desired root often is irrational and must be approximated. For example,

$$\sqrt{2} = 2^{1/2} = 1.4142$$

to four decimal places. We can approximate irrational roots by logarithms. (See Index for reference.) In this appendix, we consider only examples where roots are rational and can be determined by inspection. For example, inspecting

$$4^{3/2}$$

we first write the equivalent number

$$(4^{1/2})^3$$

to emphasize that we want the square root of 4, which is 2. Hence,

$$4^{3/2} = (4^{1/2})^3 = (2)^3 = 8.$$

Again,

$$\left(\frac{8}{125}\right)^{4/3} = \left[\left(\frac{8}{125}\right)^{1/3}\right]^4,$$

and we seek the cube roots for 8 and 125. The cube root of 8 is 2. We guess 5 as the cube root of 125 and verify our guess by noting that $5(5)(5) = 125$. Hence,

$$\left(\frac{8}{125}\right)^{4/3} = \left[\left(\frac{8}{125}\right)^{1/3}\right]^4 = \left(\frac{2}{5}\right)^4 = \frac{16}{625}.$$

Negative exponents may be changed to positive as shown in the next illustration.

$$8^{-2/3} = \frac{1}{8^{2/3}} = \frac{1}{[(8)^{1/3}]^2} = \frac{1}{2^2} = \frac{1}{4}.$$

||||||||||||||||||||||||||||||||||||||||||||||||||||||||||||||||||||||||||||||||||||||||||||||||||||||||||

Exercise. Find $(243)^{2/5}$. Answer: 9.

||||||||||||||||||||||||||||||||||||||||||||||||||||||||||||||||||||||||||||||||||||||||||||||||||||||||||

Previous rules for exponents carry over when exponents are fractional, as shown by the following examples:

$$(2^{1/3})(2^{1/2}) = 2^{5/6}$$

$$\frac{(a^{2/3})b^2}{ab} = \frac{b}{a^{1/3}}$$

$$(x^{1/3})^{-1/2} = x^{-1/6} = \frac{1}{x^{1/6}}$$

$$(a^{1/3})^2 = a^{2/3}.$$

The last expression can be written as

$$a^{2/3} = (a^{1/3})^2 = (a^2)^{1/3}.$$

Inasmuch as $a^{1/3}$ is the cube root of a, we may describe a to the two thirds as the second power of the cube root of a or as the cube root of the second power of a; that is:

$$a^{2/3} = (\sqrt[3]{a})^2 = \sqrt[3]{a^2}.$$

In general:

$$a^{m/n} = (\sqrt[n]{a})^m = \sqrt[n]{a^m}$$

A2.31 SUMMARY OF EXPONENT RULES

All the rules for exponents may now be stated in brief form:

$$a^m a^n = a^{m+n}$$

$$\frac{a^m}{a^n} = a^{m-n} = \frac{1}{a^{n-m}}$$

$$(a^m)^n = a^{mn}$$

$$a^0 = 1.$$

The rules of exponents are not restricted to terms having a single factor. We may raise a term to a power by raising each factor of the term to the power. Thus:

$$(2a)^3 = 8a^3$$

$$\left(\frac{a}{2}\right)^2 = \frac{a^2}{4}$$

$$(3x^2 y^{1/2})^3 = 27x^6 y^{3/2}.$$

||||||||||||||||||||||||||||||||||||||||||||||||||||||||||||||||||||||||||||||||||||||||||||||||||||||||||||||||||||||||||||||||||||||||

   Exercise. Express $(4y^4x^3)^{1/2}$ without parentheses. Answer: $2y^2x^{3/2}$.

||||||||||||||||||||||||||||||||||||||||||||||||||||||||||||||||||||||||||||||||||||||||||||||||||||||||||||||||||||||||||||||||||||||||

Observe, however, that in $3a^2$ the absence of parentheses means the exponent applies only to a, not to 3. Again:

$$3(2a^2)^3 = 3(8a^6) = 24a^6.$$

Exponents cannot be applied separately to terms of an expression. In the case of, say, $(a - b)^2$, we cannot simply raise each term to the second power. By definition:

$$(a - b)^2 = (a - b)(a - b).$$

According to the distributive property:

$$(a - b)(a - b) = a^2 - ab - ba + b^2 = a^2 - 2ab + b^2.$$

By way of numerical illustration:

$$(1 - 0.2)^2 = (0.8)^2 = 0.64$$

could be evaluated as

$$(1 - 0.2)(1 - 0.2) = 1^2 - 1(0.2) - 0.2(1) + 0.2^2 = 0.64.$$

This type of exercise is referred to as squaring a binomial, that is, raising the sum of two terms to the second power. We see in general that

$$(a + b)^2 = a^2 + 2ab + b^2$$

and describe the outcome as the "square of the first, plus twice the product of the two, plus the square of the second." The reader may verify that the cube of a binomial is given by

$$(a + b)^3 = a^3 + 3a^2b + 3ab^2 + b^3.$$

   Before working the problem set, the reader may wish to practice by verifying the answer stated for the following practice problem set.

A2.32 PRACTICE PROBLEM SET

   Evaluate:

1. $27^{2/3}$: $= (\sqrt[3]{27})^2 = (3)^2 = 9.$

2. $9^{-3/2}$: $= \dfrac{1}{9^{3/2}} = \dfrac{1}{3^3} = \dfrac{1}{27}.$

3. $8^{4/3}$: $= 16.$

4. $\sqrt{-4}$: not a real number.

5. $(2^{-2})(3^{-2})$: $= \dfrac{1}{4}\left(\dfrac{1}{9}\right) = \dfrac{1}{36}.$

6. $(2^{1/3})(2^{1/2})(2^{1/6})$: $= 2^{1/3+1/2+1/6} = 2.$

7. $(1 + 0.01)^2$: $= (1.01)^2 = 1.0201.$

8. $2^{-2} + 3^{-1}: = \frac{1}{4} + \frac{1}{3} = \frac{7}{12}$.

9. $2^3(1 + 2^{-3}): = 2^3 + 2^0 = 8 + 1 = 9$.

10. $(15)^{12}(15)^{-10}: = 15^2 = 225$.

11. $\frac{10^{-4}}{10^{-2}}: = \frac{1}{10^{-2+4}} = \frac{1}{100}$.

12. $\frac{1}{3}(4^{-3/2}): = \frac{1}{3}\left(\frac{1}{4^{3/2}}\right) = \left(\frac{1}{3}\right)\left(\frac{1}{8}\right) = \frac{1}{24}$.

13. $\frac{5 + 2^{-2}}{3}: = \frac{(5 + 2^{-2})2^2}{(3)(2^2)} = \frac{(5)2^2 + 2^0}{3(2^2)} = \frac{21}{12} = \frac{7}{4}$.

Combine exponents where possible, and simplify. If possible, do not leave grouping symbols, radical signs, or negative exponents in the result:

14. $a^2 x a^3 x^2: = a^5 x^3$.

15. $2x^0 + (2x)^0: = 2 + 1 = 3$.

16. $(a^2 b)(a^{-1})b^3(ab)^{-1}: = a^{2-1-1}b^{1+3-1} = b^3$.

17. $\frac{4x^2 b}{(2b)^2}: = \frac{4x^2 b}{4b^2} = \frac{x^2}{b}$.

18. $\frac{a^3}{(a^x)^2}: = \frac{a^3}{a^{2x}} = a^{3-2x}$ or $\frac{1}{a^{2x-3}}$.

19. $\frac{(2^3)(2^x)}{2^{1+x}}: = 2^{3+x-(1+x)} = 2^2 = 4$.

20. $\frac{(x^{1/3})(y^{2/3})^2}{2(xy)^{1/2}}: = \frac{x^{1/3}y^{1/3}}{2x^{1/2}y^{1/2}} = \frac{y^{5/6}}{2x^{1/6}}$.

21. $\frac{x^{1/3}\sqrt{b}}{b\sqrt{x}}: = \frac{x^{1/3}b^{1/2}}{bx^{1/2}} = \frac{1}{b^{1/2}x^{1/6}}$.

22. $(a - 2b)^2: = a^2 - 4ab + 4b^2$.

23. $\frac{1}{3}(3x)^{-2/3}: = \frac{1}{3(3)^{2/3}(x)^{2/3}} = \frac{1}{3^{5/3}x^{2/3}}$.

24. $\frac{1 + (1 + x)^{-n}}{x}: = \frac{[1 + (1 + x)^{-n}](1 + x)^n}{x(1 + x)^n} = \frac{(1 + x)^n + 1}{x(1 + x)^n}$.

A2.33 PROBLEM SET 5

Evaluate:

1. 2^4.

2. $7(2)^0$.

3. $(1 - 0.02)^{-2}$ to three decimal places.

4. $\left(\frac{3}{4}\right)^{-1}$.

5. $(-3)^{-2}(2)^{-3}$.

6. $\frac{(10^{-3})(10^5)}{(10^3)(10^{-4})}$.

7. $\sqrt{-16}$.

8. $3^{-1} + \left(\frac{2}{3}\right)^{-2}$.

9. 3^{-2}.

10. $(x - 3)^0$.

11. $16^{3/4}$.

12. $\left(\frac{1}{8}\right)^{1/3}$.

13. $\frac{10^{-4}}{10^{-5}}$.

14. $\frac{2}{3}(16)^{-3/4}$.

15. $\sqrt{\dfrac{25}{16}}$.

16. $5x^0 + (ax)^0$.

17. $(25)^4(25)^{-3}$.

18. $(1 + 0.05)^2$.

19. $(32)^{-3/5}$.

20. $\left(\dfrac{2}{3}\right)^{-1} + 2^{-3}$.

21. $\sqrt[3]{125}$.

22. $\left(\dfrac{8}{27}\right)^{-2/3}$.

23. $\dfrac{1 + 3^{-2}}{5}$.

24. $\dfrac{2^{-3} + 2^{-1}}{2^{-1}}$.

25. $3^{-2}(1 + 3^{-1})$.

Combine exponents where possible, and simplify. If possible, do not leave grouping symbols, radical signs, or negative exponents in the result:

26. a^2a.

27. $(ab)(ac)(bc)$.

28. $a^2b^3(ab^4)$.

29. $(abc^2)(a^2cb)$.

30. $(x^2a)(x^2b)$.

31. $\dfrac{a^2b^3}{ab}$.

32. $\dfrac{a^4b^2c}{a^2b^3c^2}$.

33. $\dfrac{xy^3b^3}{x^3yb}$.

34. $\dfrac{a^3(bxy)}{abx^2y}$.

35. $\dfrac{x^3y^2}{x^2y^3}$.

36. $\dfrac{a(bc)^3}{ab^2}$.

37. $\dfrac{(ab)^2(ab)}{a}$.

38. $\dfrac{a(bc)^3}{b(ac)^4}$.

39. $\dfrac{(-3b)^3(2c)^2}{12b^2c}$.

40. $\dfrac{-2b^2(-3c)^2}{-(-bc)^3}$.

41. $(xy^2)(ay^{-2})$.

42. $(ab^2c^{-1})(a^2b^{-1}c^2)$.

43. $a(ax)^{-2}$.

44. $\dfrac{2x^{-1}a^3}{(-ax)^2}$.

45. $\dfrac{x^{-2}}{ax}$.

46. $3a^{1/3}b^{1/2}a^2b$.

47. $\dfrac{\sqrt{x}\ \sqrt[3]{y}}{x^{-1}y^{1/2}}$.

48. $\dfrac{a^{2/3}(b^3)^{1/2}}{\sqrt{a}\ \sqrt[3]{b}}$.

49. $\dfrac{2}{3}(3x)^{-1/2}$.

50. $\dfrac{2x^{-1/2}\sqrt{y^3}}{3y^{-1/3}\sqrt{x^3}}$.

51. $\dfrac{2 + x^{-2}}{x + 3}$.

52. $\dfrac{(1 - x)^{-1} + 1}{x}$.

53. $\dfrac{x^{-1} + x^{-2}}{3}$.

54. $\dfrac{(ax)^2}{(a^x)^2}$.

55. $\dfrac{x^5 + 4}{x^2 + 2}$.

56. $a(a + b)$.

57. $(a - 3b)^2$.

58. $a(a - b)^2$.

59. $(x - y)(x + y)$.

60. $a(a^{-1} + 1)$.

61. $(a^{1/2} - 1)^2$.

62. $\dfrac{a^{-1} + 2}{a}$.

63. $\dfrac{(3^2)(3^a)}{3^{a-1}}$.

Mark (T) for true or (F) for false:

64. (   ) $(2^3)(3^2) = 6^5$.

65. (   ) $ab^2 = a^2b^2$.

66. (   ) $(2ab)^2 = 4a^2b^2$.

67. (   ) $\dfrac{a^2}{b^2} = \left(\dfrac{a}{b}\right)^2$.

68. (   ) $(a - b)^2 = a^2 - b^2$.

69. (   ) $(a + 1)^{1/2} = \sqrt{a} + 1$.

70. (   ) $\dfrac{\sqrt[3]{a}}{\sqrt[3]{b}} = \sqrt[3]{\dfrac{a}{b}}$.

71. (   ) $(1 + 0.1)^{-2} = \dfrac{1}{1.21}$.

72. (   ) An expression is raised to a power by raising each of its terms to the power.

73. (   ) The sum of the squares of two numbers is the same as the square of the sum of the two numbers.

74. (　) $a^{-1} + b^{-1}$ is equivalent to the sum of the reciprocals of a and b.
75. (　) $0^0 = 1$.
76. (　) $5x^0 = 1$.
77. (　) $(1 - 0.02)^{-5} = \dfrac{1}{(0.98)^5}$.
78. (　) $2x^{-1/2} = \dfrac{2}{x^2}$.
79. (　) $a^{2/3}$ is the cube root of a^2.
80. (　) $\sqrt{x^2 + y^2 + z^2} = x + y + z$.
81. (　) $\sqrt{2x} = 2^{1/2}x^{1/2}$.
82. (　) $\sqrt{-12}$ is not a real number.
83. (　) $\dfrac{2a^{1/6}}{b^{1/6}} = 2\left(\dfrac{a}{b}\right)^{1/6}$.
84. (　) $a^{1/2}$ is called the square of a.
85. (　) $(a - 2b)^2 = a^2 - 4ab + 4b^2$.

A2.34 REVIEW PROBLEMS

Reduce to a single signed number, or zero:
1. $-4 + (-2) - |+(-3)|$.
2. $-(-8) + (-4) - (+4) + 2$.
3. $-2(3) + (-3)(-4) - (1)(-5)$.
4. $-(-3) + 2\{-[-3(2)]\} - (+10)$.
5. $-5(-2)(-3) + (3)(-2)(-5)$.
6. $\dfrac{-3(-2)(-4)}{8(-5)}$.
7. $\dfrac{-(5)(3)(-4)}{2(-6)}$.
8. $\dfrac{-(-5)(20)}{-\{-[-4]\}}$.
9. $\dfrac{3(2)(5)(-6)}{(-10)(-18)}$.
10. $\dfrac{(-1)(-2)(-3)(-4)}{(-5)(6)}$.

11. Rewrite as the sum of signed numbers:
    a) $3 - 4$.
    b) $-1 + 2 - 3$.
    c) $-1 - 2 - 4 + 5$.
    d) $2 + 3$.
12. Reduce to a single signed number:
    a) $-(2 - 5) + 3(-4) - (2 + 1)$.
    b) $-[-2 - (+3)] + (-1)(1 - 3)$.
    c) $-[-(a - b) + (-b + a)]$.
13. What fundamental property or convention justifies each of the following?
    a) $x2y = 2xy$.
    b) $4(3z) = 12z$.
    c) $+2 + (-a) = 2 - a$.
    d) $3x + 6y = 3(x + 2y)$.
    e) $3a + 2 + y = 3a + y + 2$.
    f) $(a + 2) + b = a + (2 + b)$.
    g) $1 + 2 + x = 3 + x$.
    h) $x2 + y3 = 2x + 3y$.
    i) $3 + (-x) = 3 - x$.
    j) $1a = a$.
    k) $2(a + b) = 2a + 2b$.
    l) $5 - xy = -xy + 5$.
    m) $3(x + y) = 3x + 3y$.
    n) $5a + 5b = 5(a + b)$.

14. Combine like terms:
   a) $2 + 5x - 4b + 7 - (-3b)$.   b) $ab - a(2 - b) + 3a + 5$.
   c) $x(ay - 3) + 2x + 4axy - 7$.   d) $xy - 2(2 - xy) + 5xy$.

15. Starting with $3x(2y) + 3 + x + 3x$, name the fundamental property which justifies each step:
   a) $3x(2y) + 3 + x + 3x = 3x(2y) + 3 + 4x$.
   b) $= 3(2)xy + 3 + 4x$.
   c) $= 6xy + 3 + 4x$.
   d) $= 6xy + 4x + 3$.
   e) $= (6xy + 4x) + 3$.
   f) $= 2x(3y + 2) + 3$.

16. Starting with $zy + 5 + 2(3 + 4y)$, name the fundamental property which justifies each step:
   a) $zy + 5 + 2(3 + 4y) = zy + 5 + 2(3) + 2(4y)$.
   b) $= zy + 5 + 6 + 2(4y)$.
   c) $= zy + 5 + 6 + 8y$.
   d) $= zy + 11 + 8y$.
   e) $= zy + 8y + 11$.
   f) $= yz + 8y + 11$.
   g) $= (yz + 8y) + 11$.
   h) $= (z + 8)y + 11$.
   i) $= y(z + 8) + 11$.

17. Remove grouping symbols and combine like terms:
   a) $x - a\{3 - 2(1 - x)\}$.   b) $5 - 3[2 - 4(-3 + z)]$.
   c) $3 - 2[a - x + 3(2x - a)]$.   d) $(3 - a)(x - b + 2)$.
   e) $ax - a\{-2x + 3(2x - 1)\}$.

18. Reduce to a single signed number:
$$-\frac{-3 + 2[-5 - (2 - 7)]}{5 + [2 - 3(2 - 3)]}.$$

19. Factor the following:
   a) $2x + ax$.   b) $xy + 3xy + axy$.
   c) $2 + 4a + 6b$.   d) $5xy + 4x$.
   e) $3ax + 6ay + 9a$.   f) $2(x - 1) + a(x - 1)$.
   g) $a + b + ax + bx$.   h) $x^2 + x - 2$.
   i) $10x^2 + 3x - 1$.   j) $6x^2 + 7x - 20$.
   k) $8x^2 + 16x + 6$.

20. Simplify by cancellation where possible:
   a) $\dfrac{3(3 - a) + x}{2x}$.   b) $\dfrac{3xy}{2x}$.

   c) $\dfrac{x - 2xy}{3x}$.   d) $\dfrac{3}{x - y}[2(x + y)]$.

21. Multiply, leaving no grouping symbols in the answer:
   a) $\left(\dfrac{2x}{3}\right)\left(\dfrac{6a}{b}\right)$.   b) $-2\left(\dfrac{x - 3}{a}\right)$.

   c) $2(x - y)\left(\dfrac{3a}{x} - \dfrac{1}{y} - \dfrac{1}{2}\right)$.   d) $10xy\left(\dfrac{1}{5y} - \dfrac{a}{2xy}\right)$.

22. Express with a single minus sign:

   a) $\dfrac{-3x - 2y}{5 - a}$.

   b) $-\dfrac{2a - b}{x + b}$.

   c) $-\dfrac{a - b}{-5 - x}$.

   d) $\dfrac{a + b}{-5 - x}$.

23. Reduce to one simple fraction:

   a) $\dfrac{1}{3} - \dfrac{1}{6} + \dfrac{3}{4}$.

   b) $\dfrac{a}{b} + 1$.

   c) $\dfrac{3}{2x} - \dfrac{2}{x} + \dfrac{5}{4x}$.

   d) $\dfrac{y}{x} - 1 + \dfrac{y}{2a}$.

   e) $\dfrac{1}{12xy} + \dfrac{2}{3} - \dfrac{5}{x}$.

   f) $\dfrac{1}{2x - 3} + \dfrac{b}{3}$.

   g) $\dfrac{a}{x + y} - \dfrac{b}{z} + 5$.

   h) $\dfrac{3}{a} - \dfrac{b}{2} + \dfrac{c}{x + y} - d$.

   i) Three fourths of $2\frac{1}{2}$.

   j) $5\frac{1}{4} - 2\frac{1}{3}$.

   k) $1\frac{7}{8}$ divided by $2\frac{1}{2}$.

24. Reduce the following complex fractions to simple fractions by multiplication by a suitably chosen 1:

   a) $\dfrac{\frac{2}{3} - \frac{1}{4}}{\frac{1}{6} + 1}$.

   b) $\dfrac{\frac{x}{y} - 1}{\frac{2}{3} + \frac{3}{y}}$.

   c) $\dfrac{x - \frac{y - 4}{a}}{\frac{b}{2} + \frac{3}{a}}$.

   d) $\dfrac{\frac{x}{x + y} - 1}{\frac{2}{x + y}}$.

25. Evaluate:

   a) 3^3.

   b) $2x^0$.

   c) $(2x)^0$.

   d) $(1 - 0.1)^{-2}$ (to three decimals).

   e) $\left(\dfrac{2}{3}\right)^{-2}$.

   f) $(2)^{-2}(-2)^2$.

   g) $(75)^{100}(75)^{-98}$.

   h) $3^0 - (2x + 1)^0$.

   i) $\left(\dfrac{1}{16}\right)^{1/4}$.

   j) $\left(\dfrac{1}{16}\right)^{-1/4}$.

   k) $\dfrac{2}{5}(32)^{1/5}$.

   l) $(27)^{2/3}$.

   m) $(125)^{4/3}$.

   n) $(8)^{-2/3}$.

   o) $\left(\dfrac{2}{3}\right)^{-1}$.

   p) $\dfrac{(10)^{-5}(10)^2}{(10)^3(10)^{-8}}$.

   q) $2^{-3}(2^2 + 2^{-1})$.

   r) $\dfrac{3^{-1} - 3^{-2}}{3^{-3}}$.

   s) $\sqrt[3]{64}$.

   t) $(1 + 0.03)^3$.

   u) $\sqrt{-1}$.

   v) $\left(\dfrac{2}{3}\right)^{-2}\left(\dfrac{3}{2}\right)$.

   w) $\sqrt{\dfrac{4}{25}}$.

   x) $(\sqrt[3]{27})^2$.

   y) $\sqrt{16^3}$.

   z) $\sqrt{9^{-3}}$.

26. Combine exponents, where possible, and simplify. If possible, do not leave grouping symbols, radical signs, or negative exponents in the final result:

a) $a(a^2)(a^3)$.

b) $xyz(x^2y)$.

c) $(2x^2y)(3xz^2)$.

d) $(abc)^2(ab)$.

e) $\dfrac{a^2b^3}{abc}$.

f) $\dfrac{a^2yz^3}{ayz^2}$.

g) $\dfrac{(3ab)^2(2c)^3}{12ac^2}$.

h) $\dfrac{(-2x)^3y^2}{4xy}$.

i) $\dfrac{(ab^{-1}c^2)^3}{2b^2c^2}$.

j) $\dfrac{(3x^{-2}y)^2}{2x^3y^{-3}}$.

k) $\dfrac{(-xy^2)^{-3}(2z^2)}{(-2yz)^4}$.

l) $[(x\sqrt{y})^{3/2}]^2$.

m) $\dfrac{ax^{2/3}y^{-1/2}}{bx^{1/2}y^{5/3}}$.

n) $(y^{1/2})^{-3/2}$.

o) $\dfrac{3}{4}(4y)^{-3/2}$.

p) $\dfrac{\sqrt[3]{ab}\,\sqrt{xy}}{9(ab)^{-1}(xy)^{3/2}}$.

q) $\dfrac{(2a^{1/2}b^{1/3}c)^3}{4abc}$.

r) $\dfrac{1+a^{-1}}{1-a^{-1}}$.

s) $\dfrac{(a-b)^{-1}-1}{b^2}$.

t) $\dfrac{2a^{x-3}a^5}{a^2}$.

u) $\dfrac{x^{2a}}{x^a}$.

v) $a^2b^3(ab)^{-5}$.

w) $x(2-x)^2$.

x) $a^2(a^{-2}+a^{-3})$.

y) $a(a-b)(a+2b)$.

z) $(5^{x+1})(5a)^{-3}(5a^3)$.

Appendix 3

Formulas, Equations, and Graphs

A3.1 INTRODUCTION

THE DISTRIBUTIVE property asserts that

$$a(b + c) = ab + ac.$$

Inasmuch as the assertion applies for any numbers, a, b, and c, it is called an identity. On the other hand, the statement

$$x + 5 = 7$$

is true only under the condition that x is 2, and for this reason is called a conditional equality.

A three-lined symbol, $\equiv$, can be used to denote an identity. In this book, however, the two-lined symbol, $=$, is used to denote both identities and conditional equalities because the context makes clear which interpretation of the symbol is relevant. Thus, in Appendix 2 the symbol $=$ means identically equal, whereas in most of the remainder of the book the symbol denotes conditional equality. We shall refer to conditional equalities briefly as equations.

In the equation $x + 5 = 7$, the letter x can be called the unknown, and the value of the unknown which makes the statement true, the number 2, can be called the root of the equation. We may also say that a number which makes the statement of equality true satisfies the equation.

The statement $y = x + 2$ is true for various pairs of values for x and y, and we shall call the letters x and y variables. A particular set of numbers, such as

$$x = -1$$
$$y = 1$$

601

which satisfy the equation, is a solution of the equation. In this book, we use the *variable, solution* terminology (rather than unknown, root terminology) in the discussion of equations. Thus:

$$y = x + 2$$

will be called an equation in two variables. It has an unlimited number of solutions. The equation

$$x + 5 = 7$$

is an equation in one variable. It has the unique solution $x = 2$.

To solve an equation for a variable means to perform whatever operations are necessary to put the equation in a form in which the stated variable is alone, with coefficient one, on one side of the equality sign, and the expression on the other side of the equality sign does not contain the stated variable. For example:

$$x - y = 3$$

is not solved for either x or y, but

$$x = y + 3$$

is solved for x, and the equation

$$y = x - 3$$

is solved for y.

Exercise.   Fill in the blanks: The _____ $x + 10 = 4$ has a single _____, x. The _____ of the _____ is $x = -6$. On the other hand, the _____ $y - x = 7$ has two _____ and has an _____ number of solutions; if we write $y - x = 7$ in the alternate form $y = x + 7$, it is said to be _____ for _____. Answer: Equation, variable, solution, equation. Equation, variables, unlimited, solved, y.

We now turn to some elementary procedures employed when solving an equation for a variable.

A3.2 SOME AXIOMS

Axioms are statements we assume to be true. For example, we assume that if equals are added to equals, the sums will be equal. Inasmuch as an equation is a statement of equality of the numbers on either side of the equal sign, the axiom says that if we add the same number to both sides of an equation, we shall obtain another equation. For example, if we have the statement

$$x - 3 = 7$$

and we add 3 to each side, we obtain

$$x - 3 + 3 = 7 + 3$$

from which we find $x = 10$, and we have solved the equation for x.

In similar fashion, we assume that if equals are subtracted from equals, the differences are equal; if equals are multiplied by equals, the products are equal; and if equals are divided by equals (divison by zero excluded), the quotients are equal. In the context of equations, we say that the solutions of an equation are not altered if the same number is added to both sides, or if the same number is subtracted from both sides, or if both sides are multiplied or divided by the same number.

Examples.

$$
\begin{aligned}
x - 2 &= 3 \qquad &&\text{Add 2 to both sides:}\\
x - 2 + 2 &= 3 + 2\\
x &= 5.
\end{aligned}
$$

$$
\begin{aligned}
x + 2 &= 3 \qquad &&\text{Subtract 2 from both sides:}\\
x + 2 - 2 &= 3 - 2\\
x &= 1.
\end{aligned}
$$

$$
\begin{aligned}
\frac{x}{3} &= 2 \qquad &&\text{Multiply both sides by 3:}\\
3\left(\frac{x}{3}\right) &= (3)(2)\\
x &= 6.
\end{aligned}
$$

$$
\begin{aligned}
2x &= 7 \qquad &&\text{Divide both sides by 2:}\\
\frac{2x}{2} &= \frac{7}{2}\\
x &= \frac{7}{2}.
\end{aligned}
$$

Exercise. The solution $x = 4$ is obtained by applying which operation to each of the following: $6x = 24$; $x/2 = 2$; $x - 4 = 0$; $x + 2 = 6$? Answer: Divide both sides by 6; multiply both sides by 2; add 4 to both sides; subtract 2 from both sides.

A3.3 SOLUTIONS BY ADDITION AND MULTIPLICATION, WITH INVERSES

The objective in solving an equation for a certain variable is to derive an expression which has that variable alone, coefficient one, on one side of the equal sign, and an expression not involving this variable on the other side of the equal sign. Axioms are applied to accomplish this objective. If we are asked to solve

$$ax + b = c$$

for *x*, we may proceed by subtracting *b* from both sides to obtain

$$ax = c - b.$$

Next, we divide both sides by *a* to obtain the desired solution:

$$x = \frac{c - b}{a}.$$

The steps are justified by the axioms. However, it is helpful to understand what steps are required, and this understanding is enhanced if we keep in mind the notion of *inverse* operations. Thus, in solving

$$ax + b = c$$

for *x*, we wish to remove *b* from the left side. Inasmuch as *b* is *added* on the left, we apply the inverse operation and *subtract b* from both sides to obtain

$$ax = c - b.$$

We now observe that *x* is *multiplied* by *a*, so we apply the inverse operation and *divide* both sides by *a* to obtain

$$x = \frac{c - b}{a}.$$

Inverse operations are the tools we need to manipulate equations into desired form.

The student should be prepared to justify each step taken in the solution of an equation. For brevity, he may omit the phrase "both sides of the equation" when citing justification for an operation. Thus, for example, "add 3" will be assumed to mean to add 3 to both sides of the equation. For uniformity, it is to be understood that the outcome of each operation performed is shown on the line following the description of the operation. For example:

$$x - 3 = 5 \qquad \text{Add 3:}$$
$$x = 8.$$

Example.   Solve for x, citing operations performed:

$$4x - 2 = 2x + 5 \qquad \text{Subtract } 2x:$$
$$2x - 2 = 5 \qquad \text{Add 2:}$$
$$2x = 7 \qquad \text{Divide by 2:}$$

$$x = \frac{7}{2}.$$

Exercise.   Solve for x, citing operations performed: $3 - 2x = -5x + 7$. Answer:

$$3 - 2x = -5x + 7 \qquad \text{Add } 5x:$$

$$3 + 3x = 7 \qquad \text{Subtract 3:}$$

$$3x = 4 \qquad \text{Divide by 3:}$$

$$x = \frac{4}{3}.$$

Generally, equations containing fractions can be handled most efficiently by finding the lowest common denominator of all the fractions in the equation, and then multiplying both sides by the l.c.d. (lowest common demoninator).

Example.   Solve for x, citing operations performed:

$$2x - \frac{1}{3} = \frac{x}{2} \qquad \text{Multiply by 6 (the l.c.d.):}$$

$$6(2x) - 6\left(\frac{1}{3}\right) = 6\left(\frac{x}{2}\right) \qquad \text{Carry out multiplications:}$$

$$12x - 2 = 3x \qquad \text{Add 2:}$$

$$12x = 3x + 2 \qquad \text{Subtract } 3x:$$

$$9x = 2 \qquad \text{Divide by 9:}$$

$$x = \frac{2}{9}.$$

Careful attention should be paid to the fact that each side of an equation is an expression; and to multiply an expression by a number, we must multiply every *term* of the expression by that number.

It is instructive to follow through the tactics of the last example. We observe fractions in the equation, and we know that the denominator of a fraction is canceled (replaced by one, which need not be written) if the fraction is multiplied by a term having the fraction's denominator as a factor.

Rather than treat each fraction separately, we make up a term (lowest) which has the denominators of all the fractions as factors. Then, when both sides are multiplied by this term, all denominators are canceled, and we are led to the statement

$$12x - 2 = 3x.$$

We decide to gather terms involving x on the left, other terms on the right. The term -2 is removed from the left by the inverse operation of adding 2. Similarly, the term $+3x$ on the right is removed by the inverse operation of subtracting $3x$ from both sides. We now have $9x = 2$. The 9 must be removed to obtain the desired solution. Inasmuch as 9 is multiplied by x, we apply the inverse operation and divide both sides by 9 to obtain

$$x = 2/9.$$

Exercise.   Solve for x, citing operations performed:

$$\frac{3x}{4} - 5 = \frac{x}{3}.$$

Answer:

$$\frac{3x}{4} - 5 = \frac{x}{3} \qquad \text{Multiply by 12:}$$

$$9x - 60 = 4x \qquad \text{Subtract } 4x\text{:}$$

$$5x - 60 = 0 \qquad \text{Add 60:}$$

$$5x = 60 \qquad \text{Divide by 5:}$$

$$x = 12.$$

If the variable to be solved for is inside a grouping symbol, we remove the grouping symbol, citing the distributive property as justification, and then proceed with the solution. For example:

$$3x - \frac{1}{4} = 2 - \frac{1}{6}[x - (2 - x)] \qquad \text{Multiply by 12:}$$

$$36x - 3 = 24 - 2[x - (2 - x)] \qquad \text{Distributive property:}$$

$$36x - 3 = 24 - 2[x - 2 + x] \qquad \text{Distributive property:}$$

$$36x - 3 = 24 - 2x + 4 - 2x \qquad \text{Add } 4x\text{:}$$

$$40x - 3 = 24 + 4 \qquad \text{Add 3:}$$

$$40x = 31 \qquad \text{Divide by 40:}$$

$$x = \frac{31}{40}.$$

In the next example, grouping symbols are introduced to emphasize proper procedure, and then are removed in due course.

$$\frac{2}{3(x-1)} - \frac{1}{2} = \frac{1}{4} \qquad \text{Multiply by l.c.d. } 3(4)(x-1):$$

$$3(4)(x-1)\left[\frac{2}{3(x-1)}\right] - 3(4)(x-1)\left(\frac{1}{2}\right) = 3(4)(x-1)\left(\frac{1}{4}\right).$$

The last equation simplifies to:

$$8 - 6(x-1) = 3(x-1) \qquad \text{Distributive property:}$$
$$8 - 6x + 6 = 3x - 3 \qquad \text{Add } 6x; \text{ add } 3:$$
$$17 = 9x \qquad \text{Divide by } 9:$$
$$\frac{17}{9} = x.$$

The same procedures are followed when numbers are in literal form, but the next to the last step often requires factoring. Factoring derives from the distributive property, but we shall follow convention here and cite factoring rather than the distributive property when the need for justification arises.

Example. Solve for x, citing operations performed:

$$\frac{x}{a} - b = 2a(x-b) \qquad \text{Multiply by } a:$$
$$x - ab = 2a^2(x-b) \qquad \text{Distributive property:}$$
$$x - ab = 2a^2x - 2a^2b \qquad \text{Subtract } 2a^2x; \text{ add } ab:$$
$$x - 2a^2x = ab - 2a^2b \qquad \text{Factor:}$$
$$x(1 - 2a^2) = ab - 2a^2b \qquad \text{Divide by } (1 - 2a^2):$$
$$x = \frac{ab - 2a^2b}{1 - 2a^2}.$$

Exercise. Solve for x citing operations performed:

$$\frac{2x}{a} + \frac{3}{b} = x + 4.$$

Answer:

$$\frac{2x}{a} + \frac{3}{b} = x + 4 \qquad \text{Multiply by } ab:$$
$$2xb + 3a = abx + 4ab \qquad \text{Subtract } abx:$$
$$2xb - abx + 3a = 4ab \qquad \text{Subtract } 3a:$$
$$2xb - abx = 4ab - 3a \qquad \text{Factor:}$$
$$x(2b - ab) = 4ab - 3a \qquad \text{Divide by } (2b - ab):$$
$$x = \frac{4ab - 3a}{2b - ab}.$$

Example.   Solve for x, citing operations performed:

$$a = \frac{x}{1 - nx}$$ Multiply by $(1 - nx)$:

$$a(1 - nx) = x$$ Distributive property:

$$a - anx = x$$ Add anx:

$$a = x + anx$$ Factor:

$$a = x(1 + an)$$ Divide by $(1 + an)$:

$$\frac{a}{1 + an} = x.$$

Example.   Solve for y, citing operations performed:

$$\frac{a}{y} - \frac{1}{b} = 2$$ Multiply by yb:

$$ab - y = 2by$$ Add y:

$$ab = 2by + y$$ *Factor:*

$$ab = y(2b + 1)$$ Divide by $(2b + 1)$:

$$\frac{ab}{2b + 1} = y.$$

A3.4 PROBLEM SET 1

Solve each of the following for x, citing operations performed:

1. $2x - 3 = x + 4$.

2. $\frac{x}{3} + \frac{1}{2} = 3x$.

3. $3x - 7 = 2x + 4$.

4. $\frac{x + 3}{2} = x - \frac{1}{4}$.

5. $7x - 5 = 3 - 4x$.

6. $ax + 2 - x = 0$.

7. $ax + b = cx$.

8. $ax + b = x - b$.

9. $a(x - a) = 2x$.

10. $\frac{x}{a} - \frac{1}{2} = 2x$.

11. $\frac{2}{a - x} + \frac{1}{3} = 4$.

12. $y = \frac{x}{b - cx}$.

13. $\frac{3}{4} - \frac{2x}{3} = 2x(a - 1)$.

14. $3(x - 2) = 2 - a(x + 2)$.

15. $\frac{2}{3(x - 2)} + \frac{3}{a} - \frac{1}{2} = 0$.

16. $ax - \frac{b}{2} = c + \frac{5[a - 2(b - x)]}{6}$.

17. $\frac{b}{a} - x = 2a(b - x)$.

18. $x - a(b - x) = 2x - 3$.

19. $3(b - x) = 2 + b[x - (3 - x)]$.

20. $b(a + x) = a(b + x)$, where $a \neq b$.

A3.5 TRANSPOSITION

Problem Set 1 has afforded practice in citing fundamentals as justification for steps followed in the solutions of equations. From now on, we shall lessen the writing burden by omitting step-by-step justification. The burden can be lightened further by applying the rule of *transposition,* which states that a *term* may be moved from one side of an equation to the other by changing its sign. For example:

$$x - 2a = b$$

becomes

$$x = b + 2a$$

by changing the sign of $-2a$ to $+2a$ and placing $+2a$ on the other side. Similarly:

$$x + 7 = y$$

becomes

$$x = y - 7.$$

The transposition rule is a consequence of the axiom which states that the same number may be added to (subtracted from) both sides of an equation. That is, the change from

$$x + 7 = y$$

to

$$x = y - 7$$

is, in effect, subtracting 7 from both sides of the first equation. Transposition is a handy procedure, but we must keep in mind that it applies to *terms.* It would not be correct to say that any quantity can be moved to the other side by transposition. For example, the 2 in $2x = 6$ is not subject to the rule for transposition because 2 is not a term of the left member of the equation.

Exercise. The change from $ax = b$ to $x = b/a$ is not accomplished by transposition. How is it accomplished? Answer: By dividing both sides of the equation by a.

A3.6 FORMULAS

Computational procedures are described efficiently by formulas employing the symbolism of algebra. Thus, if x is the length and y the width of a rectangle, the area, A, of the rectangle is expressed by the formula $A = xy$. On the one hand, if we are told a rectangle has a length of 8 inches and a width of 5 inches, we compute the area

$$A = (8)(5) = 40 \text{ square inches.}$$

On the other hand, if we are asked how wide a rectangle of 12 inches length should be if its area is to be 84 square inches, we substitute into the equation, obtaining $84 = 12y$. Solving this equation, we find y, the desired width, is 7 inches. The point here is that we may wish to use a formula to evaluate a variable other than the one for which the formula is solved, and to do so, we call upon the usual operations for solving equation. Actually, the formula $A = xy$ leads easily to two other formulas:

$$x = \frac{A}{y}$$

$$y = \frac{A}{x}.$$

If we have the formula $A = xy$ and are given values for A and x, we can compute y by substituting the given values directly into the equation and solving for y, or we can solve the literal equation, obtaining

$$y = \frac{A}{x}$$

and then evaluate y by substituting the given values into this equation. Generally, it is more efficient to substitute the given numbers directly into the formula if a single evaluation is to be made. If, however, several evaluations are to be made, the formula should be solved for the desired variable before substituting numbers. For example, suppose that we are given

$$y = 5$$
$$a = 2$$

and are asked to evaluate x from the formula

$$y = 2(ax - 3).$$

Substituting the given numbers directly into the formula:

$$5 = 2(2x - 3)$$
$$5 = 4x - 6$$
$$11 = 4x$$
$$\frac{11}{4} = x.$$

On the other hand, if we are asked to carry out a whole series of evaluations of x for various values of y and a, it would be more efficient to solve the formula for x; thus:

$$x = \frac{y + 6}{2a}.$$

The last formula permits rapid evaluation of x for given values of y and a.

||||||||||||||||||||||||||||||||||||||||||||||||||||||||||||||||||||||||||||||||||||||||||||||||||||||||||||||||||||

Exercise. If $y = (1 + n)/2$, what would be an efficient way to compute n for many different values of y? Answer: Substitute the values for y into the equation $n = 2y - 1$.

||||||||||||||||||||||||||||||||||||||||||||||||||||||||||||||||||||||||||||||||||||||||||||||||||||||||||||||||||||

A3.7 EXACT EVALUATIONS

Evaluation by formulas often leads to approximations. Unless care is exercised, these approximations may not be accurate enough for the purpose at hand. Suppose, for example, that we wish to compute $3\frac{1}{3}$ percent of $1 million. If we change the stated percent to 3.33 percent, and then to the decimal 0.0333, the computation is

$$0.0333(1,000,000) = 33,300.$$

For most purposes the answer obtained would not be satisfactory. We can express the answer to any desired degree of accuracy by changing the stated percent to its exact equivalent, as follows:

$$3\frac{1}{3}\% = \frac{3\frac{1}{3}}{100} = \frac{\frac{10}{3}}{100} = \frac{10}{300} = \frac{1}{30}.$$

The computation with exact numbers would be

$$(1/30)(1,000,000) = 33,333.333 \cdots$$

The point of this example is that if exact numbers are used up to the last step, the final answer can be carried accurately to as many places as desired.

||||||||||||||||||||||||||||||||||||||||||||||||||||||||||||||||||||||||||||||||||||||||||||||||||||||||||||||||||||

Exercise. Compute x exactly from $3x + 2 = 1/b$ if b is $2\frac{1}{7}$. Answer: $-23/45$.

||||||||||||||||||||||||||||||||||||||||||||||||||||||||||||||||||||||||||||||||||||||||||||||||||||||||||||||||||||

As another example, let us evaluate y, given

$$y = \frac{a + b}{1 - r}$$

$$a = \frac{1}{3}$$

$$b = \frac{1}{2}$$

$$r = 1\frac{1}{3}\%.$$

To express all numbers in the same form, r is changed to its exact fractional equivalent, $1/75$. Substituting:

$$y = \frac{\frac{1}{3} + \frac{1}{2}}{1 - \frac{1}{75}}.$$

The last is an exact expression for y. If numerator and denominator are multiplied by 150, we obtain

$$y = \frac{125}{148}$$

which again is an exact expression for y. This result could now be converted to an approximate decimal, by long division, accurate to as many places as desired.

A3.8 PROBLEM SET 2

Given that $k = 12$, $m = 5$, and $n = 4$, compute y from the following formulas:

1. $y = \dfrac{km}{n}$.

2. $y = \dfrac{k^2 m - n^3}{n^2}$.

3. $y = \dfrac{m}{\dfrac{n}{k^2}}$.

4. $y = \sqrt[3]{mn - k}$.

5. $y = (n - m)[n - k(m + n)]$.

6. $y = \dfrac{8m}{\dfrac{n^2}{k}}$.

7. $y = \left(\dfrac{kn}{3}\right)^{3/2}$.

8. $y = \left(\dfrac{4}{9k}\right)^{-2/3}$.

9. $y = \left(m^2 - \dfrac{3k}{4}\right)^{5/4}$.

10. $y = m^2 n + 3[4k - n(m^3 - k^2)]$.

In the following, find the exact value of y in the form of a fraction in lowest terms, given that

$$a = \frac{5}{3}, \; b = \frac{1}{7}, \; c = \frac{2}{9}, \; d = \frac{5}{12}, \; x = \frac{3}{8}, \; z = \frac{1}{5}.$$

11. $y = \dfrac{x}{a}$.

12. $y = \dfrac{ab}{2}$.

13. $y = \dfrac{a(b + c)}{3}$.

14. $y = d - \dfrac{1}{a}$.

15. $y = ax + bz$.

16. $y = ax^2 - \dfrac{2a}{b}$.

17. $y = \left(1 - \dfrac{a}{d}\right)^3$.

18. $y = \sqrt[3]{\dfrac{c}{2x}}$.

19. $y = \dfrac{a}{2} - b\sqrt{\dfrac{3d}{5}}$.

20. $y = \dfrac{a + 2c - d}{x - 4a}$.

21. Compute y to the nearest cent:

$$y = at + b(t - c)$$
$$a = 16\tfrac{2}{3}\%, \, b = 15\%, \, c = \$75{,}000, \text{ and } t = \$140{,}000.$$

22. Given that $a = 0.52$, $x = 4$, compute y to two decimal places:

$$y = \frac{a}{1 + \dfrac{x}{100}}.$$

23. The equation relating the cost, C, of an item to its retail price, R, is $C + pR = R$, where p percent is markup on the retail price.
    a) What markup percent on retail is achieved if an item costs $5 and sells at a retail price of $8?
    b) How much can a merchant pay for an item if he wishes to sell it for $7.50 and achieve a markup which is 20 percent of the retail price?
    c) If a merchant seeks to achieve a markup of 40 percent on retail, at what price should he sell an item which cost him $2.69?

24. The amount of interest, I, earned on P dollars at simple interest of r percent per year for t years is $I = Prt$. Assuming a year to have 360 days:
    a) How long will it take for $500 to earn interest in the amount of $20 if the interest rate per year is $4\tfrac{1}{2}$ percent?
    b) How many dollars must be invested at $3\tfrac{1}{3}$ percent per year if interest in the amount of $75 is to be earned in 50 days?

25. Temperature in degrees centigrade, C, is related to temperature in degrees Fahrenheit, F, by the formula

$$C = \tfrac{5}{9}(F - 32).$$

Find the Fahrenheit temperature corresponding to the following centigrade readings:
    a) 100.   b) 0.   c) -10.6.
    d) 28.   e) -10.   f) 50.

26. A salesman's earnings, E, amount to 20 percent of his total sales, T, plus a bonus of $12\tfrac{1}{2}$ percent of any amount he sells in excess of $50,000.
    a) Write the equation relating earnings to sales if sales exceed $50,000.
    b) What must the man's sales be if he is to earn $10,000? $15,000? $20,000?
    c) How much must he sell if his earnings are to be 25 percent of total sales?

27. An executive is to receive compensation, B, amounting to x percent of his company's net profit after taxes. Taxes amount to y percent of net profits after deduction of the executive's compensation.
    a) Letting P represent net profits before taxes, complete the following formula:
    $B = x[P - ?(? - ?)]$.
    b) If the executive receives 25 percent of net profits after taxes, and taxes amount to 52 percent of net profits after deduction of the executive's compensation, how much must net profits before taxes be if the executive is to receive $50,000?

28. A customer buys an article on the installment plan. The price of the article (or the amount he still owes after making a down payment) is B dollars. The seller adds a carrying charge of C dollars and requires that the debt be paid off by n equal payments. Letting y be the number of payments which would be made

in a year (that is, if payments are weekly, y is 52; and if payments are monthly, y is 12), and letting r be the equivalent simple interest rate being paid by the customer, a formula states

$$r = \frac{2yC}{B(n+1)}.$$

a) An article priced at $500 is purchased. The customer pays $100 down (so B is $400). The seller adds $42 as a carrying charge and requires that the debt be repaid in ten equal monthly installments of $44.20 each. What rate of interest is the customer paying? (Find r, and convert it to percent form.)

b) If the simple interest equivalent of the installment rate is 15 percent, what should be the carrying charge on an article priced at $100 (no down payment) if the debt is to be repaid in 20 equal weekly installments?

29. For each dollar of increase in sales, selling expense increases by p where, it is hoped, p is less than 1. The sales level B at which the company will break even if its fixed expenses are F is:

$$B = \frac{F}{1-p}.$$

a) Find B if $F = \$12{,}000$ and $p = 0.4$.

b) Fixed expenses are $10,000 and management wishes to control p so that breakeven will occur at $12,500. Find p.

c) Repeat b) if fixed expenses are $20,000 and breakeven is to occur at $32,000.

30. The average cost per unit is A when x units are made, and

$$A = \frac{100}{x} + 0.01.$$

Find the average cost per unit if:

a) 10 units are made.

b) If 20 units are made.

How many units should be made if average cost per unit is to be

c) $1.01?   d) $0.41?   e) $0.21?

A3.9 COORDINATE AXES

An axiom of mathematics states that a correspondence exists between the real numbers and the points on a straight line; that is, for each real number, there is one and only one corresponding point on a line, and each point on a line corresponds to one and only one real number. Hence, if we draw a straight line and mark upon it an arbitrary point, zero, all negative real numbers can be represented by points to the left of zero, all positive numbers by points to the right of zero. According to the axiom, each number corresponds to a unique point on the line, and the totality of real numbers corresponds to the line itself.

A pair of rectangular coordinate axes is obtained by drawing two intersecting perpendicular lines, one horizontal, the other vertical. The intersection of the horizontal and vertical axes so obtained is taken as the zero point on each line and is called the *origin*. On the vertical, points above the origin are used to represent positive numbers, points below to represent negative numbers. On the horizontal axis, positive numbers are to the right of the origin, negative numbers to the left.

Any point in the plane of the axes can be located by means of a number pair called the *coordinates* of the point. Conventionally, coordinates are written in the form (*a, b*), where the first number in the parentheses is the horizontal coordinate, or *abscissa,* and the second number is the vertical coordinate, or *ordinate.* Thus, the point (2, 3) is located by moving two units to the right of the origin, and then three units vertically upward. The procedure is analogous to the method of directing a person from a particular spot (origin) in a city by telling him the point he wishes to reach can be found by going two blocks east, and then three blocks north.

Exercise.   What are the abscissa and the ordinate of a point? Answer: The abscissa is the horizontal coordinate and the ordinate is the vertical coordinate of the point.

Figure A3–1 shows the point (2, 3) just mentioned, together with other aspects of the coordinate system. The axes divide the plane into *quadrants,* which we number counterclockwise as I, II, III, and IV. In quadrant I, both abscissa and ordinate are positive. In II the abscissa is negative, the ordinate

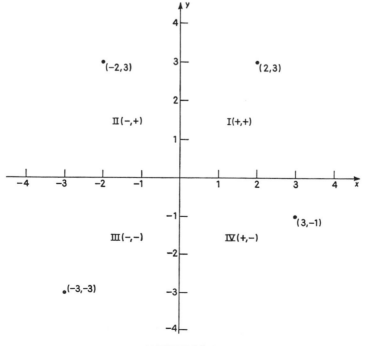

FIGURE A3–1

positive. In III both coordinates are negative. In IV the abscissa is positive, the ordinate negative.

Exercise. In what quadrants are the points $(-2, 3)$, $(3, 5)$, $(2, -3)$, $(-5, -4)$? Answer: II, I, IV, III.

A3.10 PLOTTING OBSERVATIONAL DATA

By observational data we mean numbers collected in real-world situations—numbers such as costs, sales, units produced, prices, and so on. Frequently, we wish to display such data in a graphical manner which helps to show relationships which may exist between two variables. Consider the data in Table A3–1.

TABLE A3–1

Production Cost per Unit of Product,
by Number of Units Produced

Lot Number	Number of Units in Lot (x)	Cost per Unit (y)
1	8	$11
2	20	4
3	5	13
4	10	9
5	15	8
6	25	5

Figure A3–2 is a graphic representation (or, more simply, a graph) of the data of Table A3–1. The numbers of units produced have been assigned

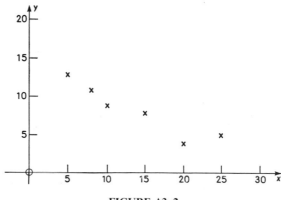

FIGURE A3–2

the general designation x, the costs per unit the general designation y. Following convention, the numbers x are assigned to the horizontal axis, the numbers y to the vertical. By definition, the vertical variable is called the *dependent* variable, and the horizontal variable is called the *independent* variable. In some cases, these definitions make real-world sense. In the present illustration, for example, it is natural to think of production cost as depending on number of units produced, so that cost would be the dependent variable and units produced the independent variable. In many circumstances, however, reference to the vertical axis as the axis of the dependent variable is purely a matter of convention.

The interpretation of Figure A3–2 is simply that unit cost *tends* to decrease as the number of units made increases. It is a graphical illustration of the economies of mass production. We refer to this set of isolated points as a *discrete* set. If we wish to illustrate the relationship by means of a smooth curve, we may sketch in freehand a curve which comes close to the points, although it does not pass through all of the points. The points on the curve are not, of course, records of observations, because real-world observations are limited in number, whereas a segment of a smooth curve has an unlimited number of points. However, we might wish to use the smooth curve to estimate costs per unit for numbers of units other than those observed.

A3.11 PLOTTING EQUATIONS IN TWO VARIABLES: STRAIGHT LINES

Given the equation

$$y = x + 2,$$

we know that for every number x, there is a number y such that the number pair (x, y) satisfies the equation. The entire solution set is an infinite set of ordered pairs. We can generate some members (obviously not all) of the solution set by arbitrarily assigning a value for x, then computing the corresponding value for y. For example,

$$\text{if } x = 1 \text{ then } y = 1 + 2 = 3$$

and $(1, 3)$ is a member of the solution set. Similarly, if $x = 2$, $y = 4$ and $(2, 4)$ is another member of the solution set.

Exercise. Find the points in the solution set corresponding to $x = 0$ and $x = 5$. Answer: $(0, 2)$; $(5, 7)$.

We *plot the graph* of an equation by finding and plotting points which satisfy the equation, then sketching in a smooth curve as suggested by the plotted points. In the case at hand, we may tabulate some of the points as in Table A3–2.

TABLE A3–2

$y = x + 2$	
x	y
0	2
1	3
2	4
3	5
4	6
5	7

The points of Table A3–2 are shown in Figure A3–3, where it is clear that all fall on the same straight line. We would also obtain a straight line if we made the graph of

$$y = 3x + 5 \quad \text{or} \quad y = -2x + 7.$$

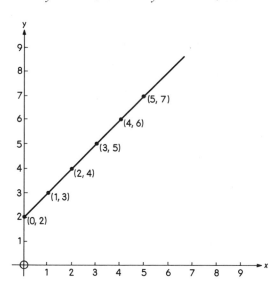

FIGURE A3–3

In general, an equation that can be made into the form

$$y = mx + b,$$

where m and b are constant numbers, has a straight line as its graph and is called a *linear equation.* The graph of such an equation is easily sketched because *two points* suffice to fix a straight line.

‖‖

Exercise. How would you plot the graph of $y = 2x - 3$? Answer: Pick two arbitrary values of x and find the corresponding values of y to obtain two points. Plot the points and use a straight-edge to draw a line through them.

‖‖

A3.12 VERTICAL PARABOLAS

In the equation

$$y = x^2 - 5$$

if we let

$$x = 0, 1, 2, 3, 4, 5$$

in succession, we find the corresponding values of y to be $-5, -4, -1, 4,$ 11, and 20. In this equation the succession of negatives

$$x = -1, -2, -3, -4, -5$$

leads to the same sequence of y's. In tabular form:

TABLE A3–3

$y = x^2 - 5$	
x	y
-5	20
-4	11
-3	4
-2	-1
-1	-4
0	-5
1	-4
2	-1
3	4
4	11
5	20

The general shape of the curve is shown in Figure A3–4. It is clear that the

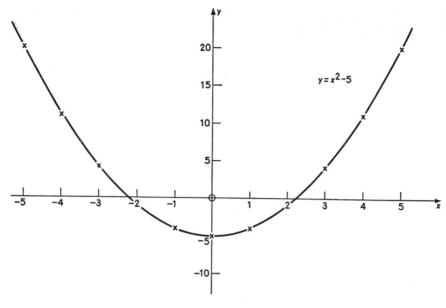

FIGURE A3–4

curve will continue to rise at both ends because if we substitute x values of 6, 7, and so on (or -6, -7, and so on), the term x^2 in

$$y = x^2 - 5$$

will overpower the number -5 and make y become larger.

Exercise.   Find the coordinates of the points on $y = x^2 - 5$ where $x = 2.5$ and $x = -2.5$. Check the graph in Figure A3–4 by plotting these points. Answer: The points are (2.5, 1.25) and (-2.5, 1.25).

The graph shown in Figure A3–4 is known as a *vertical parabola.* ·If we plotted points for

$$y = -2x^2 + 5x - 6 \quad \text{or} \quad y = 3x^2 + 5x - 4,$$

we would again obtain vertical parabolas. In general, any equation which can be made into the form

$$y = ax^2 + bx + c,$$

where a, b, c are constant numbers and $a \neq 0$, is a vertical parabola. The "nose" of the parabola is called the *vertex.* In the case of Figure A3–4, the vertex lies on the vertical axis. This is always the case if $b = 0$ in $y = ax^2 + bx + c$, as in the plotted equation

$$y = x^2 - 5.$$

Proper positioning of the vertex of a parabola is of key importance in drawing an accurate sketch. This can be done by use of the formula[1] which states that the x-coordinate of the vertex of a vertical parabola occurs at

$$x = -\frac{b}{2a}.$$

Thus, comparing the parabola $y = x^2 - 5$ with $y = ax^2 + bx + c$, we have $a = 1$, $b = 0$, $c = -5$, so

$$x = -\frac{0}{2} = 0$$

is the x-coordinate of the vertex. The corresponding y is -5. The vertex is at $(0, -5)$, as shown in Figure A3–4.

As another example, in

$$y = -2x^2 + 12x - 10$$

we have $a = -2$, $b = 12$. We find

$$x = -\frac{12}{-4} = 3$$

and

$$y = -2(9) + 12(3) - 10 = 8,$$

so $(3, 8)$ is the vertex in this case.

‖‖

Exercise.   Find the coordinates of the vertex of $y = 3x^2 - 12x + 5$. Answer: $(2, -7)$.

‖‖

In general, peaks and valleys in curves are points of critical importance in sketching accurate graphs. We shall learn more about methods of determining such points in Chapter 9, and about the important interpretations attached to them in Chapter 10. We shall now sketch the graph of the equation of the last exercise,

$$y = 3x^2 - 12x + 5.$$

---

[1]Because a vertical parabola is symmetrical with respect to a vertical line through its vertex at $x = v$ (where v stands for vertex) it follows that the y value is the same at $x + v$ and $x - v$. That is, $a(x + v)^2 + b(x + v) + c = a(x - v)^2 + b(x - v) + c$. If we square, collect terms, factor, and solve, we find $x = -b/2a$.

We found the vertex to be $(2, -7)$, and we may compute some other points as shown in Table A3–4.

TABLE A3–4

$y = 3x^2 - 12x + 5$	
x	y
-1	20
0	5
1	-4
2	-7
3	-4
4	5
5	20

The desired graph is shown in Figure A3–5.

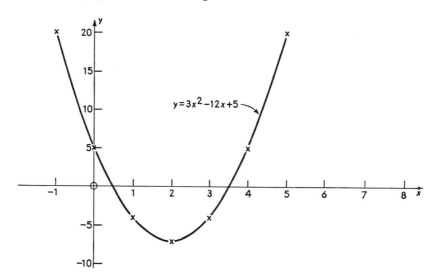

FIGURE A3–5

Exercise. Plot $y = x^2 - 8x + 15$ on Figure 5. Answer: This parabola rises on either side of the vertex at $(4, -1)$, cutting the x-axis at 3 and 5. Other points are $(2, 3)$, $(6, 3)$, $(0, 15)$, and $(8, 15)$.

As another example, the parabola $y = -4x^2 + 7x - 3$ is found to have its vertex at $(7/8, 1/16)$ and to fall on either side of the vertex. Again, $y = x^2$ is a parabola with vertex at the origin which rises on either side of the vertex.

A3.13 QUADRATIC EQUATIONS

A quadratic equation is of the form

$$ax^2 + bx + c = 0,$$

where a, b, and c are constants, with $c \neq 0$. Thus,

$$3x^2 - 12x + 5 = 0 \qquad (1)$$

is a quadratic equation with $a = 3$, $b = -12$, and $c = 5$. The solution set consists of the values for x which make the equation a true statement. If we relate this equation to

$$y = 3x^2 - 12x + 5, \qquad (2)$$

we see that the solution set of (1) consists of the values of x which make y, in (2), equal to zero. By reference to the graph in Figure A3–5, we see that the graph intersects the x-axis at about $x = 0.5$ and $x = 3.5$, and at these points y, or $3x^2 - 12x + 5$, is zero. An exact determination of the solution set can be found by the *quadratic formula* which states:

$$x = \frac{-b \pm \sqrt{b^2 - 4ac}}{2a}.$$

In the present case,

$$x = \frac{-(-12) \pm \sqrt{144 - 4(3)(5)}}{2(3)} = \frac{12 \pm \sqrt{84}}{6}.$$

By logarithms, $\sqrt{84}$ is about 9.17 so

$$x = \frac{12 \pm 9.17}{6}$$

$$= \frac{21.17}{6} \quad \text{or} \quad \frac{2.83}{6}$$

$$= 3.53 \quad \text{or} \quad 0.47$$

so the solution set may be written as $x_1 = 0.47$, $x_2 = 3.53$, or $\{0.47, 3.53\}$.

Three cases can arise in the solution of a quadratic. The graphical nature of these cases can be understood by reference to Figure A3–5. In its present state, the graph shows *two different real solutions*. If now we hold the axes fixed and move the curve upward until the vertex is tangent to the x-axis (touches it at a single point), we have the case of *two equal real* solutions. Finally, if we again move the parabola upward, it will not intersect the x-axis, and we have the case of *no real solutions*. Algebraically, these cases relate to the quantity $b^2 - 4ac$ in the quadratic formula,

$$x = \frac{-b \pm \sqrt{b^2 - 4ac}}{2a}.$$

If $b^2 - 4ac$ is *positive* (greater than zero), we have a positive and a negative value for the square root, and there are *two different real* solutions. If $b^2 - 4ac$ is *zero*, we have only one value for x and there are *two equal real* solutions. Finally, if $b^2 - 4ac$ is *negative*, its square root is not a real number, and we have *no real solutions*.

Exercise.   Apply the quadratic formula to find the solutions for $x^2 - 9 = 0$. Answer: With $a = 1, b = 0, c = -9$, we find the two different roots, $x = 3$, and $x = -3$. The same result can be obtained by writing $x^2 = 9$ and noting that the square of $+3$ and -3 equal 9.

If we apply the formula to

$$3x^2 - 5x + 10 = 0,$$

we have,

$$x = \frac{5 \pm \sqrt{25 - 120}}{6} = \frac{5 \pm \sqrt{-95}}{6}$$

and find that this equation has no real solutions.

Exercise.   Solve $x^2 - 4x + 4 = 0$ by the quadratic formula. Answer: $x_1 = x_2 = 2$. We have two real equal solutions.

Solution by Factoring.   The equation

$$(x - 3)(x + 2) = 0$$

is true if either the factor $x - 3 = 0$ or the factor $x + 2 = 0$ because 0 times any number equals zero. Hence, the solutions are $x_1 = 3, x_2 = -2$. Consequently, we can easily find the solutions if the quadratic is factorable. For example,

$$2x^2 + 7x - 15 = 0$$

can be factored to yield

$$(2x - 3)(x + 5) = 0.$$

Setting $2x - 3 = 0$ and $x + 5 = 0$, we have the solutions

$$x_1 = \frac{3}{2}, x_2 = -5.$$

Exercise.   Solve $6x^2 + x - 2 = 0$ by factoring. Answer: $x_1 = 1/2, x_2 = -2/3$.

Finally, note that the factoring is elementary if the quadratic has no constant term. Thus,

$$2x^2 - 5x = 0$$

factors to

$$x(2x - 5) = 0,$$

so the solutions are $x_1 = 0$ and $x_2 = 5/2$.

A3.14 PROBLEM SET 3

Graph each of the following:

1. $y = 3x - 1$.
2. $y = x$.
3. $y = -5x + 6$.
4. $2y - 3x = 6$.
5. $y = \frac{x}{2} + 3$.
6. $y = 3x^2$.
7. $y = x^2 - 7$.
8. $y = -2x^2 + 10$.
9. $y = 2x^2 - x$.
10. $y = x^2 + 2x - 6$.

Solve by the quadratic formula:

11. $x^2 - 10x + 3 = 0$.
12. $2x^2 + 5x - 2 = 0$.
13. $4x^2 - 12x + 9 = 0$.
14. $3x^2 + 2x + 5 = 0$.
15. $x^2 - 2x = 0$.

Solve by factoring:

16. $x^2 - 3x + 2 = 0$.
17. $3x^2 - 6x = 0$.
18. $6x^2 + 7x - 20 = 0$.
19. $x^2 - 25 = 0$.
20. $2x^2 + 5x - 7 = 0$.

A3.15 REVIEW PROBLEMS

Solve for x, citing the operations performed:

1. $3x - 2 = 2x + 5$.
2. $\frac{2x}{3} - 1 = 5x$.
3. $\frac{3 + 2x}{4} - 2 = x$.
4. $\frac{2(a - x)}{b} + 3x = c$.
5. $2a = \frac{x}{3 - cx}$.
6. $\frac{3}{a(x - 1)} + \frac{1}{2} - b = 0$.
7. $5(x - b) = a + b[3 - 2(x + 1)]$.
8. $\frac{2}{3} - \frac{3x}{4} = 4x(2 - b)$.
9. $x - \frac{c}{d} = a - \frac{3[2 - b(x - 1)]}{4}$.
10. $\frac{1}{1 - x} + 1 = b$.

In the following compute x, given that $a = 6$, $b = 2$, and $c = 10$:

11. $x = \frac{ab - c}{b}$.
12. $x = \frac{a^2b - b}{c}$.
13. $x = \frac{\frac{c^2}{b}}{a}$.
14. $x = \sqrt{c^2 - a^2}$.

15. $x = \left(3c - \dfrac{a}{2}\right)^{1/3}.$

16. $x = (3ab)^{-3/2}.$

17. $x = (c - a - b)[2b - a(c - 5b)].$

18. $x = \dfrac{\dfrac{a^2}{b} - c}{b^2}.$

In the following, compute the exact value of y in the form of a fraction in the lowest terms, given that:

$$a - \tfrac{2}{7};\ b - \tfrac{1}{3};\ c - \tfrac{4}{9};\ d - \tfrac{5}{12};\ x = \tfrac{1}{8};\ z = \tfrac{1}{5}.$$

19. $y = \dfrac{x}{a}.$

20. $y = \dfrac{ab}{2}.$

21. $y = d - \dfrac{1}{a}.$

22. $y = ax + bz.$

23. $y = \left(1 - \dfrac{b}{d}\right)^3.$

24. $y = (1 + 19x)^{1/3}.$

25. $y = \sqrt{c - b}.$

26. $y = \dfrac{\dfrac{b}{2} + 3x}{1 + a}.$

27. Compute y to the nearest cent:

$$y = at + b(t - c)$$
$$a = 4\tfrac{1}{2}\%;\ b = 22\%;\ c = \$18,000;\ t = \$28,000.$$

28. The equation relating the cost, C, of an item to its retail price, R, is $C + pR = R$, where p is the percent markup on the retail price.
   a) What markup percent on retail is achieved if an item costs $10 and sells at a retail price of $15?
   b) How much can a merchant pay for an item if he wishes to sell it for $20 and achieve a markup which is 30 percent of the retail price?
   c) If the merchant seeks to achieve a markup which is 35 percent of the retail price, at what price should he sell an item which cost him $13?

29. Compute y accurate to three decimals if $x = 0.05$:

$$y = \dfrac{(1 + x)^{-3} + 1}{x}.$$

30. The amount of interest, I, earned on P dollars at simple interest of r percent per year for t years is $i = Prt$. Assuming a year to have 360 days,
   a) How many days will it take for $2000 to earn $50 interest at 4 percent?
   b) How many dollars must be invested at $5\tfrac{1}{3}$ percent if interest in the amount of $100 is to be earned in 90 days?

31. Temperature in degrees Fahrenheit, F, is related to temperature in degrees centigrade, C, by the formula

$$F = \tfrac{9}{5}C + 32.$$

Find the centigrade temperatures corresponding to the following Fahrenheit readings:
   a) 0.   b) 100.   c) 51.   d) −32.

32. When an asset is bought, its initial book value is A. The book value will

decline over the N years of the asset's life to the salvage value, $\$S$. The ratio of salvage value to initial book value is $s = S/A$. In the straight-line method of computing book value from year to year, the appropriate formula is

$$B = A\left[1 - \frac{t}{N}(1 - s)\right]$$

where B is the book value after t years.

a) An asset with an initial book value of $1000 and a salvage value of $100 after 10 years of life will have what book value after 3 years?

b) Solve the formula for s.

c) If an asset with an initial book value of $1000 and a life of 10 years is quoted as having a book value of $650 after 5 years, find the ratio of the salvage value to the initial value, and find the salvage value.

d) What will the depreciation formula for book value be if the asset has no salvage value?

33. A customer buys an article on the installment plan. The price of the article (or the amount he still owes after making a down payment) is B dollars. The seller adds a carrying charge of C dollars and requires that the debt be paid off by n equal payments. Letting y be the number of payments which would be made in a year (that is, if payments are weekly, y is 52; and if payments are monthly, y is 12), and letting r be the equivalent simple interest rate being paid by the customer, a formula states

$$r = \frac{2yC}{B(n + 1)}.$$

a) An article priced at $200 is purchased. The customer pays $50 down. The seller adds a carrying charge of $28.50 and requires the debt to be repaid in 18 equal monthly installments. What is the amount of the installment and what is the interest rate being charged?

b) If the simple interest equivalent of the installment rate is to be 30 percent, what should be the carrying charge on an article priced at $300 if the debt is to be paid off in 12 monthly installments and no down payment is made?

34. If D (demand) is the number of units of an item which can be sold when the item is priced at p cents per unit, and if

$$D = \frac{20}{p - 1} - 1$$

for values of p greater than 1 but less than 21, at what price will the demand be:

a) 19 units.   b) 9 units.   c) 3 units.   d) 1 unit.

35. Graph the following equations:

a) $y = 2x + 1$.

b) $5y + 4x - 20 = 0$.

c) $y = \frac{x}{3}$.

d) $x - y = 1$.

e) $y = 8 - x^2$.

f) $y = x^2 - 10x + 15$.

36. Solve by the quadratic formula:

a) $x^2 - 2x - 2 = 0$.

b) $4x^2 - 4x + 1 = 0$.

c) $x^2 + 2x + 5 = 0$.

37. Solve by factoring:

a) $6x^2 - 5x + 1 = 0$.

b) $16x^2 - 40x + 25 = 0$.

c) $3x^2 - 48x = 0$.

d) $x^2 - 100 = 0$.

Answers to Problem Sets

CHAPTER 1

Problem Set 1

1. See Figure A. $AB = 6$, $AC = 3$, $DB = 4$.

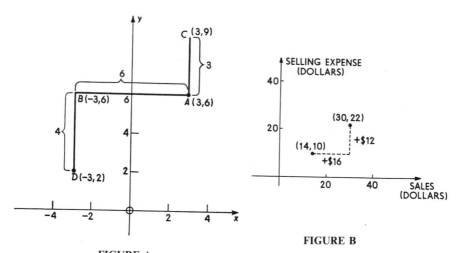

FIGURE A

FIGURE B

2. $AB = 2$, $AC = 24$, $AD = 17$, $BD = 15$.
3. AB is vertical, CD is horizontal.
4. $y_2 = y_1$, $x_1 = x_3$.
5. $y_2 = y_1$, $y_4 = y_3$.
6. a) 40 miles.　　　　b) 70 miles.
7. a) $16.　　　　b) $12.　　　　c) See Figure B.

8. Subscript notation preserves the letters x and y to mean abscissa and ordinate for *all* points.
9. 12 blocks.

Problem Set 2

1. *a*) (3, 8), (7, 5), 5.
   b) (−1, 6), (5, −2), 10.
   c) (−1, −14), (−10, −2), 15.
2. *a*) 10.   *b*) 15.   *c*) 5.10.
   d) 10.   *e*) 3.   *f*) 2.
3. *a*) (50, 25).   *b*) 55.9 feet.   *c*) 50 feet.
4. *a*) See Figure C.   *b*) 25 miles.
5. $AB + BC + CA = 140 + 150 + 130 = 420$ miles.

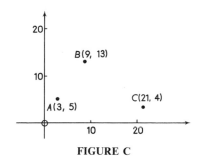

FIGURE C

Problem Set 3

1. *a*) The difference of the second (y) coordinates, $y_2 - y_1$.
   b) The difference of the first (x) coordinates, $x_2 - x_1$.
2. *a*) The steepest line is vertical.
   b) As we go from one point to another on the line, the run is zero so the slope ratio has a denominator of zero. A ratio with a denominator of zero is not a number.
3. 2/3.
4. Vertical. The plane evidently crashed.
5. 1.
6. *a*) 8.   *b*) $8.
7. The slope is 0.96 and $1 - 0.96 = 0.04$. The first is the marginal propensity to consume and the second the marginal propensity to save. The numbers 0.96 and 0.04 indicate that for an extra $1 of income, consumers spend $0.96 and save $0.04.
8. 1.   9. −1.   10. −1/8.
11. No slope number because segment is vertical.
12. 0.   13. 0.
14. No slope number because segment is vertical.
15. 1.   16. 21/32.   17. −1.   18. −11/9.

19. No slope number because segment is vertical.

20. 1.   21. $-1/13$.   22. See text.

23. $\dfrac{b-2}{a-3}, \dfrac{2-b}{3-a}$.

24. See Figure D.

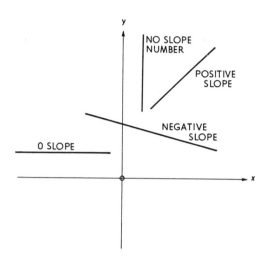

FIGURE D

25. If a is not zero, there is no number x such that $a/0 = x$; if a is 0, x is ambiguous because it could have any value.

26. F.   27. T.   28. T.

29. T.   30. T.   31. T.

32. T.   33. F.   34. T.

Problem Set 4

1. $3x - y = 5$.   2. $x + y = 1$.   3. $2x + 3y = 14$.

4. $x - 2y = 9$.   5. $y = -8$.   6. $x + 6y = 44$.

7. $13x - y = -57$.   8. $y = 0$.   9. $x = 5$.

10. $x = 0$.   11. $y = 4$.   12. $5x - y = 23$.

13. $x + 7y = 46$.   14. $6x - 7y = -51$.   15. $x - y = 0$.

16. $9x - y = -14$.   17. $3x - 2y = 0$.   18. $y = 4$.

19. $3x + 2y = 0$.   20. $x = -7$.   21. $x - 2y = -7$.

22. $y = 5$.   23. $y = 0$.   24. $5x + 7y = -29$.

25. $y = 0$.   26. $x = 0$.   27. $-24/7$.

28. -2.   29. $x = -6$.

30. *a*) $y = 3$.
    b) The tangent is the horizontal line $y = 500$ and shows the maximum (highest) profit is 500.

31. *a*) $y = 5$.
    b) $p = 2$ is a horizontal line and indicates the price is $2 per unit for any num-

ber of units demanded. $q = 500$ is vertical and indicates that 500 units are demanded at any price per unit.

32. $4y - x = 200$. 33. $y - 3x = 50$. 34. $y = 3x + 10$.

35. *a)* $y = 0.5 + 0.8x$.
   b) $E = 50 + 0.1V$. The slope is the rate of commission.

36. F. 37. T. 38. T.

39. F. 40. T. 41. T.

42. See Figure E. 43. See Figure F. 44. See Figure G.

45. See Figure H. 46. See Figure I. 47. 3/2.

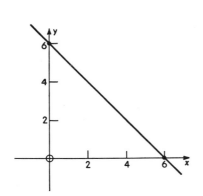

FIGURE E

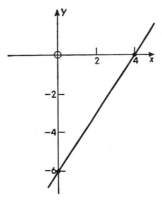

FIGURE F

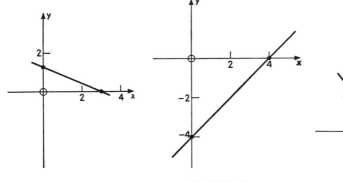

FIGURE G

FIGURE H

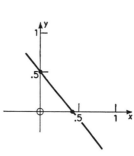

FIGURE I

48. -1. 49. 1/3. 50. 1.

51. *a)* $8x + 12y = 96$. *b)* 1.5 pounds of A per pound of B.
   c) 2/3 pounds of B per pound of A.

52. $y = 2x - 6$.

53. If the equation is solved for y, the coefficient of x, which is the slope, turns out to be $-a/b$.

54. y/x is cost per unit; $y/x = 3$ or $y = 3x$ is the equation; slope is 3, and intercepts zero (line passes through the origin).

55. *a*) $3x - 2y + 8 = 0$.   *b*) $y + 3x + 12 = 0$.

56. With p on the vertical, the vertical shift is from 40 to 35, a downward shift of 5, indicating that price is lower by $5 per unit at any level of demand. The horizontal shift is from 400 to 350, or 50 to the left, indicating that demand is 50 units less at each level of price.

57. $y = 2x$.   58. $y = 4x/5$.

59. *a*) D is a right angle; hence, angle 1 is $90° -$ angle BDC. DBC is a right triangle; hence, angle 2 is $90° -$ angle BDC. Therefore, angle $1 =$ angle 2, and triangles ABD and DBC are similar because they have equal angles. It follows that corresponding sides are in proportion, so that $DB/AB = BC/DB = 1/(DB/BC)$, where DB/AB is the slope of l_2 and DB/BC is the slope of l_1.

    b) A horizontal and a vertical line do not have reciprocal slopes because $1/0$ is not a number.

60. *a*) 1.25 miles to the right of the origin.

    b) Because the shortest distance from a point to a line is on a perpendicular to the line.

61. The coordinates of the origin $(0, 0)$, satisfy any equation of the form $ax + by = 0$.

62. F.   63. T.   64. T.   65. T.   66. T.

67. T.   68. T.   69. T.   70. T.   71. T.

Problem Set 5

1. T.   2. T.   3. F.   4. T.   5. F.

6. T.   7. T.   8. T.   9. T.   10. T.

11. *a*) 150.   *b*) 170.   *c*) 3.   *d*) 3.40.   *e*) 3.

12. *a*), (*b*), (*c*) See Figure J.

    d) $RS = 25 =$ fixed cost; $ST = 20 =$ variable cost for 10 units; $RT = 45 =$ total cost for 10 units.

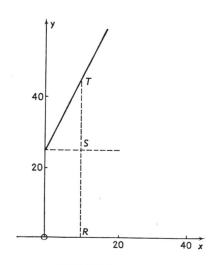

FIGURE J

13. *a*) *AB*.   *b*) *AD*.   *c*) *BD*.   *d*) *BD/OA*.
    e) *AD/OA*.   *f*) *AB/OA*.   *g*) *EB*.   *h*) *BD/OA*.

14. *a*) Setup cost, or fixed cost.
    b) Variable cost if *OM* units are made.
    c) Variable cost per unit.
    d) Average cost per unit if *OM* units are made.

Problem Set 6

1. T.   2. T.   3. T.   4. T.   5. T.
6. T.   7. F.   8. F.   9. F.   10. T.

11. *a*) $60,000.   *b*) $y = 22,800 + 0.62x$
    c) $5700   *d*) See Figure K.

12. *a*) $80,000.   *b*) $y = 36,000 + 0.55x$.
    c) −$2250.   *d*) See Figure L.

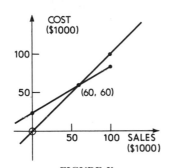

FIGURE K   FIGURE L

13. *a*) $0.47.   *b*) $29,786.   *c*) $63,626.
    d) $56,200.   *e*) $12,614.

14. *a*) $12,800.   *b*) $23,800.   *c*) $35,000.
    d) $3400 loss.

15. $1000.

16. *a*), (*b*) See Figures M and N.

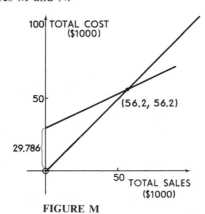

FIGURE M

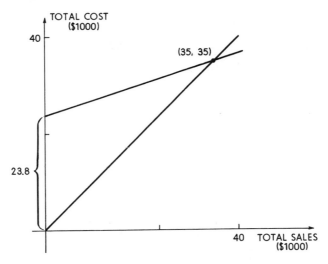

FIGURE N

Problem Set 7

1. F.	2. T.	3. T.	4. T.	5. F.
6. T.	7. T.	8. T.	9. F.	10. T.
11. T.	12. F.	13. T.	14. T.	15. T.
16. T.	17. T.	18. T.	19. T.	20. T.

21. See Figures O through Q.

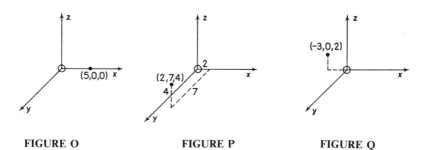

FIGURE O   FIGURE P   FIGURE Q

22. $x + 0y + 3z = 6$.   23. See Figures R through W.

24. *a*) $w + 3x + 2y + 4z = 500$.
    b) $0.5w + 0.3x + 1.2y + 0.8z = 310$.

25. x arbitrary, $y = (6 - 3x)/2$; or y arbitrary, $x = (6 - 2y)/3$.

26. x and y arbitrary, $z = (2x + 3y - 18)/2$.
    x and z arbitrary, $y = (18 + 2z - 2x)/3$.
    y and z arbitrary, $x = (18 + 2z - 3y)/2$.

27. The number of pairs of values which satisfy the equation is without limit.

28. $3x + 4y = 12$.

29. $(4, 0)$ and $(0, 3)$.

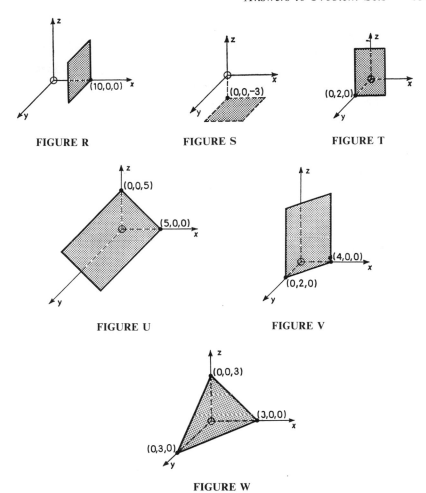

FIGURE R   FIGURE S   FIGURE T

FIGURE U   FIGURE V

FIGURE W

30. See Figure X.

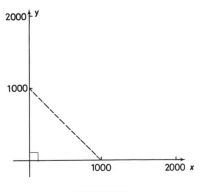

FIGURE X

31. $x = 600$, $y = 400$ gallons.
32. If per gallon costs are not equal, and there is no restriction on the relative amounts of A and B in the tank.
33. $10x + 8y + 7z$.
34. *a*) $10x + 8y + 7z = 560$.
    b) The numbers must be positive integers, or 0.
35. The set of ordered pairs, (x, y), such that $y - 3x = 6$.
36. The set of ordered triples, (x, y, z), such that $2x + y - 3z = 15$.
37. The set of ordered quadruples, (w, x, y, z), such that $2w + x - y + 3z = 30$.
38. $\{(x, y, z): y = -2x + 3z + 15\}$.
39. $\{(w, x, y, z): y = 2w + x + 3z - 30\}$.
40. w, y, and z arbitrary; $x = 30 - 2w + y - 3z$.

CHAPTER 2

Problem Set 1

1. $(2, 3)$.   2. $(-1, 4)$.   3. $(1, 1)$.   4. $(3/2, 2)$.   5. $(1/2, 2/3)$.
6. $(1/2, 3/2)$.
7. 375 gallons of X and 625 gallons of Y.
8. Invest \$20,000 in X and \$30,000 in Y.
9. Make 15 units of X and 10 units of Y.
10. Make 5 units of X and 50 units of Y.

Problem Set 2

1. $D'D'$ is $p = 90 - 0.25q$.   2. $S'S'$ is $p = 6 + 0.15q$.
3. $D'D'$ is $p = 70 - 0.75q$.
4. *a*) 200 units at \$42 per unit.   *b*) 176 units at \$50.40 per unit.
5. *a*) 1900 units at \$48 per unit.   *b*) 1820 units at \$46.40 per unit.

Problem Set 3

1. *a*) A system in which the number of equations is m and the number of variables is n.
   b) A system in which the number of equations is the same as the number of variables.
   c) A system of three equations in two variables.
   d) The total number of variables appearing in the equations of the system.
   e) A linear combination of two equations is an equation formed by taking the sum of p times one equation plus q times the other, where p and q are any numbers.
2. *a*) x arbitrary, $y = (3 - 3x)/4$.
   b) y and z arbitrary, $x = 15 - 2y + 4z$.
3. *a*) A false statement, such as $0 = 4$.
   b) An identity, such as $0 = 0$.
4. See text.

5. See text.
6. See text.
7. *a*) (3, −2). See Figure A.   *b*) (3, −2). See Figure B.
   c) (32/17, 3/17). See Figure C.

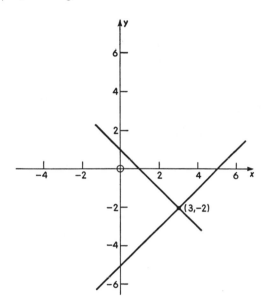

FIGURE A

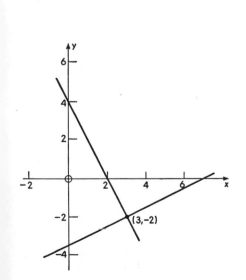

FIGURE B

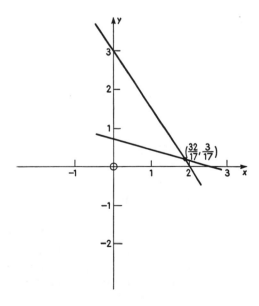

FIGURE C

8. *a*) Parallel lines, no solutions.   *b*) Lines have same graph.
9. $(1/2, 1, 3/2)$.
10. x arbitrary, $y = x - 2$, $z = 5 - x$.
11. x arbitrary, $y = 3 - 2x$, $z = 5 - 3x$.
12. $(2/5, -3, 3/5)$.
13. No solutions.
14. No solutions.
15. x arbitrary, $y = 17 - 7x$, $z = 22 - 9x$.
16. $(1, 2, 3)$.
17. No solutions.
18. $(1/2, 1/3, 1/4, -2)$.
19. x and y arbitrary, $w = x + 2$, $z = y - x - 3$.
20. $(-10, -2, -5, 8)$.
21. $(1, 2, -2, -1)$.
22. y arbitrary, $w = 2y + 6$, $x = y + 3$, $z = 2y + 7$.
23. x, y arbitrary, $z = 2y - 3x - 4$, $w = y - x - 3$.
24. $(-3, 5)$.   25. $(1, 2)$.   26. $(2, -1, 4)$.   27. $(3, 0, 2)$.
Problems 28–31. See Answers for Problems 24–27, in order.

Problem Set 4

1. No solutions.
2. y arbitrary, $x = y + 2$, $z = 2y - 1$.
3. $(1, -1, 2/3)$.
4. No solutions.
5. No solutions.
6. $(2, 0, 4)$.
7. x arbitrary, $y = 2 - x$, $z = 2x$.
8. y, z arbitrary, $x = y + z + 1$, $w = 2y - z + 3$.
9. x and z arbitrary, $y = 2x + 3z - 3$.
10. y arbitrary, $x = 2 - y$, $z = 4 - 2y$.
11. No solutions.
12. x arbitrary, $y = 1 - 2x$, $z = 3$.
13. No solutions.
14. y arbitrary, $x = y + 2$, $w = 7 + 2y$, $z = -3y - 5$.
15. x and z arbitrary, $w = 3x + z - 2$, $y = 5x + 2z - 6$.
16. x arbitrary, $w = x + 1$, $y = x$, $z = 3 - 2x$.
17. y, z arbitrary, $x = y + z$, $w = y - z$.
18. The inclusive intervals are 0 to 2 for x and y, 0 to 4 for z.
19. If the solution is stated with y arbitrary, we have $z = -3y - 5$. Consequently, for any nonnegative y, z is negative so the system has no solution if nonnegativity is imposed.
20. The inclusive intervals are: 1 to 5/2 for w; 0 to 3/2 for x and y; 0 to 3 for z.

Problem Set 5

1. $x = 10, y = 15, z = 20.$
2. z arbitrary in the range from 19 to $155/7$, $x = 10z - 190$, $y = 155 - 7z$.
3. *a)* z arbitrary in the range from 0 to $3/2$, $x = 4 + z$, $y = 3 - 2z$.
   b) Minimum is $11.50, using 5.5 pounds of A, 1.5 pounds of C, and no B.
4. x arbitrary; $w = 40 - x$, $z = 70 - x$, $y = x - 10$.
5.

$$
\begin{array}{llll}
w + x + y & = 90 & & \\
u + v & + z = 60 & x, y \text{ arbitrary} \\
u \quad + w & = 30 & w = 90 - x - y \\
v \quad + x & = 45 & u = x + y - 60 \\
y + z & = 75. & v = 45 - x \\
& & z = 75 - y.
\end{array}
$$

6. 10 grade A boxes, 15 grade B boxes.
7. The solution is z arbitrary, $x = (11z - 22)/2$, $y = (66 - 19z)/2$. It follows from these that z must be at least 2, but less than $66/19$, so that $z = 2, 3$ are possibilities; however, if $z = 3$, x and y are not integers. The single solution is $z = 2, x = 0, y = 14$; that is, 0 boxes of grade A, 14 boxes of grade B, and 2 boxes of grade C.
8. The solution is y arbitrary, $x = 2 + y$, $z = 10 - 3y$; however, y must be between 0 and 3, inclusive.
9. $x = 2, y = 0, z = 10.$
10. The solution requires that $y = -2$, which is not permissible. The problem has no solution.
11. *a)* z units of C, arbitrary from 5 through 8; x units of A where $x = 80 - 10z$; y units of B where $y = z - 5$.
    b) Maximum is $800, making 30 units of A, no B, and 5 units of C.
12. *a)* z units of C, arbitrary from 0 through 4; y units of B, where $y = z$; x units of A, where $x = 40 - 10z$.
    b) Maximum is $800, making 40 units of A and no B or C.
13. *a)* $p_1 = 415, q_1 = 65$ units; $p_2 = 445, q_2 = 65$ units.
    b) $p_1 = 474, q_1 = 96$ units; $p_2 = 472, q_2 = 64$ units.
14. *a)* $p_1 = 800, q_1 = 100$ units; $p_2 = 400, q_2 = 100$ units.
    b) $p_1 = 708, q_1 = 68$ units; $p_2 = 372, q_2 = 104$ units.

CHAPTER 3

Problem Set 1

1. F.	2. T.	3. F.
4. F.	5. F.	6. F.
7. T.	8. F.	9. T.
10. T.	11. T.	12. Left, above.
13. Left, above.	14. Right, above.	15. Left, below.
16. Left, below.	17. $x \le 3$.	18. $x < -3/2$.

19. $x \geq 1/2$.

20. See Figure A.

21. See Figure B.

22. See Figure C.

23. See Figure D.

24. See Figure E.

25. See Figure F.

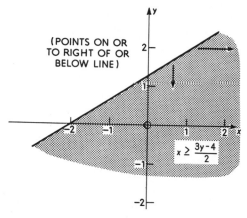

(POINTS ON OR TO RIGHT OF OR BELOW LINE)

$x \geq \dfrac{3y-4}{2}$

FIGURE A

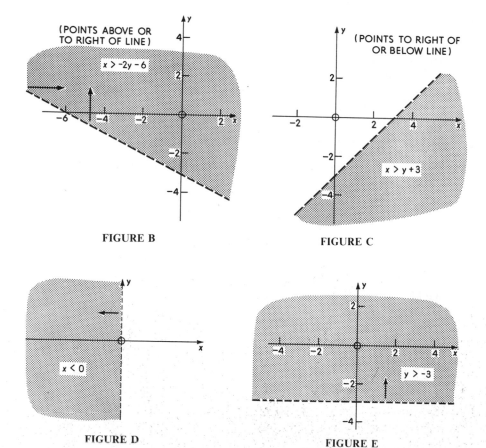

(POINTS ABOVE OR TO RIGHT OF LINE)

$x > -2y - 6$

FIGURE B

(POINTS TO RIGHT OF OR BELOW LINE)

$x > y + 3$

FIGURE C

$x < 0$

FIGURE D

$y > -3$

FIGURE E

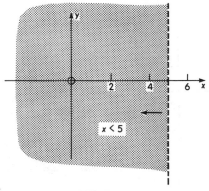

FIGURE F

Problem Set 2

1. See Figure G.   2. See Figure H.   3. See Figure I.
4. See Figure J.   5. See Figure K.

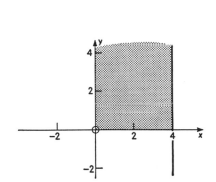

FIGURE G

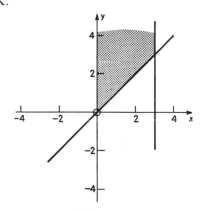

FIGURE H

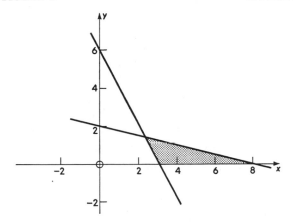

FIGURE I

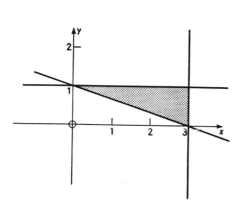

FIGURE J   FIGURE K

6. If $2 \le x \le 30/7$, $0 \le y \le (x-2)/2$; if $30/7 \le x \le 6$, $0 \le y \le (12-2x)/3$.

7. $0 \le x \le 52/19$, $(8-x)/5 \le y \le 12-4x$.

8. If $52/19 \le x \le 3$, $12-4x \le y \le (8-x)/5$; if $3 \le x \le 8$, $0 \le y \le (8-x)/5$.

9. If $0 \le x \le 52/19$, $y \ge 12-4x$; if $52/19 \le x \le 8$, $y \ge (8-x)/5$; if $x \ge 8$, $y \ge 0$.

10. *a)* $h \le (76-L)/2$; $0 < L < 76$.
    b) $x \le (108-L)/4$; $0 < L < 108$.

11. $0 \le x \le 10$, $8-0.5x \le y \le (45-3x)/5$.

12. $x = 0$, $y = 8$.

13. If $0 \le x \le 320$, $0 \le y \le (2400-3x)/8$; if $320 \le x \le 500$, $0 \le y \le 500-x$.

14. *a)* $0 \le x \le 2$, $0 \le y \le 5-x$; $2 \le x \le 7/2$, $0 \le y \le 7-2x$.

    b)

x	y
0	0, 1, 2, 3, 4, 5
1	0, 1, 2, 3, 4
2	0, 1, 2, 3
3	0, 1

15. Letting x be the number of units of A, and y the number of units of B, $0 \le x \le 9/2$; $0 \le y \le 2x$; $9/2 \le x \le 18$, $0 \le y \le (36-2x)/3$.

16. $0 \le x \le 13$, $x/4 \le y \le 4x$; $13 \le x \le 88$, $x/4 \le y \le (286-2x)/5$.

17. $0 \le x \le 20$, $0 \le y \le (40-x)/2$; $20 \le x \le 27$, $0 \le y \le (190-6x)/7$; $27 \le x \le 30$, $0 \le y \le (120-4x)/3$.

18. $20 \le x \le 24$, $(190-6x)/7 \le y \le (40-x)/2$; $24 \le x \le 27$, $(190-6x)/7 \le y \le (120-4x)/3$.

CHAPTER 4

Problem Set 1

1. *a*) 7.   *b*) 16.5.   *c*) 18.
2. *a*) 12.   *b*) 19 and 3/7.
3. 20.5.
4. 28.1.
5. *a*) 2.3.   *b*) 7.   *c*) 15.
6. Maximum is 2.5. See Figure A.

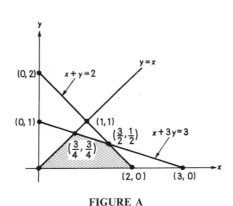

FIGURE A

7. Each would be found by solving a 3 by 3 system.
8. See text.

Problem Set 2

1. 1000 units of A, 3000 units of B.
2. 9000 units of A, no units of B.
3. 800 each of types A and B.
4. $x = 30$ units of A, $y = 15$ units of B, $\theta_{max} = \$270$.
5. $x = 20$ units of A, $y = 20$ units of B, $\theta_{max} = \$200$.
6. $x = 30$ units of A, $y = 15$ units of B, $\theta_{max} = \$270$.
7. Because no part of $x + 2y = 60$ is part of the boundary of the solution space.
8. $\theta_{min} = \$27.3$ at 0 quarts of A and 9.1 quarts of B.
9. The new constraint is $y \geq 3x$. It is not binding because $y = 3x$ is not part of the boundary of the solution space.

Problem Set 3

1. $\theta_{max} = 580$ at $x = 520$, $y = 320$, $z = 0$.
2. $\theta_{min} = 1200$.
3. $\theta_{max} = 9$ and 27/82.

Problem Set 4

1. No units of A or B, 1000 units of C.
2. 40/13 units of A, 10/13 units of B, 46/13 units of C.
3. Maximum is $190 with no units of A, 40 units of B, and 75 units of C.

CHAPTER 5

Problem Set 1

1. T. 2. F. 3. F. 4. T. 5. T.

6. T. 7. F. 8. F. 9. F. 10. T.

11. $\begin{matrix} 1/2 & 1/4 \\ -3/2 & 1/4. \end{matrix}$ 12. $\begin{matrix} -3/4 & 1/4 \\ -1/2 & 1/2. \end{matrix}$ 13. $\begin{matrix} 5 & -1 \\ 7 & -2. \end{matrix}$

14. $\begin{matrix} 17/6 & 2/3 \\ 5/3 & 1/3. \end{matrix}$ 15. $\begin{matrix} 3/4 & -3 \\ -5/4 & -3. \end{matrix}$

16.

	y	
x	3	−2
w	5	−3*

	w	
x	−1/3	2/3
y	5/3	−1/3

$x = -1/3 + (2/3)w$
$y = 5/3 - (1/3)w.$

17.

	z	
v	10	4*
y	5	−1

	v	
z	−5/2	1/4
y	15/2	−1/4

$z = -5/2 + (1/4)v$
$y = 15/2 - (1/4)v.$

18.

	x	
y	8	−2
w	2	−1*

	w	
y	4	2
x	2	−1

$y = 4 + 2w$
$x = 2 - w.$

19.

	v	
u	5	1*
w	1	−1

	u	
v	−5	1
w	6	−1

$v = -5 + u$
$w = 6 - u.$

20.

	z	
x	3	2
y	4	−3*

	y	
x	17/3	−2/3
z	4/3	−1/3

$x = 17/3 - (2/3)y$
$z = 4/3 - (1/3)y.$

Problem Set 2

1. a)

		w	z
y	−1/2	1/2	−1/2
x	7/2	3/2	−7/2.

b)

		x	z
w	−7/3	2/3	7/3
y	−5/3	1/3	2/3.

c)

		y	w
z	−1	−2	1
x	7	7	−2.

d)

		y	x
w	7/2	7/2	−1/2
z	5/2	3/2	−1/2.

2. a)

		w	u	y
x	−3	−2	1	−2
v	−1	−5	2	−7.

b)

		v	x	y
u	13	-2	5	-4
w	5	-1	2	-3.

c)

		u	x	y
w	$-3/2$	$1/2$	$-1/2$	-1
v	$13/2$	$-1/2$	$5/2$	-2.

d)

		w	x	u
y	$-3/2$	-1	$-1/2$	$1/2$
v	$19/2$	2	$7/2$	$-3/2$.

e)

		w	v	y
u	$1/2$	$5/2$	$1/2$	$7/2$
x	$-5/2$	$1/2$	$1/2$	$3/2$.

f)

		w	x	v
u	$19/3$	$4/3$	$7/3$	$-2/3$
y	$5/3$	$-1/3$	$2/3$	$-1/3$.

3. a)

		v	s	x	y
r	16	2	-2	8	6
w	3	1	$-1/2$	2	0
t	-10	-6	3	-10	-4.

b) $r = \quad 16 + 2v - \quad 2s + \ 8x + 6y$
$\quad w = \quad 3 + \ v - (1/2)s + \ 2x$
$\quad t = -10 - 6v + \quad 3s - 10x - 4y.$

4. a)

		w	x	y
u	4	3	2	1
v	2	-1	3^*	-2.

b)

		w	v	y
u	$8/3$	$11/3$	$2/3$	$7/3$
x	$-2/3$	$1/3$	$1/3$	$2/3$

$$u = \quad 8/3 + (11/3)w + (2/3)v + (7/3)y$$
$$x = -2/3 + \ (1/3)w + (1/3)v + (2/3)y.$$

Problem Set 3

1.

		x	y
θ	0	2	1
r	20	-3	-2
s	12	-1	-1.

2.

		x	y	z
θ	0	1	2	1
r	8	-1	-1	-1
s	12	-2	-3	-1
t	10	-1	-2	-3.

3.

		w	x	y	z
θ	0	3	2.8	1	2
r	150	-2	-1	-2	-3
s	100	0	-2	-1	-2
t	200	-3	-1	-4	0.

Problem Set 4

NOTE: In the Problem Set, constants have been chosen so that hand calculations will not be excessively tedious. As a consequence, the values of x, y, and z which lead to the optimal θ may not be unique. The answers given have been obtained by the method developed in this chapter. Other methods will lead to the same θ, but the x, y, and z values may differ from those shown.

1. $\theta_{max} = 60$ at $x = 0$, $y = 20$, $z = 0$.

2. $\theta_{max} = 35$ at $x = 0$, $y = 50/7$, $z = 15/7$.

3. $\theta_{max} = 125/2$ at $x = 3/2$, $y = 29/2$, $z = 1/2$.

4. $\theta_{max} = 35$ at $x = 0$, $y = 40/7$, $z = 5/7$.

5. $\theta_{max} = 27$ at $x = 0$, $y = 3$, $z = 1$.

6. $\theta_{max} = 54/5$ at $x = 0$, $y = 4$, $z = 2/5$.

7. $\theta_{max} = 335/12$ at $x = 0$, $y = 5/6$, $z = 95/12$.

8. $\theta_{max} = 22$ at $x = 2$, $y = 6$, $z = 0$.

9. $\theta_{max} = 9$ at $x = 0$, $y = 3$, $z = 3$.

10. $\theta_{max} = 79/7$ at $x = 61/7$, $y = 5/7$, $z = 13/7$.

Problem Set 5

1. 7, 7 by 7, 3, 4.

2. 7, 7 by 7, 4, 3.

3. 8, 8 by 8, 5, 3.

4. 18, 18 by 18, 7, 11.

5. 9, 9 by 9, 6, 3.

6. a) $\theta = 20 + 2x - r$.
   b) $x = 0$, $r = 0$.
   c) Using r as the pivot column would lead to *increasing r* from its present value of 0, thereby *decreasing θ*, because, as part (a) shows, r has a negative coefficient.
   d) 2.
   e) If we increase x to more than 4, leaving r at 0, $s = 4 - x - 2r$ will be negative; hence, the s row is the limiting row.
   f) Increasing x from 0 to 4 will increase θ by $2(4) = 8$.
   g) The element is positive. This means that as the variable at the head of the pivot column increases, a variable having a positive row element in the pivot column increases and cannot become negative, so the row in question cannot be a limiting factor.

7. a) $c \leq 2$. Thus, profit per unit of Z could decline by any amount or increase by at most 2, to $6.00.
   b) $-2 \leq c \leq 2/3$. $\theta' = 150 + cx = 150 + 15c$.
   c) $-1/2 \leq c \leq 3$. $\theta' = 150 + cy = 150 + 20c$.
   d) $\theta' = \$142$ at $y = 4$, $x = 11$, $t = 3$, $s = 2$, $z = r = 0$.

Problem Set 6

1.		r	s
θ	0	20	10
x	5	-4	-6
y	3	-2	-1.

2.		r	s
θ	0	10	20
x	4	-3	-1
y	2	-2	-2.

3.		r	s	t
θ	0	2	4	6
x	1	-3	-2	-4
y	1	-3	-1	-2
z	1	-1	0	-2.

4.		r	s	t
θ	0	10	15	20
x	2	-1	-2	-3
y	3	-1	-3	-1
z	1	-1	-1	-2.

5. $\theta_{min} = 10$ at $x = 3$, $y = 2$.

6. $\theta_{min} = 50$ at $x = 5$, $y = 0$, $z = 7$.

7. $\theta_{min} = 27.3$ at $x = 0$, $y = 9.1$ ounces.

8. $\theta_{min} = 92/3$ at $x = 8/3$, $y = 0$, and $z = 4$.

CHAPTER 6

Problem Set 1

1. $(3 \quad 1 \quad 7)$.

2. $\begin{pmatrix} -2 \\ 2 \end{pmatrix}$.

3. $\begin{pmatrix} 10 \\ 13 \end{pmatrix}$.

4. $(8 \quad 0 \quad 4)$.

5. $\begin{pmatrix} 21 \\ 6 \end{pmatrix}$.

6. $(-10 \quad 18 \quad -6)$.

7. 41.

8. 5.

9. $\begin{pmatrix} 4 & -4 & 0 \\ -2 & 6 & 5 \end{pmatrix}$.

10. $\begin{pmatrix} 3 & 9 \\ 2 & -7 \end{pmatrix}$.

11. $\begin{pmatrix} -2 & -27 & 17 \\ 17 & -2 & -9 \end{pmatrix}$.

12. $\begin{pmatrix} 5 & -4 & 5 \\ 14 & -10 & 15 \end{pmatrix}$.

13. $(26 \quad 35)$.

14. $\begin{pmatrix} 3 & 18 \\ 3 & -30 \end{pmatrix}$.

15. $\begin{pmatrix} 13 & 16 \\ 8 & 7 \\ -5 & -9 \end{pmatrix}$.

16 and 17. $\begin{pmatrix} 1 & 4 \\ 2 & 5 \\ 3 & 6 \end{pmatrix}$.

18. $(0.06 \quad 0.07 \quad 0.08) \begin{pmatrix} 3000 \\ 2000 \\ 4000 \end{pmatrix} = \$640$.

19. a) $(10 \quad 20) \begin{pmatrix} 8 & 2 & 5 \\ 4 & 8 & 3 \end{pmatrix} = (160 \quad 180 \quad 110)$.

The first component of the right vector, 160, is 10 batches of Superior times 8 pounds of beef per batch, plus 20 batches of Regular times 4 pounds of beef per batch, for a total of 160 pounds of beef to make the required number of batches. Similarly, we see that 180 pounds of pork and 110 pounds of lamb will be required.

648   *Mathematics*

b) $\begin{pmatrix} 8 & 2 & 5 \\ 4 & 8 & 3 \end{pmatrix}\begin{pmatrix} 1.50 \\ 0.80 \\ 1.00 \end{pmatrix} = \begin{pmatrix} 18.60 \\ 15.40 \end{pmatrix}.$

The first component in the answer, 18.60, is 8 pounds of beef times $1.50 per pound, plus 2 pounds of pork times $0.80 per pound, plus 5 pounds of lamb times $1.00 per pound, so the cost of a batch of Superior is $18.60. Similarly, the cost of a batch of Regular is $15.40.

Problem Set 2

1. $\begin{pmatrix} 2 & 3 \\ 1 & 2 \end{pmatrix}\begin{pmatrix} x_1 \\ x_2 \end{pmatrix} = \begin{pmatrix} 5 \\ 3 \end{pmatrix}.$

2. $\begin{pmatrix} 1 & 2 \\ 2 & 3 \end{pmatrix}\begin{pmatrix} x_1 \\ x_2 \end{pmatrix} + \begin{pmatrix} y_1 \\ y_2 \end{pmatrix} = \begin{pmatrix} 10 \\ 12 \end{pmatrix}.$

3. $\begin{pmatrix} 3 & 0 & 1 & -1 \\ 2 & 1 & 0 & -5 \\ 0 & 1 & 3 & 1 \end{pmatrix}\begin{pmatrix} x_1 \\ x_2 \\ x_3 \\ x_4 \end{pmatrix} \le \begin{pmatrix} 5 \\ 10 \\ 8 \end{pmatrix}.$

4. $3x_1 + x_2 + 2x_3 = 5$
   $x_1 + 4x_2 + x_3 = 4.$

5. $2x_1 + 3x_2 = 5$
   $4x_1 + 6x_2 = 10$
   $x_1 + 7x_2 = 6.$

6. $3x_1 + x_2 = 2$
   $5x_1 + 4x_2 = 6.$

7. $p_{11}x_1 + p_{12}x_2 + p_{13}x_3 + p_{14}x_4 = q_1$
   $p_{21}x_1 + p_{22}x_2 + p_{23}x_3 + p_{24}x_4 = q_2$
   $p_{31}x_1 + p_{32}x_2 + p_{33}x_3 + p_{34}x_4 = q_3.$

8. $2x_1 + x_2 + 5x_3 + y_1 = 10$
   $4x_1 + 6x_2 + 2x_3 + y_2 = 5.$

9. Maximize $3x_1 + 2x_2 + 4x_3$
   subject to: $2x_1 + x_2 + 4x_3 \le 10$
   $3x_1 + 5x_2 + 2x_3 \le 15$
   $x_1 \ge 0$
   $x_2 \ge 0$
   $x_3 \ge 0.$

10. Maximize $2x_1 + 5x_2 + 4x_3 + 3x_4$
    subject to: $x_1 + 3x_2 + 2x_3 + s_1 = 12 \quad x_1 \ge 0$
    $x_2 + 3x_3 + 2x_4 + s_2 = 15 \quad x_2 \ge 0$
    $5x_1 + 3x_2 + 2x_4 + s_3 = 10 \quad x_3 \ge 0$
    $x_1 + 4x_2 + 6x_3 + s_4 = 20 \quad x_4 \ge 0$
    $s_1 \ge 0$
    $s_2 \ge 0$
    $s_3 \ge 0$
    $s_4 \ge 0.$

11. a) y must be q by 1 and g must be p by 1.
    b) $Ay = g.$

Problem Set 3

1. *a)* 60% of customers who last purchased Eager will purchase Eager again and 40% will switch to Beaver.
   b) (0.55 0.45) and (0.555 0.445).   *c)* (5/9 Eager 4/9 Beaver).
2. *a)* (0.29 spender 0.71 saver).   *b)* (1/8 spender 7/8 saver).
3. *a)* (0.01099 users 0.98901 non-users).
   b) This is an absorbing chain. The steady state is 100% users or the state vector (1 0).
4. The new state vector is the original (*a* *b*) because the ones in the transition matrix mean that all in state #1 remain in #1 and all in #2 remain in #2.
5. (0.3 0.1 0.6); (0.1 0.6 0.3); (0.6 0.3 0.1).

Problem Set 4

1. $\begin{pmatrix} -3/2 & 1/2 \\ 1 & 0 \end{pmatrix}$.

2. $\begin{pmatrix} 1 & -4 \\ -2 & 9 \end{pmatrix}$.

3. $\begin{pmatrix} 5/4 & -1/2 \\ -3/4 & 1/2 \end{pmatrix}$.

4. The matrix has no inverse.

5. $\begin{pmatrix} 2 & 1 \\ 1 & 1 \end{pmatrix}$.

6. $\begin{pmatrix} 0 & 1/2 \\ 1/3 & -1/6 \end{pmatrix}$.

7. $\begin{pmatrix} 1 & -3 \\ -2 & 7 \end{pmatrix}$.

8. The matrix has no inverse.

9. $\begin{pmatrix} 1 & 2 & 1 \\ 4 & 5 & -3 \\ 3 & 4 & -2 \end{pmatrix}$.

10. The matrix has no inverse.

11. $\begin{pmatrix} 4 & 0 & 3 \\ 0 & 1 & 0 \\ 5 & -6 & 4 \end{pmatrix}$.

12. $\begin{pmatrix} 3/2 & -3 & -1/2 \\ 2 & -3 & -1 \\ -2 & 4 & 1 \end{pmatrix}$.

13. The matrix has no inverse.

14. $\begin{pmatrix} 2 & 2 & 3 \\ 0 & 1 & 1 \\ 1 & 1 & 1 \end{pmatrix}$.

Problem Set 5

1. *a)* $A = \begin{pmatrix} 8 & 5 \\ 3 & 2 \end{pmatrix}$; $x = \begin{pmatrix} x_1 \\ x_2 \end{pmatrix}$; $b = \begin{pmatrix} 2 \\ 1 \end{pmatrix}$.

   b) $A^{-1} = \begin{pmatrix} 2 & -5 \\ -3 & 8 \end{pmatrix}$.

   c) $\begin{pmatrix} x_1 \\ x_2 \end{pmatrix} = \begin{pmatrix} 2 & -5 \\ -3 & 8 \end{pmatrix}\begin{pmatrix} 2 \\ 1 \end{pmatrix}$.

   d) (−1 2).
   e) (1) (2 −3).   (2) (−5 8).   (3) (−3 5).
       (4) (−14 23).   (5) (−11 17).

2. *a)* $A = \begin{pmatrix} 4 & 3 \\ 9 & 7 \end{pmatrix}$; $x = \begin{pmatrix} x_1 \\ x_2 \end{pmatrix}$; $b = \begin{pmatrix} 2 \\ 3 \end{pmatrix}$.

   b) $A^{-1} = \begin{pmatrix} 7 & -3 \\ -9 & 4 \end{pmatrix}$.

   c) $\begin{pmatrix} x_1 \\ x_2 \end{pmatrix} = \begin{pmatrix} 7 & -3 \\ -9 & 4 \end{pmatrix}\begin{pmatrix} 2 \\ 3 \end{pmatrix}$.

   d) $(5 \quad -6)$.

   e) (1) $(7 \quad -9)$.   (2) $(-3 \quad 4)$.   (3) $(4 \quad -5)$.
       (4) $(11 \quad -14)$.   (5) $(-13 \quad 17)$.

3. *a)* $A = \begin{pmatrix} 6 & 8 \\ 2 & 3 \end{pmatrix}$; $x = \begin{pmatrix} x_1 \\ x_2 \end{pmatrix}$; $b = \begin{pmatrix} 3 \\ 1 \end{pmatrix}$.

   b) $A^{-1} = \begin{pmatrix} 3/2 & -4 \\ -1 & 3 \end{pmatrix}$.

   c) $\begin{pmatrix} x_1 \\ x_2 \end{pmatrix} = \begin{pmatrix} 3/2 & -4 \\ -1 & 3 \end{pmatrix}\begin{pmatrix} 3 \\ 1 \end{pmatrix}$.

   d) $(1/2 \quad 0)$.

   e) (1) $(-5/2 \quad 2)$.   (2) $(-4 \quad 3)$.   (3) $(3/2 \quad -1)$.
       (4) $(-9 \quad 7)$.   (5) $(-11/2 \quad 4)$.

4. $\begin{pmatrix} 1 & 7/5 \\ 1 & 8/5 \end{pmatrix}$.

5. *a)* $A = \begin{pmatrix} 3 & 0 & 5 \\ 2 & 2 & 5 \\ 0 & 1 & 1 \end{pmatrix}$; $x = \begin{pmatrix} x_1 \\ x_2 \\ x_3 \end{pmatrix}$; $b = \begin{pmatrix} 3 \\ 7 \\ 2 \end{pmatrix}$.

   b) $A^{-1} = \begin{pmatrix} -3 & 5 & -10 \\ -2 & 3 & -5 \\ 2 & -3 & 6 \end{pmatrix}$.

   c) $\begin{pmatrix} x_1 \\ x_2 \\ x_3 \end{pmatrix} = \begin{pmatrix} -3 & 5 & -10 \\ -2 & 3 & -5 \\ 2 & -3 & 6 \end{pmatrix}\begin{pmatrix} 3 \\ 7 \\ 2 \end{pmatrix}$.

   d) $(6 \quad 5 \quad -3)$.

   e) (1) $(-3 \quad -1 \quad 2)$.   (2) $(-31 \quad -15 \quad 19)$.   (3) $(7 \quad 3 \quad -4)$.
       (4) $(7 \quad 6 \quad -4)$.   (5) $(34 \quad 21 \quad -20)$.

6. *a)* $A = \begin{pmatrix} 7 & 3 & 0 \\ 0 & 3 & 5 \\ 1 & 1 & 1 \end{pmatrix}$, $x = \begin{pmatrix} x_1 \\ x_2 \\ x_3 \end{pmatrix}$, $b = \begin{pmatrix} 1 \\ 2 \\ 3 \end{pmatrix}$.

   b) $A^{-1} = \begin{pmatrix} -2 & -3 & 15 \\ 5 & 7 & -35 \\ -3 & -4 & 21 \end{pmatrix}$.

   c) $\begin{pmatrix} x_1 \\ x_2 \\ x_3 \end{pmatrix} = \begin{pmatrix} -2 & -3 & 15 \\ 5 & 7 & -35 \\ -3 & -4 & 21 \end{pmatrix}\begin{pmatrix} 1 \\ 2 \\ 3 \end{pmatrix}$.

d) $(37 \quad -86 \quad 52)$.

e) (1) $(16 \quad -37 \quad 22)$.   (2) $(-13 \quad 31 \quad -18)$.   (3) $(-34 \quad 79 \quad -47)$.

    (4) $(-2 \quad 8 \quad -5)$.   (5) $(3 \quad -7 \quad 4)$.

7. $\begin{pmatrix} 3/2 & -3 & -1/2 \\ 2 & -3 & -1 \\ -2 & 4 & 1 \end{pmatrix}$.

8. b) (1) $(-3 \quad 0 \quad 6)$.   (2) $(0 \quad 3 \quad 0)$.

9. b) (1) $(1 \quad 0 \quad 0 \quad 0)$.   (2) $(2 \quad -2 \quad 2 \quad -1)$.

10. a) $x_1 + \qquad\quad 2x_3 = H_1$

        $x_2 + 3x_3 = H_2$

    $x_1 + 2x_2 \qquad\quad = H_3$.

b) $A = \begin{pmatrix} 1 & 0 & 2 \\ 0 & 1 & 3 \\ 1 & 2 & 0 \end{pmatrix}$. $A^{-1} = \frac{1}{8}\begin{pmatrix} 6 & -4 & 2 \\ -3 & 2 & 3 \\ 1 & 2 & -1 \end{pmatrix}$.

c) $(130 \; 35 \; 15)$.   d) $(200 \; 100 \; 100)$.   e) $(208 \; 24 \; 56)$.

Problem Set 6

1. a) $A = \begin{pmatrix} \dfrac{1}{5} & \dfrac{3}{10} \\ \dfrac{3}{5} & \dfrac{1}{10} \end{pmatrix}$.

b) One dollar of output of industry No. 1 requires $\frac{1}{5}$ dollars of its own output and $\frac{3}{5}$ dollars of industry No. 2 output.

c) $(I - A) = \begin{pmatrix} \dfrac{4}{5} & -\dfrac{3}{10} \\ -\dfrac{3}{5} & \dfrac{9}{10} \end{pmatrix}$.

d) $(I - A)^{-1} = \begin{pmatrix} \dfrac{5}{3} & \dfrac{15}{27} \\ \dfrac{10}{9} & \dfrac{40}{27} \end{pmatrix}$.

e) $X = \begin{pmatrix} \dfrac{5}{3} & \dfrac{15}{27} \\ \dfrac{10}{9} & \dfrac{40}{27} \end{pmatrix}\begin{pmatrix} d_1 \\ d_2 \end{pmatrix}$.

f) $x_1 = 675$, $x_2 = 900$.   g) $x_1 = 1350$, $x_2 = 1800$.

2. a) $\begin{pmatrix} x_1 \\ x_2 \end{pmatrix} = \begin{pmatrix} \dfrac{40}{27} & \dfrac{10}{9} \\ \dfrac{5}{9} & \dfrac{5}{3} \end{pmatrix}\begin{pmatrix} d_1 \\ d_2 \end{pmatrix}$.   b) $x_1 = 270$, $x_2 = 270$.

c) $x_1 = 540$, $x_2 = 540$.   d) $x_1 = 510$, $x_2 = 405$.

3. a) $x_1 = 2700$, $x_2 = 2720$.   b) $x_1 = 4500$, $x_2 = 4440$.

4. a) a_{ij} is the amount of industry i output required for \$1 output by industry j.

   b) a_{13} is the amount of industry No. 1 output required for \$1 output by industry No. 3.

c) $A = \begin{pmatrix} a_{11} & a_{12} & a_{13} \\ a_{21} & a_{22} & a_{23} \\ a_{31} & a_{32} & a_{33} \end{pmatrix}$

d) $\begin{pmatrix} x_1 \\ x_2 \\ x_3 \end{pmatrix} = \left[\begin{pmatrix} 1 & 0 & 0 \\ 0 & 1 & 0 \\ 0 & 0 & 1 \end{pmatrix} - \begin{pmatrix} a_{11} & a_{12} & a_{13} \\ a_{21} & a_{22} & a_{23} \\ a_{31} & a_{32} & a_{33} \end{pmatrix} \right]^{-1} \begin{pmatrix} d_1 \\ d_2 \\ d_3 \end{pmatrix}$

or

$\begin{pmatrix} x_1 \\ x_2 \\ x_3 \end{pmatrix} = \begin{pmatrix} 1 - a_{11} & -a_{12} & -a_{13} \\ -a_{21} & 1 - a_{22} & -a_{23} \\ -a_{31} & -a_{32} & 1 - a_{33} \end{pmatrix}^{-1} \begin{pmatrix} d_1 \\ d_2 \\ d_3 \end{pmatrix}$

Problem Set 7

1. 22.

2. 15.

3. 230.

4. 45.

5. 36.

6. $\sum_{j=1}^{q} j.$

7. $\sum_{j=1}^{n} j^2.$

8. $\sum_{j=1}^{4} j^3.$

9. $\sum_{i=1}^{4} (i + 2).$

10. $\sum_{i=1}^{5} (3i).$

Problem Set 8

1. $y_1 + y_2 + y_3 + y_4 + y_5.$

2. $c_1 x_1 + c_2 x_2 + c_3 x_3.$

3. $b_1 y_1 + b_2 y_2 + \cdots + b_n y_n.$

4. $p_1 x_1 + p_2 x_2 + p_3 x_3 + p_4 x_4 + p_5 x_5 = 10.$

5. $a_1 x_1 + a_2 x_2 + \cdots + a_n x_n = c.$

6. $\sum_{i=1}^{4} x_i.$

7. $\sum_{i=1}^{6} a_i x_i = b.$

8. $\sum_{i=1}^{9} c_i x_i.$

9. $\sum_{i=1}^{q} a_i x_i.$

10. $n = 4, a_1 = 1, a_2 = 0, a_3 = 2, a_4 = 5, b = 7.$

Problem Set 9

1. $a_{11}x_1 + a_{12}x_2 + a_{13}x_3 + a_{14}x_4 = b_1$
   $a_{21}x_1 + a_{22}x_2 + a_{23}x_3 + a_{24}x_4 = b_2.$

2. Maximize $c_1x_1 + c_2x_2 + c_3x_3$
   subject to: $a_{11}x_1 + a_{12}x_2 + a_{13}x_3 \leq b_1$
   $a_{21}x_1 + a_{22}x_2 + a_{23}x_3 \leq b_2$
   $a_{31}x_1 + a_{32}x_2 + a_{33}x_3 \leq b_3$
   $a_{41}x_1 + a_{42}x_2 + a_{43}x_3 \leq b_4$
   $x_1 \geq 0$
   $x_2 \geq 0$
   $x_3 \geq 0.$

3. a) $a_{11}x_1 + a_{12}x_2 + \cdots + a_{1q}x_q = b_1$
   $a_{21}x_1 + a_{22}x_2 + \cdots + a_{2q}x_q = b_2$
   $$\vdots \qquad \vdots \qquad\qquad \vdots \qquad \vdots$$
   $a_{p1}x_1 + a_{p2}x_2 + \cdots + a_{pq}x_q = b_p.$

   b) $\sum_{j=1}^{q} a_{ij}x_j = b_i; \quad i = 1, 2, \cdots, p.$

4. a) Maximize $c_1x_1 + c_2x_2 + \cdots + c_qx_q$
   subject to: $a_{11}x_1 + a_{12}x_2 + \cdots + a_{1q}x_q \leq b_1$
   $a_{21}x_1 + a_{22}x_2 + \cdots + a_{2q}x_q \leq b_2$
   $$\cdot \qquad \cdot \qquad\qquad \cdot$$
   $a_{p1}x_1 + a_{p2}x_2 + \cdots + a_{pq}x_q \leq b_p$
   $x_1 \geq 0$
   $x_2 \geq 0$
   $$\cdot \quad \cdot$$
   $x_q \geq 0.$

   b) Maximize
   $$\sum_{j=1}^{q} c_jx_j$$
   subject to:
   $$\sum_{j=1}^{q} a_{ij}x_j \leq b_i; \quad i = 1, 2, \cdots, p$$
   and $x_j \geq 0$ for all j.

5. $\sum_{j=1}^{8} a_{ij}x_j = b_i; \quad i = 1, 2, \cdots, 6.$

6. a) m equations, n variables; $m = 3, n = 4.$
   b) $a_{11} = 2, a_{12} = 3, a_{24} = 8, a_{23} = 9, a_{33} = 7, a_{34} = 6.$
   c) $a_{13} = 0, b_1 = 20, a_{31} = 5, a_{22} = 0, a_{32} = 0.$

7. *a*) $C = x_1 d_1 + x_2 d_2 + \cdots + x_{10} d_{10}.$

   b) $C = \sum_{i=1}^{10} x_i d_i.$

8. *a*) $C = x_1 d_1 + x_2 d_2 + \cdots + x_p d_p.$

   b) $C = \sum_{i=1}^{p} x_i d_i.$

Problem Set 10

1. *a*) 12.   *b*) 62.   *c*) 4.
   d) 14.   *e*) 16.   *f*) 62/3.

2. *a*) 24.   *b*) 218.   *c*) 8.
   d) 26.   *e*) 64.   *f*) 218/3.

3. *a*) 164.   *b*) 560.   *c*) 126.
   d) 400.   *e*) 24.

4. *a*) 44.   *b*) 108.   *c*) 35.
   d) 81.   *e*) 8.

5. *a*) See Figure A.   *b*) $y = 1.9x - 4.5.$ *c*) See Figure A.

6. *a*) See Figure B.   *b*) $y = 4x/3$ *c*) See Figure B.

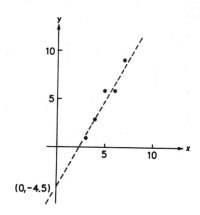

FIGURE A

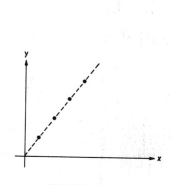

FIGURE B

7. $r = \dfrac{3(51) - 5(25)}{\sqrt{[3(13) - (5)^2][3(227) - (25)^2]}} = \dfrac{28}{\sqrt{784}} = 1.$

8. $r = 0.90.$

9. $\mu = 20,\ \sigma^2 = 4,\ \sigma = 2.$

10. $\mu = 10,\ \sigma^2 = 9,\ \sigma = 3.$

11. $\mu = 6,\ \sigma^2 = 100/9,\ \sigma = 10/3.$

12. *a)*

x	$x - \bar{x}$
3	-2
4	-1
8	3
15	0

$$\bar{x} = \frac{15}{3} = 5. \quad \sum (x - \bar{x}) = 0.$$

b) $\displaystyle\sum_{i=1}^{n} (x_i - \bar{x}) = \sum x_i - \sum \bar{x} = \sum x_i - n\left(\frac{\sum x_i}{n}\right) = 0.$

13. *a)* $\displaystyle\sum_{i=1}^{n} (x_i - 5)^2 = \sum (x_i^2 - 10x_i + 25)$

$$= \sum x_i^2 - \sum 10x_i + \sum 25$$

$$= \sum x_i^2 - 10\sum x_i + 25n \quad \text{because} \quad \sum_{i=1}^{n} 25 = 25n$$

$$= \sum x_i^2 - 10(n\bar{x}) + 25n \quad \text{because} \quad \sum x_i = n\bar{x}.$$

b) $(5 - 5)^2 + (4 - 5)^2 + (9 - 5)^2 = 25 + 16 + 81 - 10(3)(6) + 25(3)$

$$0 + 1 + 16 = 122 - 180 + 75$$

$$17 = 17.$$

14. *a)* $\bar{x} - b.$ *b)* $60 - 7 = 53.$

15. $\sum x_i^2 + 2n\bar{x} - 15n.$

16. At least 24/25 or 96 percent of the grades were in the inclusive interval 57 to 87 percent.

17. At least 84 percent of the workers earn between \$4 and \$6, inclusive.

CHAPTER 7

Problem Set 1

1. 1.5391.	2. 0.5391.	3. 0.5391 − 1.
4. 0.5391 − 2.	5. 4.5391.	6. 0.9557 − 1.
7. 0.0043.	8. 2.4116.	9. −2.
10. 0.0969 − 2.	11. 3.	12. −1.
13. 0.786.	14. 132.	15. 2.3456.
16. 3.3692.	17. 0.6794 − 3.	18. 0.
19. 0.9943.	20. 2.7810.	21. 0.5051 − 2.
22. 0.8451 − 6.	23. 4.3892.	24. 0.7782.
25. −1.	26. 1.	27. 0.3075.
28. 2.7543.	29. 0.0043 − 4.	30. 0.5159.
31. −4.	32. 1.7559.	33. 0.6990.
34. 1.2041.	35. 0.2041.	36. 3.0934.
37. 2.0453.	38. 1.1931.	39. 1.0792.
40. 0.6021 − 2.	41. 0.1430 − 1.	42. 0.2304 − 1.
43. 0.2430.	44. 2.9186.	45. 0.7528 − 2.

46. 5.1903.
49. 0.7218.
52. 6 and 7.
55. 5.
58. 0.5.

47. 0.3010 − 3.
50. 0.6794 − 9.
53. 0 and 1.
56. 2.

48. 0.1367 − 1.
51. 3 and 4.
54. 2 and 3.
57. 4.

Problem Set 2

1. 80.2.
4. 3.09.
7. 0.505.
10. 1.95.
13. 0.00281.
16. 0.000906.
19. 8.26.
22. 5.07.
25. 1.

2. 0.553.
5. 0.00309.
8. 90,300.
11. 0.4624 − 1.
14. 1.77.
17. 707.
20. 0.00308.
23. 1.01.

3. 0.0908.
6. 0.6902 − 1.
9. 0.6646 − 1.
12. 295.
15. 2.3945.
18. 6.04.
21. 0.8482 − 1.
24. 10.1.

Problem Set 3

1. 1.3918.
4. 3.5148.
7. 3.935.
10. 0.05544.
13. 0.1357.
16. 0.0149.
19. 4.093.
22. 0.6868.
25. 0.00838.
28. 1.0022.
31. 0.000008372.
34. 0.03644.
37. 0.6206.
40. 0.6906.

2. 1.9536.
5. 0.6930 − 3.
8. 3.062.
11. 0.7839 − 1.
14. 23.28.
17. 10.84.
20. 0.6791 − 1.
23. 72,780.
26. 0.7499.
29. 5517.
32. 15.40.
35. 3.440.
38. 0.6600.

3. 0.5018 − 2.
6. 2.378.
9. 0.08072.
12. 4.055.
15. 0.002328.
18. 2.5413.
21. 3.004.
24. 0.5713 − 3.
27. 0.7068.
30. 0.002367.
33. 0.3239.
36. 0.2320.
39. 0.3953.

Problem Set 4

1. See text.
4. Because $10^{-3} = 0.001$.
7. T.
10. F.
13. T.
16. F.
19. 47.36.
22. 145.3.
25. 63.91.
28. 65.77.

2. 4.
5. See text.
8. T.
11. T.
14. F.
17. 934.
20. 1.454.
23. 1.673.
26. 6.730.
29. 0.8898.

3. 2 and 3.
6. F.
9. F.
12. T.
15. F.
18. 2.356.
21. 0.00006611.
24. 0.002978.
27. 50.57.
30. 0.00004106.

31. 2.154.   32. 49.47.   33. 5.773.
34. 1.817.   35. 5.241.   36. 0.9318.
37. 5.648.   38. 66.60.   39. 0.4810.
40. 0.004507.   41. 0.007792.   42. 7.905.
43. 0.6491.   44. 2.432.   45. 6.374.
46. 0.5409.

Problem Set 5

1. $x = \dfrac{\log b}{\log a}$.

2. $x = \dfrac{\log (2d + 3)}{\log b}$.

3. $x = \dfrac{\log (k - d) - \log b}{\log c}$.

4. $x = \dfrac{\log c}{\log (ab)}$.

5. $x = \dfrac{\log c}{\log (a + b)}$.

6. $x = \dfrac{\log d - \log a}{\log (b + c)}$.

7. $x = 0$.

8. $x = \dfrac{\log 6 + 3 \log a}{\log a}$.

9. $x = \dfrac{\log c}{\log (ab)}$.

10. $x = \dfrac{\log (1 - ad) - \log (bd)}{\log c}$.

11. $x = \dfrac{\log ab}{\log c}$.

12. $x = \dfrac{\log (ab)^2}{\log c}$.

13. $x = \dfrac{\log c - (2 \log b + \log 2)}{\log c}$.

14. $x = \dfrac{\log \left(\dfrac{a^2}{b}\right)}{\log \sqrt{c}}$.

Problem Set 6

1. *a*) 1.82.
   b) $x =$ antilog $[(1/5) \log a]$.
2. *a*) $x = 1.357$.
   b) $x =$ antilog $[(1/3) \log (a/2)]$.
3. *a*) 2704.
   b) $x =$ antilog $[2 \log (a + 2)]$.
4. *a*) 3.107.
   b) $x =$ antilog $[(1/a) \log b]$.
5. *a*) 2.702.
   b) antilog $[(1/b)(\log c - \log a)]$.
6. *a*) 0.431.
   b) $x = -1 +$ antilog $[(1/n)(\log b - \log a)]$.
7. *a*) 0.017.
   b) $x = -1 +$ antilog $[(1/n)(\log a - \log b)]$.
8. *a*) By logs, 399.9.
   b) $x =$ antilog $[c(\log b)]$.
9. *a*) 1.856.
   b) $x =$ antilog $(1/n)[\log (b - c) - \log a]$.
10. *a*) Antilog $1.3263 = 21.2$.
    b) $x =$ antilog $(1/c)[\log (1 - ad) - \log (bd)]$.

Problem Set 7

1. 4.64.
2. −0.631.
3. −3.32.
4. 0.317.
5. *a*) 36.76.   *b*) 2.   *c*) 0.571.
6. *a*) 140/3.   *b*) 0.0954.   *c*) 7.50.
7. 123.
8. 23.1.
9. 23.5.
10. 2.17.

CHAPTER 8

Problem Set 1

1. *a*) $P = \text{antilog } [\log S - n \log (1 + i)]$ or $P = \dfrac{S}{(1 + i)^n}$.

   b) $n = \dfrac{\log S - \log P}{\log (1 + i)}$.

   c) $i = \text{antilog } \left[\dfrac{\log S - \log P}{n}\right] - 1$.

2. $98.65.   3. $147.8.   4. $134.3, $34.3.
5. 219.   6. 376.   7. $73.12.
8. $61.37.   9. 78.3.   10. $380.2.
11. $90.16.   12. 28 years.   13. 12.3 percent.
14. 9.3 percent.   15. $395.4.   16. About 6.4 years.
17. $2035.   18. 55,100.   19. About 34 years.

Problem Set 2

1. $1258.   2. $1060.   3. $79.49.
4. $92.59.   5. 14.2.   6. 8.9 or 9.
7. $1055.   8. *a*) 121/81. *b*) 364.

Problem Set 3

1. $216.3.   2. $579.   3. 23.4.   4. 16.7.
5. $64.7.   6. $231.2.   7. $190.39.

Problem Set 4

1. $858.30.   2. $142.46.   3. $1104.02.   4. $417.43.
5. $728.45.   6. $3860.87.   7. $240.73.   8. $5.22.
9. $12,080.40.   10. $5197.67.   11. $1987.24.   12. $136.25.
13. Monthly payment, $232.22; amount of interest, $7866.4.
14. Monthly payment, $286.70; amount of interest, $21,606.
15. 8.6 years.   16. 7.3 percent.   17. 7.3 payments.

Problem Set 5

1. *a*) $115,573.85.   *b*) $74,120.70.   *c*) $41,453.15.   *d*) $871.32.
2. $1318.95.   3. $1719.45.   4. $2209.72.
5. $161.12.   6. $2065.84.

Problem Set 6

1. $52.57.   2. $222.55.   3. $1.65.   4. $2008.55.
5. $2718.30.   6. $73.89.   7. $74.08.   8. $46.16.
9. $1.00.   10. $3.68.   11. $740.80.   12. $36.79.
13. $704, $738.91, $34.91.
14. 4, 5.06, 6.19, 7.24. Difference is 0.15.
15. *a*) Ab   years.   *b*) Almost 14 years.   *c*) Almost 10 years.
16. $n = \dfrac{1}{i}$.

CHAPTER 9

Problem Set 1

1. $A(x) = x^2$.
2. *a*) $I(t) = Prt$.   *b*) $I(r) = Prt$.
3. *a*) -2.   *b*) 1.   *c*) 1.3010.   *d*) 0.   *e*) 1.
    f) ½.   *g*) $\sqrt{2} = 1.4142$.
4. *a*) 6.021.   *b*) 4.   *c*) 0.0212.   *d*) 1.23.   *e*) 0.0143.
5. $g(a) = a^2y - y^2$.   6. $f(a) = ax^2 - a^2$.
7. *a*) $g(4)$.   *b*) $g(b)$.   *c*) $g(x + 1) - g(x)$.
8. 6.
9. *a*) 7.   *b*) $4y - 1$.
10. *a*) $0.02x + 3.01$.   *b*) $5.01.   *c*) 400 units.   *d*) $1500.
11. *a*) $0.2x + 1.1$.   *b*) $3.10.   *c*) 120 units.   *d*) $1390.
12. *a*) $\dfrac{h(x + \Delta x) - h(x)}{\Delta x}$.   *b*) $\dfrac{3^{x+\Delta x} - 3^x}{\Delta x}$.   *c*) $\dfrac{\log(x + \Delta x) - \log x}{\Delta x}$.
13. *a*) 2.   *b*) 0.1.   *c*) 0.171.   *d*) 1.44.
14. *a*) 2.   *b*) 0.01.

Problem Set 2

1. Approaching, but never touching.
2. $f(x) = 2^{-x}$ means

$$f(x) = \frac{1}{2^x}.$$

Hence, when x takes on values 1, 2, 3, 4, 5, $\longrightarrow\infty$, $f(x)$ is ½, ¼, ⅛, ¹⁄₁₆, ¹⁄₃₂, approaching but never becoming zero.

3. See Figure A.   4. See Figure B.   5. See Figure C.

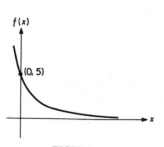

FIGURE A

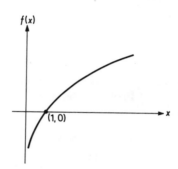

FIGURE B

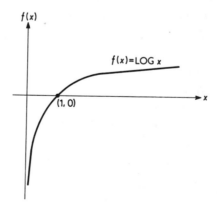

FIGURE C

6. *a*) Remember this: The logarithm of 1 to any base is 0, so log 1 = 0.
   b) 1, 2, 3.   *c*) −1, −2, −3.   *d*) See Figure D.
7. See text.

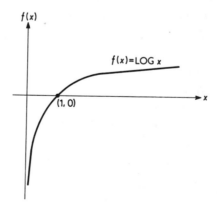

FIGURE D

8. The curve can be drawn with a continuous movement, without lifting our pencil from the paper. It has no sharp changes in direction, so that it has a unique tangent at every point for which $x > 0$. It is continuous and smooth for $x > 0$. Note: We would have to plot many more points than the few used in Problem 6 to make these statements intuitively reasonable.

9. The curve has discontinuities at $x = 2$ and $x = -2$ because these numbers make the denominator of the function zero.

10. See Figure E.

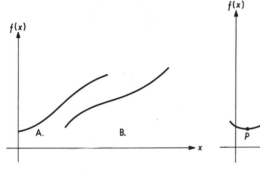

FIGURE E

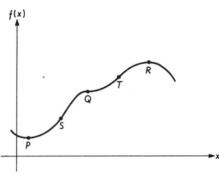

FIGURE F

11. *a*) See Figure F.
    b) There are three inflection points: *S*, *Q*, and *T*.

12. *a*) We must have a stationary point (horizontal tangent) and downward concavity.
    b) We must have a stationary point and upward concavity.
    c) We have a stationary inflection point.

13. *a*) See Figure G.

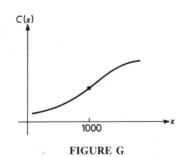

FIGURE G

b) Marginal cost is increasing up to 1000 units output, and decreasing thereafter.
c) Economists argue that the efficiencies of mass production are not realized at low output rates, so marginal cost for low output tends to be high, then decreases as output increases up to a point where operations are near the full capacity level. At this point, overtime operations and other costly steps are needed to increase output, and marginal cost rises. A curve consonant with this argument would start concave down and change to concave up.

Problem Set 3

1. See Figure H.   2. See Figure I.   3. See Figure J.   4. See Figure K.

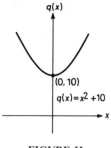

FIGURE H
(not to scale)

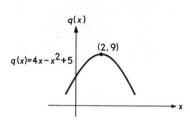

FIGURE I
(not to scale)

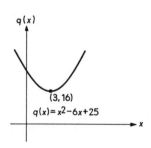

FIGURE J
(not to scale)

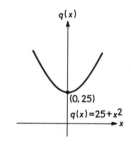

FIGURE K
(not to scale)

5. A parabola, opening upward, with vertex (minimum) at (2, 3).
6. A parabola, opening downward, with vertex (maximum) at (4, 20).
7. At $x = 40$ units, the minimum is \$34 per unit.
8. At $x = 100$ the minimum is \$25 per unit.
9. See Figure L.   10. See Figure M.

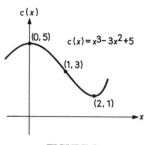

FIGURE L
(not to scale)

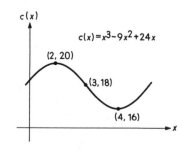

FIGURE M
(not to scale)

11. See Figure N.

12. See Figure O.

13. $x = 50$ units.

14. $x = 20$ units.

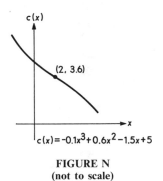

FIGURE N
(not to scale)

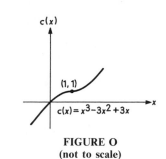

FIGURE O
(not to scale)

15. See Figure P.

16. See Figure Q.

17. See Figure R.

18. See Figure S.

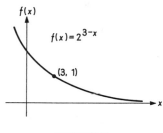

FIGURE P
(not to scale)

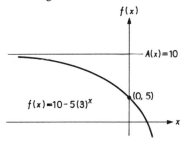

FIGURE Q
(not to scale)

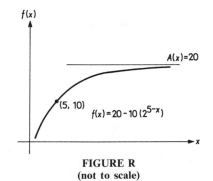

FIGURE R
(not to scale)

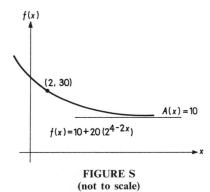

FIGURE S
(not to scale)

19. *a*) 8.

    b) ⅛.

    c) 7.3891.

    d) 0.1353.

    e) 0.9900.

    f) 4.4817.

20. *a*) 6.7%.   *b*) 44.2%.   *c*) 70%.

21. *a*) 1500.   *b*) 1184.   *c*) 1000.

22. 5.13 percent.   23. 10.52 percent.   24. About $19/_7$.

25. *a*) 0.   *b*) 0.   *c*) $0.6990 - 1 = -0.301$.
    d) -0.6931.   *e*) 1.3802.   *f*) 3.1781.
    g) 1.   *h*) 1.   *i*) 2.3026.

26. *a*) 2.3026.   *b*) $1/2.3026 = 0.4343$.
    c) 3.6243.   *d*) -0.9809.

27. *a*) $\ln 12 + \ln 3$.   *b*) $2(\ln 6)$.   *c*) $\ln 0.5 - \ln 10 = -2.9957$.

28. *a*) $3(\ln x)$.   *b*) $\ln(x/y)$.   *c*) $\ln(b^{1/a})$.
    d) $\ln a$.   *e*) $-\ln x$.

29. 8.959.   30. 0.1116.   31. About four years.

32. *a*) 5.6 years.   *b*) $1/1.75 = 0.57$ or 57 cents.

33. About eight years.   34. About 5.8 years.

Problem Set 4

1. 2, 5/4, 10/9, 17/16, 26/25 . Limit is 1.
2. $-9, -3, -1/3, 3/2, 3$. The limit as $n \longrightarrow \infty$ does not exist.
3. 2, $(3/2)^2$, $(4/3)^3$, $(5/4)^4$, $(6/5)^5$. Limit is *e*.
4. 3, 9, 27, 81, 243. Limit does not exist as $n \longrightarrow \infty$.
5. 1/2, 1/4, 1/8, 1/16, 1/32 . Limit is 0.
6. 1/2, 1, 9/8, 1, 25/32 . Limit is 0.
7. Starting with 3 as the first approximation, the sequence is 3, 19/6, 721/228, and $721/228 = 3.16228+$.
8. Starting with 2 as the first approximation, the sequence is 2, 5/2, 49/20, and $49/20 = 2.45$.
9. 211/81.   10. 255.   11. 2.
12. 3.   13. $12,500.   14. $5000.
15. *a*) *Ai*.   *b*) $80.
16. $R[(1/i) + 1]$.   17. *a*) $75,000.   *b*) $81,000.
18. *a*) $5093.
    b) B is the better buy because its capitalized cost, $4,657, is less than that of A.
19. *a*) See Text.   *b*) $2.6666 \cdots$ rounds to 2.7.
20. *a*) See Text.   *b*) 0.82.
21. No. The limit of $C(x)$ as $x \longrightarrow 50$ on $C(x) = 10 + 2x$ is $110, but as $x \longrightarrow 50$ on $C(x) = 10 + 3x$, the limit is a different value, $160. To be continuous at $x = 50$, the limit of $C(x)$ must be the same for any sequence of values approaching 50. The function has a discontinuity at $x = 50$.
22. Yes. On either $10 + 3x$ or $60 + 2x$, the limit as $x \longrightarrow 50$ is $160, and $C(50) = 160. Hence, $\lim_{x \to 50} C(x) = C(50) = 160$, so the requirement for continuity is fulfilled.
23. Because the limit can then be determined by substitution as $f(a)$.
24. 5.3891.   25. 1.9207.   26. 0.   27. 32.   28. 0.
29. The limit does not exist.

30. 0.   31. 4.   32. $2a$.

33. $f'(x) = \lim\limits_{\Delta x \to 0} \dfrac{(x + \Delta x)^n - x^n}{\Delta x}$.

34. $g'(x) = \lim\limits_{\Delta x \to 0} \dfrac{e^{x + \Delta x} - e^x}{\Delta x}$.

35. $h'(x) = \lim\limits_{\Delta x \to 0} \dfrac{\ln(x + \Delta x) - \ln x}{\Delta x}$.

36. If $f(x) = x^n$, $f'(x) = nx^{n-1}$.

37. If $g(x) = e^x$, $g'(x) = e^x$.

38. If $h(x) = \ln x$, $h'(x) = 1/x$.

CHAPTER 10

Problem Set 1

1. $3x^2$.

2. $3/x$.

3. $3^x \ln 3$.

4. $2e^x$.

5. $6x + 2$.

6. $15x^2 - 8x + 3$.

7. 4.

8. 0.

9. 1.

10. $1 + \dfrac{1}{x^2}$.

11. $-\dfrac{1}{x^2} + \dfrac{2}{x^3}$.

12. $(5/3)x^{2/3}$.

13. $x^{-2/3}$.

14. $\dfrac{1}{2}x^{-1/2}$.

15. $-\dfrac{1}{2}x^{-3/2}$.

16. $-x^{-3/2} - x^{-4/3}$.

17. $(0.9)^x \ln(0.9) = (0.9)^x(-0.1054)$.

18. $12x^3 + x^{-1/2}$.

19. $3(2)^x(\ln 2) = 2.0793(2)^x$.

20. $e^x + 5/x$.

21. $(\frac{1}{2}, \frac{3}{4})$; slope is 3.

22. $(4, 9)$; slope is 3.

23. $(0, 6)$; slope is -2.

24. $(\frac{1}{4}, \frac{1}{2})$; slope is 1.

25. $(1, 5)$; slope is 3.

26. $(2, 2)$; slope is 3.

27. $(1, 0.9)$; slope is -0.09486.

28. $(0, -2)$; slope is -1.0793.

29. $(2, 1.3069)$; slope is $\frac{1}{2}$.

30. $(0.05, 3.1539)$; slope is 3.1539.

Problem Set 2

1. *a*) 50 units. $A''(x) = 20{,}000/x^3$ is positive at $x = 50$.
   b) \$500 per unit.

2. *a*) 25 feet. $A''(x) = -2$ is negative, proving $x = 25$ yields a maximum.
   b) 25' by 25', a square.   *c*) 625 square feet.

3. *a*) $p = 0.5$. $V''(p) = -\dfrac{2}{n}$ is negative since n is a positive constant (number of respondents). Hence, we have a maximum.
   b) $V(0.5) = 0.025$ and $V(0.1) = 0.009$.

4. *a*) $x = 100$ units per hour.
   b) $C''(100) = [200/(100)^3] - [1/(100)^2] = +1/10{,}000$. $C''(x)$ is positive so the stationary point at $x = 100$ is a minimum.
   c) $C(100) = -3 + \ln 100 = -3 + 2.3026(\log 100) = \$1.6052$.

5. No maximum nor minimum.

6. Maximum at $(5/2, 25/4)$.

7. Maximum at $(0, 0)$; minimum at $(8/3, -256/27)$.
8. Minimum at $(2, 0)$.
9. Maximum at $(0, 12)$; minimum at $(8, -244)$.
10. Maximum at $(0, 2)$; minimum at $(2, -2)$.
11. Minimum at $(4, 16 - 32 \ln 4)$ or $(4, -28.36)$.
12. Minimum at $(4, 48)$.
13. Maximum at $(0, -1)$. Remember that $e^0 = 1$.
14. Minimum at $(1, 1)$.
15. Maximum at $(3, 2e^3)$ or $(3, 40.17)$.
16. $f'(x) = 2^x \ln 2$. Since ln 2 is positive and 2 to any power is positive, $f'(x)$ can never be zero. No maximum nor minimum exists.
17. Maximum at $(0, 12)$; minimum at $(8, -244)$; inflection at $(4, -116)$.
18. **Minimum at $(1/2, -3/8$.**

19.

x	$f'(x)$	$f''(x)$	*Direction*	*Concavity*
-4	$+$	$-$	Rising	Downward
-3	0	$-$	Stationary	Downward
-2	$-$	$-$	Falling	Downward
-1	$-$	0	Falling	Inflection
0	$-$	$+$	Falling	Upward
1	0	$+$	Stationary	Upward
2	$+$	$+$	Rising	Upward
3	$+$	$+$	Rising	Upward

20.

			Rising			*Falling*		
x	*Maximum*	*Minimum*	*Concave Up*	*Concave Down*	*Inflection*	*Concave Up*	*Concave Down*	*Inflection*
-6	....	....	....	....	....	X	....	....
-5	....	X	....	....	....	....	....	....
-4	....	....	X	....	....	....	....	....
-3	....	....	....	....	X	....	....	....
-2	....	....	....	X	....	....	....	....
-1	....	....	....	X	....	....	....	....
0	X	....	....	....	....	....	....	....
1	....	....	....	....	....	....	X	....
$5/3$	....	....	....	....	....	....	....	X
2	....	....	....	....	....	X	....	....
3	....	X	....	....	....	....	....	....
4	....	....	X	....	....	....	....	....
5	....	....	X	....	....	....	....	....

21. *a*) $c''(x) = 6tx + 2s$. $c''(x) = 0$ when $x = -s/3t$.
    b) Here $t = 1, s = -9$; $x = -s/3t = 9/3 = 3$.

Problem Set 3

1. *a*) 50 by 66 feet.   *b*) \$33,000.
2. *a*) 75 by 99 feet.   *b*) 7425 square feet.

3. *a)* $C(x) = \left(2x + 2\dfrac{A}{x}\right)e + ix.$ *b)* $x = \sqrt{\dfrac{2Ae}{i + 2e}}.$

4. *a)* $A(x) = x\left(\dfrac{D - 2ex - ix}{2e}\right).$ *b)* $x = \dfrac{D}{2(2e + i)}.$

5. *a)* 90 by 150 feet.   *b)* \$90,000.

6. \$210.   7. 500 square feet.

8. 18×12 inches; area $= 216$ square inches.

9. 1100 by 660 feet; increase would be 10,000 square feet.

10. *a)* $A = 2x^2 + 4xL.$ *b)* $A(x) = 432x - 14x^2.$
    c) $x = 108/7$; $A''(x) = -28$ is always negative, proving we have a maximum.
    d) 108/7 by 108/7 by 3(108/7), or about 15.4 by 15.4 by 46.3 inches.

11. $x = M/7$; $L = 3M/7.$

12. A cube 4 by 4 inches; area $= 96$ square inches.

13. 10 by 10 by 5 inches; area $= 300$ square inches.

14. 5 by 5 by 10 inches; minimum cost $= \$300.$

15. \$2.785.

16. One-inch squares.

17. 10/3 by 10/3 inches.   18. $r = 20$ feet; $h = 40$ feet.

19. $r = 24$ feet; $h = 250/9 = 27.8$ feet.

20. Maximum group charge \$1512.50 for 55 persons or more.

21. *a)* $80r^2.$ *b)* $24 - 2r.$ *c)* $T(r) = 80r^2(24 - 2r).$
    d) $r = 8$ miles.   *e)* \$40,960.

22. *a)* $N/Q.$ *b)* $cN/Q.$ *c)* $uQ.$
    d) $p(uQ)/2.$ *e)* $S(Q) = [cN/Q] + [p(uQ)/2].$
    f) $Q = \sqrt{2cN/pu}.$ *g)* 240 units.

23. *a)* $6Kx^2.$ *b)* $W/x^3.$ *c)* $(W/x^3)(6x^2d) = 6Wd/x.$
    d) $C(x) = 6Kx^2 + 6Wd/x.$

    e) $x = \sqrt[3]{\dfrac{Wd}{2K}}.$

    f) $x = 8$ feet.

Problem Set 4

1. $30(3x - 2)^9.$ 2. $12(2x^3 - 5x)^3(6x^2 - 5).$

3. $3e^{3x+2}.$ 4. $-e^{-x}.$

5. $18x^2e^{2x^3}.$ 6. $-0.1e^{-0.05x}.$

7. $-2^{3-x}\ln 2 = -0.6931(2^{3-x}).$

8. $x[3^{(x^2)/2}]\ln 3 = 1.0986x[3^{(x^2)/2}].$

9. $12/(3x + 5).$ 10. $(4x + 3)/(2x^2 + 3x).$

11. $(3/2)(3x)^{-1/2} = 3/2(3x)^{1/2}.$

12. $1 - 2(2x - 3)^{-2} = 1 - 2/(2x - 3)^2.$

13. $(3/2)(3x - 2)^{-1/2} = 3/2(3x - 2)^{1/2}.$

14. $-3(3 - x)^2 - 4(5 - 2x).$ 15. $-xe^{(-x^2)/2} + 3e^x.$

16. $1/x.$

17. a) $\dfrac{dH(w)}{dx} = \dfrac{dH(w)}{dw}\dfrac{dw}{dx}$.   b) w is a function of x.

18. a) $\dfrac{dP(g)}{dz} = \dfrac{dP(g)}{dg}\dfrac{dg}{dz}$.   b) g is a function of z.

19. a) $4(x^{2/3})^3(2/3)x^{-1/3} = (8/3)x^{5/3}$.
    b) $(8/3)x^{5/3}$.

20. a) $(d/dx)(9 - 12x + 4x^2) = -12 + 8x$.
    b) $(d/dx)(3 - 2x)^2 = 2(3 - 2x)(-2) = -12 + 8x$.

21. Letting $P(t)$ be the profit function, $dP(t)/dt$ is zero when $e^{-0.25t} = 0.04$, so by logarithms, $-0.25t = \ln(0.04)$. Computing $\ln(0.04)$ as $2.3026 \log(0.04)$, we find $t = 12.9$ days.

22. $dy/dx = -3/4$ at $(60, 80)$ so, on the tangent, if x decreases from 60 to 0, y will increase from 80 to $80 + 45 = 125$. Hence, we have $Q(0, 125)$.

23. a) $\overline{RP} = [(x - 1)^2 + 1]^{1/2}$, $\overline{RQ} = [(x - 3)^2 + 4]^{1/2}$.
    b) $s = [(x - 1)^2 + 1]^{1/2} + [(x - 3)^2 + 4]^{1/2}$.
    c) $\dfrac{ds}{dx} = \dfrac{x - 1}{[(x - 1)^2 + 1]^{1/2}} + \dfrac{x - 3}{[(x - 3)^2 + 4]^{1/2}}$.
    d) Set $ds/dx = 0$, place one expression on each side of the equation, square both sides and simplify to obtain $4(x - 1)^2 = (x - 3)^2$. Taking both positive and negative square roots of both sides yields $x = -1$ and $x = 5/3$. Substitution in (b) shows s is larger at $x = -1$ than at $x = 5/3$. Hence, $x = 5/3$.
    e) $s = 3.61$.

24. a) $\dfrac{de^y}{dx} = e^y \cdot \dfrac{dy}{dx}$.

    b) $\dfrac{de^{-3y}}{dx} = \dfrac{de^{-3y}}{dy} \cdot \dfrac{dy}{dx} = e^{-3y}(-3)\dfrac{dy}{dx}$.

    c) $\dfrac{d}{dI}e^{-y/10} = \dfrac{de^{-y/10}}{dy} \cdot \dfrac{dy}{dI} = e^{-y/10}\left(-\dfrac{1}{10}\right)\dfrac{dy}{dI}$.

    d) $\dfrac{d}{dz}\ln x = \dfrac{1}{x}\dfrac{dx}{dz}$.

    e) $\dfrac{d}{dh}\ln(2x + 3) = \dfrac{d}{dx}\ln(2x + 3) \cdot \dfrac{dx}{dh} = \left(\dfrac{2}{2x + 3}\right)\dfrac{dx}{dh}$.

    f) $\dfrac{d}{dw}(2y + 3)^2 = \dfrac{d}{dy}(2y + 3)^2\dfrac{dy}{dw} = 2(2y + 3)(2)\dfrac{dy}{dw}$.

Problem Set 5

1. $(2x + 1)(3) + (3x + 1)(2) = 12x + 5$.
2. $x^{1/2}(6x + 2) + (3x^2 + 2x)(1/2)x^{-1/2} = (15x^2 + 6x)/2x^{1/2}$.
3. $-3(3x + 1)^{4/3}(3 - 2x)^{1/2} + 4(3 - 2x)^{3/2}(3x + 1)^{1/3}$.
4. $12x^3 + 6x^2$.
5. $x^{-2}(6)(x + 1)^5 + (x + 1)^6(-2)x^{-3} = 2x^{-3}(x + 1)^5(2x - 1)$.
6. $(2/3) + (1/3) = 1$.
7. $\dfrac{x(1) - x(1)}{x^2} = 0$.
8. $1/(1 - x)^2$.
9. $3/(x + 1)^2$.

10. $(2x^2 + 6x)/(2x + 3)^2$.

11. $\dfrac{(x + 3)(1/2)(2x + 5)^{-1/2}(2) - (2x + 5)^{1/2}}{(x + 3)^2} = -\dfrac{x + 2}{(2x + 5)^{1/2}(x + 3)^2}$.

12. $[(2x - 3)^{1/2} - x(1/2)(2x - 3)^{-1/2}(2)]/(2x - 3) = (x - 3)/(2x - 3)^{3/2}$.

13. $x^2 e^{2x}(2) + 2xe^{2x} = 2xe^{2x}(x + 1)$.

14. $\dfrac{e^x(3x^2) - x^3 e^x}{(e^x)^2} = \dfrac{x^2(3 - x)}{e^x}$.

15. $\ln x$.

16. $-2xe^{-x^2} + 2(1 - 2x)/(x - x^2)$.

17. $2^x(\ln 2)$.

18. $e^{2x}(2x - 1)/x^2$.

19. $-3xe^{-x^2/2}$.

20. $\dfrac{(\ln x)e^x - e^x\left(\dfrac{1}{x}\right)}{(\ln x)^2} = \dfrac{e^x(x \ln x - 1)}{x(\ln x)^2}$.

21. $(5^x/x) + (\ln 3x)(5^x)(\ln 5)$.

22. $6x[\ln(x^2 - 5)]^2/(x^2 - 5)$.

23. *a*) 0.   *b*) $-e^{-0.5} = -0.6065$.

24. $y'(x) = \dfrac{e^{-x}}{x} + (\ln x)(e^{-x})(-1) = e^{-x}\left(\dfrac{1}{x} - \ln x\right)$.

   a) 0.3679.   *b*) $(0.1353)(0.5 - 0.6931) = -0.02612643$.

25. The starting expression is

$$Q'(x) = \dfrac{\dfrac{(x + \Delta x)^2}{(3x + 3\Delta x)} - \dfrac{x^2}{3x}}{\Delta x}.$$

Algebraic manipulation reduces this to

$$Q'(x) = \lim_{\Delta x \to 0} \dfrac{(3x)(2x) + 3x\Delta x - x^2(3)}{3x(3x + 3\Delta x)} = \dfrac{3x(2x) - (x^2)(3)}{(3x)^2},$$

which is the form which would appear if the quotient rule were applied to $x^2/3x$.

26. $dP(p)/dp = 4p(1 - p)^3(-1) + (1 - p)^4$ by the product and chain rule. Setting this to zero and factoring out $(1 - p)^3$ yields the expected result, $p = 0.2$. Show that the second derivative is negative to prove this p yields a maximum.

Problem Set 6

1. *a*) $q = 300$ units.   *b*) Maximum profit is $880.
2. *a*) $q = 1,350$ units.   *b*) Maximum profit is $1,772.50.
3. *a*) $q = 200$ units.   *b*) Maximum profit is $18,000.
4. *a*) At $q = 10$ and $q = 70$ units.
   b) The derivative of marginal cost is $d/dq[C'(q)] = C''(q)$ which is $C''(q)$ $= 0.02q - 0.9$. $C''(10) = -0.7$ and $C''(70) = +0.5$. Hence, marginal cost is decreasing at $q = 10$ and increasing at $q = 70$.
   c) $q = 70$ because here marginal cost is increasing.
   d) At $q = 45$ marginal cost has its minimum, $26.75.

5. $q = 45$ units.

6. See text.

7. a) The derivative of average cost, $(-400/q^2) + 0.01$ has its minimum at $q^* = 200$ units. Break-even selling price is $AC(200) = \$9$ per unit.

b) Average cost equals marginal cost when $(400/q) + 5 + 0.01q = 5 + 0.02q$, yielding, as in a), $a^* = 200$ units.

8. a) $AC(20) = \$20.4$; $AC(50) = \$15$; $AC(80) = \$14.1$; $AC(100) = \$14$; $AC(200) = \$15$; $AC(400) = \$18.5$.

b) The derivative of average cost, $-(200/q^2) + 0.02$, is zero at $q = 100$ units and the break-even selling price is $AC(100) = \$14$ per unit.

c) $C'(50) = \$12$; $C'(100) = \$14$; $C'(200) = \$18$.

d) See Figure A.

e) Average cost equals marginal cost when average cost is at its minimum.

f) At $q = 250$ units.

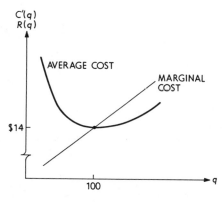

FIGURE A

9. At $q = 70$ units.   10. See text.

11. a) 0.8.

b) \$0.80 of an additional dollar of income will be spent on consumption.

c) 0.82 or 82 percent.   d) 5.

e) Income will increase at a rate of \$5 per additional dollar of investment.

12. a) 0.7788 and 0.4724 at \$20 and \$60 thousand, respectively.

b) 0.88 or 88 percent.

c) 4.5 and 1.9 at \$20 and \$60 thousand, respectively.

13. $1/(0.6 - 0.96\,Y^{-1/4})$.

Problem Set 7

1. a) $dC(i)/di = n(1 + i)^{n-1}$.

b) From Table II, $20(2.10685) = 42.137$.

c) 0.042137.   d) 2.2333.

2. a) The exponential rule.

b) $dC(n)/dn = (1 + i)^n \ln(1 + i)$. At $i = 0.05$, $n = 10$, $dC(n)/dn = (1.62889)$ $(0.04879) = 0.07947$, using Table II.

c) $(0.25)(0.07947) = 0.01987$.

d) $C(10.25) = 1.62889 + 0.01987 = 1.64876$. (Note: To five decimals, the correct figure is 1.64888.)

3. *a*) $(1 - \ln 500r)/r^2$.   *b*) -90.12925.
   c) 23.03 or 23 days.   *d*) -0.90 or about one day less.

4. *a*) $e^{-1/x}\left(\dfrac{1}{x} + 1\right)$.   *b*) 0.7358.

   c) 0.007358.   *d*) 0.3753.

5. *a*) $1/2x^{1/2}$.   *b*) 0.1.
   c) -0.01.   *d*) 4.99.

6. *a*) $p/20$ or $5 - \ln q$.
   b) $e(20) = 2.00$.
   c) At 20 units demand, a price increase of 1 percent will be accompanied by a demand decrease of about 2 percent.

7. *a*) -2.   *b*) $2p/q$ or $(400/q) - 1$.
   c) $e(100) = 3$.
   d) At demand 100 units, a price increase of 1 percent will be accompanied by a demand decrease of about 3 percent.
   e) At $q = 200$ units, $p = \$100$ per unit.
   f) \$20,000.

8. $R(q) = q(200 - 0.5q) = 200q - 0.5q^2$, and $R'(q) = 0$ when $q = 200$ units, as in 7 *e*).

Problem Set 8

1. *a*) $f_x = 2x + 3y$.   *b*) $f_y = 4y + 3x$.
   c) $f_{xx} = 2$.   *d*) $f_{yy} = 4$.
   e) $f_{xy} = 3$.   *f*) $f_{yx} = 3$.

2. *a*) $f_x = 2y^2 + 6xy - 2y$.   *b*) $f_y = 4xy + 3x^2 - 2x$.
   c) $f_{xx} = 6y$.   *d*) $f_{yy} = 4x$.
   e) $f_{xy} = 4y + 6x - 2$.   *f*) $f_{yx} = 4y + 6x - 2$.

3. Maximum of 122,500 at $x = 1000$, $y = 300$.

4. Minimum of -21.5 at $x = 4$, $y = 5$.

5. *a*) Maximum occurs at $q_1 = 180$, $p_1 = \$62.50$; $q_2 = 15$, $p_2 = \$59$.
   b) \$7065.

CHAPTER 11

Problem Set 1

1. $(x^4/4) + C$.   2. $(2/5)(x^{5/2}) + C$.
3. $2x^{1/2} + C$.   4. $3x + C$.
5. $x + C$.   6. $\ln x + C$.
7. $2 \ln x + e^x + C$.   8. $x \ln x + C$.
9. $(3^x/\ln 3) + C$.   10. $-x^{-1} + [2(5^x)/\ln 5] + C$.
11. $3 \ln\left(\dfrac{x}{1 + 6x}\right) + C$.   12. $(1/2)e^{5+2x} + C$.

13. $(2/9)(2 + 3x)^{3/2} + C.$

14. $(-1/2)e^{-2x} + C.$

15. $2x - \ln(1 + 2x) + C.$

16. $(3/4)[e^{2x}(2x - 1)] + C.$

17. $[-10(2)^{-0.1x}/\ln 2] + C.$

18. $(1 + 3x)[\ln(2 + 6x) - 1] + C.$

19. Apply the product rule.

20. Apply the logarithmic, quotient, and chain rules.

21. *a*) $C(q) = 1500 + 1.5q + 0.05q^2.$
    b) \$2150.

22. *a*) $Q(t) = 20 + 15e^{-0.06t}.$ *b*) $\lim_{t \to \infty} Q(t) = 20.$

23. $p(q) = 20 - 0.01q, \, q \le 1{,}000.$

24. $A(t) = 2{,}500(e^{0.06t} - 1).$

25. $C(q) = 139 + 2q \ln q + q.$

Problem Set 2

1. 8.   2. 5.   3. 1.6931.

4. $(1/\ln 0.8)(0.64 - 1) = 1.614.$

5. 0.9.   6. 1.   7. 3.4366.

8. 2.   9. 52/3.   10. 0.3862.

11. $31/(16 \ln 2) = 31/11.0896 = 2.795.$

12. 5.

13. 0.6931.

14. *a*) See Figure A.

b) 250/3.

15. *a*) See Figure B.

b) 9/2.

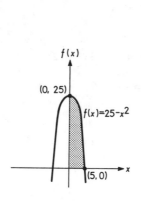

FIGURE A

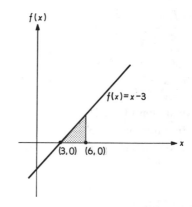

FIGURE B

16. *a*) $R'(t)\Delta t.$

b) $R'(t)\Delta t$ is (rate per year)(Δt years) which approximates the number of gallons sold in the time interval Δt years.

c) Fill the interval with area elements of width Δt, then sum these area elements.

d) By using more area elements (that is, smaller Δt).

e) The exact area is the limit of the sum in *c*) as Δt approaches 0.

f) 43.2.

g) Total sales over the 10 year period will be 43.2 thousand gallons.

17. See Figure C.

a) $M'(t)\Delta t$.

b) $M'(t)\Delta t$ is (rate per year)(Δt years) which approximates the dollars of maintenance cost during the time interval Δt.

c) Fill the interval with area elements of width Δt, and sum these area elements.

d) By using more area elements (smaller Δt).

e) The exact area is the sum in c) as Δt approaches zero.

f) 25.

g) Total maintenance cost over the 10-year period will be $25 thousand.

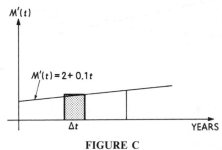

FIGURE C

18. $500 - 2000(0.3679) - [-2000] - 1235.8 = \$528.4$.

19. $\int_0^{999} [50 - 5 \ln (q + 1)] \, dq - 999[50 - 5 \ln (1000)]$

$$= 50(999) - 5(1000)(\ln 1000 - 1) - (-5)(-1) - 999(50) + 4995 \ln 1000$$
$$= 4995 - 5 \ln 1000 = 4995 - 5(2.3026)(3) = \$4960.46.$$

20. See Figure D.

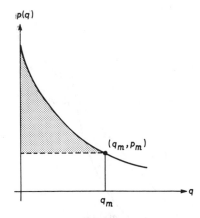

FIGURE D

21. $5(300) + 300e^3 - 5(300) - 100e^3 + 100 = \$4117.1$.

22. $190.77.

23. $p''(q) = -16/(2q + 1)^2$ which is negative for all q, so the curve is concave downward rather than upward.

24. *a*) See Figure E.
   b) $q_m = 300$ units, $p_m = \$40$ per unit.
   c) $9000.
   d) $4500.

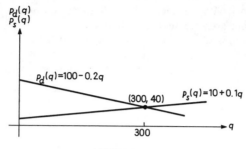

FIGURE E

25. $(38.9630)(0.5) = 19.4815.$
26. $(3.7219)(0.2) = 0.74438.$
27. $(3.2176)(0.4) = 1.28704.$
28. $(46.5)(1) = 46.5.$

Problem Set 3

1. *a*) $A_1 = \int_0^b f(x)\, dx - \int_0^b g(x)\, dx = \int_0^b [f(x) - g(x)]\, dx.$

   b) $A_2 = \int_b^d [g(x) - f(x)]\, dx.$

2. *a*) See Figure F.
   b) $(0, 0)$ and $(3, 9)$.
   c) $9/2$.

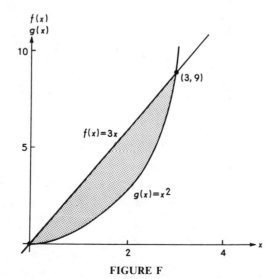

FIGURE F

3. *a*) See Figure G.
   b) (1, 12) and (12, 1).
   c) 41.6812.

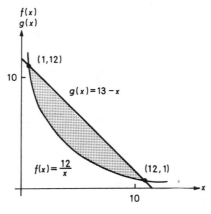

FIGURE G

4. 1/3.
5. *a*) See Figure H.
   b) $(e^x + e^{-x}) \Big|_0^2 = 5.5244.$

6. *a*) See Figure I.
   b) 6.

7. $\int_0^2 (2x + 2)\, dx + \int_2^8 (8 - x)\, dx = 26.$

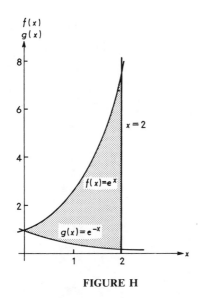

FIGURE H

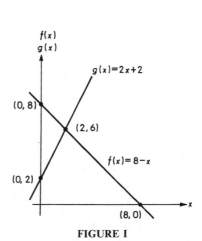

FIGURE I

8. *a*) See Figure J.
   b) $(-1, 3)$; $(2, 6)$.
   c) 4.5.

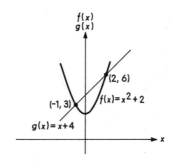

FIGURE J

9. *a*) 50 days.
   b) \$250 thousand.
10. *a*) $t = -1.6094/(-0.01)$. Round this to 161 days.
    b) \$954, taking $e^{-1.61}$ as 0.201.
11. *a*) See Figure K.
    b) \$216 thousand.

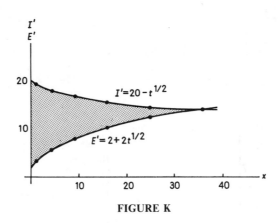

FIGURE K

12. 1/2.   13. 1.   14. 2.8856.
15. 2.   16. 10/3.

17. *a*) $\int_0^{10} (1/500)(30x - 3x^2)\, dx = 1$.
    b) 5.
18. $440\,E(x) = 440(45/22) = \$900$.

19. $-2\left(\dfrac{0.5x + 1}{e^{0.5x}}\right)\Big|_0^{\infty} = -2(0 - 1) = 2$.

Problem Set 4

1. $y = x + C$.
2. $y^2/2 = (x^3/3) + C$.
3. $\ln x = -(1/y) + C$.
4. $y^2/2 = \ln x + C$.
5. $y = (x^2/2) - e^{-x} + C$.
6. $(1/2) \ln (1 - x^2) = \ln y + C$.
7. $y^2/2 = (x^2/2) - x + C$.
8. $\ln y = -x^2 + C$.
9. $P = 10e^t$.
10. $\ln (y + 1) = \ln x$, or $y = x - 1$.
11. $P = 100e^{0.03t}$, so $P = 135$ at time $t = 10$.
12. *a*) $dP(h) = (25 + 5e^{-0.01h+1})\, dh$.
    b) $P(h) = 25h + 500(1 - e^{-0.01h+1})$.
    c) \$5316 thousand.
13. *a*) $dA(t) = 0.062A(t) + 500$.

    b) $A(t) = \dfrac{500(e^{0.062t} - 1)}{0.062}$.

    c) \$6927. (to the nearest dollar).

CHAPTER 12

Problem Set 1

1. *a*) 0.30.	*b*) 0.55.	*c*) 0.	*d*) 0.125.	*e*) 0.
f) 0.03.	*g*) 0.03.	*h*) 0.15.	*i*) 0.20.	*j*) 0.15.
k) 0.50.	*l*) 0.50.	*m*) 0.775.	*n*) 0.47.	*o*) 0.85.

2. *a*) $P(M|A) = 0.3 = P(M)$.
   b) M and A are independent in the probability sense.
   c) Independent because $P(A|L) = 0.15 = P(A)$, or $P(L|A) = 0.20 = P(L)$.
   d) Not independent because $P(B|L) = 0.75$ and $P(B) = 0.55$, so $P(B|L) \neq P(B)$.
   e) $P(A \cap B) = 0$. A and B are mutually exclusive.

3. *a*) Independent. Inasmuch as the first item is *replaced*, the probability that the second item will be a G does not depend upon what occurred when the first item was selected.
   b) D and G are not mutually exclusive because the drawing of D the first time does not exclude the drawing of G the second time. That is, $P(G \cap D) \neq 0$.

4. *a*) Not independent. The probability of drawing G the second time depends upon whether or not D was drawn the first time. That is, if D is drawn the first time, there remain $DGGG$ so the probability of G the second time is $3/4$; however, if G is drawn the first time, leaving $DDGG$, the probability of G the second time is $1/2$.
   b) See Answer to 3 *b*).

5. Mutually exclusive because DG means the first is D and the second is G which excludes the possibility of GD where the first is G and the second is D.

Problem Set 2

1.

Area	L	M	S	
A	0.03	0.045	0.075	0.15
B	0.15	0.125	0.275	0.55
C	0.02	0.130	0.150	0.30
	0.20	0.30	0.50	1.00

2. *a*)

	F	F'	
C	0.24	0.36	0.60
C'	0.16	0.24	0.40
	0.40	0.60	1.00

b) 0.36.   *c*) 0.76.   *d*) 0.
e) 0.40.   *f*) 0.60.   *g*) 0.
h) $P(CF) = 0.24$ and $P(C)P(F) = 0.60(0.40) = 0.24$. Hence, $P(CF) = P(C)P(F)$, so C and F are independent.
i) F and F' are complementary events.

3. *a*) 0.79.
b) As a consequence of these probability assignments, $P(A \cup B) = 1.09$, which is greater than 1, so the candidate should be advised to reassess her subjective probabilities.

4. *a*) $7/15 = 0.467$.   *b*) $1/15 = 0.0667$.   *c*) $7/15 = 0.467$.
d) $1 - P(0 \text{ defective}) = 1 - P(GGG) = 1 - (7/10)(6/9)(5/8) = 0.708$.

5. $P(CS) = 0.35$ because $P(CS) = P(C)P(S|C) = 0.5(0.7) = 0.35$.

6. $P(G|H) = 0.90$, where G means good job and H means honors. This follows from the rule $P(G|H) = P(GH)/P(H) = 0.09/0.10 = 0.90$.

7. *a*) 0.9025.   *b*) 0.0025.   *c*) 0.9975.
d) 0.095.   *e*) 0.95 because of independence.

8. *a*) *HHH, THH, HTH, HHT, TTH, THT, HTT, TTT.*
b) 0, 1, 2, 3.
c)

Event	HHH	THH	HTH	HHT	TTH	THT	HTT	TTT.
Probability	0.125	0.125	0.125	0.125	0.125	0.125	0.125	0.125.

d)

Event	0	1	2	3
Probability	0.125	0.375	0.375	0.125.

e) 0.875.   *f*) 0.50.   *g*) 0.375.   *h*) 0.375.

9. *a*)

Event	W_1M	W_2M	W_1W_2	MW_1	MW_2	W_2W_1.
Probability	1/6	1/6	1/6	1/6	1/6	1/6.

b) 1/3.   *c*) 2/3.   *d*) 1/3.

10. *a*) *WM.*
    b) $P(WM) = P(W)P(M|W) = (2/3)(1/2) = 1/3.$
    c) *WW.*
    d) $P(WW) = P(W)P(W|W) = (2/3)(1/2) = 1/3.$

Problem Set 3

1. *a*) $T_1T_2, T_1F_1, T_1F_2, T_1F_3, T_2F_1, T_2F_2, T_2F_3, F_1F_2, F_1F_3, F_2F_3.$
   b) The events in *a*) are equally likely, probability $\frac{1}{10}$ each. The only way not to have an event on Thursday and an event on Friday is to postpone both Thursday events (T_1T_2). The remaining nine events all provide an event on both days, so the desired probability is $\frac{9}{10}$.
   c) $P(T_1T_2) = P(T_1)P(T_2|T_1) = (2/5)(1/4) = 1/10 = 0.10.$ If T_1T_2 does not occur, there will be an event on both days. Hence, the desired probability is $1 - (1/10) = 9/10.$

2. *a*) $2/5 = 0.40.$ *b*) The complement of *a*), $3/5 = 0.60.$

3. $5/6 = 0.833.$

4. $P(X|Y) = 1/4; P(Y|X) = 1/6.$

5. $P(X) = 1/2; P(Y) = 2/5.$

6. 0.67.

7. *a*) $1/2.$ *b*) $4/7.$ *c*) $23/70.$ *d*) $47/70.$

8. *a*) 0.35.   *b*) 0.60.   *c*) 0.55.   *d*) 0.45.

9. *a*) 0.48.   *b*) 0.16.

10. *a*) $P(CCCC) = (1/3)(1/3)(1/3)(1/3) = 1/81 = 0.0123.$
    b) $P(3 \text{ correct or } 4 \text{ correct})$
       $= P(CCCW) + P(CCWC) + P(CWCC) + P(WCCC) + P(CCCC)$
       $= (2/81) + (2/81) + (2/81) + (2/81) + (1/81) = 1/9 = 0.1111.$
    c) $1 - P(0 \text{ correct}) = 65/81 = 0.8025.$
    d) $4P(CWWW) = 4(8/81) = 32/81 = 0.3951.$

11. *a*) $P(SC) = 0.5; P(S)P(C) = 0.48; P(SC) \neq P(S)P(C).$
    b) 0.12.

12. *a*) $1/4000.$ *b*) $3871/4000.$ *c*) $129/4000.$ *d*) $128/4000.$

13. *a*) $1/12.$ *b*) $7/12.$ *c*) $1/4.$ *d*) $23/36.$

14. 0.141.

15.

	H	M	L	
S	0.08	0.12	0.20	0.40
HP	0.12	0.18	0.30	0.60
	0.20	0.30	0.50	

It is seen that each joint probability equals the product of the corresponding probabilities in the margins of the table.

16.

	H	M	L
Female	20	160	20
Male	10	80	10.

17. 0.56.   18. 0.01099.   19. 0.07831.

20. 0.0034.   21. 0.98.   22. 0.857.

23. 0.664.

Problem Set 4

1. *a*) 0.441.   *b*) 0.657.   *c*) 0.3087.   *d*) 0.0513.
2. 0.992.   3. 0.913.
4. *a*) 0.886.   *b*) 0.655.
5. *a*) 0.760.   *b*) 0.0837.
6. 5.
7. *a*) $1/32 = 0.03125$.   *b*) $3/16 = 0.1875$.
   c) 5/16, 0.3125.   *d*) $1/2 = 0.50$.
   e) $3/16 = 0.1875$.
8. *a*) $1/1024 = 0.000977$.   *b*) $1/64 = 0.0156$.
   c) $45/512 = 0.0879$.   *d*) $53/512 = 0.104$.

9. *a*) $\sum_{x=0}^{15} C_x^{100}(0.05)^x(0.95)^{100-x}$ *b*) $\sum_{x=11}^{200} C_x^{200}(0.05)^x(0.95)^{200-x}$.

10. *a*) 0.993.   *b*) 0.421.
11. *a*) 0.642.   *b*) 0.027.
12. *a*) 0.983.   *b*) 0.006.   *c*) 0.579.
    d) 0.062.   *e*) 0.377.   *f*) 0.617.
13. *a*) 0.115.   *b*) 0.788.   *c*) 1.000 to 3 decimals.
    d) 0.032.   *e*) 0.210.   *f*) 0.002.

Problem Set 5

1. 0.   2. $0.8.
3. A_2.   4. A_4.
5. Make three. EMV = $8.35.
6. *a*) $42.75.
   b) $P(0 \text{ loss}) = 0.99825$ is not stated.
7. The author would choose A_1 in each case, even though A_2 presents no chance
   of loss. Individual readers may differ with the choice of A_1, especially in *c*) where
   there is a 50 percent chance of a loss of $1000. However, $E(A_1)$ in *c*) is $1000,
   compared to $E(A_2) = 250.

CHAPTER 13

Problem Set 1

1. *a*) $\int_0^{10} f(x)\,dx = \int_0^{10} x\,dx = 50$; hence, $p(x) = x/50$.
   b) 0.09, 0.24, 0.09.

2. *a*) $\int_0^9 x^{1/2}\,dx = 18$; hence, $p(x) = x^{1/2}/18$.
   b) 7/27.
3. 19/216.
4. $k = 3/124$.
5. 0.462.

Problem Set 2

1. *a*) 0.1915
   b) 0.1915
   c) 0.1587
   d) 0.1498
   e) 0.2266
   f) 0.3721
   g) 0.1554
   h) 0.1464.

2. *a*) 0.8400
   b) 0.9544
   c) 0.9974
   d) 0.0456
   e) 0.7881
   f) 0.2295
   g) 0.8413
   h) 0.1151.

3. *a*) 0.9772
   b) 0.3023
   c) 0.3085
   d) 0.3413
   e) 0.1498
   f) 0.6826
   g) 0.2266
   h) 0.6247
   i) 0.0013.

Problem Set 3

1. $\bar{x} = 7, s = 2.$
3. $\bar{x} = 11.2, s = 2.17.$
5. $\bar{x} = 0.58, s = 0.0334.$

2. $\bar{x} = 15, s = 2.94.$
4. $\bar{x} = 3.5, s = 2.07.$

Problem Set 4

1. *a*) 52.2.
   d) 46.6 to 53.4.

   b) 56.410.
   e) 58.225.

   c) 41.775.

2. *a*) 0.1587.
   d) 4.56 percent.
   f) 2.2484.

   b) 0.9772.
   e) 1.5730 to 1.5770.
   g) $53.44.

   c) 23 parts.

3. *a*) 0.3085.
   d) 10.56 percent.

   b) 69.15 percent.
   e) 16.465.

   c) 134 cans.

4. *a*) 1908 to 2092.
   c) 1868.

   b) 10.56 percent.
   d) 0.62 percent.

5. *a*) 0.13 percent.
   d) 159.

   b) 5.48 percent.
   e) 99th percentile.

   c) 183 to 217.

APPENDIX 1

Problem Set 1

1. *a*) "*B* is the set whose elements are 2, 4, and 6."
   b) "*C* is the set whose element is 0."
   c) "*Q* is the set whose members are the odd numbers 1 through 29."
   d) "*F* is the set whose elements are the lower case English vowels."
   e) "*L* is the set whose members are George and Charles."
   f) "*K* is the set of positive even integers starting with 2."

2. *a*) $F = \{$Lower case English vowels$\}$.
   b) $S = \{$Positive integral multiples of 3$\}$.
   c) $P = \{$First five positive integers$\}$.
   d) $G = \{$First five upper case English letters$\}$.
   e) $R = \{$Integers greater than 99$\}$.
   f) $J = \{$Negative odd integers from -1 through $-101\}$.
      $= \{$Negative odd integers greater than $-102\}$.

3. *a*) {1, 3, 5, 7}.
   b) {January, February, March, April}.
   c) {14, 28, 42, . . .}.
   d) {5, 10, 15, . . . , 995}.
   e) ϕ.
   f) {1^2, 2^2, 3^2, . . . , 10^2}.

4. ϕ has no elements, whereas {0} has the element 0.

5. *a*) $1/2 \notin$ {1, 2, 3, . . .}.   *b*) $64 \in$ {4, 8, 12, . . .}.
   c) $a \in$ {a, e, i, o, u}.   *d*) $4 \notin$ {1, 3, 5, . . .}.
   e) $\$ \notin$ {$a, b, c, . . . , z$}.   *f*) $g \notin$ {a, b, c, d, e, f}.

6. *a*) {1, 3, 4, 7, 12}.   *b*) {3, 7}.
   c) ϕ.   *d*) {1, 4}.
   e) {1}.   *f*) ϕ.

7. *a*) {$y : y + 2 = 10$} = {8}.
   b) The set of y's such that y plus 5 equals 8 is the set with the single member, 3.

8. *a*) {$m : m - 6 = 4$} = {10}.
   b) The set of q's such that two times q is six is the set with the single member, 3.

9. *a*) The relation is a function because whenever we replace x by a number, there is one and only one corresponding value for y.
   b) Multiply x by five and add four.
   c) The set of ordered pairs, {(x, y)}, whose elements make the sentence $y = 5x + 4$ a true sentence.
   d) The solution set is an infinite set.
   e) (1, 9); (2, 14); (3, 19); (4, 24).

10. ϕ.

11. The relation is not a function because the vertical line intersects the circle at two points. Thus, for any first point, there is not one and only one second point.

Problem Set 2

1. *a*) {a, b, c, d, e, f, g, h}.   *b*) {a, b, c, d, e, i, j}.
   c) {d, e}.   *d*) ϕ.
   e) {d, e, f, g, h, i, j}.   *f*) ϕ.

2. *a*) {0, 1, 2, 3, . . .}.   *b*) ϕ.
   c) {5, 7, 9}.   *d*) ϕ.
   e) {1, 3, 5, . . .}.

3. $A \cup B$ = {face cards}; $A \cap B$ = {queens}.

4. Assuming an ace is a face card, and using the notation $3D$, jC to represent the 3 of diamonds and the jack of clubs, and so on:
   $A \cup B$ = {Face card, $2H$, $3H$, . . . , $10H$, $2D$, $3D$, . . . , $10D$}.
   $A \cap B$ = {jH, qH, kH, aH, jD, qD, kD, aD}.
   $A \cap C$ = ϕ.
   $B \cap C$ = {$9H$, $9D$}.

5. *a*) ϕ.
   b) $M \cap N$ is the point at which the lines intersect.

6. *a*) Rain tomorrow, or warmer tomorrow, or rain and warmer tomorrow.
   b) Rain and warmer tomorrow.

APPENDIX 2

Problem Set 1

1. -14. 2. $+13$. 3. -12. 4. $+3$.
5. $+5.34$. 6. -24. 7. -23. 8. $+3$.
9. -4. 10. -4. 11. $+15$. 12. -3.
13. $-7/2$. 14. $+3/2$. 15. -32. 16. $-28/5$.
17. $+3 + (-4) + (+5) + (-2) + (+6)$.
18. $+3 + (-2) + (+4) + (-5)$.
19. $-1.5 + (-3) + (-2.5)$.
20. $-3 + (+7) + (-4) + (+2)$.
21. -16.
22. 23.
23. -2.
24. 3.
25. 16.

Problem Set 2

1. ab means a times b; cxy means c times x times y.
2. $100a + 10b + c$.
3. All.
4. If a and b are any odd integers, then $a + b$ is an even integer.
5. Whenever we add two clock numbers (in the time sense), the sum is another clock number.
6. By convention, it is agreed that three times the sum of a and c will be indicated by parentheses; thus, $3(a + c)$.
7. Any (whole) number.
8. In $a(b + c)$, we have the product of the number a by the number $(b + c)$. The distributive property says this product equals the sum of ab and ac.
9. Commutative, multiplication.
10. Commutative, addition.
11. Distributive.
12. Commutative, addition.
13. Convention; no sign means assume a plus sign.
14. Commutative, multiplication.
15. Convention; indicate addition of negative by minus sign.
16. Commutative, multiplication.
17. Associative, addition.
18. Commutative, multiplication.
19. Associative, multiplication.
20. Convention; b means $+b$; indicate addition of negative by minus sign.
21. Associative, addition.
22. Distributive property (inverse).

23. Commutative, addition.
24. Commutative, addition.
25. Associative, multiplication.
26. $-b$.
27. $7a - 5b$.
28. $5abc - 2d$.
29. $7 + x$.
30. $8abc$.
31. $8 + 2a - b$.
32. $\quad -b + a$
$= \quad a - b \qquad$ Commutative, addition.
33. $\quad 2b(3a)$
$= 2b3(a) \qquad$ Associative, multiplication
$= 2(3)ba \qquad$ Commutative, multiplication
$= 6ba \qquad$ Associative, multiplication
$= 6ab \qquad$ Commutative, multiplication.
34. $\quad 3 + xy + \ 5 + 3(2ab)$
$= 3 + \ 5 + xy + 3(2ab) \qquad$ Commutative, addition
$= \qquad 8 + xy + 3(2ab) \qquad$ Associative, addition
$= \qquad 8 + xy + 6ab \qquad$ Associative, multiplication.
35. $\quad 3 + a + 2$
$= 3 + 2 + a \qquad$ Commutative, addition
$= \qquad 5 + a \qquad$ Associative, addition.
36. $\quad a + 3 + c + 2$
$= a + c + 3 + 2 \qquad$ Commutative, addition
$= a + c + 5 \qquad$ Associative, addition.
37. $\quad acdb$
$= adcb \qquad$ Commutative, multiplication
$= adbc \qquad$ Commutative, multiplication.
38. $a + b$.
39. $a + (b + c)$.
40. $(a + b) + 2(cd)$.
41. $2(a + b) - 3(c + 2d)$.
42. $2d(a + b + c)$.

Problem Set 3

1. $2abc - 4ab$.
2. $ab + a - 2b - 2$.
3. 11.
4. $ac - bc + 3c + 2a - 2b + 6$.
5. $3a + 3b - 1$.
6. $15a - 62$.
7. $13a + 54$.
8. $3ax - 6bx - 18x - 2a + 4b + 12$.
9. $x + 2b - a$.
10. $2x + 3y$.
11. $a - 3b$.
12. $a + 3x + 2$.
13. $7b - 2ab$.
14. $abc - 2abx + 10ab$.
15. $ax - 2ab + 2b$.
16. $3a - ac - 3bx + bcx$.

17. $ax + bx + x + ay + by + y$.

18. $ab + a - b - 1$.

19. -2.

20. $-2/5$.

21. $+4298.936$.

22. $-4/9$.

23. $+2333$.

24. See text.

25. $b(a - 2)$.

26. $a(3 + 5) = a(8) = 8a$.

27. $2a(2bc - b + 3)$.

28. $x(a - b + 1)$.

29. $a(3d - 5c + 1)$.

30. $2(2uv - xv + 1)$.

31. $ab(x + y - 1)$.

32. $2a(x - 3y + 2z)$.

33. $x(2 + a + b)$.

34. $-a(b + 3c + 1)$.

35. $(x + 1)(a + b)$.

36. $(x + 1)(1 + y)$.

37. $(x + y)(2 - a)$.

38. $(x + y)(a + b)$.

39. $(x - 2)(x + 1)$.

40. $(2x + 1)(x - 5)$.

41. $(4x - 3)(3x - 4)$.

42. $(5x - 1)(2x + 3)$.

43. $(x - 3)(x + 2)$.

44. $(x + 3)(x - 3)$.

45. $(x - y)(x + y)$.

46. $(2x + 3y)(2x - 3y)$.

47. F.

48. T.

49. T.

50. T.

51. F.

52. T.

53. F.

54. F.

55. T.

56. T.

57. F.

58. T.

59. F.

60. T.

Problem Set 4

1. F.

2. F.

3. T.

4. T.

5. T.

6. T.

7. F.

8. F.

9. T.

10. T.

11. T.

12. The product of a number and its reciprocal is one.

13. $3a/2c$.

14. Cancellation not possible as the expression stands.

15. 1.

16. $a + 3$.

17. $6c$.

18. Cancellation not possible as the expression stands.

19. $(2y + 6a + 1)/4y$.

20. $8x$.

21. See text.

22. $3ab/8$.

23. $\dfrac{2a + 2b}{15}$.

24. $-3a/4$.

25. $\dfrac{2ab + 4a}{21}$.

26. $\dfrac{-2}{3a + 3b}$.

27. $a + 6b + 2$.

28. $4a - 6b$.

29. $b(c - d)/2$.

30. $(b - 2)/(a + 3)$.

31. $[2 + (c - b)]/2a$.

32. $[2 + (c - b) + a]/2x(a + b)$.

33. $(x + y)/(y - x)$.

34. $1/3$.

35. $(5a - 6)/10$.

36. $(45b - 5ab + 12a)/30ab$.

37. $(a + 24)/6a$.

38. $7/6$.

39. $(a - 2ab + 4b + bx - 2)/b(a - 2)$.

40. $(x - 2ab + 6)/2a$.

41. $(18abx - 3ab + 1)/6ab$.

42. $(11a + 5b - 3ab - 5)/12(b - 1)$.

43. $-(12x + 17)/4(x + 3)$.

44. $91/12$.

45. $27/176$.

46. $7/6$.

47. $(6b + 2ab)/(5b - 3a)$.

48. $(3abc - 6b + 18)/(2bc - 6c)$.

49. $(4ac - 6b + 12bc)/(bc - 12)$.

50. $(6a + 4b - 2c)/(12b - 3a)$.

Problem Set 5

1. 16.

2. 7.

3. 1.041.

4. 4/3.

5. 1/72.

6. 1000.

7. Not a real number.

8. 31/12.

9. 1/9.

10. 1.

11. 8.

12. 1/2.

13. 10.

14. 1/12.

15. 5/4.

16. 6.

17. 25.

18. 1.1025.

19. 1/8.

20. 13/8.

21. 5.

22. 9/4.

23. 2/9.

24. 5/4.

25. 4/27.

26. a^3.

27. $a^2b^2c^2$.

28. a^3b^7.

29. $a^3b^2c^3$.

30. x^4ab.

31. ab^2.

32. a^2/bc.

33. y^2b^2/x^2.

34. a^2/x.

35. x/y.

36. bc^3.

37. a^2b^3.

38. b^2/a^3c.

39. $-9bc$.

40. $-18/bc$.

41. ax.

42. a^3bc.

43. $1/ax^2$.

44. $2a/x^3$.

45. $1/ax^3$.

46. $3a^{7/3}b^{3/2}$.

47. $x^{3/2}/y^{1/6}$.

48. $a^{1/6}b^{7/6}$.

49. $2/3^{3/2}x^{1/2}$.

50. $2y^{11/6}/3x^2$.

51. $\dfrac{2x^2 + 1}{x^3 + 3x^2}$.

52. $\dfrac{2 - x}{x - x^2}$.

53. $\dfrac{x + 1}{3x^2}$.

54. $\dfrac{x^2}{a^{2x-2}}$.

55. Exponents in this expression cannot be combined.

56. $a^2 + ab$.

57. $a^2 - 6ab + 9b^2$.

58. $a^3 - 2a^2b + ab^2$.

59. $x^2 - y^2$.

60. $1 + a$.

61. $a - 2a^{1/2} + 1$.

62. $\dfrac{1 + 2a}{a^2}$.

63. 27.

64. F.

65. F.

66. T.

67. T.

68. F.

69. F.

70. T.

71. T.

72. F.

73. F.

74. T.

75. F.

76. F.

77. T.

78. F.

79. T.

80. F.

81. T.

82. T.

83. T.

84. F.

85. T.

APPENDIX 3

Problem Set 1

1. $2x - 3 = x + 4$ Subtract x
   $x - 3 = 4$ Add 3
   $x = 7$.

2. $x/3 + 1/2 = 3x$ Multiply by 6
   $2x + 3 = 18x$ Subtract $18x$
   $-16x + 3 = 0$ Subtract 3
   $-16x = -3$ Divide by -16
   $x = 3/16$.

3. $3x - 7 = 2x + 4$ Subtract $2x$
     $x - 7 = 4$ Add 7
        $x = 11.$

4. $(x + 3)/2 = x - 1/4$ Multiply by 4
    $2(x + 3) = 4x - 1$ Apply distributive property
    $2x + 6 = 4x - 1$ Subtract $4x$
   $-2x + 6 = -1$ Subtract 6
       $-2x = -7$ Divide by -2
         $x = 7/2.$

5. $7x - 5 = 3 - 4x$ Add $4x$
  $11x - 5 = 3$ Add 5
     $11x = 8$ Divide by 11
       $x = 8/11.$

6. $ax + 2 - x = 0$ Subtract 2
     $ax - x = -2$ Factor
   $x(a - 1) = -2$ Divide by $a - 1$
        $x = -2/(a - 1)$ Change signs of numerator and denominator

        $x = 2/(1 - a).$

7. $ax + b = cx$ Subtract cx
  $ax - cx + b = 0$ Subtract b
    $ax - cx = -b$ Factor
    $x(a - c) = -b$ Divide by $a - c$,
        $x = b/(c - a).$ change signs

8. $ax + b = x - b$ Subtract b
       $ax = x - 2b$ Subtract x
   $ax - x = -2b$ Factor
 $x(a - 1) = -2b$ Divide by $a - 1$
       $x = -2b/(a - 1)$ Change signs of numerator and denominator

       $x = 2b/(1 - a).$

9. $a(x - a) = 2x$ Apply distributive property
   $ax - a^2 = 2x$ Add a^2
      $ax = 2x + a^2$ Subtract $2x$
 $ax - 2x = a^2$ Factor
$x(a - 2) = a^2$ Divide by $a - 2$
      $x = a^2/(a - 2).$

10. $x/a - 1/2 = 2x$ Multiply by $2a$
    $2x - a = 4ax$ Add a
      $2x = 4ax + a$ Subtract $4ax$
$2x - 4ax = a$ Factor
$x(2 - 4a) = a$ Divide by $2 - 4a$
       $x = a/(2 - 4a).$

11. $2/(a - x) + 1/3 = 4$ Multiply by $3(a - x)$

      $6 + a - x = 12(a - x)$ Apply distributive property

      $6 + a - x = 12a - 12x$ Add $12x$

      $6 + a + 11x = 12a$ Subtract $6 + a$

          $11x = 11a - 6$ Divide by 11

            $x = (11a - 6)/11.$

12. $y = x/(b - cx)$ Multiply by $b - cx$

     $y(b - cx) = x$ Apply distributive property

      $yb - ycx = x$ Add ycx

         $yb = x + ycx$ Factor

         $yb = x(1 + yc)$ Divide by $1 + yc$

   $yb/(1 + yc) = x$

or

          $x = yb/(1 + yc).$

13. $3/4 - 2x/3 = 2x(a - 1)$ Multiply by 12

      $9 - 8x = 24x(a - 1)$ Apply distributive property

      $9 - 8x = 24ax - 24x$ Add $8x$

          $9 = 24ax - 16x$ Factor

          $9 = x(24a - 16)$ Divide by $24a - 16$

   $9/(24a - 16) = x$

or

          $x = 9/(24a - 16).$

14. $3(x - 2) = 2 - a(x + 2)$ Apply distributive property

      $3x - 6 = 2 - ax - 2a$ Add ax

   $3x + ax - 6 = 2 - 2a$ Add 6

     $3x + ax = 8 - 2a$ Factor

     $x(3 + a) = 8 - 2a$ Divide by $3 + a$

          $x = (8 - 2a)/(3 + a).$

15. $2/3(x - 2) + 3/a - 1/2 = 0$ Multiply by $6a(x - 2)$

  $4a + 18(x - 2) - 3a(x - 2) = 0$ Apply distributive property

  $4a + 18x - 36 - 3ax + 6a = 0$ Subtract $10a$

         $18x - 36 - 3ax = -10a$ Add 36

           $18x - 3ax = 36 - 10a$ Factor

          $x(18 - 3a) = 36 - 10a$ Divide by $18 - 3a$

               $x = (36 - 10a)/$

                   $(18 - 3a).$

16. $ax - b/2 = c + 5[a - 2(b - x)]/6$ Multiply by 6

     $6ax - 3b = 6c + 5[a - 2(b - x)]$ Apply distributive property

     $6ax - 3b = 6c + 5[a - 2b + 2x]$ Apply distributive property

     $6ax - 3b = 6c + 5a - 10b + 10x$ Subtract $10x$

  $6ax - 10x - 3b = 6c + 5a - 10b$ Add $3b$

     $6ax - 10x = 6c + 5a - 7b$ Factor

    $x(6a - 10) = 6c + 5a - 7b$ Divide by $6a - 10$

           $x = (6c + 5a - 7b)/$

              $(6a - 10).$

17. $b/a - x = 2a(b - x)$ Multiply by a
        $b - ax = 2a^2(b - x)$ Apply distributive property
        $b - ax = 2a^2b - 2a^2x$ Add $2a^2x$
$2a^2x + b - ax = 2a^2b$ Subtract b
     $2a^2x - ax = 2a^2b - b$ Factor
     $x(2a^2 - a) = 2a^2b - b$ Divide by $2a^2 - a$
              $x = (2a^2b - b)/(2a^2 - a)$.

18. $x - a(b - x) = 2x - 3$ Apply distributive property
      $x - ab + ax = 2x - 3$ Subtract $2x$
      $-x - ab + ax = -3$ Add ab
         $-x + ax = -3 + ab$ Factor
        $x(-1 + a) = -3 + ab$ Divide by $(-1 + a)$
               $x = (-3 + ab)/(-1 + a)$
               $x = (ab - 3)/(a - 1)$.

19. $3(b - x) = 2 + b[x - (3 - x)]$ Apply distributive property
         $3b - 3x = 2 + b[x - 3 + x]$ Apply distributive property
         $3b - 3x = 2 + bx - 3b + bx$ Subtract $2bx$
 $3b - 3x - 2bx = 2 - 3b$ Subtract $3b$
    $-3x - 2bx = 2 - 6b$ Factor
    $x(-3 - 2b) = 2 - 6b$ Divide by $-3 - 2b$
            $x = (2 - 6b)/(-3 - 2b)$ Change signs of numerator and
                                                 denominator

            $x = (6b - 2)/(2b + 3)$.

20. $b(a + x) = a(b + x)$ Apply distributive property
         $ba + bx = ab + ax$ Subtract ax
    $ba + bx - ax = ab$ Subtract ba
       $bx - ax = 0$ Factor
       $x(b - a) = 0$ Divide by $b - a$
            $x = 0$.

Problem Set 2

1. 15.   2. 41.
3. 180.   4. 2.
5. 104.   6. 30.
7. 64.   8. 9.
9. 32.   10. 472.
11. 9/40.   12. 5/42.
13. 115/567.   14. $-11/60$.
15. 183/280.   16. $-4435/192$.
17. -27.   18. 2/3.
19. 16/21.   20. $-122/453$.
21. $33,083.33.   22. 0.50.
23. *a*) 37.5 percent.
    b) $6.00.
    c) $4.48.

24. *a*) 320 days.
    b) $16,200.
25. *a*) 212°F.
    b) 32°F.
    c) 12.92°F.
    d) 82.4°F.
    e) 14°F.
    f) 122°F.
26. *a*) $E = 0.2T + 0.125(T - 50,000)$.
    b) $50,000; $65,384.62; $80,769.23.
    c) $83,333.33.
27. *a*) $B = x[P - y(P - B)]$.
    b) $362,500.
28. *a*) 22.9 percent.
    b) $3.03.
29. *a*) $20,000.   *b*) 0.2.   *c*) 0.375.
30. *a*) $10.01. *b*) $5.01.   *c*) 100.   *d*) 250.   *e*) 500.

Problem Set 3

Problems 1–10: See Figures A through J.

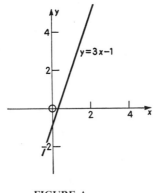

FIGURE A

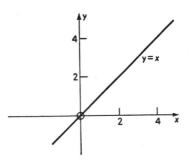

FIGURE B

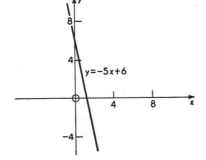

FIGURE C

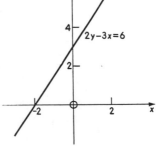

FIGURE D

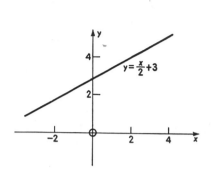

FIGURE E

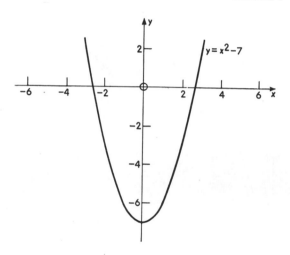

FIGURE F

FIGURE G

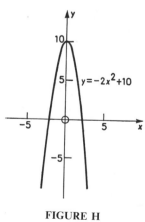

FIGURE H

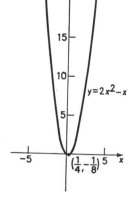

FIGURE I

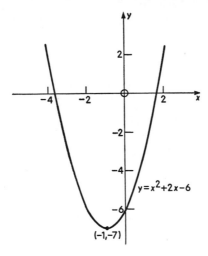

FIGURE J

11. $x_1 = 9.69$; $x_2 = 0.31$.
12. $x_1 = 0.35$; $x_2 = -2.85$.
13. $x_1 = x_2 = 3/2$.
14. No real solutions.
15. $x_1 = 0$; $x_2 = 2$.
16. $x_1 = 2$; $x_2 = 1$.
17. $x_1 = 0$; $x_2 = 2$.
18. $x_1 = -5/2$; $x_2 = 4/3$.
19. $x_1 = 5$; $x_2 = -5$.
20. $x_1 = -7/2$; $x_2 = 1$.

Tables

TABLE I
Common Logarithms: 100–549

N	0	1	2	3	4	5	6	7	8	9
10	0000	0043	0086	0128	0170	0212	0253	0294	0334	0374
11	0414	0453	0492	0531	0569	0607	0645	0682	0719	0755
12	0792	0828	0864	0899	0934	0969	1004	1038	1072	1106
13	1139	1173	1206	1239	1271	1303	1335	1367	1399	1430
14	1461	1492	1523	1553	1584	1614	1644	1673	1703	1732
15	1761	1790	1818	1847	1875	1903	1931	1959	1987	2014
16	2041	2068	2095	2122	2148	2175	2201	2227	2253	2279
17	2304	2330	2355	2380	2405	2430	2455	2480	2504	2529
18	2553	2577	2601	2625	2648	2672	2695	2718	2742	2765
19	2788	2810	2833	2856	2878	2900	2923	2945	2967	2989
20	3010	3032	3054	3075	3096	3118	3139	3160	3181	3201
21	3222	3243	3263	3284	3304	3324	3345	3365	3385	3404
22	3424	3444	3464	3483	3502	3522	3541	3560	3579	3598
23	3617	3636	3655	3674	3692	3711	3729	3747	3766	3784
24	3802	3820	3838	3856	3874	3892	3909	3927	3945	3962
25	3979	3997	4014	4031	4048	4065	4082	4099	4116	4133
26	4150	4166	4183	4200	4216	4232	4249	4265	4281	4298
27	4314	4330	4346	4362	4378	4393	4409	4425	4440	4456
28	4472	4487	4502	4518	4533	4548	4564	4579	4594	4609
29	4624	4639	4654	4669	4683	4698	4713	4728	4742	4757
30	4771	4786	4800	4814	4829	4843	4857	4871	4886	4900
31	4914	4928	4942	4955	4969	4983	4997	5011	5024	5038
32	5051	5065	5079	5092	5105	5119	5132	5145	5159	5172
33	5185	5198	5211	5224	5237	5250	5263	5276	5289	5302
34	5315	5328	5340	5353	5366	5378	5391	5403	5416	5428
35	5441	5453	5465	5478	5490	5502	5514	5527	5539	5551
36	5563	5575	5587	5599	5611	5623	5635	5647	5658	5670
37	5682	5694	5705	5717	5729	5740	5752	5763	5775	5786
38	5798	5809	5821	5832	5843	5855	5866	5877	5888	5899
39	5911	5922	5933	5944	5955	5966	5977	5988	5999	6010
40	6021	6031	6042	6053	6064	6075	6085	6096	6107	6117
41	6128	6138	6149	6160	6170	6180	6191	6201	6212	6222
42	6232	6243	6253	6263	6274	6284	6294	6304	6314	6325
43	6336	6345	6355	6365	6375	6385	6395	6405	6415	6425
44	6435	6444	6454	6464	6474	6484	6493	6503	6513	6522
45	6532	6542	6551	6561	6571	6580	6590	6599	6609	6618
46	6628	6637	6646	6656	6665	6675	6684	6693	6702	6712
47	6721	6730	6739	6749	6758	6767	6776	6785	6794	6803
48	6812	6821	6830	6839	6848	6857	6866	6875	6884	6893
49	6902	6911	6920	6928	6937	6946	6955	6964	6972	6981
50	6990	6998	7007	7016	7024	7033	7042	7050	7059	7067
51	7076	7084	7093	7101	7110	7118	7126	7135	7143	7152
52	7160	7168	7177	7185	7193	7202	7210	7218	7226	7235
53	7243	7251	7259	7267	7275	7284	7292	7300	7308	7316
54	7324	7332	7340	7348	7356	7364	7372	7380	7388	7396

Taken by permission from Ernest Kurnow, Gerald J. Glasser, and Frederick R. Ottman, *Statistics for Business Decisions* (Homewood, Ill.: Richard D. Irwin, Inc., 1959), pp. 507–8.

TABLE I—(*continued*)
Common Logarithms: 550–999

N	0	1	2	3	4	5	6	7	8	9
55	7404	7412	7419	7427	7435	7443	7451	7459	7466	7474
56	7482	7490	7497	7505	7513	7520	7528	7536	7543	7551
57	7559	7566	7574	7582	7589	7597	7604	7612	7619	7627
58	7634	7642	7649	7657	7664	7672	7679	7686	7694	7701
59	7709	7716	7723	7731	7738	7745	7752	7760	7767	7774
60	7782	7789	7796	7803	7810	7818	7825	7832	7839	7846
61	7853	7860	7868	7875	7882	7889	7896	7903	7910	7917
62	7924	7931	7938	7945	7952	7959	7966	7973	7980	7987
63	7993	8000	8007	8014	8021	8028	8035	8041	8048	8055
64	8062	8069	8075	8082	8089	8096	8102	8109	8116	8122
65	8129	8136	8142	8149	8156	8162	8169	8176	8182	8189
66	8195	8202	8209	8215	8222	8228	8235	8241	8248	8254
67	8261	8267	8274	8280	8287	8293	8299	8306	8312	8319
68	8325	8331	8338	8344	8351	8357	8363	8370	8376	8382
69	8388	8395	8401	8407	8414	8420	8426	8432	8439	8445
70	8451	8457	8463	8470	8476	8482	8488	8494	8500	8506
71	8513	8519	8525	8531	8537	8543	8549	8555	8561	8567
72	8573	8579	8585	8591	8597	8603	8609	8615	8621	8627
73	8633	8639	8645	8651	8657	8663	8669	8675	8681	8686
74	8692	8698	8704	8710	8716	8722	8727	8733	8739	8745
75	8751	8756	8762	8768	8774	8779	8785	8791	8797	8802
76	8808	8814	8820	8825	8831	8837	8842	8848	8854	8859
77	8865	8871	8876	8882	8887	8893	8899	8904	8910	8915
78	8921	8927	8932	8938	8943	8949	8954	8960	8965	8971
79	8976	8982	8987	8993	8998	9004	9009	9015	9020	9025
80	9031	9036	9042	9047	9053	9058	9063	9069	9074	9079
81	9085	9090	9096	9101	9106	9112	9117	9122	9128	9133
82	9138	9143	9149	9154	9159	9165	9170	9175	9180	9186
83	9191	9196	9201	9206	9212	9217	9222	9227	9232	9238
84	9243	9248	9253	9258	9263	9269	9274	9279	9284	9289
85	9294	9299	9304	9309	9315	9320	9325	9330	9335	9340
86	9345	9350	9355	9360	9365	9370	9375	9380	9385	9390
87	9395	9400	9405	9410	9415	9420	9425	9430	9435	9440
88	9445	9450	9455	9460	9465	9469	9474	9479	9484	9489
89	9494	9499	9504	9509	9513	9518	9523	9528	9533	9538
90	9542	9547	9552	9557	9562	9566	9571	9576	9581	9586
91	9590	9595	9600	9605	9609	9614	9619	9624	9628	9633
92	9638	9643	9647	9652	9657	9661	9666	9671	9675	9680
93	9685	9689	9694	9699	9703	9708	9713	9717	9722	9727
94	9731	9736	9741	9745	9750	9754	9759	9763	9768	9773
95	9777	9782	9786	9791	9795	9800	9805	9809	9814	9818
96	9823	9827	9832	9836	9841	9845	9850	9854	9859	9863
97	9868	9872	9877	9881	9886	9890	9894	9899	9903	9908
98	9912	9917	9921	9926	9930	9934	9939	9943	9948	9952
99	9956	9961	9965	9969	9974	9978	9983	9987	9991	9996

TABLE II
Compound Amount of $1
$$(1 + i)^n$$

n	1%	2%	3%	4%
1 . . . . . 1.01000	1.02000	1.03000	1.04000
2 . . . . . 1.02010	1.04040	1.06090	1.08160
3 . . . . . 1.03030	1.06121	1.09273	1.12486
4 . . . . . 1.04060	1.08243	1.12551	1.16986
5 . . . . . 1.05101	1.10408	1.15927	1.21665
6 . . . . . 1.06152	1.12616	1.19405	1.26532
7 . . . . . 1.07214	1.14869	1.22987	1.31593
8 . . . . . 1.08286	1.17166	1.26677	1.36857
9 . . . . . 1.09369	1.19509	1.30477	1.42331
10 . . . . . 1.10462	1.21899	1.34392	1.48024
11 . . . . . 1.11567	1.24337	1.38423	1.53945
12 . . . . . 1.12683	1.26824	1.42576	1.60103
13 . . . . . 1.13809	1.29361	1.46853	1.66507
14 . . . . . 1.14947	1.31948	1.51259	1.73168
15 . . . . . 1.16097	1.34587	1.55797	1.80094
16 . . . . . 1.17258	1.37279	1.60471	1.87298
17 . . . . . 1.18430	1.40024	1.65285	1.94790
18 . . . . . 1.19615	1.42825	1.70243	2.02582
19 . . . . . 1.20811	1.45681	1.75351	2.10685
20 . . . . . 1.22019	1.48595	1.80611	2.19112
21 . . . . . 1.23239	1.51567	1.86029	2.27877
22 . . . . . 1.24472	1.54598	1.91610	2.36992
23 . . . . . 1.25716	1.57690	1.97359	2.46472
24 . . . . . 1.26973	1.60844	2.03279	2.56330
25 . . . . . 1.28243	1.64061	2.09378	2.66584
26 . . . . . 1.29526	1.67342	2.15659	2.77247
27 . . . . . 1.30821	1.70689	2.22129	2.88337
28 . . . . . 1.32129	1.74102	2.28793	2.99870
29 . . . . . 1.33450	1.77584	2.35657	3.11865
30 . . . . . 1.34785	1.81136	2.42726	3.24340
31 . . . . . 1.36133	1.84759	2.50008	3.37313
32 . . . . . 1.37494	1.88454	2.57508	3.50806
33 . . . . . 1.38869	1.92223	2.65234	3.64838
34 . . . . . 1.40258	1.96068	2.73191	3.79432
35 . . . . . 1.41660	1.99989	2.81386	3.94609
36 . . . . . 1.43077	2.03989	2.89828	4.10393
37 . . . . . 1.44508	2.08069	2.98523	4.26809
38 . . . . . 1.45953	2.12230	3.07478	4.43881
39 . . . . . 1.47412	2.16474	3.16703	4.61637
40 . . . . . 1.48886	2.20804	3.26204	4.80102

TABLE II—(*continued*)

Compound Amount of $1

$$(1 + i)^n$$

n	5%	6%	7%	8%
1	1.05000	1.06000	1.07000	1.08000
2	1.10250	1.12360	1.14490	1.16640
3	1.15762	1.19102	1.22504	1.25971
4	1.21551	1.26248	1.31080	1.36049
5	1.27628	1.33823	1.40255	1.46933
6	1.34010	1.41852	1.50073	1.58687
7	1.40710	1.50363	1.60578	1.71382
8	1.47746	1.59385	1.71819	1.85093
9	1.55133	1.68975	1.83846	1.99900
10	1.62889	1.79085	1.96715	2.15892
11	1.71034	1.89830	2.10485	2.33164
12	1.79586	2.01220	2.25219	2.51817
13	1.88565	2.13293	2.40985	2.71962
14	1.97993	2.26090	2.57853	2.93719
15	2.07893	2.39656	2.75903	3.17217
16	2.18287	2.54035	2.95216	3.42594
17	2.29202	2.69277	3.15882	3.70002
18	2.40662	2.85434	3.37993	3.99602
19	2.52695	3.02560	3.61653	4.31570
20	2.65330	3.20714	3.86968	4.66096
21	2.78596	3.39956	4.14056	5.03383
22	2.92526	3.60354	4.43040	5.43654
23	3.07152	3.81975	4.74053	5.87146
24	3.22510	4.04893	5.07237	6.34118
25	3.38635	4.29187	5.42743	6.84848
26	3.55567	4.54938	5.80735	7.39635
27	3.73346	4.82235	6.21387	7.98806
28	3.92013	5.11169	6.64884	8.62711
29	4.11614	5.41839	7.11426	9.31727
30	4.32194	5.74349	7.61226	10.06266
31	4.53804	6.08810	8.14511	10.86767
32	4.76494	6.45339	8.71527	11.73708
33	5.00319	6.84059	9.32534	12.67605
34	5.25335	7.25103	9.97811	13.69013
35	5.51602	7.68609	10.67658	14.78534
36	5.79182	8.14725	11.42394	15.96817
37	6.08141	8.63608	12.22362	17.24563
38	6.38548	9.15425	13.07927	18.62528
39	6.70475	9.70351	13.99482	20.11530
40	7.03999	10.28572	14.97446	21.72452

TABLE III
Present Value of $1
$$(1 + i)^{-n}$$

n	1%	2%	3%	4%
1	0.990099	0.980392	0.970874	0.961538
2	0.980296	0.961169	0.942596	0.924556
3	0.970590	0.942322	0.915142	0.888996
4	0.960980	0.923845	0.888487	0.854804
5	0.951466	0.905731	0.862609	0.821927
6	0.942045	0.887971	0.837484	0.790315
7	0.932718	0.870560	0.831092	0.759918
8	0.923483	0.853490	0.789409	0.730690
9	0.914340	0.836755	0.766417	0.702587
10	0.905287	0.820348	0.744094	0.675564
11	0.896324	0.804263	0.722421	0.649581
12	0.887449	0.788493	0.701380	0.624597
13	0.878663	0.773033	0.680951	0.600574
14	0.869963	0.757875	0.661118	0.577475
15	0.861349	0.743015	0.641862	0.555265
16	0.852821	0.728446	0.623167	0.533908
17	0.844377	0.714163	0.605016	0.513373
18	0.836017	0.700159	0.587395	0.493628
19	0.827740	0.686431	0.570286	0.474642
20	0.819544	0.672971	0.553676	0.456387
21	0.811430	0.659776	0.537549	0.438834
22	0.803396	0.646839	0.521893	0.421955
23	0.795442	0.634156	0.506692	0.405726
24	0.787566	0.621721	0.491934	0.390121
25	0.779768	0.609531	0.477606	0.375117
26	0.772048	0.597579	0.463695	0.360689
27	0.764404	0.585862	0.450189	0.346817
28	0.756836	0.574375	0.437077	0.333477
29	0.749342	0.563112	0.424346	0.320651
30	0.741923	0.552071	0.411987	0.308319
31	0.734577	0.541246	0.399987	0.296460
32	0.727304	0.530633	0.388337	0.285058
33	0.720103	0.520229	0.377026	0.274094
34	0.712973	0.510028	0.366045	0.263552
35	0.705914	0.500028	0.355383	0.253415
36	0.698925	0.490223	0.345032	0.243669
37	0.692005	0.480611	0.334983	0.234297
38	0.685153	0.471187	0.325226	0.225285
39	0.678370	0.461948	0.315754	0.216621
40	0.671653	0.452890	0.306557	0.208289

TABLE III–(*continued*)
Present Value of $1

$$(1 + i)^{-n}$$

n	5%	6%	7%	8%
1	0.952381	0.943396	0.934579	0.925926
2	0.907029	0.889996	0.873439	0.857339
3	0.863838	0.839619	0.816298	0.793832
4	0.822702	0.792094	0.762895	0.735030
5	0.783526	0.747258	0.712986	0.680583
6	0.746215	0.704961	0.666342	0.630170
7	0.710681	0.665057	0.622750	0.583490
8	0.676839	0.627412	0.582009	0.540269
9	0.644609	0.591898	0.543934	0.500249
10	0.613913	0.558395	0.508349	0.463193
11	0.584679	0.526788	0.475093	0.428883
12	0.556837	0.496969	0.444012	0.397114
13	0.530321	0.468839	0.414964	0.367698
14	0.505068	0.442301	0.387817	0.340461
15	0.481017	0.417265	0.362446	0.315242
16	0.458112	0.393646	0.338735	0.291890
17	0.436297	0.371364	0.316574	0.270269
18	0.415521	0.350344	0.295864	0.250249
19	0.395734	0.330513	0.276508	0.231712
20	0.376889	0.311805	0.258419	0.214548
21	0.358942	0.294155	0.241513	0.198656
22	0.341850	0.277505	0.225713	0.183941
23	0.325571	0.261797	0.210947	0.170315
24	0.310068	0.246979	0.197147	0.157699
25	0.295303	0.232999	0.184249	0.146018
26	0.281241	0.219810	0.172195	0.135202
27	0.267848	0.207368	0.160930	0.125187
28	0.255094	0.195630	0.150402	0.115914
29	0.242946	0.184557	0.140563	0.107328
30	0.231377	0.174110	0.131367	0.099377
31	0.220359	0.164255	0.122773	0.092016
32	0.209866	0.154957	0.114741	0.085200
33	0.199873	0.146186	0.107235	0.078889
34	0.190355	0.137912	0.100219	0.073045
35	0.181290	0.130105	0.093663	0.067635
36	0.172657	0.122741	0.087535	0.062625
37	0.164436	0.115793	0.081809	0.057986
38	0.156605	0.109239	0.076457	0.053690
39	0.149148	0.103056	0.071455	0.049713
40	0.142046	0.097222	0.066780	0.046031

TABLE IV
Amount of $1 per Period

$$s_{\overline{n}|i} = \frac{(1 + i)^n - 1}{i}$$

n	1%	2%	3%	4%
1	1.00000	1.00000	1.00000	1.00000
2	2.01000	2.02000	2.03000	2.04000
3	3.03010	3.06040	3.09090	3.12160
4	4.06040	4.12161	4.18363	4.24646
5	5.10101	5.20404	5.30914	5.41632
6	6.15202	6.30812	6.46841	6.63298
7	7.21354	7.43428	7.66246	7.89829
8	8.28567	8.58297	8.89234	9.21423
9	9.36853	9.75463	10.15911	10.58280
10	10.46221	10.94972	11.46388	12.00611
11	11.56683	12.16872	12.80780	13.48635
12	12.68250	13.41209	14.19203	15.02581
13	13.80933	14.68033	15.61779	16.62684
14	14.94742	15.97394	17.08632	18.29191
15	16.09690	17.29342	18.59891	20.02359
16	17.25786	18.63929	20.15688	21.82453
17	18.43044	20.01207	21.76159	23.69751
18	19.61475	21.41231	23.41444	25.64541
19	20.81090	22.84056	25.11687	27.67123
20	22.01900	24.29737	26.87037	29.77808
21	23.23919	25.78332	28.67649	31.96920
22	24.47159	27.29898	30.53678	34.24797
23	25.71630	28.84496	32.45288	36.61789
24	26.97346	30.42186	34.42647	39.08260
25	28.24320	32.03030	36.45926	41.64591
26	29.52563	33.67091	38.55304	44.31174
27	30.82089	35.34432	40.70963	47.08421
28	32.12910	37.05121	42.93092	49.96758
29	33.45039	38.79223	45.21885	52.96629
30	34.78489	40.56808	47.57542	56.08494
31	36.13274	42.37944	50.00268	59.32834
32	37.49407	44.22703	52.50276	62.70147
33	38.86901	46.11157	55.07784	66.20953
34	40.25770	48.03380	57.73018	69.85791
35	41.66028	49.99448	60.46208	73.65222
36	43.07688	51.99437	63.27594	77.59831
37	44.50765	54.03425	66.17422	81.70225
38	45.95272	56.11494	69.15945	85.97034
39	47.41225	58.23724	72.23423	90.40915
40	48.88637	60.40198	75.40126	95.02552

TABLE IV—(*continued*)
Amount of $1 per Period

$$s_{\overline{n}|i} = \frac{(1 + i)^n - 1}{i}$$

n	5%	6%	7%	8%
1	1.00000	1.00000	1.00000	1.00000
2	2.05000	2.06000	2.07000	2.08000
3	3.15250	3.18360	3.21490	3.24640
4	4.31012	4.37462	4.43994	4.50611
5	5.52563	5.63709	5.75074	5.86660
6	6.80191	6.97532	7.15329	7.33593
7	8.14201	8.39384	8.65402	8.92280
8	9.54911	9.89747	10.25980	10.63663
9	11.02656	11.49132	11.97799	12.48756
10	12.57789	13.18079	13.81645	14.48656
11	14.20679	14.97164	15.78360	16.64549
12	15.91713	16.86994	17.88845	18.97713
13	17.71298	18.88214	20.14064	21.49530
14	19.59863	21.01507	22.55049	24.21492
15	21.57856	23.27597	25.12902	27.15211
16	23.65749	25.67253	27.88805	30.32428
17	25.84037	28.21288	30.84022	33.75023
18	28.13238	30.90565	33.99903	37.45024
19	30.53900	33.75999	37.37896	41.44626
20	33.06595	36.78559	40.99549	45.76196
21	35.71925	39.99273	44.86518	50.42292
22	38.50521	43.39229	49.00574	55.45676
23	41.43048	46.99583	53.43614	60.89330
24	44.50200	50.81558	58.17667	66.76476
25	47.72710	54.86451	63.24904	73.10594
26	51.11345	59.15638	68.67647	79.95442
27	54.66913	63.70577	74.48382	87.35077
28	58.40258	68.52811	80.69769	95.33883
29	62.32271	73.63980	87.34653	103.96594
30	66.43885	79.05819	94.46079	113.28321
31	70.76079	84.80168	102.07304	123.34587
32	75.29883	90.88978	110.21815	134.21354
33	80.06377	97.34316	118.93343	145.95062
34	85.06696	104.18375	128.25876	158.62667
35	90.32031	111.43478	138.23688	172.31680
36	95.83632	119.12087	148.91346	187.10215
37	101.62814	127.26812	160.33740	203.07032
38	107.70955	135.90421	172.56102	220.31595
39	114.09502	145.05846	185.64029	238.94122
40	120.79977	154.76197	199.63511	259.05652

TABLE V
Present Value of $1 per Period

$$a_{\overline{n}|i} = \frac{1 - (1 + i)^{-n}}{i}$$

n	1%	2%	3%	4%
1	0.99010	0.98039	0.97087	0.96154
2	1.97040	1.94156	1.91347	1.88609
3	2.94099	2.88388	2.82861	2.77509
4	3.90197	3.80773	3.71710	3.62990
5	4.85343	4.71346	4.57971	4.45182
6	5.79548	5.60143	5.41719	5.24214
7	6.72819	6.47199	6.23028	6.00205
8	7.65168	7.32548	7.01969	6.73274
9	8.56602	8.16224	7.78611	7.43533
10	9.47130	8.98259	8.53020	8.11090
11	10.36763	9.78685	9.25262	8.76048
12	11.25508	10.57534	9.95400	9.38507
13	12.13374	11.34837	10.63496	9.98565
14	13.00370	12.10625	11.29607	10.56312
15	13.86505	12.84926	11.93794	11.11839
16	14.71787	13.57771	12.56110	11.65230
17	15.56225	14.29187	13.16612	12.16567
18	16.39827	14.99203	13.75351	12.65930
19	17.22601	15.67846	14.32380	13.13394
20	18.04555	16.35143	14.87747	13.59033
21	18.85698	17.01121	15.41502	14.02916
22	19.66038	17.65805	15.93692	14.45112
23	20.45582	18.29220	16.44361	14.85684
24	21.24339	18.91393	16.93554	15.24696
25	22.02316	19.52346	17.41315	15.62208
26	22.79520	20.12104	17.87684	15.98277
27	23.55961	20.70690	18.32703	16.32959
28	24.31644	21.28127	18.76411	16.66306
29	25.06579	21.84438	19.18845	16.98371
30	25.80771	22.39646	19.60044	17.29203
31	26.54229	22.93770	20.00043	17.58849
32	27.26959	23.46833	20.38877	17.87355
33	27.98969	23.98856	20.76579	18.14765
34	28.70267	24.49859	21.13184	18.41120
35	29.40858	24.99862	21.48722	18.66461
36	30.10751	25.48884	21.83225	18.90828
37	30.79951	25.96945	22.16724	19.14258
38	31.48466	26.44064	22.49246	19.36786
39	32.16303	26.90259	22.80822	19.58448
40	32.83469	27.35548	23.11477	19.79277

TABLE V—(*continued*)
Present Value of $1 per Period

$$a_{\overline{n}|\,i} = \frac{1 - (1 + i)^{-n}}{i}$$

n	5%	6%	7%	8%
1	0.95238	0.94340	0.93458	0.92593
2	1.85941	1.83339	1.80802	1.78326
3	2.72325	2.67301	2.62432	2.57710
4	3.54595	3.46511	3.38721	3.31213
5	4.32948	4.21236	4.10020	3.99271
6	5.07569	4.91732	4.76654	4.62288
7	5.78637	5.58238	5.38929	5.20637
8	6.46321	6.20979	5.97130	5.74664
9	7.10782	6.80169	6.51523	6.24689
10	7.72173	7.36009	7.02358	6.71008
11	8.30641	7.88687	7.49867	7.13896
12	8.86325	7.38384	7.94269	7.53608
13	9.39357	8.85268	8.35765	7.90378
14	9.89864	9.29498	8.74547	8.24424
15	10.37966	9.71225	9.10791	8.55948
16	10.83777	10.10590	9.44665	8.85137
17	11.27407	10.47726	9.76322	9.12164
18	11.68959	10.82760	10.05909	9.37189
19	12.08532	11.15812	10.33560	9.60360
20	12.46221	11.46992	10.59401	9.81815
21	12.82115	11.76408	10.83553	10.01680
22	13.16300	12.04158	11.06124	10.20074
23	13.48857	12.30338	11.27219	10.37106
24	13.79864	12.55036	11.46933	10.52876
25	14.09394	12.78336	11.65358	10.67478
26	14.37519	13.00317	11.82578	10.80998
27	14.64303	13.21053	11.98671	10.93516
28	14.89813	13.40616	12.13711	11.05108
29	15.14107	13.59072	12.27767	11.15841
30	15.37245	13.76483	12.40904	11.25778
31	15.59281	13.92909	12.53181	11.34980
32	15.80268	14.08404	12.64656	11.43500
33	16.00255	14.23023	12.75379	11.51389
34	16.19290	14.36814	12.85401	11.58693
35	16.37419	14.49825	12.94767	11.65457
36	16.54685	14.62099	13.03521	11.71719
37	16.71129	14.73678	13.11702	11.77518
38	16.86789	14.84602	13.19347	11.82887
39	17.01704	14.94907	13.26593	11.87858
40	17.15909	15.04630	13.33171	11.92461

TABLE VI
Per Period Equivalent of $1 Present Value

$$\frac{1}{a_{\overline{n}|\,i}} = \frac{i}{1 - (1 + i)^{-n}}$$

n	1%	2%	3%	4%
1	1.010000	1.020000	1.030000	1.040000
2	0.507512	0.515050	0.522611	0.530196
3	0.340022	0.346755	0.353530	0.360349
4	0.256281	0.262624	0.269027	0.275490
5	0.206040	0.212158	0.218355	0.224627
6	0.172548	0.178526	0.184598	0.190762
7	0.148628	0.154512	0.160506	0.166610
8	0.130690	0.136510	0.142456	0.148528
9	0.116740	0.122515	0.128434	0.134493
10	0.105582	0.111327	0.117231	0.123291
11	0.096454	0.102178	0.108077	0.114149
12	0.088848	0.094560	0.100462	0.106552
13	0.082415	0.088118	0.094030	0.100144
14	0.076901	0.082602	0.088526	0.094669
15	0.072124	0.077825	0.083767	0.089941
16	0.067945	0.073650	0.079611	0.085820
17	0.064258	0.069970	0.075953	0.082199
18	0.060982	0.066702	0.072709	0.078993
19	0.058052	0.063782	0.069814	0.076139
20	0.055415	0.061157	0.067216	0.073582
21	0.053031	0.058785	0.064872	0.071280
22	0.050864	0.056631	0.062747	0.069199
23	0.048886	0.054668	0.060814	0.067309
24	0.047073	0.052871	0.059047	0.065587
25	0.045407	0.051220	0.057428	0.064012
26	0.043869	0.049699	0.055938	0.062567
27	0.042446	0.048293	0.054564	0.061239
28	0.041124	0.046990	0.053293	0.060013
29	0.039895	0.045778	0.052115	0.058880
30	0.038748	0.044650	0.051019	0.057830
31	0.037676	0.043596	0.049999	0.056855
32	0.036671	0.042611	0.049047	0.055949
33	0.035727	0.041687	0.048156	0.055104
34	0.034840	0.040819	0.047322	0.054315
35	0.034004	0.040002	0.046539	0.053577
36	0.033214	0.039233	0.045804	0.052887
37	0.032468	0.038507	0.045112	0.052240
38	0.031761	0.037821	0.044459	0.051632
39	0.031092	0.037171	0.043844	0.051061
40	0.030456	0.036556	0.043262	0.050523

TABLE VI–(*continued*)
Per Period Equivalent of $1 Present Value

$$\frac{1}{a_{\overline{n}|i}} = \frac{i}{1 - (1 + i)^{-n}}$$

n	5%	6%	7%	8%
1	1.050000	1.060000	1.070000	1.080000
2	0.537805	0.545437	0.553092	0.560769
3	0.367209	0.374110	0.381052	0.388034
4	0.282012	0.288591	0.295228	0.301921
5	0.230975	0.237396	0.243891	0.250456
6	0.197017	0.203363	0.209796	0.216315
7	0.172820	0.179135	0.185553	0.192072
8	0.154722	0.161036	0.167468	0.174015
9	0.140690	0.147022	0.153486	0.161080
10	0.129505	0.135868	0.142378	0.149029
11	0.120389	0.126793	0.133357	0.140076
12	0.112825	0.119277	0.125902	0.132695
13	0.106456	0.112960	0.119651	0.126522
14	0.101024	0.107585	0.114345	0.121297
15	0.096342	0.102963	0.109795	0.116830
16	0.092270	0.098952	0.105858	0.112977
17	0.088699	0.095445	0.102425	0.109629
18	0.085546	0.092357	0.099412	0.106702
19	0.082745	0.089621	0.096753	0.104128
20	0.080243	0.087185	0.094393	0.101852
21	0.077996	0.085005	0.092289	0.099832
22	0.075971	0.083046	0.090406	0.098032
23	0.074137	0.081278	0.088714	0.096422
24	0.072471	0.079679	0.087189	0.094978
25	0.070952	0.078227	0.085811	0.093679
26	0.069564	0.076904	0.084561	0.092507
27	0.068292	0.075697	0.083426	0.091448
28	0.067123	0.074593	0.082392	0.090489
29	0.066046	0.073580	0.081449	0.089619
30	0.065051	0.072649	0.080586	0.088827
31	0.064132	0.071792	0.079797	0.088107
32	0.063280	0.071002	0.079073	0.087451
33	0.062490	0.070273	0.078408	0.086852
34	0.061755	0.069598	0.077797	0.086304
35	0.061072	0.068974	0.077234	0.085803
36	0.060434	0.068395	0.076715	0.085345
37	0.059840	0.067857	0.076237	0.084924
38	0.059284	0.067358	0.075795	0.084539
39	0.058765	0.066894	0.075387	0.084185
40	0.058278	0.066462	0.075009	0.083860

TABLE VII
Per Period Equivalent of $1 Future Value

$$\frac{1}{s_{\overline{n}|i}} = \frac{i}{(1 + i)^n - 1} = \frac{1}{a_{\overline{n}|i}} - i$$

n	1%	2%	3%	4%
1	1.0000000	1.0000000	1.0000000	1.0000000
2	0.4975124	0.4950495	0.4926108	0.4901961
3	0.3300221	0.3267547	0.3235304	0.3203485
4	0.2462811	0.2426238	0.2390270	0.2354900
5	0.1960398	0.1921584	0.1883546	0.1846271
6	0.1625484	0.1585258	0.1545975	0.1507619
7	0.1386283	0.1345120	0.1305064	0.1266096
8	0.1206903	0.1165098	0.1124564	0.1085278
9	0.1067404	0.1025154	0.0984339	0.0944930
10	0.0955821	0.0913265	0.0872305	0.0832909
11	0.0864541	0.0821779	0.0780774	0.0741490
12	0.0788488	0.0745596	0.0704621	0.0665522
13	0.0724148	0.0681184	0.0640295	0.0601437
14	0.0669012	0.0626020	0.0585263	0.0546690
15	0.0621238	0.0578255	0.0537666	0.0499411
16	0.0579446	0.0536501	0.0496108	0.0458200
17	0.0542581	0.0499698	0.0459525	0.0421985
18	0.0509820	0.0467021	0.0427087	0.0389933
19	0.0480518	0.0437818	0.0398139	0.0361386
20	0.0454153	0.0411567	0.0372157	0.0335818
21	0.0430308	0.0387848	0.0348718	0.0312801
22	0.0408637	0.0366314	0.0327474	0.0291988
23	0.0388858	0.0346681	0.0308139	0.0273091
24	0.0370735	0.0328711	0.0290474	0.0255868
25	0.0354068	0.0312204	0.0274279	0.0240120
26	0.0338689	0.0296992	0.0259383	0.0225674
27	0.0324455	0.0282931	0.0245642	0.0212385
28	0.0311244	0.0269897	0.0232932	0.0200130
29	0.0298950	0.0257784	0.0221147	0.0188799
30	0.0287481	0.0246499	0.0210193	0.0178301
31	0.0276757	0.0235963	0.0199989	0.0168554
32	0.0266709	0.0226106	0.0190466	0.0159486
33	0.0257274	0.0216865	0.0181561	0.0151036
34	0.0248400	0.0208187	0.0173220	0.0143148
35	0.0240037	0.0200022	0.0165393	0.0135773
36	0.0232143	0.0192329	0.0158038	0.0128869
37	0.0224680	0.0185068	0.0151116	0.0122396
38	0.0217615	0.0178206	0.0144593	0.0116319
39	0.0210916	0.0171711	0.0138439	0.0110608
40	0.0204556	0.0165557	0.0132624	0.0105235

TABLE VII–(*continued*)
Per Period Equivalent of $1 Future Value

$$\frac{1}{s_{\overline{n}|i}} = \frac{i}{(1 + i)^n - 1} = \frac{1}{a_{\overline{n}|i}} - i$$

n	5%	6%	7%	8%
1	1.0000000	1.0000000	1.0000000	1.0000000
2	0.4878049	0.4854369	0.4830918	0.4807692
3	0.3172086	0.3141098	0.3110517	0.3080335
4	0.2320118	0.2285915	0.2252281	0.2219208
5	0.1809748	0.1773964	0.1738907	0.1704565
6	0.1470175	0.1433626	0.1397958	0.1363154
7	0.1228198	0.1191350	0.1155532	0.1120724
8	0.1047218	0.1010359	0.0974678	0.0940148
9	0.0906901	0.0870222	0.0834865	0.0800797
10	0.0795046	0.0758680	0.0723775	0.0690295
11	0.0703889	0.0667929	0.0633569	0.0600763
12	0.0628254	0.0592770	0.0559020	0.0526950
13	0.0564558	0.0529601	0.0496508	0.0465218
14	0.0510240	0.0475849	0.0443449	0.0412969
15	0.0463423	0.0429628	0.0397946	0.0368295
16	0.0422699	0.0389521	0.0358576	0.0329769
17	0.0386991	0.0354448	0.0324252	0.0296294
18	0.0355462	0.0323565	0.0294126	0.0267021
19	0.0327450	0.0296209	0.0267530	0.0241276
20	0.0302426	0.0271846	0.0243929	0.0218522
21	0.0279961	0.0250045	0.0222890	0.0198323
22	0.0259705	0.0230456	0.0204058	0.0180321
23	0.0241368	0.0212785	0.0187139	0.0164222
24	0.0224709	0.0196790	0.0171890	0.0149780
25	0.0209525	0.0182267	0.0158105	0.0136788
26	0.0195643	0.0169043	0.0145610	0.0125071
27	0.0182919	0.0156972	0.0134257	0.0114481
28	0.0171225	0.0145926	0.0123919	0.0104889
29	0.0160455	0.0135796	0.0114487	0.0096185
30	0.0150514	0.0126489	0.0105864	0.0088274
31	0.0141321	0.0117922	0.0097969	0.0081073
32	0.0132804	0.0110023	0.0090729	0.0074508
33	0.0124900	0.0102729	0.0084081	0.0068516
34	0.0117554	0.0095984	0.0077967	0.0063041
35	0.0110717	0.0089739	0.0072340	0.0058033
36	0.0104345	0.0083948	0.0067153	0.0053447
37	0.0098398	0.0078574	0.0062368	0.0049244
38	0.0092842	0.0073581	0.0057951	0.0045389
39	0.0087646	0.0068938	0.0053868	0.0041851
40	0.0082782	0.0064615	0.0050091	0.0038602

TABLE VIII
Areas under the Normal Curve

Normal Deviate, z	.00	.01	.02	.03	.04	.05	.06	.07	.08	.09
0.0	.0000	.0040	.0080	.0120	.0160	.0199	.0239	.0279	.0319	.0359
0.1	.0398	.0438	.0478	.0517	.0557	.0596	.0636	.0675	.0714	.0753
0.2	.0793	.0832	.0871	.0910	.0948	.0987	.1026	.1064	.1103	.1141
0.3	.1179	.1217	.1255	.1293	.1331	.1368	.1406	.1443	.1480	..1517
0.4	.1554	.1591	.1628	.1664	.1700	.1736	.1772	.1808	.1844	.1879
0.5	.1915	.1950	.1985	.2019	.2054	.2088	.2123	.2157	.2190	.2224
0.6	.2257	.2291	.2324	.2357	.2389	.2422	.2454	.2486	.2517	.2549
0.7	.2580	.2611	.2642	.2673	.2704	.2734	.2764	.2794	.2823	.2852
0.8	.2881	.2910	.2939	.2967	.2995	.3023	.3051	.3078	.3106	.3133
0.9	.3159	.3186	.3212	.3238	.3264	.3289	.3315	.3340	.3365	.3389
1.0	.3413	.3438	.3461	.3485	.3508	.3531	.3554	.3577	.3599	.3621
1.1	.3643	.3665	.3686	.3708	.3729	.3749	.3770	.3790	.3810	.3830
1.2	.3849	.3869	.3888	.3907	.3925	.3944	.3962	.3980	.3997	.4015
1.3	.4032	.4049	.4066	.4082	.4099	.4115	.4131	.4147	.4162	.4177
1.4	.4192	.4207	.4222	.4236	.4251	.4265	.4279	.4292	.4306	.4319
1.5	.4332	.4345	.4357	.4370	.4382	.4394	.4406	.4418	.4429	.4441
1.6	.4452	.4463	.4474	.4484	.4495	.4505	.4515	.4525	.4535	.4545
1.7	.4554	.4564	.4573	.4582	.4591	.4599	.4608	.4616	.4625	.4633
1.8	.4641	.4649	.4656	.4664	.4671	.4678	.4686	.4693	.4699	.4706
1.9	.4713	.4719	.4726	.4732	.4738	.4744	.4750	.4756	.4761	.4767
2.0	.4772	.4778	.4783	.4788	.4793	.4798	.4803	.4808	.4812	.4817
2.1	.4821	.4826	.4830	.4834	.4838	.4842	.4846	.4850	.4854	.4857
2.2	.4861	.4864	.4868	.4871	.4875	.4878	.4881	.4884	.4887	.4890
2.3	.4893	.4896	.4898	.4901	.4904	.4906	.4909	.4911	.4913	.4916
2.4	.4918	.4920	.4922	.4925	.4927	.4929	.4931	.4932	.4934	.4936
2.5	.4938	.4940	.4941	.4943	.4945	.4946	.4948	.4949	.4951	.4952
2.6	.4953	.4955	.4956	.4957	.4959	.4960	.4961	.4962	.4963	.4964
2.7	.4965	.4966	.4967	.4968	.4969	.4970	.4971	.4972	.4973	.4974
2.8	.4974	.4975	.4976	.4977	.4977	.4978	.4979	.4979	.4980	.4981
2.9	.4981	.4982	.4982	.4983	.4984	.4984	.4985	.4985	.4986	.4986
3.0	.4987	.4987	.4987	.4988	.4988	.4989	.4989	.4989	.4990	.4990

Adapted by permission from Ernest Kurnow, Gerald J. Glasser, and Frederick R. Ottman, *Statistics for Business Decisions* (Homewood, Ill.: Richard D. Irwin, Inc., 1959), p. 501.

TABLE IX
e^x and e^{-x}

x	e^x	e^{-x}	x	e^x	e^{-x}	x	e^x	e^{-x}	x	e^x	e^{-x}
0.01	1.0101	0.9900	0.31	1.3634	0.7334	0.61	1.8404	0.5434	0.91	2.4843	0.4025
0.02	1.0202	0.9802	0.32	1.3771	0.7261	0.62	1.8589	0.5379	0.92	2.5093	0.3985
0.03	1.0305	0.9704	0.33	1.3910	0.7189	0.63	1.8776	0.5326	0.93	2.5345	0.3946
0.04	1.0408	0.9608	0.34	1.4049	0.7118	0.64	1.8965	0.5273	0.94	2.5600	0.3906
0.05	1.0513	0.9512	0.35	1.4191	0.7047	0.65	1.9155	0.5220	0.95	2.5857	0.3867
0.06	1.0618	0.9418	0.36	1.4333	0.6977	0.66	1.9348	0.5169	0.96	2.6117	0.3829
0.07	1.0725	0.9324	0.37	1.4477	0.6907	0.67	1.9542	0.5117	0.97	2.6379	0.3791
0.08	1.0833	0.9231	0.38	1.4623	0.6839	0.68	1.9739	0.5066	0.98	2.6645	0.3753
0.09	1.0942	0.9139	0.39	1.4770	0.6771	0.69	1.9937	0.5016	0.99	2.6912	0.3716
0.10	1.1052	0.9048	0.40	1.4918	0.6703	0.70	2.0138	0.4966	1.0	2.7183	0.3679
0.11	1.1163	0.8958	0.41	1.5068	0.6637	0.71	2.0340	0.4916	1.1	3.0042	0.3329
0.12	1.1275	0.8869	0.42	1.5220	0.6570	0.72	2.0544	0.4868	1.2	3.3201	0.3012
0.13	1.1388	0.8781	0.43	1.5373	0.6505	0.73	2.0751	0.4819	1.3	3.6693	0.2725
0.14	1.1503	0.8694	0.44	1.5527	0.6440	0.74	2.0959	0.4771	1.4	4.0552	0.2466
0.15	1.1618	0.8607	0.45	1.5683	0.6376	0.75	2.1170	0.4724	1.5	4.4817	0.2231
0.16	1.1735	0.8521	0.46	1.5841	0.6313	0.76	2.1383	0.4677	1.6	4.9530	0.2019
0.17	1.1853	0.8437	0.47	1.6000	0.6250	0.77	2.1598	0.4630	1.7	5.4739	0.1827
0.18	1.1972	0.8353	0.48	1.6161	0.6188	0.78	2.1815	0.4584	1.8	6.0496	0.1653
0.19	1.2092	0.8270	0.49	1.6323	0.6126	0.79	2.2034	0.4538	1.9	6.6859	0.1496
0.20	1.2214	0.8187	0.50	1.6487	0.6065	0.80	2.2255	0.4493	2.0	7.3891	0.1353
0.21	1.2337	0.8106	0.51	1.6653	0.6005	0.81	2.2479	0.4449	2.1	8.1662	0.1225
0.22	1.2461	0.8025	0.52	1.6820	0.5945	0.82	2.2705	0.4404	2.2	9.0250	0.1108
0.23	1.2586	0.7945	0.53	1.6989	0.5886	0.83	2.2933	0.4360	2.3	9.9742	0.1003
0.24	1.2712	0.7866	0.54	1.7160	0.5827	0.84	2.3164	0.4317	2.4	11.0232	0.0907
0.25	1.2840	0.7788	0.55	1.7333	0.5769	0.85	2.3396	0.4274	2.5	12.1825	0.0821
0.26	1.2969	0.7711	0.56	1.7507	0.5712	0.86	2.3632	0.4232	2.6	13.4637	0.0743
0.27	1.3100	0.7634	0.57	1.7683	0.5655	0.87	2.3869	0.4190	2.7	14.8797	0.0672
0.28	1.3231	0.7558	0.58	1.7860	0.5599	0.88	2.4109	0.4148	2.8	16.4446	0.0608
0.29	1.3364	0.7483	0.59	1.8040	0.5543	0.89	2.4351	0.4107	2.9	18.1741	0.0550
0.30	1.3499	0.7408	0.60	1.8221	0.5488	0.90	2.4596	0.4066	3.0	20.0855	0.0498

TABLE X
Natural (Napierian) Logarithm of N

N	.0	.1	.2	.3	.4	.5	.6	.7	.8	.9
0	...	−2.3026	−1.6094	−1.2040	−0.9163	−0.6931	−0.5108	−0.3567	−0.2231	−0.1054
1	0.0000	0.0953	0.1823	0.2624	0.3365	0.4055	0.4700	0.5306	0.5878	0.6419
2	0.6931	0.7419	0.7885	0.8329	0.8755	0.9163	0.9555	0.9933	1.0296	1.0647
3	1.0986	1.1314	1.1632	1.1939	1.2238	1.2528	1.2809	1.3083	1.3350	1.3610
4	1.3863	1.4110	1.4351	1.4586	1.4816	1.5041	1.5261	1.5476	1.5686	1.5892
5	1.6094	1.6292	1.6487	1.6677	1.6864	1.7047	1.7228	1.7405	1.7579	1.7750
6	1.7918	1.8083	1.8245	1.8405	1.8563	1.8718	1.8871	1.9021	1.9169	1.9315
7	1.9459	1.9601	1.9741	1.9879	2.0015	2.0149	2.0281	2.0412	2.0541	2.0669
8	2.0794	2.0919	2.1041	2.1163	2.1282	2.1401	2.1518	2.1633	2.1748	2.1861
9	2.1972	2.2083	2.2192	2.2300	2.2407	2.2513	2.2618	2.2721	2.2824	2.2925
10	2.3026	2.3125	2.3224	2.3321	2.3418	2.3514	2.3609	2.3702	2.3795	2.3888
11	2.3979	2.4069	2.4159	2.4248	2.4336	2.4423	2.4510	2.4596	2.4681	2.4765
12	2.4849	2.4932	2.5014	2.5096	2.5177	2.5257	2.5337	2.5416	2.5494	2.5572
13	2.5649	2.5726	2.5802	2.5878	2.5953	2.6027	2.6101	2.6174	2.6247	2.6319
14	2.6391	2.6462	2.6532	2.6603	2.6672	2.6741	2.6810	2.6878	2.6946	2.7014
15	2.7081	2.7147	2.7213	2.7279	2.7344	2.7408	2.7473	2.7537	2.7600	2.7663
16	2.7726	2.7788	2.7850	2.7912	2.7973	2.8034	2.8094	2.8154	2.8214	2.8273
17	2.8332	2.8391	2.8449	2.8507	2.8565	2.8622	2.8679	2.8736	2.8792	2.8848
18	2.8904	2.8959	2.9014	2.9069	2.9124	2.9178	2.9232	2.9285	2.9339	2.9392
19	2.9444	2.9497	2.9549	2.9601	2.9653	2.9704	2.9755	2.9806	2.9857	2.9907
20	2.9957	3.0007	3.0057	3.0106	3.0155	3.0204	3.0253	3.0301	3.0350	3.0397
21	3.0445	3.0493	3.0540	3.0587	3.0634	3.0681	3.0727	3.0773	3.0819	3.0865
22	3.0910	3.0956	3.1001	3.1046	3.1091	3.1135	3.1179	3.1224	3.1268	3.1311
23	3.1355	3.1398	3.1442	3.1485	3.1527	3.1570	3.1612	3.1655	3.1697	3.1739
24	3.1781	3.1822	3.1864	3.1905	3.1946	3.1987	3.2027	3.2068	3.2108	3.2149
25	3.2189	3.2279	3.2268	3.2308	3.2347	3.2387	3.2426	3.2465	3.2504	3.2542

TABLE XI
Monthly Payment for $1 Mortgage

$$\frac{\frac{r}{1200}}{1-\left(1+\frac{r}{1200}\right)^{-n}}$$

Annual Rate r%	Number of Months, n					
	120	180	240	300	360	420
7.00	.0116108479	.0089982827	.0077529894	.0070677920	.0066530250	.0063885636
7.25	.0117401041	.0091286288	.0079037598	.0072280686	.0068217628	.0065646724
7.50	.0118701769	.0092701236	.0080559319	.0073899118	.0069921451	.0067424260
7.75	.0120010631	.0094127575	.0082094856	.0075532876	.0071641225	.0069217594
8.00	.0121327594	.0095565208	.0083644007	.0077181622	.0073376457	.0071026088
8.25	.0122652625	.0097014036	.0085206565	.0078845013	.0075126660	.0072849114
8.50	.0123985689	.0098473956	.0086782324	.0080522708	.0076891348	.0074686057
8.75	.0125326750	.0099944865	.0088371071	.0082214364	.0078670041	.0076536314
9.00	.0126675774	.0101426658	.0089972596	.0083919636	.0080462262	.0078399297
9.25	.0128032722	.0102919229	.0091586683	.0085638184	.0082267543	.0080274432
9.50	.0129397558	.0104422468	.0093213119	.0087369666	.0084085421	.0082161160
9.75	.0130770242	.0105936266	.0094851685	.0089113742	.0085915441	.0084058939
10.00	.0132150737	.0107460512	.0096502165	.0090870075	.0087757157	.0085967243
10.25	.0133539002	.0108995092	.0098164339	.0092638328	.0089610130	.0087885561
10.50	.0134934997	.0110539892	.0099837989	.0094441817	.0091473929	.0089813402
10.75	.0136338680	.0112094798	.0101522895	.0096209272	.0093348136	.0091750290
11.00	.0137750011	.0113659693	.0103218839	.0098011308	.0095232340	.0093695765

TABLE XII
Derivatives and Integrals[1]

1. $\frac{d}{dx}[f(x) \pm g(x)] = \frac{d}{dx}f(x) \pm \frac{d}{dx}g(x).$

2. $\frac{d}{dx}[cf(x)] = c\frac{d}{dx}f(x).$

3. $\frac{d}{dx}x^n = nx^{n-1}.$

4. $\frac{d}{dx}[f(x)]^n = n[f(x)]^{n-1}\frac{d}{dx}f(x).$

5. $\frac{d}{dx}e^x = e^x.$

6. $\frac{d}{dx}a^x = a^x \ln a.$

7. $\frac{d}{dx}e^{f(x)} = e^{f(x)}\frac{d}{dx}f(x).$

8. $\frac{d}{dx}a^{f(x)} = a^{f(x)}(\ln a)\frac{d}{dx}f(x).$

9. $\frac{d}{dx}\ln x = \frac{1}{x}.$

10. $\frac{d}{dx}\ln[f(x)] = \frac{1}{f(x)}\frac{d}{dx}f(x).$

11. $\frac{d}{dx}\log_{10} x = \frac{1}{x(\ln 10)} = \frac{0.4343}{x}.$

12. $\frac{d}{dx}\log_{10} f(x) = \frac{0.4343}{f(x)} \cdot \frac{d}{dx}f(x).$

13. Chain rule: $\frac{d}{dx}g[f(x)] = \frac{dg(f)}{df} \cdot \frac{df(x)}{dx}.$

14. Product rule: $\frac{d}{dx}[f(x)g(x)] = f(x) \cdot \frac{d}{dx}g(x) + g(x)\frac{d}{dx}f(x).$

15. Quotient rule: $\frac{d}{dx}\left[\frac{f(x)}{g(x)}\right] = \frac{g(x)\frac{d}{dx}f(x) - f(x)\frac{d}{dx}g(x)}{[g(x)]^2}.$

16. $\int cf(x)\, dx = c \int f(x)\, dx.$

17. $\int [f(x) \pm g(x)]\, dx = \int f(x)\, dx \pm \int g(x)\, dx.$

18. $\int x^n\, dx = \frac{x^{n+1}}{n + 1} + C, n \neq -1.$

19. $\int x^{-1}\, dx = \int \frac{dx}{x} = \ln x + C.$

20. $\int (a + bx)^n\, dx = \frac{(a + bx)^{n+1}}{b(n + 1)} + C, n \neq -1.$

21. $\int (a + bx)^{-1}\, dx = \int \frac{dx}{a + bx} = \frac{\ln(a + bx)}{b} + C.$

22. $\int \frac{x\, dx}{a + bx} = \frac{x}{b} - \frac{a}{b^2}\ln(a + bx) + C.$

23. $\int \frac{dx}{x(a + bx)} = \frac{1}{a}\ln\left(\frac{x}{a + bx}\right) + C.$

[1]Note: $a, b, c,$ and n are constants.

TABLE XII—(continued)
Derivatives and Integrals[1]

24. $\int e^x \, dx = e^x + C.$

25. $\int a^x \, dx = \dfrac{a^x}{\ln a} + C.$

26. $\int e^{b+cx} \, dx = \dfrac{e^{b+cx}}{c} + C.$

27. $\int a^{b+cx} \, dx = \dfrac{a^{b+cx}}{c \ln a} + C.$

28. $\int xe^{bx} = \dfrac{e^{bx}(bx - 1)}{b^2} + C.$

29. $\int \ln x \, dx = x \ln x - x + C = x(\ln x - 1) + C.$

30. $\int \ln(a + bx) \, dx = \dfrac{(a + bx)[\ln(a + bx) - 1]}{b} + C.$

[1]Note: a, b, c, and n are constants.

TABLE XIII
Cumulative Binomial Distribution for $n = 25$

Tabulated values are $\displaystyle\sum_0^x C_x^{25} \, p^x q^{25-x}$

Values of x						Values of p							
	0.01	0.05	0.10	0.20	0.30	0.40	0.50	0.60	0.70	0.80	0.90	0.95	0.99
0	.778	.277	.072	.004	.000	.000	.000	.000	.000	.000	.000	.000	.000
1	.974	.642	.271	.027	.002	.000	.000	.000	.000	.000	.000	.000	.000
2	.998	.873	.537	.098	.009	.000	.000	.000	.000	.000	.000	.000	.000
3	1.000	.966	.764	.234	.033	.002	.000	.000	.000	.000	.000	.000	.000
4	1.000	.993	.902	.421	.090	.009	.000	.000	.000	.000	.000	.000	.000
5	1.000	.999	.967	.617	.193	.029	.002	.000	.000	.000	.000	.000	.000
6	1.000	1.000	.991	.780	.341	.074	.007	.000	.000	.000	.000	.000	.000
7	1.000	1.000	.998	.891	.512	.154	.022	.001	.000	.000	.000	.000	.000
8	1.000	1.000	1.000	.953	.677	.274	.054	.004	.000	.000	.000	.000	.000
9	1.000	1.000	1.000	.983	.811	.425	.115	.013	.000	.000	.000	.000	.000
10	1.000	1.000	1.000	.994	.902	.586	.212	.034	.002	.000	.000	.000	.000
11	1.000	1.000	1.000	.998	.956	.732	.345	.078	.006	.000	.000	.000	.000
12	1.000	1.000	1.000	1.000	.983	.846	.500	.154	.017	.000	.000	.000	.000
13	1.000	1.000	1.000	1.000	.994	.922	.655	.268	.044	.002	.000	.000	.000
14	1.000	1.000	1.000	1.000	.998	.966	.788	.414	.098	.006	.000	.000	.000
15	1.000	1.000	1.000	1.000	1.000	.987	.885	.575	.189	.017	.000	.000	.000
16	1.000	1.000	1.000	1.000	1.000	.996	.946	.726	.323	.047	.000	.000	.000
17	1.000	1.000	1.000	1.000	1.000	.999	.978	.846	.488	.109	.002	.000	.000
18	1.000	1.000	1.000	1.000	1.000	1.000	.993	.926	.659	.220	.009	.000	.000
19	1.000	1.000	1.000	1.000	1.000	1.000	.998	.971	.807	.383	.033	.001	.000
20	1.000	1.000	1.000	1.000	1.000	1.000	1.000	.991	.910	.579	.098	.007	.000
21	1.000	1.000	1.000	1.000	1.000	1.000	1.000	.998	.967	.766	.236	.034	.000
22	1.000	1.000	1.000	1.000	1.000	1.000	1.000	1.000	.991	.902	.463	.127	.002
23	1.000	1.000	1.000	1.000	1.000	1.000	1.000	1.000	.998	.973	.729	.358	.026
24	1.000	1.000	1.000	1.000	1.000	1.000	1.000	1.000	1.000	.996	.928	.723	.222

Index

*This book has been set in 10 and 9 point Times
Roman, leaded 2 points. Chapter numbers are 48
point Times New Roman and chapter titles are 21
point Times New Roman. The size of the type page
is 27 $\times$ 46½ picas.*